Legal Notice

This book is copyright 2020 with all rights reserved. It is illegal to copy, distribute, or create derivative works from this book in whole or in part or to contribute to the copying, distribution, or creating of derivative works of this book.

For information on bulk purchases and licensing agreements, please email

support@SATPrepGet800.com

ISBN-13: 978-1-951619-06-0

Also Available from Dr. Steve Warner

CONNECT WITH DR. STEVE WARNER

www.facebook.com/SATPrepGet800

www.youtube.com/TheSATMathPrep

www.twitter.com/SATPrepGet800

www.linkedin.com/in/DrSteveWarner

www.pinterest.com/SATPrepGet800

Real Analysis for Beginners

A Rigorous Introduction to Set Theory, Functions, Topology, Limits, Continuity, Differentiation, Riemann Integration, Sequences, and Series

Dr. Steve Warner

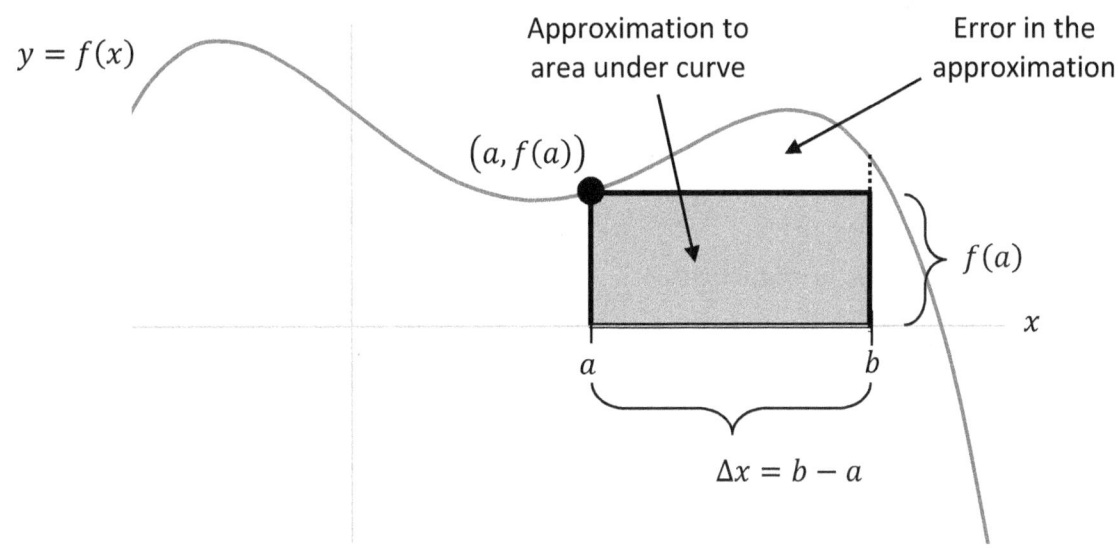

© 2020, All Rights Reserved

Table of Contents

Introduction — 7
    For students — 7
    For instructors — 8

Lesson 1 – Basic Set Theory — 9
    Describing Sets — 9
    Subsets and Proper Subsets — 14
    Basic Theorems Involving Subsets — 16
    Power Sets — 18
    Transitivity of the Subset Relation — 19
    Equality of Sets — 19
    Cartesian Products — 21
    Basic Set Operations — 24
    Properties of Unions and Intersections — 27
    Arbitrary Unions and Intersections — 28
    Problem Set 1 — 31

Lesson 2 – Relations and Partitions — 35
    Binary Relations — 35
    n**-ary Relations** — 37
    Orderings — 37
    Intervals — 40
    Equivalence Relations — 43
    Partitions — 45
    Problem Set 2 — 48

Lesson 3 – Functions — 50
    Functions — 50
    Injections, Surjections, and Bijections — 52
    Inverse Functions — 56
    Composite Functions — 55
    Identity Functions — 56
    Images and Inverse images — 57
    Sequences — 57
    Subsequences — 59
    Convergent and Cauchy Sequences — 60
    Real-valued Functions — 62
    Problem Set 3 — 72

Lesson 4 – Number Systems and Induction — 75
    The Natural Numbers — 75
    Well Ordering and the Principle of Mathematical Induction — 77
    The Integers — 83
    The Rational Numbers — 84
    The Real Numbers — 85
    The Complex Numbers — 87
    Problem Set 4 — 89

Lesson 5 – Equinumerosity	92
Basic Definitions and Examples	92
Countable and Uncountable Sets	93
Problem Set 5	99
Lesson 6 – Algebraic Structures and Completeness	101
Binary Operations and Closure	101
Groups	103
Fields	105
Ordered Rings and Fields	108
Why Isn't $\mathbb{Q}$ Enough?	111
Completeness	114
Problem Set 6	119
Lesson 7 – Basic Topology of $\mathbb{R}$	122
Absolute Value and Distance	122
Neighborhoods	123
Open and Closed Sets in $\mathbb{R}$	124
Compactness	132
Connectedness	134
Problem Set 7	136
Lesson 8 – Limits and Continuity	138
Strips and Rectangles in $\mathbb{R}$	138
Limits and Continuity in $\mathbb{R}$	141
Equivalent Definitions of Limits and Continuity in $\mathbb{R}$	145
Basic Examples in $\mathbb{R}$	147
Limit Theorems in $\mathbb{R}$	151
Limits in $\mathbb{R}$ Involving Infinity	153
One-sided Limits in $\mathbb{R}$	155
Limits of Sequences	156
Problem Set 8	160
Lesson 9 – Topological Continuity	163
Topological Treatment of Continuous Functions	163
Continuity on Compact Sets	165
Continuity on Connected Sets	166
Homeomorphisms	167
Problem Set 9	168
Lesson 10 – Differentiation	170
Linear Functions and Slope	170
Secant Lines	172
Tangent Lines	176
The Derivative	178
Differentiation Rules	180
The Mean Value Theorem	185
L'Hôpital's Rule	188
Problem Set 10	193

Lesson 11 – Riemann Integration — 195
    Area under a Curve — 195
    The Riemann Integral — 206
    Integration Rules — 211
    Fundamental Theorem of Calculus — 216
    Problem Set 11 — 222

Lesson 12 – Logarithmic and Exponential Functions — 225
    The Natural Logarithmic Function — 225
    The Natural Exponential Function — 230
    General Exponential and Logarithmic Functions — 233
    Logarithmic Differentiation — 234
    Problem Set 12 — 239

Lesson 13 – Improper Integration — 241
    Improper Integrals — 241
    Integration by Parts — 244
    Comparison Tests — 249
    Absolute and Conditional Convergence — 251
    Problem Set 13 — 253

Lesson 14 – Sequences — 256
    Sequences of Real Numbers — 256
    Monotone Sequences — 258
    Limit Superior and Limit Inferior — 260
    Problem Set 14 — 262

Lesson 15 – Series — 264
    Series of Real Numbers — 264
    The Integral Test — 267
    The Alternating Series Test — 272
    Dirichlet's and Abel's Tests — 276
    Comparison Tests — 280
    Absolute and Conditional Convergence — 283
    Tests for Absolute Convergence — 286
    Problem Set 15 — 290

Lesson 16 – Sequences and Series of Functions — 293
    Sequences of Functions — 293
    The Uniform Norm — 296
    Preservation Properties of Uniform Convergence — 298
    Series of Functions — 301
    Power Series — 304
    Taylor Series — 312
    Problem Set 16 — 319

Index — 322
About the Author — 327
Books by Dr. Steve Warner — 328

INTRODUCTION
REAL ANALYSIS

This book was written to provide a basic but rigorous introduction to real analysis.

For students: There are no prerequisites for this book. The content is completely self-contained. Students with a bit of mathematical knowledge may have an easier time getting through some of the material, but no such knowledge is necessary to read this book.

More important than mathematical knowledge is "mathematical maturity." Although there is no single agreed upon definition of mathematical maturity, one reasonable way to define it is as "one's ability to analyze, understand, and communicate mathematics." A student with a higher level of mathematical maturity will be able to move through this book more quickly than a student with a lower level of mathematical maturity.

Whether your level of mathematical maturity is low or high, if you are just starting out in real analysis, then you're in the right place. If you read this book the "right way," then your level of mathematical maturity will continually be increasing. This increased level of mathematical maturity will not only help you to succeed in advanced math courses, but it will improve your general problem solving and reasoning skills. This will make it easier to improve your performance in college, in your professional life, and on standardized tests.

So, what is the "right way" to read this book? Simply reading each lesson from end to end without any further thought and analysis is not the best way to read the book. You will need to put in some effort to have the best chance of absorbing and retaining the material. When a new theorem is presented, don't just jump right to the proof and read it. Think about what the theorem is saying. Try to describe it in your own words. Do you believe that it is true? If you do believe it, can you give a convincing argument that it is true? If you do not believe that it is true, try to come up with an example that shows it is false, and then figure out why your example does not contradict the theorem. Pick up a pen or pencil. Draw some pictures, come up with your own examples, and try to write your own proof.

You may find that this book goes into more detail than other real analysis books when explaining examples, discussing concepts, and proving theorems. This was done so that any student can read this book, and not just students that are naturally gifted in mathematics. So, it is up to you as the student to try to answer questions before they are answered for you. When a new definition is given, try to think of your own examples before looking at those presented in the book. And when the book provides an example, do not just accept that it satisfies the given definition. Convince yourself. Prove it.

Each lesson is followed by a Problem Set. The problems in each Problem Set have been organized into five levels of difficulty, followed by one or more Challenge Problems. Level 1 problems are the easiest and Level 5 problems are the most difficult, except for the Challenge Problems. If you want to get just a small taste of real analysis, then you can work on the easier problems. If you want to achieve a deeper understanding of the material, take some time to struggle with the harder problems.

For instructors: This book can be used as an undergraduate text or an introductory graduate text in real analysis. The subject is developed slowly with an emphasis early on of developing skill with proof writing.

Lessons 1 through 3 provide a rigorous treatment of the basic set theory that anyone studying real analysis should know. All the basics of sets, subsets, set operations, relations, partitions, and functions are included. If students have already seen this material in a prerequisite to your course, then this material can be reviewed quickly or skipped altogether.

Lesson 4 covers induction and a formal treatment of the natural numbers, integers, rational numbers, real numbers, and complex numbers. The formal treatments of these number systems can be included or omitted, depending on the taste of each individual instructor. On the other hand, I would spend at least a little time reviewing induction, as it can be a useful tool in many mathematical proofs.

Lesson 5 covers equinumerosity of sets. Countable sets, uncountable sets, Cantor's Theorem, and the Cantor-Schroeder-Bernstein Theorem are all discussed in detail.

Lesson 6 provides an introduction to groups, rings, and fields, with an emphasis on the specific structures that will serve as examples throughout this book. Ordered rings and fields and the completeness of the reals are covered in detail as well.

Lesson 7 covers the basic topology of the real numbers. Concepts such as distance, openness, compactness, and connectedness are all discussed.

Lesson 8 introduces limits and continuity of real-valued functions and Lesson 9 goes into detail on the topological properties of continuous functions.

Lesson 10 covers differentiation including the Mean Value Theorem and L'Hôpital's rule, while Lesson 11 covers Riemann integration and the Fundamental Theorem of Calculus.

Lesson 12 discusses logarithmic and exponential functions as well as logarithmic differentiation. Lesson 13 covers improper integration and integration by parts. Comparison tests and the notions of absolute and conditional convergence of improper integrals are also introduced in Lesson 13.

Lessons 14 and 15 cover sequences and series of real numbers, respectively. Finally, in Lesson 16, we provide an introduction to sequences and series of functions.

Students will have access to solutions to all problems in the Problem Sets at the end of each lesson, except for the Challenge Problems. These Challenge Problems can be used for graded assignments. I would recommend giving students at least several days to a week to work on each one. Most students will find them quite difficult.

The author welcomes all feedback from instructors. Any suggestions will be considered for future editions of the book. The author would also love to hear about the various courses that are created using these lessons. Feel free to email Dr. Steve Warner with any feedback at

steve@SATPrepGet800.com

LESSON 1
BASIC SET THEORY

Describing Sets

A **set** is simply a collection of "objects." These objects can be numbers, letters, colors, animals, funny quotes, or just about anything else you can imagine. We will usually refer to the objects in a set as the **members** or **elements** of the set.

If a set consists of a small number of elements, we can describe the set simply by listing the elements in the set in curly braces, separating elements by commas. We call this method of describing a set the **roster method**.

Example 1.1:

1. {fire, water} is the set consisting of two elements: *fire* and *water*.

2. {pear, lion, tower, black, poster, chair} is the set consisting of six elements: *pear, lion, tower, black, poster,* and *chair*.

3. $\{0, 2, 4, 6, 8, 10\}$ is the set consisting of six elements: 0, 2, 4, 6, 8, and 10. The elements in this set happen to be *numbers*.

A set is determined by its elements and not the order in which the elements are presented. For example, the set $\{2, 6, 4, 0, 10, 8\}$ is the same as the set $\{0, 2, 4, 6, 8, 10\}$.

Also, the set $\{0, 0, 2, 4, 4, 6, 6, 6, 8, 8, 10\}$ is the same as the set $\{0, 2, 4, 6, 8, 10\}$. If we are describing a set by listing its elements, the most natural way to do this is to list each element just once.

We will usually name sets using capital letters such as $A, B, C,...$, and so on. For example, we might write $A = \{a, b, c, d, e\}$. So, A is the set consisting of the elements $a, b, c, d,$ and e.

Example 1.2: Consider the sets $X = \{x, y, z\}$, $Y = \{y, z, x\}$, $Z = \{x, y, x, z, x\}$. Then $X, Y,$ and Z all represent the same set. We can write $X = Y = Z$.

We use the symbol $\in$ for the membership relation (we will define the term "relation" more carefully in Lesson 2). So, $x \in A$ means "x is an element of A," whereas $x \notin A$ means "x is **not** an element of A." We will often simply say "x is in A," and "x is not in A," respectively.

Example 1.3: Let $B = \{c, z, 5, \Delta, \Gamma\}$. Then $c \in B, z \in B, 5 \in B, \Delta \in B,$ and $\Gamma \in B$.

If a set consists of many elements, we can use **ellipses** (...) to help describe the set. For example, the set consisting of the natural numbers between 1 and 80, inclusive, can be written $\{1, 2, 3, ..., 79, 80\}$ ("inclusive" means that we include 1 and 80). The ellipses between 3 and 79 are there to indicate that there are elements in the set that we are not explicitly mentioning.

Ellipses can also be used to help describe **infinite sets**. The set of **natural numbers** can be written $\mathbb{N} = \{0, 1, 2, 3, ...\}$, and the set of **integers** can be written $\mathbb{Z} = \{..., -4, -3, -2, -1, 0, 1, 2, 3, 4, ...\}$.

Notes: (1) Some mathematicians exclude 0 from the set of natural numbers. In this book, 0 will always be included. Symbolically, $0 \in \mathbb{N}$.

(2) Notice how we use special character symbols to represent the natural numbers and integers. The characters $\mathbb{N}$ and $\mathbb{Z}$ are said to be **doublestruck** or in **blackboard bold**. In general, important sets are written using doublestruck character symbols. The natural numbers and integers are two such examples. We will see several more shortly.

Example 1.4:

1. The even natural numbers can be written $\mathbb{E} = \{0, 2, 4, 6, ...\}$.
2. The odd natural numbers can be written $\mathbb{O} = \{1, 3, 5, ...\}$.
3. The even integers can be written $2\mathbb{Z} = \{..., -6, -4, -2, 0, 2, 4, 6, ...\}$.
4. The positive integers can be written $\mathbb{Z}^+ = \{1, 2, 3, 4, ...\}$ or $\mathbb{N}^+ = \{1, 2, 3, 4, ...\}$ (the positive integers and the positive natural numbers describe the same set).

A set can also be described by a certain property P that all its elements have in common. There are endless possibilities for what a property could be. For example, suppose that P is the property of being an insect. Then "mosquito" satisfies the property P, whereas "computer" does **not** satisfy the property P.

If we wish to describe a set with a property P, then we can use the **set-builder notation** $\{x|P(x)\}$. The expression $\{x|P(x)\}$ can be read "the set of all x such that the property $P(x)$ is true." Note that the symbol "|" is read as "such that."

The letter x in the set-builder notation is called a **variable**. The choice of x is completely arbitrary. We could have used any other letter or symbol. In other words, $\{x|P(x)\}$, $\{t|P(t)\}$, $\{\square|P(\square)\}$, and $\{?|P(?)\}$ all have exactly the same meaning. The idea here is that we substitute objects in for the variable. If that object has the given property, then it is in the set. If it does not have the given property, then it is not in the set. Sticking with our example above, let $A = \{x \mid x \text{ is an insect}\}$. Then we have mosquito $\in A$ ("mosquito" is an element of the set A) because when we substitute "mosquito" in for the variable x, we get a true statement. Indeed, a mosquito is an insect. On the other hand, computer $\notin A$ ("computer" is **not** an element of the set A) because when we substitute "computer" in for the variable x, we get a false statement. Indeed, the statement "a computer is an insect" is false.

As another simple example, let's consider the property of being a beverage. Formally, we can let $P(x)$ be the statement "x is a beverage." Symbolically, we can write $\{x|P(x)\} = \{x \mid x \text{ is a beverage}\}$. In words, this set can be described as "the set of all x such that x is a beverage." Water is an element of this set because water is a beverage. In other words, if we replace x by "water," then $P(x)$ is a true statement. Symbolically, we can write "water $\in \{x \mid x \text{ is a beverage}\}$." Tiger is not an element of this set because a tiger is not a beverage. Symbolically, we can write "tiger $\notin \{x \mid x \text{ is a beverage}\}$." However, tiger would be in the set $\{x \mid x \text{ is an animal}\}$. Tiger would also be in the set $\{x \mid x \text{ is a cat that roars}\}$.

Let's now turn to an example involving numbers.

Example 1.5: Let's look at a few different ways that we can describe the set $\{0, 2, 4, 6, 8, 10\}$. We have already seen that reordering and/or repeating elements does not change the set. For example, $\{2, 2, 0, 4, 10, 8, 8, 8, 6\}$ describes the same set. Here are a few more descriptions using set-builder notation:

- $\{n \mid n$ is an even natural number less than or equal to $10\}$
- $\{n \mid n$ is an even integer between 0 and 10, inclusive$\}$
- $\{2t \mid t \in \mathbb{N}$ and $0 \leq t \leq 5\}$
- $\{2k \mid k = 0, 1, 2, 3, 4, 5\}$

The first expression in the bulleted list can be read "the set of n such that n is an even natural number less than or equal to 10."

The second expression can be read "the set of n such that n is an even integer between 0 and 10, inclusive. Recall that the word "inclusive" means that we include 0 and 10.

The third expression can be read "the set of $2t$ such that t is a natural number and t is between 0 and 5, including both 0 and 5. Note that the abbreviation "$t \in \mathbb{N}$" can be read "t is in the set of natural numbers," or more briefly, "t is a natural number." We used the letter "t" for the variable here (as opposed to the letter "n" that was used in the first two expressions). Once again, we can use any variable name we like, as long as it doesn't lead to confusion.

The fourth expression can be read "the set of $2k$ such that k is 0, 1, 2, 3, 4 or 5."

For easier readability, we may include a **bounding set** when using set-builder notation. If we wish to describe a set with a property P and a bounding set A, then the corresponding set-builder notation is $\{x \in A \mid P(x)\}$. As an example, consider the set $\{7, 8, 9\}$. Since every element of this set is a natural number, we can use the set of natural numbers, $\mathbb{N}$, as a bounding set. So, instead of writing $\{k \mid k \in \mathbb{N} \wedge 7 \leq k \leq 9\}$, we can use the friendlier notation $\{k \in \mathbb{N} \mid 7 \leq k \leq 9\}$. Notice how the bounding set appears to the **left** of the vertical line.

Example 1.5 continued: Let's look at a couple more ways that we can describe the set $\{0, 2, 4, 6, 8, 10\}$, this time using bounding sets in the notation:

- $\{n \in \mathbb{N} \mid n$ is even and $n \leq 10\}$
- $\{n \in \mathbb{Z} \mid n$ is even, $0 \leq n < 12\}$

In addition to the sets $\mathbb{N}$ (the natural numbers) and $\mathbb{Z}$ (the integers), let's look at a few more sets that will show up throughout this book.

Example 1.6:

1. The set of **rational numbers** is $\mathbb{Q} = \{\frac{a}{b} \mid a, b \in \mathbb{Z} \text{ and } b \neq 0\}$. In words, $\mathbb{Q}$ is "the set of quotients a over b such that a and b are integers and b is not zero." Some examples of rational numbers are $\frac{0}{5}, \frac{1}{3}, \frac{2}{5}$, and $\frac{-6}{7}$. We identify rational numbers $\frac{a}{b}$ and $\frac{c}{d}$ whenever $ad = bc$. For example, $\frac{1}{2} = \frac{3}{6}$ because $1 \cdot 6 = 2 \cdot 3$. We also abbreviate the rational number $\frac{a}{1}$ as a. In this way, we can think of every integer as a rational number. For example, we have $\frac{0}{3} = \frac{0}{1}$ (because $0 \cdot 1 = 3 \cdot 0$), and therefore, we can abbreviate $\frac{0}{3}$ as 0. Similarly, we can abbreviate $\frac{15}{3}$ as 5 (because $\frac{15}{3} = \frac{5}{1}$).

   If a and b are integers and $b \neq 0$, then the expression $\frac{a}{b}$ is called a **fraction**. So, the set of rational numbers, $\mathbb{Q}$, can also be referred to as the set of fractions. Each fraction can also be represented in another way, namely as a **decimal**. We will discuss this a bit more after defining the real numbers.

2. In this book we will be most interested in the set of **real numbers**, $\mathbb{R}$. There are many equivalent ways of formally defining the real numbers and we will look at one such method in Lesson 4.

   For now, let's go with a naïve definition of the real numbers. We first define a **digit** to be one of the symbols 0, 1, 2, 3, 4, 5, 6, 7, 8, or 9. We then define $\mathbb{R}$ to be the set of numbers of the form $x.y$, where $x \in \mathbb{Z}$ and y is an infinite "string" of digits without a **tail of 9's** (meaning there are infinitely many digits in the string that are **not** 9). Symbolically, we have

   $$\mathbb{R} = \{x.y \mid x \in \mathbb{Z} \text{ and } y \text{ is an infinite string of digits without a tail of 9's}\}.$$

   Some examples of real numbers are $0.000\ldots$, $0.333\ldots$, $-16.000\ldots$, and $1.010010001\ldots$ We will generally delete tails of 0's. So, we would write $0.000\ldots$ as 0 and $-16.000\ldots$ as -16. We will not consider $53.023999999\ldots$ to be a real number because of the tail of 9's (an alternative approach would be to identify $53.023999999\ldots$ with 53.024).

   We will sometimes refer to the real number $x.y$ as a **decimal**. The dot between x and y is called a **decimal point**.

   We can visualize the set of real numbers with the **real line**.

   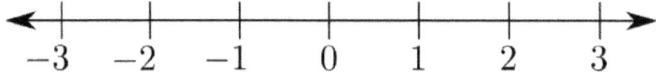

   There is a fairly simple algorithm that allows us to identify every rational number as a real number. For example, the rational number $\frac{3}{2}$ can be represented as the real number 1.5 and the rational number $\frac{2}{3}$ can be represented as the real number $0.66666\ldots$ We may abbreviate this last number by using the notation $0.\overline{6}$. The "bar" over the 6 indicates that the 6 repeats forever. As another example of this notation, the number $0.12\overline{345}$ abbreviates $0.12345345345\ldots$ Notice how the 1 and 2 appear just once (because the bar is not over those digits), whereas the 3, 4, and 5 repeat forever. I leave the details of this algorithm for the interested reader to explore (see Problem 30 below).

12

Every rational number can be represented as a decimal that either **terminates** (has a tail of 0's) or **repeats** (has a tail with a finite repeating pattern). Any decimal (real number) that does **not** terminate or repeat is called an **irrational number**.

3. The set of **complex numbers** is defined as $\mathbb{C} = \{a + bi \mid a, b \in \mathbb{R}\}$. In words, $\mathbb{C}$ is "the set of $a + bi$ such that a and b are real numbers." Some examples of complex numbers are $0 + 0i$, $-2 + 0i$, $2.3 - 5i = 2.3 + (-5)i$, and $4.235235235\ldots + 51.2020020002\ldots i$. We will abbreviate $0 + 0i$ as 0, $a + 0i$ as a, and $0 + bi$ as bi. For example, $-2 + 0i = -2$. By identifying $a + 0i$ as a, we can think of every real number as a complex number. Complex numbers of the form bi are called **pure imaginary numbers**.

If we identify $1 = 1 + 0i$ with the ordered pair $(1,0)$, and we identify $i = 0 + 1i$ with the ordered pair $(0,1)$, then it is natural to write the complex number $a + bi$ as the point (a, b). Here is a reasonable justification for this: $a + bi = a(1,0) + b(0,1) = (a, 0) + (0, b) = (a, b)$

In this way, we can visualize a complex number as a point in the **Complex Plane**. A portion of the Complex Plane is shown to the right with several complex numbers displayed as points of the form (x, y).

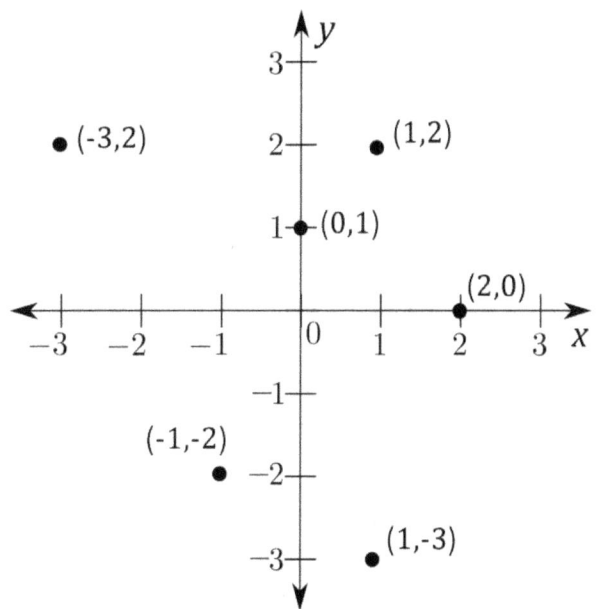

The Complex Plane is formed by taking two copies of the real line and placing one horizontally and the other vertically. The horizontal copy of the real line is called the x-axis or the **real axis** (labeled x in the figure) and the vertical copy of the real line is called the y-axis or **imaginary axis** (labeled y in the figure). The two axes intersect at the point $(0, 0)$. This point is called the **origin**.

The **empty set** is the unique set with no elements. We use the symbol $\emptyset$ to denote the empty set (some authors use the symbol $\{\ \}$ instead).

If A is a finite set, we define the **cardinality** of A, written $|A|$, to be the number of elements of A. For example, $|\{a, b\}| = 2$.

Example 1.7: Let $A = \{\text{painting, diamond, starfish}\}$, $B = \{c, d, d\}$, $C = \{25, 26, 27, \ldots, 3167, 3168\}$, $D = \{\{x\}, \{x, x\}, \{x, x, x\}\}$, and $E = \emptyset$. Then $|A| = 3$, $|B| = 2$, $|C| = 3144$, $|D| = 1$, and $|E| = 0$.

Notes: (1) The set A consists of the three elements "painting," "diamond," and "starfish."

(2) The set B consists of just two elements: c and d. Remember that $\{c, d, d\} = \{c, d\}$.

(3) The number of consecutive integers from m to n, inclusive, is $\boldsymbol{n - m + 1}$. For set C, we have $m = 25$ and $n = 3168$. Therefore, $|C| = 3168 - 25 + 1 = 3144$.

(4) I call the formula "$n - m + 1$" the **fence-post formula**. If you construct a 3-foot fence by placing a fence-post every foot, then the fence will consist of 4 fence-posts ($3 - 0 + 1 = 4$).

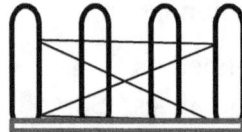

(5) Since $\{x, x\} = \{x\}$ and $\{x, x, x\} = \{x\}$, it follows that $D = \{\{x\}, \{x\}, \{x\}\} = \{\{x\}\}$. So, D consists of the single element $\{x\}$.

(6) Remember that $\emptyset$ (pronounced "the empty set") is the unique set with no elements.

Subsets and Proper Subsets

We say that a set A is a **subset** of a set B, written $A \subseteq B$, if every element of A is an element of B.

Example 1.8:

1. Let $A = \{x, y\}$ and $B = \{x, y, z\}$. The only elements of A are x and y. Since x and y are also elements of B, we see that $A \subseteq B$.

    Notice that $B \nsubseteq A$ (B is **not** a subset of A) because $z \in B$, but $z \notin A$.

2. Let $\mathbb{N} = \{0, 1, 2, 3, \ldots\}$ be the set of natural numbers and let $\mathbb{Z} = \{\ldots, -3, -2, -1, 0, 1, 2, 3, 4, \ldots\}$ be the set of integers. Since every natural number is an integer, $\mathbb{N} \subseteq \mathbb{Z}$.

3. As we saw in Example 1.6, by making appropriate identifications, we have the following sequence of inclusions:

    $$\mathbb{N} \subseteq \mathbb{Z} \subseteq \mathbb{Q} \subseteq \mathbb{R} \subseteq \mathbb{C}$$

    We will see in Theorem 1.14 below that if $A \subseteq B$ and $B \subseteq C$, then $A \subseteq C$ (we say that $\subseteq$ is a **transitive** relation). In this way we see that we have many other inclusions such as $\mathbb{N} \subseteq \mathbb{Q}$, $\mathbb{N} \subseteq \mathbb{R}$,..., and so on.

4. Consider the sets $2\mathbb{Z} = \{\ldots, -6, -4, -2, 0, 2, 4, 6, \ldots\}$ and $4\mathbb{Z} = \{\ldots, -12, -8, -4, 0, 4, 8, 12, \ldots\}$. Then $4\mathbb{Z} \subseteq 2\mathbb{Z}$. Note that the opposite inclusion is false. That is, $2\mathbb{Z} \nsubseteq 4\mathbb{Z}$. To see this, we just need a single **counterexample** (a counterexample is an example that is used to show that a statement is false). Well, we have $2 \in 2\mathbb{Z}$, but $2 \notin 4\mathbb{Z}$.

5. Consider the sets $2\mathbb{Z} = \{\ldots, -6, -4, -2, 0, 2, 4, 6, \ldots\}$ and $3\mathbb{Z} = \{\ldots, -9, -6, -3, 0, 3, 6, 9, \ldots\}$. Neither of these sets is a subset of the other. To see that $2\mathbb{Z} \nsubseteq 3\mathbb{Z}$, observe that $2 \in 2\mathbb{Z}$, whereas $2 \notin 3\mathbb{Z}$. To see that $3\mathbb{Z} \nsubseteq 2\mathbb{Z}$, observe that $3 \in 3\mathbb{Z}$, whereas $3 \notin 2\mathbb{Z}$.

To the right we see a physical representation of $A \subseteq B$. This figure is called a **Venn diagram**. These types of diagrams are very useful to help visualize relationships among sets. Notice that set A lies completely inside set B. We assume that all the elements of A and B lie in some **universal set** U.

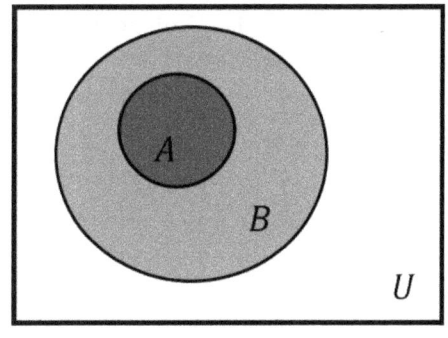

$A \subseteq B$

As an example, let's let U be the set of all species of animals. If we let A be the set of species of cats and we let B be the set of species of mammals, then we have $A \subseteq B \subseteq U$, and we see that the Venn diagram to the right gives a visual representation of this situation. (Note that every cat is a mammal and every mammal is an animal.)

We say that A is a **proper subset** of B, written $A \subset B$ (or sometimes $A \subsetneq B$), if $A \subseteq B$, but $A \neq B$. For example, $\mathbb{N} \subset \mathbb{Z}$, whereas $\mathbb{N} \not\subset \mathbb{N}$ (although $\mathbb{N} \subseteq \mathbb{N}$).

Note: The definition of proper subset is not very important. It just gives us a convenient way to discuss all the subsets of a specific set except for the set itself. For example, it is quite cumbersome to say "Find all subsets of A, but exclude the set A." It's nice to be able to rephrase this as "Find all proper subsets of A."

Let's look at the definition of $\subseteq$ (subset) in a bit more detail.

Once again, we write $A \subseteq B$ if every element of A is an element of B. That is, $A \subseteq B$ if, for every x, $x \in A$ implies $x \in B$. Symbolically, we can write the following:

$$\forall x(x \in A \rightarrow x \in B)$$

Notes: (1) The symbol $\forall$ is called a **universal quantifier**, and it is pronounced "For all."

(2) The logical expression $\forall x(x \in A \rightarrow x \in B)$ can be translated into English as "For all x, if x is an element of A, then x is an element of B."

(3) To show that a set A is a subset of a set B, we need to show that the expression $\forall x(x \in A \rightarrow x \in B)$ is true. If the set A is finite and the elements are listed, we can just check that each element of A is also an element of B. However, if the set A is described by a property, say $A = \{x|P(x)\}$, we may need to craft an argument more carefully. We can begin by taking an **arbitrary but specific element** a from A and then arguing that this element a is in B.

What could we possibly mean by an arbitrary but specific element? Aren't the words "arbitrary" and "specific" antonyms? Well, by arbitrary, we mean that we don't know which element we are choosing – it's just some element a that satisfies the property P. So, we are just assuming that $P(a)$ is true. However, once we choose this element a, we use this same a for the rest of the argument, and that is what we mean by it being specific.

(4) "$p \rightarrow q$" is an example of a **statement** in **propositional logic**. It is usually read as "if p, then q" or "p implies q." The letters p and q are called **propositional variables**, and we generally assign a truth value of T (for true) or F (for false) to each propositional variable. Formally, we define a **truth assignment** of a list of propositional variables to be a choice of T or F for each propositional variable in the list.

The symbol → is called a **conditional** or **implication**. It is one example of a **logical connective** (it *connects* two propositional variables). The rules for determining the truth value for $p \to q$ are given by the following truth table:

p	q	$p \to q$
T	T	T
T	F	F
F	T	T
F	F	T

For example, if p and q are both assigned the truth value T, then the truth value of $p \to q$ is also T, as can be seen by the first row of the above truth table. We can write T → T ≡ T. The symbol "≡" can be read **"is logically equivalent to."** Similarly, we have

$$T \to F \equiv F \qquad F \to T \equiv T \qquad F \to F \equiv T$$

Observe that the only time $p \to q$ can be false is if p is true and q is false. So, one way to prove that $p \to q$ is true is to assume that p is true and then provide a logically correct argument that q must also be true.

If we let p represent the statement "$x \in A$" and we let q represent the statement "$x \in B$," then $p \to q$ represents the statement "if $x \in A$, then $x \in B$." As stated in the last paragraph, one way to prove that this statement is true is to assume that $x \in A$ is true and then provide a logically correct argument that $x \in B$ must also be true.

Basic Theorems Involving Subsets

Let's try to prove our first theorem using the definition of a subset together with Note 3 above about arbitrary but specific elements.

Theorem 1.9: Every set A is a subset of itself.

Analysis: Before writing the proof, let's think about our strategy. We want to prove $A \subseteq A$. In other words, we want to show $\forall x (x \in A \to x \in A)$. So, we will take an arbitrary but specific $a \in A$ and then argue that $a \in A$. But that's pretty obvious, isn't it? In this case, the property we're describing is precisely the conclusion we are looking for. Here are the details.

Proof of Theorem 1.9: Let A be a set and let $a \in A$. Then $a \in A$. So, $a \in A \to a \in A$ is true. Since a was an arbitrary element of A, $\forall x (x \in A \to x \in A)$ is true. Therefore, $A \subseteq A$. □

Notes: (1) The proof begins with the **opening statement** "Let A be a set and let $a \in A$." In general, the opening statement states what is given in the problem and/or fixes any arbitrary but specific objects that we will need.

(2) The proof ends with the **closing statement** "Therefore, $A \subseteq A$." In general, the closing statement states the result.

(3) Everything between the opening statement and the closing statement is known as the **argument**.

(4) We place the symbol □ at the end of the proof to indicate that the proof is complete.

(5) Consider the logical statement $p \to p$. This statement is always true (T $\to$ T $\equiv$ T and F $\to$ F $\equiv$ T). $p \to p$ is an example of a tautology. A **tautology** is a statement that is true for every possible truth assignment of the propositional variables.

(6) If we let p represent the statement $a \in A$, by Note 5, we see that $a \in A \to a \in A$ is always true.

Alternate proof of Theorem 1.9: Let A be a set and let $a \in A$. Since $p \to p$ is a tautology, we have that $a \in A \to a \in A$ is true. Since a was arbitrary, $\forall x(x \in A \to x \in A)$ is true. Therefore, $A \subseteq A$. □

Let's prove another basic but important theorem.

Theorem 1.10: The empty set is a subset of every set.

Analysis: This time we want to prove $\emptyset \subseteq A$. In other words, we want to show $\forall x(x \in \emptyset \to x \in A)$. Since $x \in \emptyset$ is always false (the empty set has no elements), $x \in \emptyset \to x \in A$ is always true.

In general, if p is a false statement, then we say that $p \to q$ is **vacuously true**.

Proof of Theorem 1.10: Let A be a set. The statement $x \in \emptyset \to x \in A$ is vacuously true for any x, and so, $\forall x(x \in \emptyset \to x \in A)$ is true. Therefore, $\emptyset \subseteq A$. □

Note: The opening statement is "Let A be a set," the closing statement is "Therefore, $\emptyset \subseteq A$," and the argument is everything in between.

Example 1.11: Let $C = \{a, b, c\}$, $D = \{a, c\}$, $E = \{b, c\}$, $F = \{b, d\}$, and $G = \emptyset$. Then $D \subseteq C$ and $E \subseteq C$. Also, since *the empty set is a subset of every set*, we have $G \subseteq C$, $G \subseteq D$, $G \subseteq E$, $G \subseteq F$, and $G \subseteq G$. *Every set is a subset of itself*, and so, $C \subseteq C$, $D \subseteq D$, $E \subseteq E$, and $F \subseteq F$.

Note: Below are possible Venn diagrams for this problem. The diagram on the left shows the relationship between the sets C, D, E, and F. Notice how D and E are both subsets of C, whereas F is not a subset of C. Also, notice how D and E overlap, E and F overlap, but there is no overlap between D and F (they have no elements in common). The diagram on the right shows the proper placement of the elements. Here, I chose the universal set to be $U = \{a, b, c, d, e, f, g\}$. This choice for the universal set is somewhat arbitrary. Any set containing $\{a, b, c, d\}$ would do.

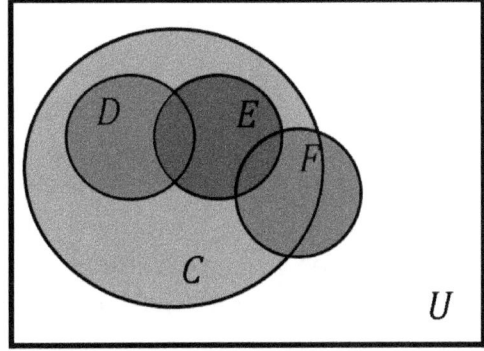
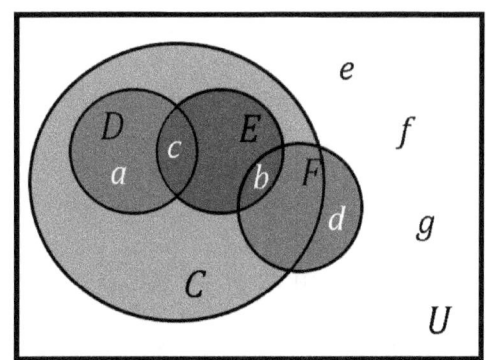

Power Sets

If A is a set, then the **power set** of A, written $\mathcal{P}(A)$, is the set of all subsets of A. In set-builder notation, we write $\mathcal{P}(A) = \{B \mid B \subseteq A\}$.

Example 1.12: The set $A = \{a, b\}$ has 2 elements and 4 subsets. The subsets of A are $\emptyset$, $\{a\}$, $\{b\}$, and $\{a, b\}$. It follows that $\mathcal{P}(A) = \{\emptyset, \{a\}, \{b\}, \{a, b\}\}$.

The set $B = \{a, b, c\}$ has 3 elements and 8 subsets. The subsets of B are $\emptyset$, $\{a\}$, $\{b\}$, $\{c\}$, $\{a, b\}$, $\{a, c\}$, $\{b, c\}$, and $\{a, b, c\}$. It follows that $\mathcal{P}(B) = \{\emptyset, \{a\}, \{b\}, \{c\}, \{a, b\}, \{a, c\}, \{b, c\}, \{a, b, c\}\}$.

Let's draw a **tree diagram** for the subsets of each of the sets A and B.

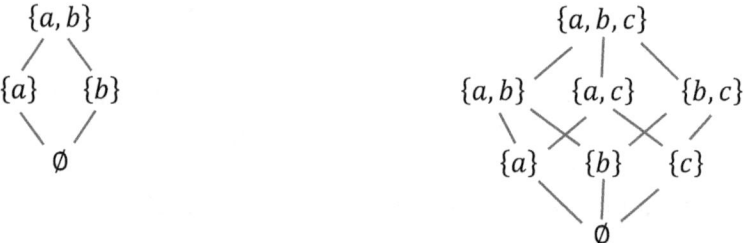

The tree diagram on the left is for the subsets of the set $A = \{a, b\}$. We start by writing the set $A = \{a, b\}$ at the top. On the next line we write the subsets of cardinality 1 ($\{a\}$ and $\{b\}$). On the line below that we write the subsets of cardinality 0 (just $\emptyset$). We draw a line segment between any two sets when the smaller (lower) set is a subset of the larger (higher) set. So, we see that $\emptyset \subseteq \{a\}$, $\emptyset \subseteq \{b\}$, $\{a\} \subseteq \{a, b\}$, and $\{b\} \subseteq \{a, b\}$. There is actually one more subset relationship, namely $\emptyset \subseteq \{a, b\}$ (and of course each set displayed is a subset of itself). We didn't draw a line segment from $\emptyset$ to $\{a, b\}$ to avoid unnecessary clutter. Instead, we can simply trace the path from $\emptyset$ to $\{a\}$ to $\{a, b\}$ (or from $\emptyset$ to $\{b\}$ to $\{a, b\}$). We are using a property called **transitivity** here (see Theorem 1.14 below).

The tree diagram on the right is for the subsets of $B = \{a, b, c\}$. Observe that from top to bottom we write the subsets of B of cardinality 3, then 2, then 1, and then 0. We then draw the appropriate line segments, just as we did for $A = \{a, b\}$.

How many subsets does a set of cardinality n have? Let's start by looking at some examples.

Example 1.13: A set with 0 elements must be $\emptyset$, and this set has exactly 1 subset (the only subset of the empty set is the empty set itself).

A set with 1 element has 2 subsets, namely $\emptyset$ and the set itself.

In the last example, we saw that a set with 2 elements has 4 subsets, and we also saw that a set with 3 elements has 8 subsets.

Do you see the pattern yet? $1 = 2^0$, $2 = 2^1$, $4 = 2^2$, $8 = 2^3$. So, we see that a set with 0 elements has 2^0 subsets, a set with 1 element has 2^1 subsets, a set with 2 elements has 2^2 subsets, and a set with 3 elements has 2^3 subsets.

A reasonable guess would be that a set with n elements has 2^n subsets. You will be asked to prove this result later (Problem 16 in Problem Set 4). We can also say that if $|A| = n$, then $|\mathcal{P}(A)| = 2^n$.

Transitivity of the Subset Relation

Let's get back to the transitivity mentioned above in Example 1.8 and in our discussion of tree diagrams.

Theorem 1.14: Let A, B, and C be sets such that $A \subseteq B$ and $B \subseteq C$. Then $A \subseteq C$.

Proof: Suppose that A, B, and C are sets with $A \subseteq B$ and $B \subseteq C$, and let $a \in A$. Since $A \subseteq B$ and $a \in A$, it follows that $a \in B$. Since $B \subseteq C$ and $a \in B$, it follows that $a \in C$. Since a was an arbitrary element of A, we have shown that every element of A is an element of C. That is, $\forall x(x \in A \rightarrow x \in C)$ is true. Therefore, $A \subseteq C$. □

Note: To the right we have a Venn diagram illustrating Theorem 1.14.

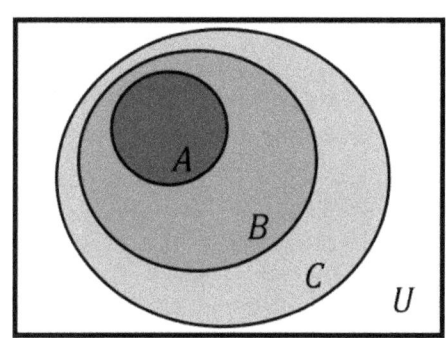

$A \subseteq B \subseteq C$

Theorem 1.14 tells us that the relation $\subseteq$ is **transitive**. Since $\subseteq$ is transitive, we can write things like $A \subseteq B \subseteq C \subseteq D$, and without explicitly saying it, we know that $A \subseteq C$, $A \subseteq D$, and $B \subseteq D$.

Example 1.15: The membership relation $\in$ is an example of a relation that is **not** transitive. For example, let $A = \{0\}$, $B = \{0, t, \{0\}\}$, and $C = \{x, y, \{0, t, \{0\}\}\}$. Observe that $A \in B$ and $B \in C$, but $A \notin C$.

Notes: (1) The set A has just 1 element, namely 0.

(2) The set B has 3 elements, namely 0, t, and $\{0\}$. But wait! $A = \{0\}$. So, $A \in B$. The set A is circled twice in the above image.

(3) The set C also has 3 elements, namely, x, y, and $\{0, t, \{0\}\}$. But wait! $B = \{0, t, \{0\}\}$. So, $B \in C$. The set B has a rectangle around it twice in the above image.

(4) Since $A \neq x$, $A \neq y$, and $A \neq \{0, t, \{0\}\}$, we see that $A \notin C$.

(5) Is it clear that $\{0\} \notin C$? $\{0\}$ is in a set that's in C (namely, B), but $\{0\}$ is not itself in C.

(6) Here is a more basic example showing that $\in$ is not transitive: $\emptyset \in \{\emptyset\} \in \{\{\emptyset\}\}$, but $\emptyset \notin \{\{\emptyset\}\}$ The only element of $\{\{\emptyset\}\}$ is $\{\emptyset\}$.

Equality of Sets

Two sets A and B are **equal**, written $A = B$, if they have the same elements. Symbolically, we can write the following:

$$\forall x(x \in A \leftrightarrow x \in B)$$

Notes: (1) "$p \leftrightarrow q$" is another example of a **statement** in **propositional logic**. It is usually read as "p if and only if q."

The logical connective $\leftrightarrow$ is called a **biconditional**. The rules for determining the truth value for $p \leftrightarrow q$ are given by the following truth table:

p	q	$p \leftrightarrow q$
T	T	T
T	F	F
F	T	F
F	F	T

In other words, $p \leftrightarrow q$ is true when p and q have the same truth value (both T or both F) and false when p and q have opposite truth values (one T and the other F).

If we let p represent the statement "$x \in A$" and we let q represent the statement "$x \in B$," then $p \leftrightarrow q$ represents the statement "$x \in A$ if and only if $x \in B$."

(2) In addition to the conditional ($\rightarrow$) and biconditional ($\leftrightarrow$), there are three additional commonly used logical connectives: the **conjunction** ($\wedge$), the **disjunction** ($\vee$), and the **negation** ($\neg$). They have the following truth tables:

p	q	$p \wedge q$
T	T	T
T	F	F
F	T	F
F	F	F

p	q	$p \vee q$
T	T	T
T	F	T
F	T	T
F	F	F

p	$\neg p$
T	F
F	T

We can use statements in place of propositional variables to form compound statements. For example, given the statements $p \rightarrow q$ and $q \rightarrow p$, we can form the compound statement $(p \rightarrow q) \wedge (q \rightarrow p)$. If A and B are arbitrary statements and we know the truth values of A and B, then we can find the truth value of A $\wedge$ B in the same way as if A and B were propositional variables. The same holds true when applying other logical connectives to compound statements. We will use this in Note 3 below.

(3) We say that two statements are **logically equivalent** if every truth assignment of the propositional variables appearing in either statement (or both statements) leads to the same truth value for both statements.

It is easy to verify that $p \leftrightarrow q$ is logically equivalent to $(p \rightarrow q) \wedge (q \rightarrow p)$. To see this, we check that all possible truth assignments for p and q lead to the same truth value for the two statements. For example, if p and q are both true, then

$$p \leftrightarrow q \equiv T \leftrightarrow T \equiv T \quad \text{and} \quad (p \rightarrow q) \wedge (q \rightarrow p) \equiv (T \rightarrow T) \wedge (T \rightarrow T) \equiv T \wedge T \equiv T.$$

As another example, if p is true and q is false, then

$$p \leftrightarrow q \equiv T \leftrightarrow F \equiv F \quad \text{and} \quad (p \rightarrow q) \wedge (q \rightarrow p) \equiv (T \rightarrow F) \wedge (F \rightarrow T) \equiv F \wedge T \equiv F.$$

The reader should check the other two truth assignments for p and q.

(4) Letting p be the statement $x \in A$, letting q be the statement $x \in B$, and replacing $p \leftrightarrow q$ by the logically equivalent statement $(p \to q) \wedge (q \to p)$ gives us

$$\forall x(x \in A \leftrightarrow x \in B) \text{ if and only if } \forall x\big((x \in A \to x \in B) \wedge (x \in B \to x \in A)\big).$$

(5) It is also true that $\forall x\big(p(x) \wedge q(x)\big)$ is logically equivalent to $\forall x\big(p(x)\big) \wedge \forall x\big(q(x)\big)$. And so, we have

$$\forall x(x \in A \leftrightarrow x \in B) \text{ if and only if } \forall x(x \in A \to x \in B) \text{ and } \forall x(x \in B \to x \in A).$$

In other words, to show that $A = B$, we can instead show that $A \subseteq B$ and $B \subseteq A$.

The statement "$A = B$ if and only if $A \subseteq B$ and $B \subseteq A$" is usually called the **Axiom of Extensionality**. It is often easiest to prove that two sets are equal by showing that each one is a subset of the other.

Example 1.16: Let $A = \{n \in \mathbb{N} \mid n < 100\}$ and let $B = \{n \in \mathbb{Z} \mid 0 \leq n \leq 99\}$. Let's use the Axiom of Extensionality to prove that $A = B$.

Note: We are assuming that $<$ and $\leq$ are defined in the usual way on $\mathbb{N}$ and $\mathbb{Z}$. In the proof, we will freely use the properties of these relations that we know to be true. For example, 0 is the least element of $\mathbb{N}$ with respect to $<$ and $\leq$. As another example, in both $\mathbb{N}$ and $\mathbb{Z}$, we have $n < k + 1$ if and only if $n \leq k$.

Proof: We first prove that $A \subseteq B$. Let $n \in A$. Then $n \in \mathbb{N}$ and $n < 100$. Since $n \in \mathbb{N}$ and $\mathbb{N} \subseteq \mathbb{Z}$, it follows that $n \in \mathbb{Z}$. Since $n \in \mathbb{N}$ and the least element of $\mathbb{N}$ is 0, $0 \leq n$. Since $n < 100$ and $n \in \mathbb{N}$, we have $n \leq 99$. Therefore, $n \in B$. Since $n \in A$ was arbitrary, $\forall n(n \in A \to n \in B)$. Therefore, we have shown $A \subseteq B$.

We now prove that $B \subseteq A$. Let $n \in B$. Then $n \in \mathbb{Z}$ and $0 \leq n \leq 99$. Since $n \in \mathbb{Z}$ and $0 \leq n$, it follows that $n \in \mathbb{N}$. Since $n \leq 99$ and $99 < 100$, it follows that $n < 100$. Therefore, $n \in A$. Since $n \in B$ was arbitrary, $\forall n(n \in B \to n \in A)$. Therefore, we have shown $B \subseteq A$.

Since $A \subseteq B$ and $B \subseteq A$, we have $A = B$. □

Cartesian Products

An **unordered pair** is a set with 2 elements. Recall, that a set doesn't change if we write the elements in a different order or if we write the same element multiple times. For example, $\{x, y\} = \{y, x\}$ and $\{x, x\} = \{x\}$.

We now define the **ordered pair** (x, y) in such a way that (y, x) will **not** be the same as (x, y). The simplest way to define a set with this property is as follows:

$$(x, y) = \{\{x\}, \{x, y\}\}$$

Let's show that with this definition, the ordered pair behaves as we would expect.

Theorem 1.17: $(x, y) = (z, w)$ if and only if $x = z$ and $y = w$.

Part of the proof of this theorem is a little trickier than expected. Assuming that $(x, y) = (z, w)$, there are actually two cases to consider: $x = y$ and $x \neq y$. If $x = y$, then (x, y) is a set with just one element. Indeed, $(x, x) = \{\{x\}, \{x, x\}\} = \{\{x\}, \{x\}\} = \{\{x\}\}$. So, the only element of (x, x) is $\{x\}$. Watch carefully how this plays out in the proof.

Proof of Theorem 1.17: First suppose that $x = z$ and $y = w$. Then by direct substitution, $\{x\} = \{z\}$ and $\{x, y\} = \{z, w\}$. So, $(x, y) = \{\{x\}, \{x, y\}\} = \{\{z\}, \{z, w\}\} = (z, w)$.

Conversely, suppose that $(x, y) = (z, w)$. Then $\{\{x\}, \{x, y\}\} = \{\{z\}, \{z, w\}\}$. There are two cases to consider.

Case 1: If $x = y$, then $\{\{x\}, \{x, y\}\} = \{\{x\}\}$. So, $\{\{x\}\} = \{\{z\}, \{z, w\}\}$. It follows that $\{z\} = \{x\}$ and $\{z, w\} = \{x\}$. Since $\{z, w\} = \{x\}$, we must have $z = x$ and $w = x$. Therefore, $x, y, z,$ and w are all equal. In particular, $x = z$ and $y = w$.

Case 2: If $x \neq y$, then $\{x, y\}$ is a set with two elements. So, $\{x, y\}$ cannot be equal to $\{z\}$ (because $\{z\}$ has just one element). Therefore, we must have $\{x, y\} = \{z, w\}$. It then follows that $\{x\} = \{z\}$. So, we have $x = z$. Since $x = z$ and $\{x, y\} = \{z, w\}$, we must have $y = w$. □

Note: (x, y) is an abbreviation for the set $\{\{x\}, \{x, y\}\}$. In the study of set theory, every object can be written as a set like this. It's often convenient to use abbreviations, but we should always be aware that if necessary, we can write any object in its unabbreviated form.

We can extend the idea of an ordered pair to an **ordered k-tuple**. An ordered 3-tuple (also called an **ordered triple**) is defined by $(x, y, z) = ((x, y), z)$, an ordered 4-tuple is $(x, y, z, w) = ((x, y, z), w)$, and so on. For a general k-tuple, we will use a single letter with subscripts for the variable names. For example, using the letter x, we can write a k-tuple as $(x_1, x_2, \ldots, x_k)$.

The dedicated reader may want to verify that $(x, y, z) = (u, v, w)$ if and only if $x = u$, $y = v$, and $z = w$. This result extends in the obvious way to k-tuples in general.

Example 1.18: Let's write the ordered triple (x, y, z) in its unabbreviated form.

$$(x, y, z) = ((x, y), z) = \{\{(x, y)\}, \{(x, y), z\}\} = \{\{\{\{x\}, \{x, y\}\}\}, \{\{\{x\}, \{x, y\}\}, z\}\}$$

The **Cartesian product** of the sets A and B, written $A \times B$ is the set of ordered pairs (a, b) with $a \in A$ and $b \in B$. Symbolically, we have

$$A \times B = \{(a, b) \mid a \in A \land b \in B\}.$$

Observe that if A and B are finite sets with $|A| = m$ and $|B| = n$, then $|A \times B| = mn$.

Example 1.19:
1. Let $A = \{x, y\}$ and $B = \{0, 1, 2\}$. Then $A \times B = \{(x, 0), (x, 1), (x, 2), (y, 0), (y, 1), (y, 2)\}$. Note that $|A| = 2$, $|B| = 3$, and $|A \times B| = 2 \cdot 3 = 6$.
2. Let $C = \emptyset$ and $D = \{2, 4, 6, 8, 10\}$. Then $C \times D = \emptyset$ (since there are no elements in C, there can be no elements in $C \times D$). Note that $|C| = 0$, $|D| = 5$, and $|C \times D| = 0 \cdot 5 = 0$.

3. $\mathbb{N} \times \mathbb{Z} = \{(m, n) \mid m \in \mathbb{N} \wedge n \in \mathbb{Z}\}$. For example, $(5, -3) \in \mathbb{N} \times \mathbb{Z}$, whereas $(-3, 5) \notin \mathbb{N} \times \mathbb{Z}$ (although it is in $\mathbb{Z} \times \mathbb{N}$). We can visualize $\mathbb{N} \times \mathbb{Z}$ as follows:

$$\ldots, (0, -3), (0, -2), (0, -1), (0, 0), (0, 1), (0, 2), (0, 3), \ldots$$
$$\ldots, (1, -3), (1, -2), (1, -1), (1, 0), (1, 1), (1, 2), (1, 3), \ldots$$
$$\ldots, (2, -3), (2, -2), (2, -1), (2, 0), (2, 1), (2, 2), (2, 3), \ldots$$
$$\vdots \qquad \vdots \qquad \vdots$$

4. $\mathbb{R} \times \mathbb{R} = \{(x, y) \mid x, y \in \mathbb{R}\}$. We can visualize elements of $\mathbb{R} \times \mathbb{R}$ as points in the **Cartesian plane.** A portion of the Cartesian plane is shown to the right. The elements $(3, 2)$ and $(-1, -2)$ of $\mathbb{R} \times \mathbb{R}$ are displayed as points.

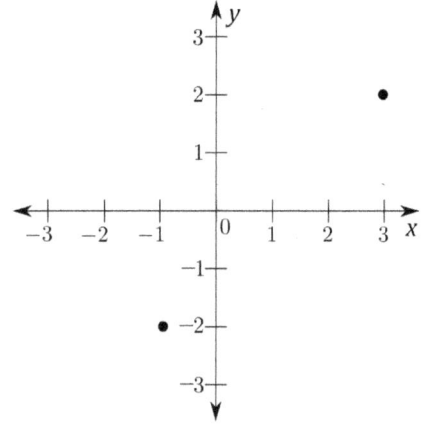

We form the Cartesian plane by taking two copies of the real line and placing one horizontally and the other vertically, exactly as we did for the Complex Plane in part 3 of Example 1.6. The horizontal copy of the real line is called the x-axis (labeled x in the figure) and the vertical copy of the real line is called the y-axis (labeled y in the figure). The two axes intersect at the point $(0, 0)$. This point is called the **origin**.

Notice that visually the Cartesian plane $\mathbb{R} \times \mathbb{R}$ is indistinguishable from the Complex Plane.

We can extend the definition of the Cartesian product to more than two sets in the obvious way:
$$A \times B \times C = \{(a, b, c) \mid a \in A \wedge b \in B \wedge c \in C\}$$
$$A \times B \times C \times D = \{(a, b, c, d) \mid a \in A \wedge b \in B \wedge c \in C \wedge d \in D\}$$

Observe that if A, B, and C are finite sets with $|A| = m$, $|B| = n$, and $|C| = k$, then we have $|A \times B \times C| = mnk$. Similarly, if A, B, C, and D are finite sets with $|A| = m$, $|B| = n$, $|C| = k$, and $|D| = j$, then we have $|A \times B \times C \times D| = mnkj$.

In general, the **Cartesian product** of the sets $A_1, A_2, \ldots, A_n$ is
$$A_1 \times A_2 \times \cdots \times A_n = \{(a_1, a_2, \ldots, a_n) \mid a_1 \in A_1 \wedge a_2 \in A_2 \wedge \cdots \wedge a_n \in A_n\}.$$

Note: Recall that $\wedge$ is just a symbol representing the word "and."

If $A_1, A_2, \ldots, A_n$ are finite, then the cardinality of $A_1 \times A_2 \times \cdots \times A_n$ is the product of the cardinalities of $A_1, A_2, \ldots, A_n$. Symbolically, we have $|A_1 \times A_2 \times \cdots \times A_n| = |A_1| \cdot |A_2| \cdots |A_n|$.

Example 1.20:

1. $\{a\} \times \{b\} \times \{c\} \times \{d\} = \{(a, b, c, d)\}$.

   Note that $|\{a\} \times \{b\} \times \{c\} \times \{d\}| = |\{a\}| \cdot |\{b\}| \cdot |\{c\}| \cdot |\{d\}| = 1 \cdot 1 \cdot 1 \cdot 1 = 1$.

2. $\{0\} \times \{0, 1\} \times \{1\} \times \{0, 1\} \times \{0\} = \{(0, 0, 1, 0, 0), (0, 0, 1, 1, 0), (0, 1, 1, 0, 0), (0, 1, 1, 1, 0)\}$.

   Note that $|\{0\} \times \{0, 1\} \times \{1\} \times \{0, 1\} \times \{0\}| = 1 \cdot 2 \cdot 1 \cdot 2 \cdot 1 = 4$.

3. $\mathbb{Z} \times \mathbb{N} \times \mathbb{R} = \{(m, n, x) \mid m \in \mathbb{Z} \land n \in \mathbb{N} \land x \in \mathbb{R}\}$.

We abbreviate Cartesian products of sets with themselves using exponents.

$$A^2 = A \times A \qquad A^3 = A \times A \times A \qquad A^4 = A \times A \times A \times A \qquad A^n = \underbrace{A \times A \times \cdots \times A}_{n \text{ times}}$$

Example 1.21:

1. $\mathbb{Z}^2 = \mathbb{Z} \times \mathbb{Z} = \{(x, y) \mid x, y \in \mathbb{Z}\}$ is the set of ordered pairs of integers. A few sample elements in $\mathbb{Z}^2$ are $(0, 0), (-1, 2), (15, -106)$ and $(-53, -53)$.

2. $\mathbb{N}^5 = \mathbb{N} \times \mathbb{N} \times \mathbb{N} \times \mathbb{N} \times \mathbb{N} = \{(a, b, c, d, e) \mid a, b, c, d, e \in \mathbb{N}\}$ is the set of ordered 5-tuples of natural numbers. A few sample elements in $\mathbb{N}^5$ are $(0, 0, 0, 0, 0), (1, 1, 2, 3, 3), (0, 1, 17, 86, 0)$, and $(1000, 2529, 8, 900, 106)$.

3. $\{0, 1\}^2 = \{0, 1\} \times \{0, 1\} = \{(0, 0), (0, 1), (1, 0), (1, 1)\}$.

4. $\{0, 1\}^3 = \{0, 1\} \times \{0, 1\} \times \{0, 1\}$
   $= \{(0, 0, 0), (0, 0, 1), (0, 1, 0), (0, 1, 1), (1, 0, 0), (1, 0, 1), (1, 1, 0), (1, 1, 1)\}$.

5. $\mathbb{R}^2 = \mathbb{R} \times \mathbb{R} = \{(x, y) \mid x, y \in \mathbb{R}\}$ was discussed in part 4 of Example 1.19 above. We will also be interested in larger Cartesian products of $\mathbb{R}$ such as $\mathbb{R}^3 = \{(x, y, z) \mid x, y, z \in \mathbb{R}\}$, and more generally, $\mathbb{R}^n = \{(x_1, x_2, \ldots, x_n) \mid x_1, x_2, \ldots, x_n \in \mathbb{R}\}$.

6. $\mathbb{C}^n = \{(z_1, z_2, \ldots, z_n) \mid z_1, z_2, \ldots, z_n \in \mathbb{C}\}$ is the set of ordered n-tuples of complex numbers. For example, $\left(3 + 2i, -1 + \frac{5}{7}i\right) \in \mathbb{C}^2$ and $\left(1 - i, 2 + \frac{1}{2}i, -5i, 0.253\right) \in \mathbb{C}^4$.

Basic Set Operations

The **union** of the sets A and B, written $A \cup B$, is the set of elements that are in A or B (or both).

$$A \cup B = \{x \mid x \in A \text{ or } x \in B\}$$

The **intersection** of A and B, written $A \cap B$, is the set of elements that are simultaneously in A and B.

$$A \cap B = \{x \mid x \in A \text{ and } x \in B\}$$

The following Venn diagrams for the union and intersection of two sets can be useful for visualizing these operations.

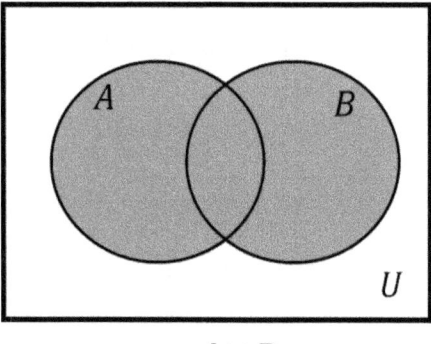

$A \cup B$

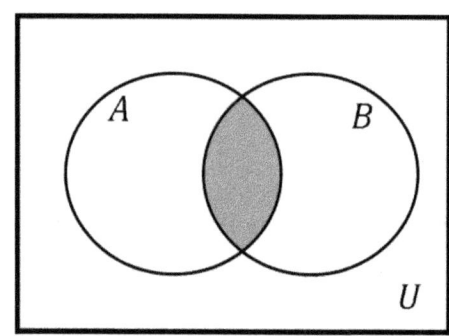

$A \cap B$

The **difference** $A \setminus B$ is the set of elements that are in A and not in B.
$$A \setminus B = \{x \mid x \in A \text{ and } x \notin B\}$$

The **symmetric difference** between A and B, written $A \triangle B$, is the set of elements that are in A or B, but not both.
$$A \triangle B = (A \setminus B) \cup (B \setminus A)$$

Let's also look at Venn diagrams for the difference and symmetric difference of two sets.

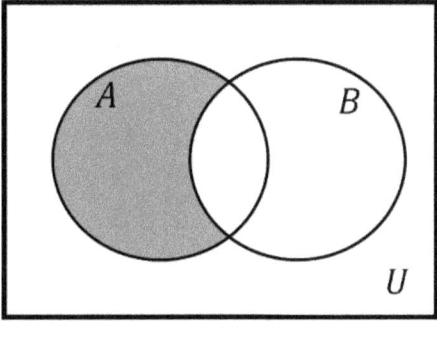

$A \setminus B$

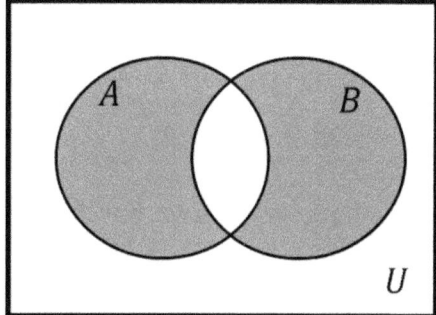

$A \triangle B$

Example 1.22: Let $A = \{0, 1, 2, 3, 4\}$ and $B = \{3, 4, 5, 6\}$. We have

1. $A \cup B = \{0, 1, 2, 3, 4, 5, 6\}$.
2. $A \cap B = \{3, 4\}$.
3. $A \setminus B = \{0, 1, 2\}$.
4. $B \setminus A = \{5, 6\}$.
5. $A \triangle B = \{0, 1, 2\} \cup \{5, 6\} = \{0, 1, 2, 5, 6\}$.

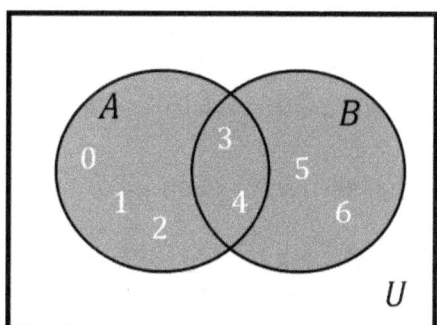

Example 1.23: Recall that the set of natural numbers is $\mathbb{N} = \{0, 1, 2, 3, \ldots\}$ and the set of integers is $\mathbb{Z} = \{\ldots, -4, -3, -2, -1, 0, 1, 2, 3, 4, \ldots\}$. Observe that in this case, $\mathbb{N} \subseteq \mathbb{Z}$. We have

1. $\mathbb{N} \cup \mathbb{Z} = \mathbb{Z}$.
2. $\mathbb{N} \cap \mathbb{Z} = \mathbb{N}$.
3. $\mathbb{N} \setminus \mathbb{Z} = \emptyset$.
4. $\mathbb{Z} \setminus \mathbb{N} = \{\ldots, -4, -3, -2, -1\} = \mathbb{Z}^-$. ($\mathbb{Z}^-$ is "the set of negative integers.")
5. $\mathbb{N} \triangle \mathbb{Z} = \emptyset \cup \mathbb{Z}^- = \mathbb{Z}^-$.

Note: Whenever A and B are sets and $B \subseteq A$, then $A \cup B = A$, $A \cap B = B$, and $B \setminus A = \emptyset$. We will prove the first and third of these three facts in Theorems 1.27 and 1.28 below, respectively. You will be asked to prove the second in Problem 21 below.

Example 1.24: Let $\mathbb{E} = \{0, 2, 4, 6, \ldots\}$ be the set of even natural numbers and let $\mathbb{O} = \{1, 3, 5, 7, \ldots\}$ be the set of odd natural numbers. We have

1. $\mathbb{E} \cup \mathbb{O} = \{0, 1, 2, 3, 4, 5, 6, 7, \ldots\} = \mathbb{N}$.
2. $\mathbb{E} \cap \mathbb{O} = \emptyset$.
3. $\mathbb{E} \setminus \mathbb{O} = \mathbb{E}$.
4. $\mathbb{O} \setminus \mathbb{E} = \mathbb{O}$.
5. $\mathbb{E} \triangle \mathbb{O} = \mathbb{E} \cup \mathbb{O} = \mathbb{N}$.

In general, we say that sets A and B are **disjoint** or **mutually exclusive** if $A \cap B = \emptyset$. To the right is a Venn diagram for disjoint sets.

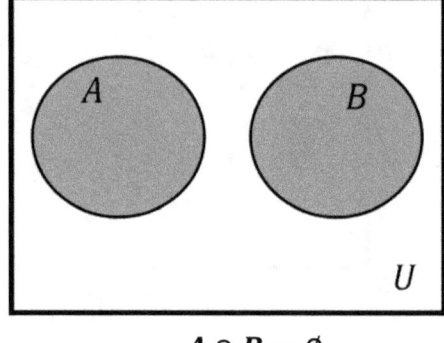

$A \cap B = \emptyset$

In Example 1.24 above, we saw that the sets $\mathbb{E}$ and $\mathbb{O}$ are disjoint.

Example 1.25: Consider the sets $A = \{a + bi \in \mathbb{C} \mid a, b \in \mathbb{Z}\}$ and $B = \{a + bi \in \mathbb{C} \mid a \notin \mathbb{Q}\}$. Then A and B are disjoint. To see this, suppose that $a + bi \in A \cap B$. Then $a + bi \in A$, and so, $a \in \mathbb{Z}$. Since $\mathbb{Z} \subseteq \mathbb{Q}$, $a \in \mathbb{Q}$. Also, $a + bi \in B$. So, $a \notin \mathbb{Q}$. Since we cannot have both $a \in \mathbb{Q}$ and $a \notin \mathbb{Q}$, we must have $A \cap B = \emptyset$.

Let's prove some theorems involving unions of sets. You will be asked to prove the analogous results for intersections of sets in Problems 18 and 21 below.

Theorem 1.26: If A and B are sets, then $A \subseteq A \cup B$.

Before going through the proof, look once more at the Venn diagram above for $A \cup B$ and convince yourself that this theorem should be true.

Proof of Theorem 1.26: Suppose that A and B are sets and let $x \in A$. Then $x \in A$ or $x \in B$. Therefore, $x \in A \cup B$. Since x was an arbitrary element of A, we have shown that every element of A is an element of $A \cup B$. That is, $\forall x(x \in A \to x \in A \cup B)$ is true. Therefore, $A \subseteq A \cup B$. □

Note: If p is a true statement, then $p \vee q$ (p or q) is true no matter what the truth value of q is. You can see this by looking at the truth table for $\vee$ above (see Note 2 before Example 1.16). In the second sentence of the proof above, we are using this fact with p being the statement $x \in A$ and q being the statement $x \in B$.

We will use this same reasoning in the second paragraph of the next proof as well.

Theorem 1.27: $B \subseteq A$ if and only if $A \cup B = A$.

Before going through the proof, it's a good idea to draw a Venn diagram for $B \subseteq A$ and convince yourself that this theorem should be true.

Proof of Theorem 1.27: Suppose that $B \subseteq A$ and let $x \in A \cup B$. Then $x \in A$ or $x \in B$. If $x \in A$, then $x \in A$ (trivially). If $x \in B$, then since $B \subseteq A$, it follows that $x \in A$. Since x was an arbitrary element of $A \cup B$, we have shown that every element of $A \cup B$ is an element of A. That is, $\forall x(x \in A \cup B \rightarrow x \in A)$ is true. Therefore, $A \cup B \subseteq A$. By Theorem 1.26, $A \subseteq A \cup B$. Since $A \cup B \subseteq A$ and $A \subseteq A \cup B$, it follows that $A \cup B = A$.

Now, suppose that $A \cup B = A$ and let $x \in B$. Since $x \in B$, it follows that $x \in A$ or $x \in B$. Therefore, $x \in A \cup B$. Since $A \cup B = A$, we have $x \in A$. Since x was an arbitrary element of B, we have shown that every element of B is an element of A. That is, $\forall x(x \in B \rightarrow x \in A)$. Therefore, $B \subseteq A$. □

Theorem 1.28: Let A and B be sets. If $B \subseteq A$, then $B \setminus A = \emptyset$.

We will use an **indirect proof** to prove Theorem 1.28. Specifically, we will use a **proof by contrapositive**.

The contrapositive of the conditional statement $p \rightarrow q$ is the statement $\neg q \rightarrow \neg p$. These two statements are logically equivalent. To see this, we check that all possible truth assignments for p and q lead to the same truth value for the two statements. For example, if p and q are both true, then $p \rightarrow q \equiv T \rightarrow T \equiv T$ and $\neg q \rightarrow \neg p \equiv F \rightarrow F \equiv T$. The reader should check the other three truth assignments for p and q.

The contrapositive of the statement "If $B \subseteq A$, then $B \setminus A = \emptyset$" is "If $B \setminus A \neq \emptyset$, then $B \nsubseteq A$." So, we will prove Theorem 1.28 by assuming that $B \setminus A \neq \emptyset$ and using this to show that $B \nsubseteq A$.

Proof of Theorem 1.28: Let A and B be sets such that $B \setminus A \neq \emptyset$. Since $B \setminus A \neq \emptyset$, there is $a \in B \setminus A$. Then $a \in B$ and $a \notin A$. So, $a \in B$ is true and $a \in A$ is false. So, $a \in B \rightarrow a \in A$ is false (because $T \rightarrow F \equiv F$). It follows that $\forall x(x \in B \rightarrow x \in A)$ is false, and therefore, $B \nsubseteq A$. □

Properties of Unions and Intersections

Unions, intersections, and set differences have many nice algebraic properties such as the following:

1. **Commutativity:** $A \cup B = B \cup A$ and $A \cap B = B \cap A$.
2. **Associativity:** $(A \cup B) \cup C = A \cup (B \cup C)$ and $(A \cap B) \cap C = A \cap (B \cap C)$.
3. **Distributivity:** $A \cap (B \cup C) = (A \cap B) \cup (A \cap C)$ and $A \cup (B \cap C) = (A \cup B) \cap (A \cup C)$.
4. **De Morgan's Laws:** $C \setminus (A \cup B) = (C \setminus A) \cap (C \setminus B)$ and $C \setminus (A \cap B) = (C \setminus A) \cup (C \setminus B)$.
5. **Idempotent Laws:** $A \cup A = A$ and $A \cap A = A$.

As an example, let's prove that the operation of forming unions is associative. You will be asked to prove that the other properties hold in the problems below.

Theorem 1.29: The operation of forming unions is associative.

Note: Before beginning the proof, let's draw Venn diagrams of the situation to convince ourselves that the theorem is true.

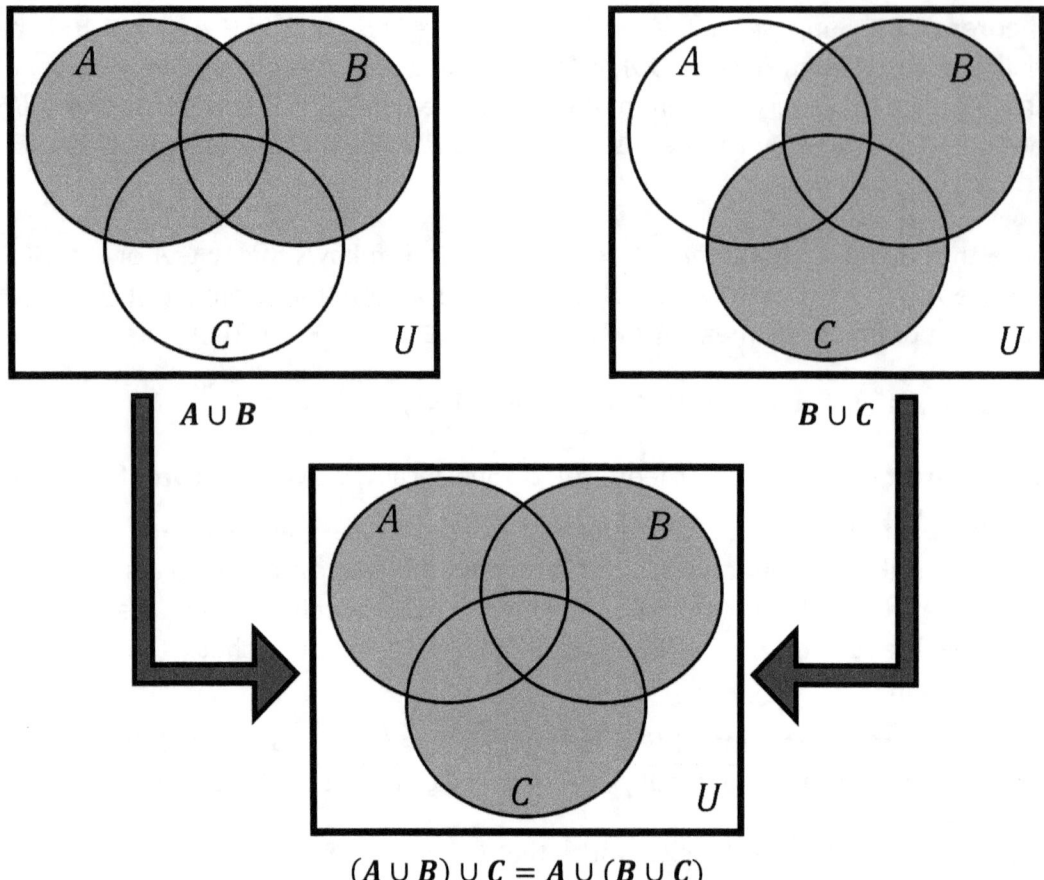

$(A \cup B) \cup C = A \cup (B \cup C)$

Proof of Theorem 1.29: Let A, B, and C be sets, and let $x \in (A \cup B) \cup C$. Then $x \in A \cup B$ or $x \in C$. If $x \in C$, then $x \in B$ or $x \in C$. So, $x \in B \cup C$. Then $x \in A$ or $x \in B \cup C$. So, $x \in A \cup (B \cup C)$. If, on the other hand, $x \in A \cup B$, then $x \in A$ or $x \in B$. If $x \in A$, then $x \in A$ or $x \in B \cup C$. So, $x \in A \cup (B \cup C)$. If $x \in B$, then $x \in B$ or $x \in C$. So, $x \in B \cup C$. Then $x \in A$ or $x \in B \cup C$. So, $x \in A \cup (B \cup C)$. Since x was arbitrary, we have shown $\forall x(x \in (A \cup B) \cup C \rightarrow x \in A \cup (B \cup C))$. Therefore, we have shown that $(A \cup B) \cup C \subseteq A \cup (B \cup C)$.

A similar argument can be used to show $A \cup (B \cup C) \subseteq (A \cup B) \cup C$ (the reader should write out the details).

Since $(A \cup B) \cup C \subseteq A \cup (B \cup C)$ and $A \cup (B \cup C) \subseteq (A \cup B) \cup C$, $(A \cup B) \cup C = A \cup (B \cup C)$, and therefore, the operation of forming unions is associative. □

Associativity allows us to drop parentheses. So, we can now simply write $A \cup B \cup C$ when taking the union of the three sets A, B, and C.

Arbitrary Unions and Intersections

Many students find the definitions given in this section difficult to understand at first. If you have trouble grasping the material here, I wouldn't worry too much at first. I would suggest coming back and rereading this section as examples of infinite unions and intersections come up throughout the book. Try to understand it a little better each time and eventually it will all become clear.

We will often be interested in taking unions and intersections of more than two sets. Therefore, we make the following more general definitions.

Let X be a nonempty set of sets.

$$\cup X = \{y \mid \text{there is } Y \in X \text{ with } y \in Y\} \quad \text{and} \quad \cap X = \{y \mid \text{for all } Y \in X, y \in Y\}.$$

If you're having trouble understanding what these definitions are saying, you're not alone. The notation probably looks confusing, but the ideas behind these definitions are very simple. You have a whole bunch of sets (possibly infinitely many). To take the union of all these sets, you simply throw all the elements together into one big set. To take the intersection of all these sets, you take only the elements that are in every single one of those sets.

Example 1.30:

1. Let A and B be sets and let $X = \{A, B\}$. Then

   $\cup X = \{y \mid \text{there is } Y \in X \text{ with } y \in Y\} = \{y \mid y \in A \text{ or } y \in B\} = A \cup B.$
   $\cap X = \{y \mid \text{for all } Y \in X, y \in Y\} = \{y \mid y \in A \text{ and } y \in B\} = A \cap B.$

2. Let A, B, and C be sets, and let $X = \{A, B, C\}$. Then

   $\cup X = \{y \mid \text{there is } Y \in X \text{ with } y \in Y\} = \{y \mid y \in A, y \in B, \text{or } y \in C\} = A \cup B \cup C.$
   $\cap X = \{y \mid \text{for all } Y \in X, y \in Y\} = \{y \mid y \in A, y \in B, \text{and } y \in C\} = A \cap B \cap C.$

3. Let $X = \{\{-n, \ldots, -3, -2, -1, 0, 1, 2, 3, 4, \ldots, n\} \mid n \in \mathbb{N}\}$. The sets in X look as follows:

   $\{0\}, \quad \{-1, 0, 1\}, \quad \{-2, -1, 0, 1, 2\}, \quad \{-3, -2, -1, 0, 1, 2, 3\}, \quad \{-4, -3, -2, -1, 0, 1, 2, 3, 4\}, \ldots$

   We have
   $$\cup X = \{y \mid \text{there is } Y \in X \text{ with } y \in Y\}$$
   $$= \{y \mid \text{there is } n \in \mathbb{N} \text{ with } y \in \{-n, \ldots, -3, -2, -1, 0, 1, 2, 3, 4, \ldots, n\}\} = \mathbb{Z}.$$
   $$\cap X = \{y \mid \text{for all } Y \in X, y \in Y\}$$
   $$= \{y \mid \text{for all } n \in \mathbb{N}, y \in \{-n, \ldots, -3, -2, -1, 0, 1, 2, 3, 4, \ldots, n\}\} = \{0\}.$$

Notes: (1) Examples 1 and 2 give a good idea of what $\cup X$ and $\cap X$ look like when X is finite. More generally, if $X = \{A_1, A_2, \ldots, A_n\}$, then $\cup X = A_1 \cup A_2 \cup \cdots \cup A_n$ and $\cap X = A_1 \cap A_2 \cap \cdots \cap A_n$.

(2) As a specific example of Note 1, let $A_1 = \{\ldots, -3, -2, -1, 0, 1, 2\}$, $A_2 = \{0, 1, 2, 3, 4, 5\}$, $A_3 = \{1, 2\}$, and $A_4 = \{2, 3, 4, \ldots, 98, 99\}$. Let $X = \{A_1, A_2, A_3, A_4\}$. Then

$$\cup X = A_1 \cup A_2 \cup A_3 \cup A_4 = \{\ldots, -3, -2, -1, 0, 1, 2, 3, \ldots, 98, 99\}.$$
$$\cap X = A_1 \cap A_2 \cap A_3 \cap A_4 = \{2\}.$$

If you have trouble seeing how to compute the intersection, it may help to take the intersections two at a time:

$A_1 \cap A_2 = \{\ldots, -3, -2, -1, 0, 1, 2\} \cap \{0, 1, 2, 3, 4, 5\} = \{0, 1, 2\}.$

$\{0, 1, 2\} \cap A_3 = A_3 = \{1, 2\}$ because $A_3 \subseteq \{0, 1, 2\}$.

$\{1, 2\} \cap A_4 = \{1, 2\} \cap \{2, 3, 4, \ldots, 98, 99\} = \{2\}.$

(3) Let's prove carefully that $\{y \mid \text{there is } n \in \mathbb{N} \text{ with } y \in \{-n, \ldots, -3, -2, -1, 0, 1, 2, 3, 4, \ldots, n\}\} = \mathbb{Z}$.

For convenience, let's let $A = \{y \mid \text{there is } n \in \mathbb{N} \text{ with } y \in \{-n, \ldots, -3, -2, -1, 0, 1, 2, 3, 4, \ldots, n\}\}$.

If $y \in A$, then there is $n \in \mathbb{N}$ with $y \in \{-n, \ldots, -3, -2, -1, 0, 1, 2, 3, 4, \ldots, n\}$. In particular, $y \in \mathbb{Z}$. Since $y \in A$ was arbitrary, we have shown that $A \subseteq \mathbb{Z}$.

Let $y \in \mathbb{Z}$. Then $y \in \{-n, \ldots, -3, -2, -1, 0, 1, 2, 3, 4, \ldots, n\}$, where $n = y$ if $y \geq 0$ and $n = -y$ if $y < 0$ (in other words, n is the **absolute value** of y, written $n = |y|$). So, $y \in A$. Since $y \in \mathbb{Z}$ was arbitrary, we have shown that $\mathbb{Z} \subseteq A$.

Since $A \subseteq \mathbb{Z}$ and $\mathbb{Z} \subseteq A$, it follows that $A = \mathbb{Z}$.

(4) Let's also prove carefully that $\{y \mid \text{for all } n \in \mathbb{N}, y \in \{-n, \ldots, -3, -2, -1, 0, 1, 2, 3, 4, \ldots, n\}\} = \{0\}$.

For convenience, let's let $B = \{y \mid \text{for all } n \in \mathbb{N}, y \in \{-n, \ldots, -3, -2, -1, 0, 1, 2, 3, 4, \ldots, n\}\}$.

If $y \in B$, then for all $n \in \mathbb{N}$, $y \in \{-n, \ldots, -3, -2, -1, 0, 1, 2, 3, 4, \ldots, n\}$. In particular, $y \in \{0\}$. Since $y \in B$ was arbitrary, we have shown that $B \subseteq \{0\}$.

Now, let $y \in \{0\}$. Then $y = 0$. For all $n \in \mathbb{N}$, $0 \in \{-n, \ldots, -3, -2, -1, 0, 1, 2, 3, 4, \ldots, n\}$. So, $y \in B$. It follows that $\{0\} \subseteq B$.

Since $B \subseteq \{0\}$ and $\{0\} \subseteq B$, it follows that $B = \{0\}$.

(5) Note that the empty union is empty. Indeed, we have $\bigcup \emptyset = \{y \mid \text{there is } Y \in \emptyset \text{ with } y \in Y\} = \emptyset$.

If X is a nonempty set of sets, we say that X is **disjoint** if $\bigcap X = \emptyset$. We say that X is **pairwise disjoint** if for all $A, B \in X$ with $A \neq B$, A and B are disjoint. For example, if we let $X = \{(n, n + 1) \mid n \in \mathbb{Z}\}$, then X is both disjoint and pairwise disjoint.

Are the definitions of disjoint and pairwise disjoint equivalent? You will be asked to answer this question in Problem 17 below.

Problem Set 1

Full solutions to these problems are available for free download here:
www.SATPrepGet800.com/RABQXZ

LEVEL 1

1. Determine whether each of the following statements is true or false:

    (i) $9 \in \{9\}$

    (ii) $c \in \{a, b, c\}$

    (iii) $-3 \in \{3\}$

    (iv) $\frac{2}{3} \in \mathbb{Z}$

    (v) $-18 \in \mathbb{N}$

    (vi) $\frac{3}{19} \in \mathbb{Q}$

    (vii) $\emptyset \subseteq \{0, 1, 2\}$

    (viii) $\{\delta\} \subseteq \{\delta, \Delta\}$

    (ix) $\{f, g, h\} \subseteq \{f, g, h\}$

    (x) $\{a, b, \{c, d\}\} \subseteq \{a, b, c, d\}$

2. Determine the cardinality of each of the following sets:

    (i) $\{0, 1, 2\}$

    (ii) $\{a, b, c, d, e, f\}$

    (iii) $\{1, 2, \ldots, 99\}$

    (iv) $\{\frac{1}{2}, \frac{1}{3}, \ldots, \frac{1}{11}\}$

3. List the elements of $\{k, x, t\} \times \{5, 6\}$.

4. Let $A = \{0\}$. Evaluate (i) A^2; (ii) A^3; (iii) $\mathcal{P}(A)$.

5. Let $A = \{a, b, \Delta, \delta\}$ and $B = \{b, c, \delta, \gamma\}$. Determine each of the following:

    (i) $A \cup B$

    (ii) $A \cap B$

    (iii) $A \setminus B$

    (iv) $B \setminus A$

    (v) $A \Delta B$

6. Draw Venn diagrams for $(A \setminus B) \setminus C$ and $A \setminus (B \setminus C)$. Are these two sets equal for all sets A, B, and C? If so, prove it. If not, provide a counterexample.

LEVEL 2

7. Compute the power set of each of the following sets:

    (i) $\emptyset$

    (ii) $\{2\}$

    (iii) $\{a, b\}$

    (iv) $\{\emptyset, \{\emptyset\}\}$

    (v) $\{\{\emptyset\}\}$

8. Determine whether each of the following statements is true or false:

    (i) $2 \in \emptyset$

    (ii) $\emptyset \in \{a, b\}$

    (iii) $\emptyset \in \emptyset$

    (iv) $\emptyset \in \{\emptyset, \{\emptyset\}\}$

    (v) $\{\emptyset\} \in \emptyset$

    (vi) $\{\emptyset\} \in \{\emptyset\}$

    (vii) $\emptyset \subseteq \emptyset$

    (viii) $\emptyset \subseteq \{\emptyset\}$

    (ix) $\{\emptyset\} \subseteq \emptyset$

    (x) $\{\emptyset\} \subseteq \{\emptyset\}$

    (xi) $\mathbb{Q} \subseteq \mathbb{C}$

    (xii) $3 \in \{2k \mid k = 1, 2, 3, 4, 5, 6\}$

9. Determine the cardinality of each of the following sets:

    (i) $\{a, a, b, c, d, d, d\}$

    (ii) $\{\{1, 2\}, \{3, 4, 5\}\}$

    (iii) $\{5, 6, 7, \ldots, 2122, 2123\}$

10. Compute $\{a, b\}^4$.

11. Let $A = \{\emptyset, \{\emptyset, \{\emptyset\}\}\}$ and $B = \{\emptyset, \{\emptyset\}\}$. Compute each of the following:

    (i) $A \cup B$

    (ii) $A \cap B$

    (iii) $A \setminus B$

    (iv) $B \setminus A$

    (v) $A \triangle B$

12. Prove the following:
    (i) The operation of forming unions is commutative.
    (ii) The operation of forming intersections is commutative.
    (iii) The operation of forming intersections is associative.

LEVEL 3

13. Determine the cardinality of each of the following sets:
    (i) $\{\{\{a,b\}\}\}$
    (ii) $\{\{0,1\}, 0, \{0\}, \{0, \{0,1,2\}\}\}$
    (iii) $\{a, \{a\}, \{a,a\}, \{a,a,a,a\}, \{a,a,\{a\}\}, \{a,\{a\},\{a\}\}\}$

14. How many subsets does $\{a,b,c,d\}$ have? Draw a tree diagram for the subsets of $\{a,b,c,d\}$.

15. Let $A, B, C, D,$ and E be sets such that $A \subseteq B$, $B \subseteq C$, $C \subseteq D$, and $D \subseteq E$. Prove that $A \subseteq E$.

16. Let $A, B, C,$ and D be sets with $A \subseteq B$ and $C \subseteq D$. Prove that $A \times C \subseteq B \times D$.

17. Prove or provide a counterexample:
    (i) Every pairwise disjoint set of sets is disjoint.
    (ii) Every disjoint set of sets is pairwise disjoint.

18. Let A and B be sets. Prove that $A \cap B \subseteq A$.

LEVEL 4

19. Determine whether each of the following statements is true or false:
    (i) $c \in \{a, \{c\}\}$
    (ii) $\{\Delta\} \in \{\delta, \Delta\}$
    (iii) $\{1\} \in \{1, a, 2, b\}$
    (iv) $\emptyset \in \{\{\emptyset\}\}$
    (v) $\{\{\emptyset\}\} \in \emptyset$

20. Let $A, B, C,$ and D be sets. Determine if each of the following statements is true or false. If true, provide a proof. If false, provide a counterexample.
    (i) $(A \times B) \cap (C \times D) = (A \cap C) \times (B \cap D)$
    (ii) $(A \times B) \cup (C \times D) = (A \cup C) \times (B \cup D)$

21. Prove that $B \subseteq A$ if and only if $A \cap B = B$.

22. Let A, B, and C be sets. Prove each of the following:

   (i) $A \cap (B \cup C) = (A \cap B) \cup (A \cap C)$.

   (ii) $A \cup (B \cap C) = (A \cup B) \cap (A \cup C)$.

   (iii) $C \setminus (A \cup B) = (C \setminus A) \cap (C \setminus B)$.

   (iv) $C \setminus (A \cap B) = (C \setminus A) \cup (C \setminus B)$.

LEVEL 5

23. Let A and B be sets with $B \subseteq A$. Determine if the following are true or false. If true, provide a proof. If false, provide a counterexample.

   (i) $B \in \mathcal{P}(A)$

   (ii) $B \subseteq \mathcal{P}(A)$

   (iii) $\mathcal{P}(B) \in \mathcal{P}(A)$

   (iv) $\mathcal{P}(B) \subseteq \mathcal{P}(A)$

24. Let $P(x)$ be the property $x \notin x$. Prove that $\{x | P(x)\}$ cannot be a set.

25. Let $A = \{a, b, c, d\}$, $B = \{X \mid X \subseteq A \wedge d \notin X\}$, and $C = \{X \mid X \subseteq A \wedge d \in X\}$. Show that there is a natural one-to-one correspondence between the elements of B and the elements of C. Then generalize this result to a set with $n+1$ elements for $n > 0$. (A **one-to-one correspondence** between two sets is a pairing so that each element of the first set is matched up with exactly one element of the second set, and vice versa.)

26. Let X be a nonempty set of sets. Prove the following:

   (i) For all $A \in X$, $A \subseteq \bigcup X$.

   (ii) For all $A \in X$, $\bigcap X \subseteq A$.

27. Let A be a set and let X be a nonempty set of sets. Prove each of the following:

   (i) $A \cap \bigcup X = \bigcup \{A \cap B \mid B \in X\}$

   (ii) $A \cup \bigcap X = \bigcap \{A \cup B \mid B \in X\}$

   (iii) $A \setminus \bigcup X = \bigcap \{A \setminus B \mid B \in X\}$

   (iv) $A \setminus \bigcap X = \bigcup \{A \setminus B \mid B \in X\}$.

CHALLENGE PROBLEMS

28. Let X be a nonempty set of sets. Prove that $\mathcal{P}(\bigcap X) = \bigcap \{\mathcal{P}(A) \mid A \in X\}$.

29. Let X be a nonempty set of sets. Prove that $\mathcal{P}(\bigcup X) = \bigcup \{\mathcal{P}(A) \mid A \in X\}$ if and only if $\bigcup X \in X$.

30. Prove that there is a natural one-to-one correspondence between the elements of $\mathbb{Q}$ and the elements of a proper subset of $\mathbb{R}$.

LESSON 2
RELATIONS AND PARTITIONS

Binary Relations

A **binary relation** on a set A is a subset of $A^2 = A \times A$. Symbolically, we have

R is a binary relation on A if and only if $R \subseteq A \times A$.

We will usually abbreviate $(a, b) \in R$ as aRb.

Remark: The statement $R \subseteq A \times A$ is equivalent to the statement $R \in \mathcal{P}(A \times A)$. It follows that for a finite set A, the number of binary relations on A is $|\mathcal{P}(A \times A)|$.

Example 2.1:

1. Let $R = \{(a, b) \in \mathbb{N} \times \mathbb{N} \mid a < b\}$. For example, we have $(0, 1) \in R$ because $0 < 1$. However, $(1, 1) \notin R$ because $1 \not< 1$. We abbreviate $(0, 1) \in R$ by $0R1$.

   Observe that $R \subseteq \mathbb{N} \times \mathbb{N}$, and so, R is a binary relation on $\mathbb{N}$.

   We would normally use the name $<$ for this relation R. So, we have $(0, 1) \in <$, which we abbreviate as $0 < 1$, and we have $(1, 1) \notin <$, which we abbreviate as $1 \not< 1$.

2. There are binary relations $<, \leq, >, \geq$ defined on $\mathbb{N}, \mathbb{Z}, \mathbb{Q}$, and $\mathbb{R}$. For example, if we consider $> \subseteq \mathbb{Z}^2$, we have $(13, -7) \in >$, or equivalently, $13 > -7$.

3. Let $A = \{a\}$. Since $|A| = 1$, we have $|A \times A| = 1 \cdot 1 = 1$. So, $|\mathcal{P}(A \times A)| = 2^1 = 2$. So, there are 2 binary relations on A. They are $R_1 = \emptyset$ and $R_2 = \{(a, a)\}$.

4. Let $B = \{0, 1\}$. Since $|B| = 2$, we have $|B \times B| = 2 \cdot 2 = 4$. So, $|\mathcal{P}(B \times B)| = 2^4 = 16$. So, there are 16 binary relations on B. A few examples are $R_1 = \emptyset$, $R_2 = \{(0, 0)\}$, $R_3 = \{(0, 1)\}$, and $R_4 = \{(0, 0), (0, 1)\}$. Can you list the rest of them?

5. Let A be a set and let R be the binary relation on A defined by $R = \{(a, b) \in A \times A \mid a \in b\}$. R is known as the **membership relation**, and it is usually denoted by $\in$. So, if $a, b \in A$ and a is a member of b, we can write $(a, b) \in \in$, which we will usually abbreviate as $a \in b$. As a specific example, let $A = \{\emptyset, \{\emptyset\}, \{\{\emptyset\}\}\}$. Then $(\emptyset, \{\emptyset\}) \in \in$, or equivalently, $\emptyset \in \{\emptyset\}$. Similarly, we have $(\{\emptyset\}, \{\{\emptyset\}\}) \in \in$, or equivalently, $\{\emptyset\} \in \{\{\emptyset\}\}$.

6. Let $R = \{((a, b), (c, d)) \in (\mathbb{N} \times \mathbb{N})^2 \mid a + d = b + c\}$. Then R is a binary relation on $\mathbb{N} \times \mathbb{N}$. For example, we have $(5, 0)R(6, 1)$ because $5 + 1 = 0 + 6$. However, we see that $(5, 0)\not R(6, 2)$ because $5 + 2 \neq 0 + 6$.

7. Let $R = \{((a, b), (c, d)) \in (\mathbb{Z} \times \mathbb{Z}^*)^2 \mid ad = bc\}$. (Recall that $\mathbb{Z}^*$ is the set of *nonzero* integers.) Then R is a binary relation on $\mathbb{Z} \times \mathbb{Z}^*$. For example, $(1, 2)R(2, 4)$ because $1 \cdot 4 = 2 \cdot 2$. However, $(1, 2)\not R(2, 5)$ because $1 \cdot 5 \neq 2 \cdot 2$. Compare this to the rational number system (see part 1 of Example 1.6), where we have $\frac{1}{2} = \frac{2}{4}$ because $1 \cdot 4 = 2 \cdot 2$, but $\frac{1}{2} \neq \frac{2}{5}$ because $1 \cdot 5 \neq 2 \cdot 2$.

The **domain** of a binary relation R, written dom R, is $\{x \mid \exists y(xRy)\}$. The **range** of a binary relation, written ran R, is $\{y \mid \exists x(xRy)\}$. The **field** of a binary relation R is dom $R \cup$ ran R.

Notes: (1) The symbol $\exists$ is called an **existential quantifier**, and it is pronounced "There exists" or "There is."

(2) The expression $\exists y(xRy)$ can be translated into English as "There exists a y such that xRy." Similarly, the expression $\exists x(xRy)$ can be translated into English as "There exists an x such that xRy." In general, if $P(x)$ is some property, then the expression $\exists x(P(x))$ can be translated into English as "There exists an x such that $P(x)$."

(3) The definition of field given above is not related to the algebraic structure that we will learn about in Lesson 6. Unfortunately, we use the same word for these two unrelated definitions.

Example 2.2:

1. Let $R = \{(a, b) \in \mathbb{N} \times \mathbb{N} \mid a < b\}$. Then dom $R = \mathbb{N}$, ran $R = \mathbb{N}$, and field $R = \mathbb{N} \cup \mathbb{N} = \mathbb{N}$.
2. Let $B = \{0, 1, 2, 3\}$ and $R = \{(0, 2), (0, 3), (1, 3)\}$. Then dom $R = \{0, 1\}$, ran $R = \{2, 3\}$, and field $R = \{0, 1\} \cup \{2, 3\} = \{0, 1, 2, 3\} = B$.
3. Let $R = \{((a, b), (c, d)) \in (\mathbb{N} \times \mathbb{N})^2 \mid a + d = b + c\}$. Then dom $R = \mathbb{N} \times \mathbb{N}$, ran $R = \mathbb{N} \times \mathbb{N}$, and field $R = (\mathbb{N} \times \mathbb{N}) \cup (\mathbb{N} \times \mathbb{N}) = \mathbb{N} \times \mathbb{N}$.

We say that a binary relation R on a set A is

- **reflexive** if for all $a \in A$, aRa.
- **symmetric** if for all $a, b \in A$, aRb implies bRa.
- **transitive** if for all $a, b, c \in A$, aRb and bRc imply aRc.
- **antireflexive** if for all $a \in A$, $a\not{R}a$.
- **antisymmetric** if for all $a, b \in A$, aRb and bRa imply $a = b$.

Example 2.3:

1. Let A be any set and let $R = \{(a, b) \in A^2 \mid a = b\}$. Then R is reflexive ($a = a$), symmetric (if $a = b$, then $b = a$), transitive (if $a = b$ and $b = c$, then $a = c$), and antisymmetric (trivially). If $A \neq \emptyset$, then this relation is not antireflexive because $a \neq a$ is false for any $a \in A$.

2. The binary relations $\leq$ and $\geq$ defined in the usual way on $\mathbb{Z}$ are transitive (if $a \leq b$ and $b \leq c$, then $a \leq c$, and similarly for $\geq$), reflexive ($a \leq a$ and $a \geq a$), and antisymmetric (if $a \leq b$ and $b \leq a$, then $a = b$, and similarly for $\geq$). These relations are not symmetric. For example, $1 \leq 2$, but $2 \not\leq 1$). These relations are not antireflexive. For example, $1 \leq 1$ is true.

    Any relation that is transitive, reflexive, and antisymmetric is called a **partial ordering**.

3. The binary relations $<$ and $>$ defined on $\mathbb{Z}$ are transitive (if $a < b$ and $b < c$, then $a < c$, and similarly for $>$), antireflexive ($a \not< a$ and $a \not> a$), and antisymmetric (this is vacuously true because $a < b$ and $b < a$ can never occur). These relations are not symmetric (for example, $1 < 2$, but $2 \not< 1$). These relations are not reflexive (for example, $1 < 1$ is false).

Any relation that is transitive, antireflexive, and antisymmetric is called a **strict partial ordering**.

4. Let $R = \{(0,0), (0,2), (2,0), (2,2), (2,3), (3,2), (3,3)\}$ be a binary relation on $\mathbb{N}$. Then R is clearly symmetric. R is not reflexive because $1 \in \mathbb{N}$, but $(1,1) \notin R$ (however, if we were to consider R as a relation on $\{0,2,3\}$ instead of on $\mathbb{N}$, then R **would** be reflexive). R is not transitive because we have $(0,2),(2,3) \in R$, but $(0,3) \notin R$. R is not antisymmetric because we have $(2,3),(3,2) \in R$ and $2 \neq 3$. R is not antireflexive because $(0,0) \in R$ (and also, $(2,2) \in R$ and $(3,3) \in R$).

n-ary Relations

We can extend the idea of a binary relation on a set A to an **n-ary relation** on A. For example, a 3-ary relation (or **ternary relation**) on A is a subset of $A^3 = A \times A \times A$. More generally, we have that R is an n-ary relation on A if and only if $R \subseteq A^n$. A **1-ary relation** (or **unary relation**) on A is just a subset of A.

Example 2.4:

1. $\mathbb{R}$ is a unary relation on $\mathbb{C}$ because $\mathbb{R} \subseteq \mathbb{C}$.

2. Let $R = \{(x,y,z) \in \mathbb{Z}^3 \mid x + y = z\}$. Then R is a ternary (or 3-ary) relation on $\mathbb{Z}$. We have, for example, $(1,2,3) \in R$ (because $1 + 2 = 3$) and $(1,2,4) \notin R$ (because $1 + 2 \neq 4$).

3. Let C be the set of all colors. For example, blue $\in C$, pink $\in C$, and violet $\in C$. Let $S = \{(a,b,c) \in C^3 \mid$ when a and b are combined in equal quantities, the result is $c\}$. Then S is a ternary relation on C. We have, for example, (red, yellow, orange) $\in S$.

4. Let $T = \{(a,b,c,d,e) \in \mathbb{N}^5 \mid ab + c = de\}$. Then T is a 5-ary relation on $\mathbb{N}$. We have, for example, $(1,2,8,5,2) \in T$ ($1 \cdot 2 + 8 = 5 \cdot 2$) and $(1,1,1,1,1) \notin T$ ($1 \cdot 1 + 1 \neq 1 \cdot 1$).

Orderings

A binary relation $\leq$ on a set A is a **partial ordering** on A if $\leq$ is reflexive, antisymmetric, and transitive on A. If we replace "reflexive" by "antireflexive," then we call the relation a **strict partial ordering** on A (we would normally use the symbol $<$ instead of $\leq$ for a strict partial ordering).

A **partially ordered set** (or **poset**) is a pair $(A, \leq)$, where A is a set and $\leq$ is a partial ordering on A. Similarly, a **strict poset** is a pair $(A, <)$, where A is a set and $<$ is a strict partial ordering on A.

Note: If $<$ is a strict partial ordering on a set A, then $\forall a, b \in A(a < b \rightarrow b \not< a)$. To see this, let $a < b$. Since $<$ is antireflexive, $a \neq b$. Since $<$ is antisymmetric, $b < a$ would lead to $a = b$. Since we cannot have both $a \neq b$ and $a = b$, we must have $b \not< a$.

Example 2.5:

1. The usual ordering $\leq$ on $\mathbb{Z} = \{\ldots, -3, -2, -1, 0, 1, 2, 3, \ldots\}$ is a partial ordering, and the ordering $<$ on $\mathbb{Z}$ is a strict partial ordering. See Example 2.3 (parts 2 and 3).

2. If A is a set, then $(\mathcal{P}(A), \subseteq)$ is a poset. Since every set is a subset of itself, $\subseteq$ is reflexive (see Theorem 1.9). If $X, Y \in \mathcal{P}(A)$ with $X \subseteq Y$ and $Y \subseteq X$, then $X = Y$ (see the Axiom of Extensionality right before Example 1.16). So, $\subseteq$ is antisymmetric. By Theorem 1.14, $\subseteq$ is transitive.

See the tree diagrams at the end of Example 1.12 for visual representations of this poset when $A = \{a, b\}$ and $A = \{a, b, c\}$.

Similarly, $(\mathcal{P}(A), \subset)$ is a strict poset (the relation here is the **proper subset** relation). The relation $\subset$ is antireflexive because no set is a proper subset of itself. If $X \subset Y$ and $Y \subset X$, then $X \subseteq Y$ and $Y \subseteq X$, and so, $X = Y$ (again, see the Axiom of Extensionality right before Example 1.16). Finally, suppose that $X \subset Y$ and $Y \subset Z$. Then $X \subseteq Y$ and $Y \subseteq Z$, and so, by Theorem 1.14, $X \subseteq Z$. Suppose toward contradiction that $X = Z$. Then $Z \subseteq X$ and again, by Theorem 1.14, $Z \subseteq Y$. Since $Y \subseteq Z$ and $Z \subseteq Y$, it follows that $Y = Z$, contradicting our assumption that Y is a *proper* subset of Z. So, $X \neq Z$, and therefore, $X \subset Z$.

Note that in the argument above, to show that $X \neq Z$, we assumed that $X = Z$, and then used a logically valid argument to derive the statement $Y = Z$, which we already knew was false. This is known as a proof by contradiction. We will use this kind of argument in Theorems 2.7 and 2.8 below as well. See the notes after the proof of Theorem 4.6 for a more detailed explanation as to how this type of argument works. Theorem 5.2 provides another good example.

3. Let $(A, \leq_A)$ be a partially ordered set, $B \subseteq A$, and $\leq_B = \{(x, y) \mid x, y \in B \land x \leq_A y\}$. Then $(B, \leq_B)$ is also a partially ordered set. Let's check this carefully.

    To see that $\leq_B$ is reflexive on B, let $x \in B$. Since $B \subseteq A$, $x \in A$. Since $\leq_A$ is reflexive on A, $x \leq_A x$. Since $x \in B$, $x \leq_B x$.

    To see that $\leq_B$ is antisymmetric on B, let $x, y \in B$ with $x \leq_B y$ and $y \leq_B x$. Then $x \leq_A y$ and $y \leq_A x$. Since $\leq_A$ is antisymmetric on A, $x = y$.

    Finally, to see that $\leq_B$ is transitive on B, let $x, y, z \in B$ with $x \leq_B y$ and $y \leq_B z$. Then $x \leq_A y$ and $y \leq_A z$. Since $\leq_A$ is transitive on A, $x \leq_A z$. Since $x, z \in B$, $x \leq_B z$.

4. Let $X = \{x \mid x$ is a word in the English language$\}$ and define the **dictionary order** on X as follows: $x <_D y$ if x appears before y alphabetically. Then $(X, <_D)$ is a strict poset. In this poset, we have aardvark $<_D$ antelope (because a = a and a appears before n alphabetically), we have stranger $<_D$ violin (because s appears before v alphabetically), and we have dragon $<_D$ drainage (because d = d, r = r, a = a and g appears before i alphabetically).

    We can use a similar idea to define the **dictionary order** on Cartesian products of posets (and strict posets).

    For example, the dictionary order $<_D$ can be defined on $\mathbb{Z} \times \mathbb{Q}$ by $(a, b) <_D (c, d)$ if and only if either $a <_\mathbb{Z} c$ or both $a = c$ and $b <_\mathbb{Q} d$, where $<_\mathbb{Z}$ and $<_\mathbb{Q}$ are the usual strict partial orderings on $\mathbb{Z}$ and $\mathbb{Q}$, respectively. It is easy (but a bit tedious) to verify that $(\mathbb{Z} \times \mathbb{Q}, <_D)$ is a strict poset. I leave the verification to the reader.

    In the strict poset $(\mathbb{Z} \times \mathbb{Q}, <_D)$, we have $(4, 7) <_D (9, -2)$ because $4 <_\mathbb{Z} 9$ (notice that the second coordinates are irrelevant because the first coordinates are not equal). We also have $\left(3, \frac{1}{2}\right) <_D \left(3, \frac{3}{4}\right)$ because $3 = 3$ and $\frac{1}{2} <_\mathbb{Q} \frac{3}{4}$.

    We can visualize this particular dictionary order in a Cartesian plane as solid vertical lines passing through each integer value along the x-axis. Each point is less than any point higher than it on the same vertical line and each point is also less than any point on a vertical line to the right of that point (regardless of the height).

In the figure below, we see that $(1,-2) <_D \left(1, \frac{3}{2}\right)$ because $\left(1, \frac{3}{2}\right)$ is above $(1,-2)$ on the same vertical line. We also have $(1,-2) <_D (3,-3)$ because $(3,-3)$ is to the right of $(1,-2)$ (note that it does **not** matter that $(3,-3)$ is below $(1,-2)$).

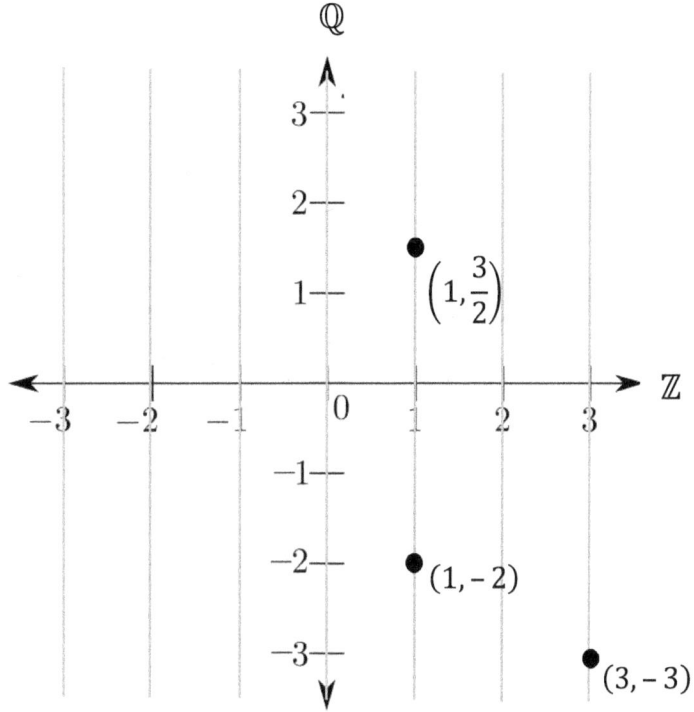

As another example, in the dictionary order of $\mathbb{R}^5$, we have $(1,3,5,2,11) <_D (1,3,5,4,10)$ because $1 = 1, 3 = 3, 5 = 5$, and $2 <_{\mathbb{R}} 4$.

As one more example, let $A = \{a,b\}$, consider the strict poset $(\mathcal{P}(A), \subset)$ (the relation here is the **proper subset** relation), and let $<_D$ be the corresponding dictionary order on $(\mathcal{P}(A))^2 = \mathcal{P}(A) \times \mathcal{P}(A)$. We have $(\emptyset, \{a\}) <_D (\{b\}, \{b\})$ because $\emptyset \subset \{b\}$. Also, we have $(\{a\}, \{b\}) <_D (\{a\}, \{a,b\})$ because $\{a\} = \{a\}$ and $\{b\} \subset \{a,b\}$.

Let $(A, \leq)$ be a poset. We say that $a, b \in A$ are **comparable** if $a \leq b$ or $b \leq a$. The poset satisfies the **comparability condition** if every pair of elements in A are comparable. A poset that satisfies the comparability condition is called a **linearly ordered set** (or **totally ordered set**). Similarly, a **strict linearly ordered set** $(A, <)$ satisfies **trichotomy**: If $a, b \in A$, then $a < b$, $a = b$, or $b < a$.

Example 2.6:

1. $(\mathbb{N}, \leq), (\mathbb{Z}, \leq), (\mathbb{Q}, \leq)$, and $(\mathbb{R}, \leq)$ are linearly ordered sets. Similarly, $(\mathbb{N}, <), (\mathbb{Z}, <), (\mathbb{Q}, <)$, and $(\mathbb{R}, <)$ are strict linearly ordered sets.

2. If A has at least two elements, then $(\mathcal{P}(A), \subset)$ is **not** a strict linearly ordered set. Indeed, if $a, b \in A$ with $a \neq b$, then $\{a\} \not\subset \{b\}$ and $\{b\} \not\subset \{a\}$. See either of the tree diagrams at the end of Example 1.12.

   Similarly, $\left((\mathcal{P}(A))^2, <_D\right)$, where $<_D$ is the dictionary order is **not** a strict linearly ordered set. Indeed, $(\{a\}, \emptyset)$ and $(\{b\}, \emptyset)$ are not comparable with respect to $<_D$ because $\{a\} \not\subset \{b\}$ and $\{b\} \not\subset \{a\}$, just like we observed above. Similarly, $(\{a\}, \{a\})$ and $(\{a\}, \{b\})$ are not comparable.

3. If $(A, <_A)$ and $(B, <_B)$ are strict linearly ordered sets, then the dictionary order $<_D$ on $A \times B$ is also a strict linearly ordered set. To see this, let $(a, b), (c, d) \in A \times B$. If $a <_A c$, then we have $(a, b) <_D (c, d)$. If $c <_A a$, then $(c, d) <_D (a, b)$. If $a = c$, then we look at the second coordinates. In this case, if $b <_B d$, then $(a, b) <_D (c, d)$, and if $d <_B b$, then $(c, d) <_D (a, b)$. Otherwise, $a = c$ and $b = d$, and therefore, $(a, b) = (c, d)$.

We can modify a partial ordering slightly to get a strict partial ordering. We do this as follows:

Theorem 2.7: Let A be a set and let $\leq_A$ be a partial ordering on A. Define the binary relation $<_A$ on A by $a <_A b$ if and only if $a \leq_A b$ and $a \neq b$. Then $<_A$ is a strict partial ordering on A.

Proof: Suppose that $a <_A b$ and $b <_A c$. Then $a \leq_A b$ and $b \leq_A c$. Since $\leq_A$ is transitive, $a \leq_A c$. Also, $a \neq b$ and $b \neq c$. Suppose toward contradiction that $a = c$. Since $b \leq_A c$, we have $b \leq_A a$. Since $\leq_A$ is antisymmetric, $a = b$, contrary to our assumption. Therefore, $a \neq c$. Since $a \leq_A c$ and $a \neq c$, we have $a <_A c$. This shows that $<_A$ is transitive.

Now, suppose that $a <_A b$ and $b <_A a$. Then $a \leq_A b$ and $b \leq_A a$. Since $\leq_A$ is antisymmetric, $a = b$. Therefore, $<_A$ is antisymmetric.

Finally, by definition, $a <_A a$ is false. Therefore, $<_A$ is antireflexive.

Since $<_A$ is transitive, antisymmetric, and antireflexive, $<_A$ is a strict partial ordering on A. □

Similarly, we can modify a strict partial ordering to get a partial ordering.

Theorem 2.8: Let A be a set and let $<_A$ be a strict partial ordering on A. Define the binary relation $\leq_A$ on A by $a \leq_A b$ if and only if $a <_A b$ or $a = b$. Then $\leq_A$ is a partial ordering on A.

Proof: Suppose that $a \leq_A b$ and $b \leq_A c$. If $a = b$, then we have $a \leq_A c$ by direct substitution. Similarly, if $b = c$, we have $a \leq_A c$ by direct substitution. If $a <_A b$ and $b <_A c$, then $a <_A c$ because $<_A$ is transitive. It follows that $a \leq_A c$. This shows that $\leq_A$ is transitive.

Now, suppose that $a \leq_A b$ and $b \leq_A a$. Assume toward contradiction that $a \neq b$. Then $a <_A b$ and $b <_A a$. Since $<_A$ is antisymmetric, $a = b$, contrary to our assumption. Therefore, $\leq_A$ is antisymmetric.

Finally, by definition, $a \leq_A a$ is true. Therefore, $\leq_A$ is reflexive.

Since $\leq_A$ is transitive, antisymmetric, and reflexive, $\leq_A$ is a partial ordering on A. □

Intervals

A set I of real numbers is called an **interval** if any real number that lies between two numbers in I is also in I. Symbolically, we can write

$$\forall x, y \in I \ \forall z \in \mathbb{R} \ (x < z < y \rightarrow z \in I).$$

The expression above can be read "For all x, y in I and all $z \in \mathbb{R}$, if x is less than z and z is less than y, then z is in I."

A simple way to think of an interval is as a set of real numbers with no "gaps" or "holes." If x and y are in the interval, then everything between x and y is in the interval as well.

Example 2.9:

1. The set $A = \{0, 1\}$ is **not** an interval. A consists of just the two real numbers 0 and 1. There are infinitely many real numbers between 0 and 1. For example, the real number $\frac{1}{2}$ satisfies $0 < \frac{1}{2} < 1$, but $\frac{1}{2} \notin A$.

2. The set $B = \{x \in \mathbb{R} \mid 0 < x < 1\}$ is an example of an **open interval**. This set consists of all real numbers between 0 and 1, exclusive (0 and 1 are excluded). We will usually write the set B using the **interval notation** $B = (0, 1)$.

3. The set $C = \{x \in \mathbb{R} \mid 0 \leq x \leq 1\}$ is an example of a **closed interval**. This set consists of all real numbers between 0 and 1, inclusive (0 and 1 are included). We will usually write the set C using the **interval notation** $C = [0, 1]$.

4. The set $D = \{x \in \mathbb{R} \mid 0 \leq x < 1\}$ is an example of a **half-open interval**. This set consists of all real numbers between 0 and 1, including 0, but excluding 1. We will usually write the set D using the **interval notation** $D = [0, 1)$.

5. The set $E = \{x \in \mathbb{R} \mid x > 1\}$ is an example of an **infinite open interval**. This set consists of all real numbers greater than 1. We will usually write the set E using the **interval notation** $E = (1, \infty)$. The symbol "∞" is pronounced "**infinity**." It is **not** a number, but rather a symbol indicating that the set E has no upper bound.

6. $\mathbb{R}$ is an interval. This follows trivially from the definition. If we replace I by $\mathbb{R}$, we get $\forall x, y \in \mathbb{R} \, \forall z \in \mathbb{R} \, (x < z < y \to z \in \mathbb{R})$. In other words, if we start with two real numbers, and take a real number between them, then that number is a real number (which we already said).

When we are thinking of $\mathbb{R}$ as an interval, we sometimes use the notation $(-\infty, \infty)$ and refer to this as **the real line**. The following picture gives the standard geometric interpretation of the real line.

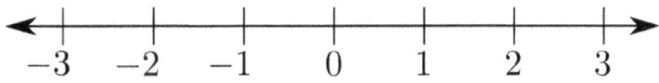

In addition to the real line, there are 8 other types of intervals.

Open Interval: $\quad (a, b) = \{x \in \mathbb{R} \mid a < x < b\}$

Closed Interval: $\quad [a, b] = \{x \in \mathbb{R} \mid a \leq x \leq b\}$

Half-open Intervals: $\quad (a, b] = \{x \in \mathbb{R} \mid a < x \leq b\} \quad [a, b) = \{x \in \mathbb{R} \mid a \leq x < b\}$

Infinite Open Intervals: $\quad (a, \infty) = \{x \in \mathbb{R} \mid x > a\} \quad (-\infty, b) = \{x \in \mathbb{R} \mid x < b\}$

Infinite Closed Intervals: $\quad [a, \infty) = \{x \in \mathbb{R} \mid x \geq a\} \quad (-\infty, b] = \{x \in \mathbb{R} \mid x \leq b\}$

Warning: It is unfortunate that the notation "(a, b)" is used for both an open interval and an ordered pair of real numbers. Most of the time it will be clear which one of these we mean. However, occasionally we may be discussing open intervals and ordered pairs of real numbers at the same time. In these instances, it is important to pay attention to the details of the discussion.

Example 2.10:

1. The half-open interval $(-2, 1] = \{x \in \mathbb{R} \mid -2 < x \leq 1\}$ has the following graph:

   The left parenthesis appearing at -2 indicates that -2 is **not** included in the set, whereas the right bracket appearing at 1 indicates that 1 is included in the set.

2. The infinite open interval $(0, \infty) = \{x \in \mathbb{R} \mid x > 0\}$ has the following graph:

   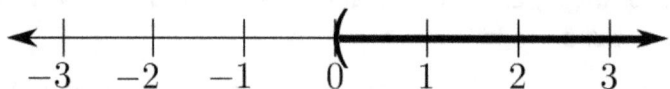

Notes: (1) If I is an interval of reals and $A \subseteq \mathbb{R}$, then we will use the notation $I \cap A$ for the corresponding interval as a subset of A. For example, $(-2, 1] \cap \mathbb{Q} = \{x \in \mathbb{Q} \mid -2 < x \leq 1\}$ is a half-open interval of rational numbers. It consists of only the rational numbers between -2 and 1, including 1 and excluding -2. We can visualize this interval of rational numbers the same way we visualize the corresponding interval of real numbers. If we wish, we may label the graph with a $\mathbb{Q}$ for extra clarification as follows:

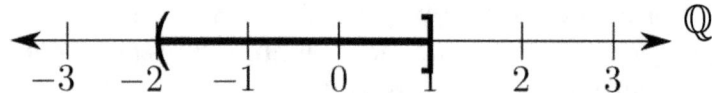

(2) It is fairly obvious that each of the eight types of sets defined above satisfies the definition of being an interval. For example, let's consider the set (a, b). If $x, y \in (a, b)$, $z \in \mathbb{R}$, and $x < z < y$, then we have $a < x < z < y < b$. Using the transitivity of the relation $<$, we have $a < z < b$. So, $z \in (a, b)$. The proofs for the other seven types are very similar.

Conversely, it turns out that every interval has one of these nine forms (the ninth form is the whole real line). This is harder to prove and requires a property of the real numbers called completeness. Completeness will be introduced in Lesson 6 and you will be asked to prove this result in Problem 21 in Problem Set 6.

Example 2.11: Let $A = (-2, 1]$ and $B = (0, \infty)$. We have

1. $A \cup B = (-2, \infty)$
2. $A \cap B = (0, 1]$
3. $A \setminus B = (-2, 0]$
4. $B \setminus A = (1, \infty)$
5. $A \triangle B = (-2, 0] \cup (1, \infty)$

Note: If you have trouble seeing how to compute these, it may be helpful to draw the graphs of A and B lined up vertically, and then draw vertical lines through the endpoints of each interval.

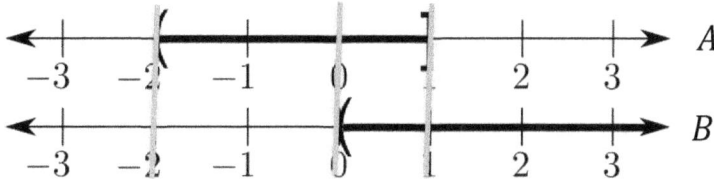

The results follow easily by combining these graphs into a single graph using the vertical lines as guides. For example, let's look at $A \cap B$ in detail. We're looking for all numbers that are in both A and B. The two rightmost vertical lines drawn passing through the two graphs above isolate all those numbers nicely. We see that all numbers between 0 and 1 are in the intersection. We should then think about the two endpoints 0 and 1 separately. $0 \notin B$ and therefore, 0 cannot be in the intersection of A and B. On the other hand, $1 \in A$ and $1 \in B$. Therefore, $1 \in A \cap B$. So, we see that $A \cap B = (0, 1]$.

Equivalence Relations

A binary relation R on a set A is an **equivalence relation** if R is reflexive, symmetric, and transitive.

Example 2.12:

1. The most basic equivalence relation on a set A is the relation $R = \{(a, b) \in A^2 \mid a = b\}$ (the **equality relation**). We already saw in part 1 of Example 2.3 that this relation is reflexive, symmetric and transitive.

2. Another obvious equivalence relation on a set A is the set A^2. Since every ordered pair (a, b) is in A^2, reflexivity, symmetry, and transitivity can never fail. We will refer to A^2 as the **trivial equivalence relation** on A.

3. We say that integers a and b have the same **parity** if they are both even or both odd. Define $\equiv_2$ on $\mathbb{Z}$ by $\equiv_2 = \{(a, b) \in \mathbb{Z}^2 \mid a \text{ and } b \text{ have the same parity}\}$. It is easy to see that $\equiv_2$ is reflexive ($a \equiv_2 a$ because every integer has the same parity as itself), $\equiv_2$ is symmetric (if $a \equiv_2 b$, then a has the same parity as b, so b has the same parity as a, and therefore, $b \equiv_2 a$), and $\equiv_2$ is transitive (if $a \equiv_2 b$ and $b \equiv_2 c$, then a, b, and c all have the same parity, and so, $a \equiv_2 c$). Therefore, $\equiv_2$ is an equivalence relation. The relation $\equiv_2$ is called **congruence modulo 2**. If $a \equiv_2 b$, then we say that a is congruent to b modulo 2. Some authors write $a \equiv b \pmod 2$ instead of $a \equiv_2 b$.

4. An integer n is **divisible** by an integer m, written $m|n$, if there is another integer k such that $n = mk$. For example, 18 is divisible by 6 because $18 = 6 \cdot 3$ and -35 is divisible by 5 because $-35 = 5 \cdot (-7)$. Another way to say that a and b have the same parity is to say that $b - a$ is divisible by 2, or equivalently, $2|b - a$. This observation allows us to generalize the notion of having the same parity. For example, $\equiv_3 = \{(a, b) \in \mathbb{Z}^2 \mid 3|b - a\}$ is an equivalence relation, and more generally, for each $n \in \mathbb{Z}^+$, $\equiv_n = \{(a, b) \in \mathbb{Z}^2 \mid n|b - a\}$ is an equivalence relation. I leave the proof that $\equiv_n$ is reflexive, symmetric, and transitive on $\mathbb{Z}$ as an exercise (see Problem 6 below). The relation $\equiv_n$ is called **congruence modulo n**. If $a \equiv_n b$, then we say that a is congruent to b modulo n. Some authors write $a \equiv b \pmod n$ instead of $a \equiv_n b$.

5. Consider the relation $R = \{((a,b),(c,d)) \in (\mathbb{N} \times \mathbb{N})^2 \mid a + d = b + c\}$ defined in part 6 of Example 2.1. Since $a + b = b + a$, we see that $(a,b)R(a,b)$, and therefore, R is reflexive. If $(a,b)R(c,d)$, then $a + d = b + c$. Therefore, $c + b = d + a$, and so, $(c,d)R(a,b)$. Thus, R is symmetric. Finally, suppose that $(a,b)R(c,d)$ and $(c,d)R(e,f)$. Then $a + d = b + c$ and $c + f = d + e$. So, $a + d + c + f = b + c + d + e$. Therefore, $a + f = b + e$, and so, we have $(a,b)R(e,f)$. So, R is transitive. Since R is reflexive, symmetric, and transitive, it follows that R is an equivalence relation.

6. Consider the relation $R = \{((a,b),(c,d)) \in (\mathbb{Z} \times \mathbb{Z}^*)^2 \mid ad = bc\}$ defined in part 7 of Example 2.1. Since $ab = ba$, we see that $(a,b)R(a,b)$, and therefore, R is reflexive. If $(a,b)R(c,d)$, then $ad = bc$. Therefore, $cb = da$, and so, $(c,d)R(a,b)$. Thus, R is symmetric. Finally, suppose that $(a,b)R(c,d)$ and $(c,d)R(e,f)$. Then $ad = bc$ and $cf = de$. So, $adcf = bcde$. Therefore, $cd(af - be) = adcf - bcde = 0$. If $a = 0$, then $bc = 0$, and so, $c = 0$ (because $b \neq 0$). So, $de = 0$, and therefore, $e = 0$ (because $d \neq 0$). So, $af = be$ (because they're both 0). If $a \neq 0$, then $c \neq 0$. Therefore, $af - be = 0$, and so, $af = be$. Since $a = 0$ and $a \neq 0$ both lead to $af = be$, we have $(a,b)R(e,f)$. So, R is transitive. Since R is reflexive, symmetric, and transitive, it follows that R is an equivalence relation.

Let $\sim$ be an equivalence relation on a set S. If $x \in S$, the **equivalence class** of x, written $[x]$, is the set

$$[x] = \{y \in S \mid x \sim y\}.$$

Example 2.13:

1. Let A be a set and let $R = \{(a,b) \in A^2 \mid a = b\}$ be the equality relation. Then for each $a \in A$, $[a] = \{a\}$.

2. Let A be a set and let $R = A^2$ be the trivial equivalence relation. Then for each $a \in A$, $[a] = A$.

3. Consider the equivalence relation $\equiv_2 = \{(a,b) \in \mathbb{Z}^2 \mid a \text{ and } b \text{ have the same parity}\}$ on $\mathbb{Z}$ (see part 3 of Example 2.12). We have $[0]_2 = \{y \in \mathbb{Z} \mid 0 \equiv_2 y\} = 2\mathbb{Z}$ (we add the subscript 2 into the notation $[0]_2$ to emphasize that this is an equivalence class for the equivalence relation $\equiv_2$). Observe that $[2]_2 = [0]_2$, and in fact, if n is any even integer, then $[n]_2 = [0]_2 = 2\mathbb{Z}$. Similarly, if n is any odd integer, then $[n]_2 = [1]_2 = 2\mathbb{Z} + 1$.

4. More generally, consider the equivalence relation $\equiv_n = \{(a,b) \in \mathbb{Z}^2 \mid n|b - a\}$ (see part 4 of Example 2.12). Then $[a]_n = n\mathbb{Z} + a$, where $n\mathbb{Z} + a = \{nk + a \mid k \in \mathbb{Z}\}$ (again, we add the subscript n into the notation $[a]_n$ to emphasize that this is an equivalence class for the equivalence relation $\equiv_n$). To see this, note that $b \in [a]_n$ if and only if $a \equiv_n b$ if and only if $n|b - a$ if and only if there is an integer k such that $b - a = nk$ if and only if there is an integer k such that $b = nk + a$ if and only if $b \in n\mathbb{Z} + a$. Since b was arbitrary, we see that $[a]_n = n\mathbb{Z} + a$.

   As a specific example, let's consider $\equiv_3$. Then $[0]_3 = 3\mathbb{Z}$, $[1]_3 = 3\mathbb{Z} + 1$, and $[2]_3 = 3\mathbb{Z} + 2$. Note that these are the only three equivalence classes. For example, we see that $[3]_3 = 3\mathbb{Z} + 3 = \{-3, -2, -1, 0, 3, 6, 9, \ldots\} = 3\mathbb{Z} = [0]_3$.

Theorem 2.14: Let $\sim$ be an equivalence relation on a set S and let $x, y \in S$. The following are equivalent:

1. $x \sim y$.
2. $[x] = [y]$.
3. $x \in [y]$.

This is the first theorem where we want to prove more than two statements equivalent. We will do this with the following chain: $1 \to 2 \to 3 \to 1$. In other words, we will assume statement 1 and use it to prove statement 2. We will then assume statement 2 and use it to prove statement 3. Finally, we will assume statement 3 and use it to prove statement 1.

Proof: ($1 \to 2$) Assume that $x \sim y$ and let $z \in [x]$. Then $x \sim z$. Since $\sim$ is symmetric, $y \sim x$. Since $\sim$ is transitive, $y \sim z$. Therefore, $z \in [y]$. Since $z \in [x]$ was arbitrary, we have shown that $[x] \subseteq [y]$. A symmetric argument shows that $[y] \subseteq [x]$. So, $[x] = [y]$.

($2 \to 3$) Assume $[x] = [y]$. Since $\sim$ is reflexive, $x \sim x$. So, $x \in [x]$. Since $x \in [x]$ and $[x] = [y]$, $x \in [y]$.

($3 \to 2$) Assume that $x \in [y]$. Then $y \sim x$. Since $\sim$ is symmetric, $x \sim y$. □

Partitions

Recall: (1) If X is a nonempty set of sets, we say that X is **pairwise disjoint** if for all $A, B \in X$ with $A \neq B$, A and B are disjoint ($A \cap B = \emptyset$).

(2) If X is a nonempty set of sets, then **union** X is defined by $\cup X = \{y \mid \text{there is } Y \in X \text{ with } y \in Y\}$.

A **partition** of a set S is a set of pairwise disjoint nonempty subsets of S whose union is S. Symbolically, X is a partition of S if and only if

$$\forall A \in X (A \neq \emptyset \land A \subseteq S) \land \forall A, B \in X (A \neq B \to A \cap B = \emptyset) \land \cup X = S.$$

Example 2.15:

1. Recall that $2\mathbb{Z} = \{2k \mid k \in \mathbb{Z}\}$ is the set of even integers and $2\mathbb{Z} + 1 = \{2k + 1 \mid k \in \mathbb{Z}\}$ is the set of odd integers. $X = \{2\mathbb{Z}, 2\mathbb{Z} + 1\}$ is a partition of $\mathbb{Z}$. We can visualize this partition as follows:
$$\mathbb{Z} = \{\ldots, -4, -2, 0, 2, 4, \ldots\} \cup \{\ldots, -3, -1, 1, 3, 5, \ldots\}$$

2. Recall that $3\mathbb{Z} = \{3k \mid k \in \mathbb{Z}\}$, $3\mathbb{Z} + 1 = \{3k + 1 \mid k \in \mathbb{Z}\}$, and $3\mathbb{Z} + 2 = \{3k + 2 \mid k \in \mathbb{Z}\}$. $X = \{3\mathbb{Z}, 3\mathbb{Z} + 1, 3\mathbb{Z} + 2\}$ is a partition of $\mathbb{Z}$. We can visualize this partition as follows:
$$\mathbb{Z} = \{\ldots, -6, -3, 0, 3, 6, \ldots\} \cup \{\ldots, -5, -2, 1, 4, 7, \ldots\} \cup \{\ldots, -4, -1, 2, 5, 8, \ldots\}$$

3. For each $n \in \mathbb{N}$, let $A_n = \{2n, 2n+1\}$. Then $X = \{A_n \mid n \in \mathbb{N}\}$ is a partition of $\mathbb{N}$. We can visualize this partition as follows:

$$\mathbb{N} = \{0,1\} \cup \{2,3\} \cup \{4,5\} \cup \{6,7\} \cup \{8,9\} \cup \cdots$$

4. For each $n \in \mathbb{Z}$, let $A_n = \{(n,m) \mid m \in \mathbb{Z}\}$. Then $X = \{A_n \mid n \in \mathbb{Z}\}$ is a partition of $\mathbb{Z} \times \mathbb{Z}$. We can visualize this partition as follows:

$$\vdots \qquad \vdots \qquad \vdots$$
$$A_{-2} = \{\ldots, (-2,-3), (-2,-2), (-2,-1), (-2,0), (-2,1), (-2,2), (-2,3), \ldots\}$$
$$A_{-1} = \{\ldots, (-1,-3), (-1,-2), (-1,-1), (-1,0), (-1,1), (-1,2), (-1,3), \ldots\}$$
$$A_0 = \{\ldots, (0,-3), (0,-2), (0,-1), (0,0), (0,1), (0,2), (0,3), \ldots\}$$
$$A_1 = \{\ldots, (1,-3), (1,-2), (1,-1), (1,0), (1,1), (1,2), (1,3), \ldots\}$$
$$A_2 = \{\ldots, (2,-3), (2,-2), (2,-1), (2,0), (2,1), (2,2), (2,3), \ldots\}$$
$$\vdots \qquad \vdots \qquad \vdots$$

5. For each $a \in \mathbb{R}$, let $X_a = \{a + bi \mid b \in \mathbb{R}\}$. Then $X = \{X_a \mid a \in \mathbb{R}\}$ is a partition of $\mathbb{C}$. This partition cannot be visualized as easily as the partition given in part 4 above. We will see in Lesson 5 that it is impossible to form a list of the real numbers and therefore, we cannot form a list of the elements of X or lists of the elements of each X_a.

6. The only partition of the one element set $\{a\}$ is $\{\{a\}\}$. The partitions of the two element set $\{a,b\}$ with $a \ne b$ are $\{\{a\},\{b\}\}$ and $\{\{a,b\}\}$.

We will now explore the relationship between equivalence relations and partitions.

Example 2.16: Consider the equivalence relation $\equiv_2$ from part 3 of Example 2.12, defined by $a \equiv_2 b$ if and only if a and b have the same parity, and the partition $\{2\mathbb{Z}, 2\mathbb{Z}+1\}$ of $\mathbb{Z}$ from part 1 of Example 2.15. For this partition, we are thinking of $\mathbb{Z}$ as the union of the even and odd integers:

$$\mathbb{Z} = \{\ldots, -4, -2, 0, 2, 4, \ldots\} \cup \{\ldots, -3, -1, 1, 3, 5, \ldots\}$$

Observe that a and b are in the same member of the partition if and only if $a \equiv_2 b$ if and only if $[a]_2 = [b]_2$. If n is any even integer, then we have $[n]_2 = [0]_2 = 2\mathbb{Z}$ and if n is any odd integer, then we have $[n]_2 = [1]_2 = 2\mathbb{Z} + 1$.

Similarly, the equivalence classes of $\equiv_3$ are $[0]_3 = 3\mathbb{Z}$, $[1]_3 = 3\mathbb{Z}+1$, and $[2]_3 = 3\mathbb{Z}+2$, and so, we see that $\{[0]_3, [1]_3, [2]_3\} = \{3\mathbb{Z}, 3\mathbb{Z}+1, 3\mathbb{Z}+2\}$ is a partition of $\mathbb{Z}$.

Example 2.17: Recall that the **power set** of A, written $\mathcal{P}(A)$, is the set consisting of all subsets of A.

$$\mathcal{P}(A) = \{X \mid X \subseteq A\}$$

For example, if $A = \{a,b,c\}$, then $\mathcal{P}(A) = \{\emptyset, \{a\}, \{b\}, \{c\}, \{a,b\}, \{a,c\}, \{b,c\}, \{a,b,c\}\}$. We can define a binary relation $\sim$ on $\mathcal{P}(A)$ by $X \sim Y$ if and only if $|X| = |Y|$ (X and Y have the same number of elements). It is easy to see that $\sim$ is an equivalence relation on $\mathcal{P}(A)$. There are four equivalence classes.

$$[\emptyset] = \{\emptyset\} \qquad\qquad [\{a\}] = \{\{a\}, \{b\}, \{c\}\}$$
$$[\{a,b\}] = \{\{a,b\}, \{a,c\}, \{b,c\}\} \qquad\qquad [\{a,b,c\}] = \{\{a,b,c\}\}$$

Notes: (1) $\{a\} \sim \{b\} \sim \{c\}$ because each of these sets has one element. It follows that $\{a\}$, $\{b\}$, and $\{c\}$ are all in the same equivalence class. Above, we chose to use $\{a\}$ as the **representative** for this equivalence class. This is an arbitrary choice. In fact, $[\{a\}] = [\{b\}] = [\{c\}]$.

Similarly, $[\{a,b\}] = [\{a,c\}] = [\{b,c\}]$.

(2) The empty set is the only subset of A with 0 elements. Therefore, the equivalence class of $\emptyset$ contains only itself. Similarly, the equivalence class of $A = \{a,b,c\}$ contains only itself.

(3) Notice that the four equivalence classes are pairwise disjoint, nonempty, and their union is $\mathcal{P}(A)$. In other words, the equivalence classes form a partition of $\mathcal{P}(A)$.

Theorem 2.18: Let P be a partition of a set S. Then there is an equivalence relation $\sim$ on S for which the elements of P are the equivalence classes of $\sim$. Conversely, if $\sim$ is an equivalence relation on a set S, then the equivalence classes of $\sim$ form a partition of S.

You will be asked to prove Theorem 2.18 in Problem 12 below.

Important note: We will sometimes want to define relations and operations on equivalence classes. When we do this, we must be careful that what we are defining is **well-defined**. If E is a set of equivalence classes on a set A and R is a relation on A, then we say that R is **well-defined** on E if for all $x, y, z, w \in A$ with $[x] = [z]$ and $[y] = [w]$, xRy if and only if zRw. If R is in fact well-defined on E, then we can define the relation R^* by $[x]R^*[y]$ if and only if xRy. We often identify R and R^* and say that R^* is well-defined on E.

For example, consider the equivalence relation $\equiv_2$ on $\mathbb{Z}$, and let $E = \{[0], [1]\}$ be the set of equivalence classes. Now, consider the relation $<$ on $\mathbb{Z}$. Is $<$ well-defined on E? Well, we have $0 \equiv_2 2$, $0 < 1$ and $2 \not< 1$. It follows that $<$ is **not** well-defined on E. Therefore, if we attempt to define the "relation" $<^*$ on E by $[x] <^* [y]$ if and only if $x < y$, we get something that does not make any sense. Therefore, this definition cannot be made.

Now, let's define the relation R on $\mathbb{Z}$ by $R = \{(x,y) \mid x \text{ is even and } y \text{ is odd}\}$. This is an example of a relation that **is** well-defined on E. Indeed, suppose that $x, y, z, w \in \mathbb{Z}$ with $[x] = [z]$ and $[y] = [w]$ and let xRy. Then x is even and y is odd. Since $[x] = [z]$, z must be even. Since $[y] = [w]$, w must be odd. Therefore, by the definition of R, zRw. A symmetrical argument shows that zRw implies xRy. It follows that we can define R^* on E by $[x]R^*[y]$ if and only if xRy and the definition of R^* makes sense.

In practice, we would usually just define R on E by $[x]R[y]$ if and only if x is even and y is odd. Of course, we should then use the argument given in the last paragraph to verify that R is well-defined.

As one more example, let's attempt to define an operation $+: X \times X \to X$ by $[x]_2 + [y]_2 = [x+y]_2$. This **is** a well-defined operation. To see this, suppose that $[x]_2 = [z]_2$ and $[y]_2 = [w]_2$. By Theorem 2.14, $x \equiv_2 z$ and $y \equiv_2 w$. So, there are integers a and b such that $z - x = 2a$ and $w - y = 2b$. Adding these equations gives us $(z-x) + (w-y) = 2a + 2b$, and so, $(z+w) - (x+y) = 2(a+b)$. Therefore, $x + y \equiv_2 z + w$. Again, by Theorem 2.14, $[x+y]_2 = [z+w]_2$.

Problem Set 2

Full solutions to these problems are available for free download here:
www.SATPrepGet800.com/RABQXZ

LEVEL 1

1. Determine if each of the following sets is an interval:
   (i) $A = \{x \in \mathbb{R} \mid 3 \leq x \leq 7\}$
   (ii) $B = \{x \in \mathbb{Q} \mid x < -205\}$
   (iii) $C = \mathbb{R}^+$
   (iv) $D = \{x \in \mathbb{R} \mid x \geq -16\}$
   (v) $E = \mathbb{R} \setminus \{0\}$

2. Let $C = (-\infty, 2]$ and $D = (-1, 3]$. Compute each of the following:
   (i) $C \cup D$
   (ii) $C \cap D$
   (iii) $C \setminus D$
   (iv) $D \setminus C$
   (v) $C \triangle D$

3. Find all partitions of the three-element set $\{a, b, c\}$ and the four-element set $\{a, b, c, d\}$.

4. Let $A = \{1, 2, 3, 4\}$ and let $R = \{(1,1), (1,3), (2,2), (2,4), (3,1), (3,3), (4,2), (4,4)\}$. Note that R is an equivalence relation on A. Find the equivalence classes of R.

LEVEL 2

5. Find the domain, range, and field of each of the following relations:
   (i) $R = \{(a,b), (c,d), (e,f), (f,a)\}$
   (ii) $S = \{(2k, 2t+1) \mid k, t \in \mathbb{Z}\}$

6. Prove that for each $n \in \mathbb{Z}^+$, $\equiv_n$ (see part 4 of Example 2.12) is an equivalence relation on $\mathbb{Z}$.

LEVEL 3

7. Prove that there do not exist sets A and B such that the relation $<$ on $\mathbb{R}$ is equal to $A \times B$.

8. Let X be a set of equivalence relations on a nonempty set A. Prove that $\bigcap X$ is an equivalence relation on A.

LEVEL 4

9. Let $R = \{(x, y) \in \mathbb{R} \times \mathbb{R} \mid x - y \in \mathbb{Z}\}$. Prove that R is an equivalence relation on $\mathbb{R}$ and describe the equivalence classes of R.

10. Let R be a relation on a set A. Determine if each of the following statements is true or false. If true, provide a proof. If false, provide a counterexample.

    (i) If R is symmetric and transitive on A, then R is reflexive on A.

    (ii) If R is antisymmetric on A, then R is not symmetric on A.

LEVEL 5

11. For $a, b \in \mathbb{N}$, we will say that a divides b, written $a|b$, if there is a natural number k such that $b = ak$. Notice that $|$ is a binary relation on $\mathbb{N}$. Prove that $(\mathbb{N}, |)$ is a partially ordered set, but it is not a linearly ordered set.

12. Let $\boldsymbol{P}$ be a partition of a set S. Prove that there is an equivalence relation $\sim$ on S for which the elements of $\boldsymbol{P}$ are the equivalence classes of $\sim$. Conversely, if $\sim$ is an equivalence relation on a set S, prove that the equivalence classes of $\sim$ form a partition of S.

CHALLENGE PROBLEMS

13. Let $a, b, c, d \in \mathbb{N}$, let A and B be finite sets with $|A| = a$ and $|B| = b$, let $C = \left(\mathcal{P}(A \times B)\right)^c$, let $D = \left(\mathcal{P}(C^2)\right)^d$, and let Z be the set of relations on $C \times D$. Evaluate $|Z|$.

14. Let R and S be binary relations on a set A. The composition of R and S, written $R \circ S$, is defined as $R \circ S = \{(a, b) \mid \exists c \in A\big((a, c) \in R \wedge (c, b) \in S\big)\}$. Suppose that R and S are equivalence relations on A. Prove that $R \circ S$ is an equivalence relation if and only if $R \circ S = S \circ R$.

Lesson 3
Functions

Functions

Let A and B be sets. f is a **function** from A to B, written $f: A \to B$, if the following two conditions hold.

1. $f \subseteq A \times B$.
2. For all $a \in A$, there is a unique $b \in B$ such that $(a, b) \in f$.

Notes: (1) A function $f: A \to B$ is a binary relation on $A \cup B$.

(2) Not every binary relation on $A \cup B$ is a function from A to B. See part 2 of Example 3.1 below.

(3) The uniqueness in the second clause in the definition of a function above is equivalent to the statement "if $(a, b), (a, c) \in f$, then $b = c$."

(4) When we know that f is a function, we will abbreviate $(a, b) \in f$ by $f(a) = b$.

If $f: A \to B$, the **domain** of f, written dom f, is the set A, and the **range** of f, written ran f, is the set $\{f(a) \mid a \in A\}$. Observe that ran $f \subseteq B$. The set B is sometimes called the **codomain** of f.

Example 3.1:

1. $f = \{(0, a), (1, a)\}$ is a function with dom $f = \{0, 1\}$ and ran $f = \{a\}$. Instead of $(0, a) \in f$, we will usually write $f(0) = a$. Similarly, instead of $(1, a) \in f$, we will write $f(1) = a$. Here is a visual representation of this function.

   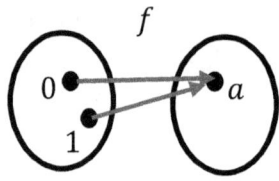

   This function $f: \{0, 1\} \to \{a\}$ is called a **constant function** because the range of f consists of a single element.

   Note also that f is a binary relation on the set $\{0, 1, a\}$.

2. If $a \neq b$, then $g = \{(0, a), (0, b)\}$ is **not** a function because it violates the second clause in the definition of being a function. It is, however, a binary relation on $\{0, a, b\}$.

   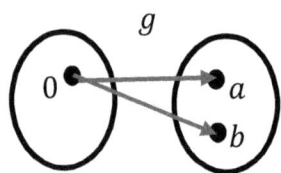

50

3. $h = \{(a,b) \mid a, b \in \mathbb{R} \wedge a > 0 \wedge a^2 + b^2 = 2\}$ is a relation on $\mathbb{R}$ that is **not** a function. $(1, 1)$ and $(1, -1)$ are both elements of h, violating the second clause in the definition of a function. See the figure below on the left. Notice how a vertical line hits the graph twice.

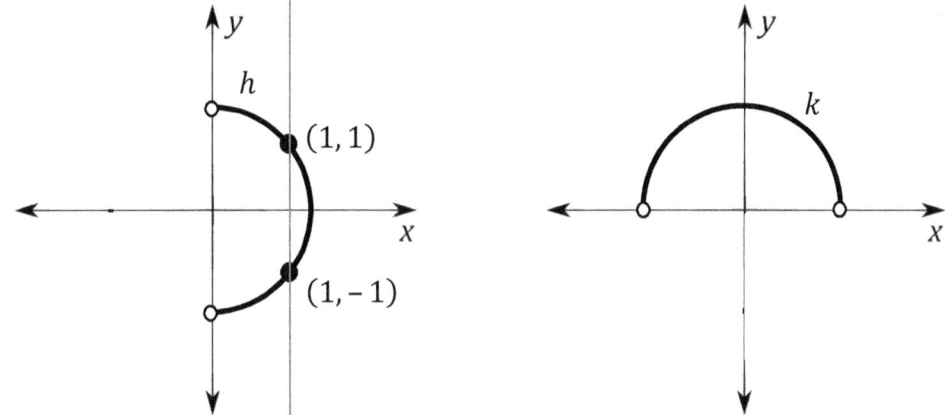

4. $k = \{(a,b) \mid a, b \in \mathbb{R} \wedge b > 0 \wedge a^2 + b^2 = 2\}$ **is** a function. See the figure above on the right. To see that the second clause in the definition of a function is satisfied, suppose that (a, b) and (a, c) are both in f. Then $a^2 + b^2 = 2$, $a^2 + c^2 = 2$, and b and c are both positive. It follows that $b^2 = c^2$, and since b and c are both positive, we have $b = c$.

   We have dom $k = \left(-\sqrt{2}, \sqrt{2}\right)$ and ran $k = \left(0, \sqrt{2}\,\right]$. So, $k\colon \left(-\sqrt{2}, \sqrt{2}\right) \to \left(0, \sqrt{2}\,\right]$.

   This function k is an example of a *real-valued function* (or $\mathbb{R}$-*valued function*) because the codomain of k consists of only real numbers.

   Note that if $a \in \mathbb{R}$, then $a^2 = a \cdot a$. Also, if $a \geq 0$ and $a^2 = b$, then $\sqrt{b} = a$.

5. $b = \{(z, w) \mid z, w \in \mathbb{C} \wedge w = z - 1\}$ is a function. By Note 4 above, we can describe b using the notation $b(z) = z - 1$.

   If we write $z = x + yi$ and $b(z) = u + vi$, then $b(x + yi) = x + yi - 1 = (x - 1) + yi$ and we see that $u(x, y) = x - 1$ and $v(x, y) = y$.

   b is an example of a complex-valued function. One way to visualize this function is to simply stay in the same plane and to analyze how a typical point moves or how a certain set is transformed. The function b takes the point (x, y) to the point $(x - 1, y)$. That is, each point is shifted one unit to the left. Similarly, if $S \subseteq \mathbb{C}$, then each point of the set S is shifted one unit to the left by the function b. Both these situations are demonstrated in the figure to the right. Observe how the point $(2, 2)$ is shifted to the point $(1, 2)$ and how each point of the rightmost rectangle is shifted one point to the left to form the leftmost rectangle.

   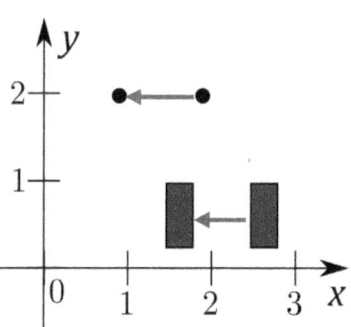

   A second way to visualize this function is to draw two separate planes: an xy-plane and a uv-plane. We can then draw a point or a set in the xy-plane and its image under b in the uv-plane. In the figure below, we do this for the same point and the same rectangle as we did in the previous figure.

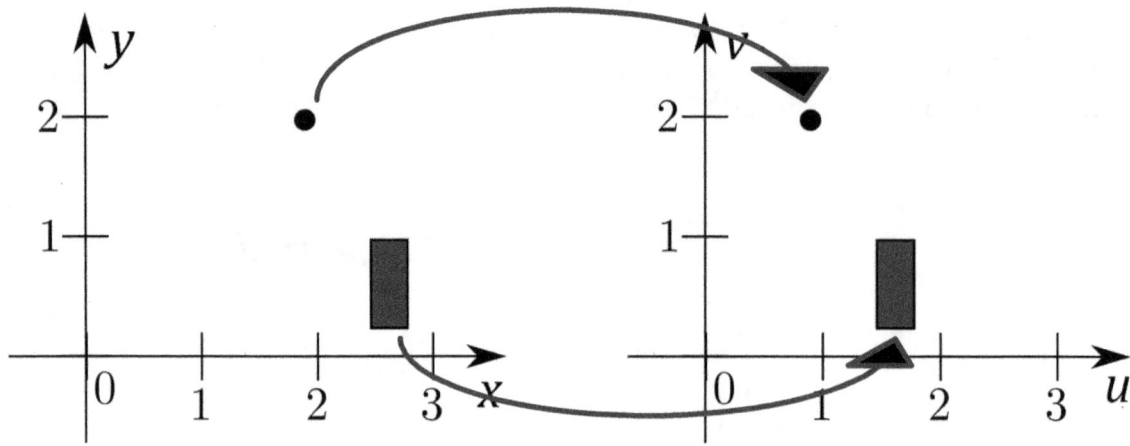

Injections, Surjections, and Bijections

A function $f: A \to B$ is **injective** (or **one-to-one**), written $f: A \hookrightarrow B$, if for all $a, b \in A$, if $a \neq b$, then $f(a) \neq f(b)$. In this case, we call f an **injection**.

Notes: (1) The contrapositive of the conditional statement $p \to q$ is the statement $\neg q \to \neg p$. These two statements are logically equivalent. See the analysis after the statement of Theorem 1.28.

(2) The contrapositive of the statement "If $a \neq b$, then $f(a) \neq f(b)$" is "If $f(a) = f(b)$, then $a = b$." So, we can say that a function $f: A \to B$ is injective if for all $a, b \in A$, if $f(a) = f(b)$, then $a = b$.

A function $f: A \to B$ is **surjective** (or **onto B**), written $f: A \mapsto B$, if for all $b \in B$, there is an $a \in A$ such that $f(a) = b$. In this case, we call f a **surjection**.

A function $f: A \to B$ is **bijective**, written $f: A \cong B$ if f is both injective and surjective. In this case, we call f a **bijection**.

Example 3.2:

1. $f = \{(0, a), (1, a)\}$ from part 1 of Example 3.1 is **not** an injective function because $f(0) = a$, $f(1) = a$, and $0 \neq 1$.

   If we think of f as $f: \{0, 1\} \to \{a\}$, then f is surjective. However, if we think of f as $f: \{0, 1\} \to \{a, b\}$, then f is **not** surjective. So, surjectivity depends upon the codomain of the function.

   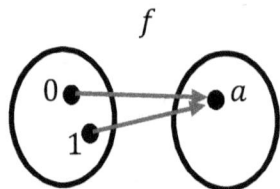

2. $k = \{(a, b) \mid a, b \in \mathbb{R} \wedge b > 0 \wedge a^2 + b^2 = 2\}$ from part 4 of Example 3.1 is **not** an injective function. For example, $(1, 1) \in k$ because $1^2 + 1^2 = 1 + 1 = 2$ and $(-1, 1) \in k$ because $(-1)^2 + 1^2 = 1 + 1 = 2$. Notice how a horizontal line hits the graph twice.

If we think of k as a function from $(-\sqrt{2},\sqrt{2})$ to $\mathbb{R}^+$, then k is **not** surjective. For example, $2 \notin \operatorname{ran} k$ because for any $a \in \mathbb{R}$, $a^2 + 2^2 = a^2 + 4 \geq 4$, and so, $a^2 + 2^2$ cannot be equal to 2. However, if instead we consider k as a function with codomain $(0,\sqrt{2}]$, that is $k:(-\sqrt{2},\sqrt{2}) \to (0,\sqrt{2}]$, then k **is** surjective. Indeed, if $0 < b \leq \sqrt{2}$, then $0 < b^2 \leq 2$, and so, $a^2 = 2 - b^2 \geq 0$. Therefore, $a = \sqrt{2-b^2}$ is a real number such that $k(a) = b$.

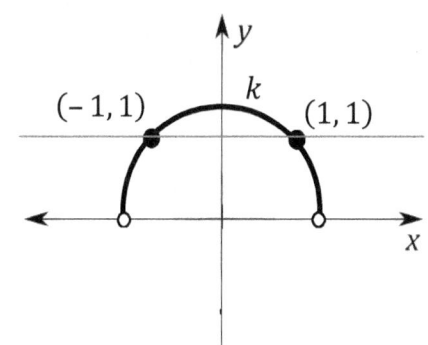

3. Define $g: \mathbb{R} \to \mathbb{R}$ by $g(x) = 7x - 3$. Then g is injective because if $g(a) = g(b)$, we then have $7a - 3 = 7b - 3$. Adding 3 to each side of this equation, we get $7a = 7b$, and then multiplying each side of this last equation by $\frac{1}{7}$, we get $a = b$ (see Lesson 6 for justification of these cancellation rules). Also, g is surjective because if $b \in \mathbb{R}$, then $\frac{b+3}{7} \in \mathbb{R}$ and

$$g\left(\frac{b+3}{7}\right) = 7\left(\frac{b+3}{7}\right) - 3 = (b+3) - 3 = b + (3-3) = b + 0 = b$$

Therefore, g is bijective. See the image to the right for a visual representation of $\mathbb{R}^2 = \mathbb{R} \times \mathbb{R}$ and the graph of the function g.

Notice that any vertical line will hit the graph of g exactly once because g is a function with domain $\mathbb{R}$. Also, any horizontal line will hit the graph exactly once because g is bijective. Injectivity ensures that each horizontal line hits the graph *at most* once and surjectivity ensures that each horizontal line hits the graph *at least* once.

This function g is an example of a *real-valued function* (or *$\mathbb{R}$-valued function*) because the codomain of g consists of only real numbers.

Inverse Functions

If $f: A \to B$ is bijective, we define $f^{-1}: B \to A$, the **inverse** of f, by $f^{-1} = \{(b,a) \mid (a,b) \in f\}$. In other words, for each $b \in B$, $f^{-1}(b) =$ "the unique $a \in A$ such that $f(a) = b$."

Notes: (1) Let $f: A \to B$ be bijective. Since f is surjective, for each $b \in B$, there is an $a \in A$ such that $f(a) = b$. Since f is injective, there is only one such value of a.

(2) The inverse of a bijective function is also bijective.

Example 3.3:

1. Define $f: \{0,1\} \to \{a,b\}$ by $f = \{(0,a),(1,b)\}$. Then f is a bijection and $f^{-1}: \{a,b\} \to \{0,1\}$ is defined by $f^{-1} = \{(a,0),(b,1)\}$. Observe that f^{-1} is also a bijection.

2. Let $\mathbb{E} = \{0,2,4,6,8,\ldots\}$ be the set of even natural numbers and let $\mathbb{O} = \{1,3,5,7,9\ldots\}$ be the set of odd natural numbers. The function $f: \mathbb{E} \to \mathbb{O}$ defined by $f(n) = n + 1$ is a bijection with inverse $f^{-1}: \mathbb{O} \to \mathbb{E}$ defined by $f(n) = n - 1$.

3. If X and Y are sets, we define ^XY to be the set of functions from X to Y. Symbolically, we have
$$^XY = \{f \mid f:X \to Y\}$$

For example, if $A = \{a, b\}$ and $B = \{0, 1\}$, then ^AB has 4 elements (each element is a function from A to B). The elements are $f_1 = \{(a, 0), (b, 0)\}$, $f_2 = \{(a, 0), (b, 1)\}$, $f_3 = \{(a, 1), (b, 0)\}$, and $f_4 = \{(a, 1), (b, 1)\}$. Here is a visual representation of these four functions.

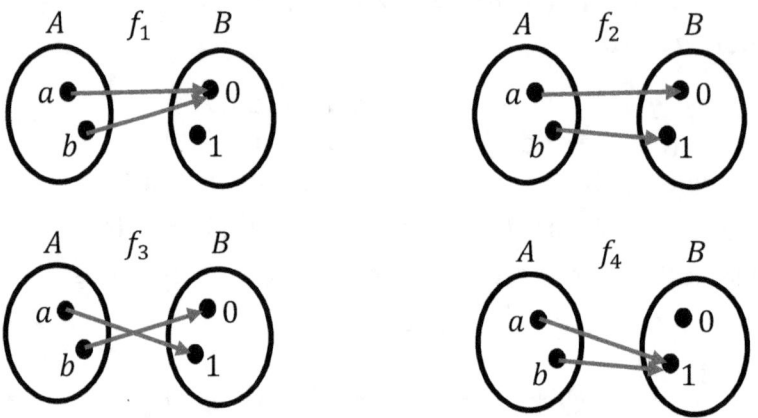

Define $F: {^AB} \to \mathcal{P}(A)$ by $F(f) = \{x \in A \mid f(x) = 1\}$.

So, $F(f_1) = \emptyset$, $F(f_2) = \{b\}$, $F(f_3) = \{a\}$, and $F(f_4) = \{a, b\}$.

Since $\mathcal{P}(A) = \{\emptyset, \{a\}, \{b\}, \{a, b\}\}$, we see that F is a bijection from ^AB to $\mathcal{P}(A)$.

The inverse of F is the function $F^{-1}: \mathcal{P}(A) \to {^AB}$ defined by $F^{-1}(C)(x) = \begin{cases} 0 & \text{if } x \notin C \\ 1 & \text{if } x \in C \end{cases}$

So, we see that $F^{-1}(\emptyset) = f_1$, $F^{-1}(\{b\}) = f_2$, $F^{-1}(\{a\}) = f_3$, and $F^{-1}(\{a, b\}) = f_4$.

4. For $A \neq \emptyset$ and $B = \{0, 1\}$, the function $F: {^AB} \to \mathcal{P}(A)$ defined by $F(f) = \{x \in A \mid f(x) = 1\}$ is always a bijection.

To see that F is injective, let $f, g \in {^AB}$ with $f \neq g$. Since f and g are different, there is some $a \in A$ such that either $f(a) = 0, g(a) = 1$ or $f(a) = 1, g(a) = 0$. **Without loss of generality**, (see Note 1 below) assume that $f(a) = 0$ and $g(a) = 1$. Since $f(a) = 0$, $a \notin F(f)$. Since $g(a) = 1$, $a \in F(g)$. So, $F(f) \neq F(g)$. Since $f \neq g$ implies $F(f) \neq F(g)$, F is injective.

To see that F is surjective, let $C \in \mathcal{P}(A)$, so that $C \subseteq A$. Define $f \in {^AB}$ by $f(x) = \begin{cases} 0 & \text{if } x \notin C \\ 1 & \text{if } x \in C \end{cases}$
Then $x \in F(f)$ if and only if $f(x) = 1$ if and only if $x \in C$. So, $F(f) = C$. Since $C \in \mathcal{P}(A)$ was arbitrary, F is surjective.

As in 3, the inverse of F is the function $F^{-1}: \mathcal{P}(A) \to {^AB}$ defined by $F^{-1}(C)(x) = \begin{cases} 0 & \text{if } x \notin C \\ 1 & \text{if } x \in C \end{cases}$

Notes: (1) In part 4 of Example 3.3, we used the expression "Without loss of generality." This expression can be used when an argument can be split up into two or more cases, and the proof of each of the cases is nearly identical. In the example above, the two cases are (i) $f(a) = 0$, $g(a) = 1$ and (ii) $f(a) = 1$, $g(a) = 0$. The argument for case (ii) is the same as the argument for case (i), essentially word for word—only the roles of f and g are interchanged.

(2) In the study of set theory, we define the natural numbers by letting $0 = \emptyset$, $1 = \{0\}$, $2 = \{0, 1\}$, $3 = \{0, 1, 2\}$,... and so on. In general, $n = \{0, 1, 2, \ldots, n-1\}$. We have just shown that for any nonempty set A, there is a bijection $f: {}^A 2 \to \mathcal{P}(A)$.

Composite Functions

Given functions $f: A \to B$ and $g: B \to C$, the **composite** (or **composition**) of f and g, written $g \circ f: A \to C$, is defined by $(g \circ f)(a) = g(f(a))$ for all $a \in A$. Symbolically, we have

$$g \circ f = \{(a, c) \in A \times C \mid \text{There is a } b \in B \text{ such that } (a, b) \in f \text{ and } (b, c) \in g\}.$$

We can visualize the composition of two functions f and g as follows.

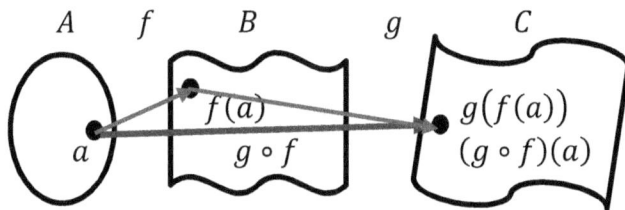

In the picture above, sets A, B, and C are drawn as different shapes simply to emphasize that they can all be different sets. Starting with an arbitrary element $a \in A$, we have an arrow showing a being mapped by f to $f(a) \in B$ and another arrow showing $f(a)$ being mapped by g to $g(f(a)) \in C$. There is also an arrow going directly from $a \in A$ to $(g \circ f)(a) = g(f(a))$ in C. However, note that the only way we know how to get from a to $(g \circ f)(a)$ is to first travel from a to $f(a)$, and then to travel from $f(a)$ to $g(f(a))$.

Example 3.4:

1. Define $f: \mathbb{R} \to \mathbb{R}$ by $f(x) = x + 3$ and $g: \mathbb{R} \to \mathbb{R}$ by $g(x) = (x - 1)^2$ Then $g \circ f: \mathbb{R} \to \mathbb{R}$ is defined by $(g \circ f)(x) = g(f(x)) = g(x + 3) = ((x + 3) - 1)^2 = (x + 2)^2$.

2. Define $f: \mathbb{Z} \to \mathbb{Q}$ by $f(n) = \frac{n}{2}$ and define $g: \mathbb{Q} \to \{0, 1\}$ by $g(x) = \begin{cases} 0 & \text{if } x \in \mathbb{Z}. \\ 1 & \text{if } x \in \mathbb{Q} \setminus \mathbb{Z} \end{cases}$ We will show that $g \circ f: \mathbb{Z} \to \{0, 1\}$ is defined by $(g \circ f)(n) = \begin{cases} 0 & \text{if } n \text{ is even.} \\ 1 & \text{if } n \text{ is odd.} \end{cases}$

   To see this, observe that if $n \in \mathbb{Z}$ is even, then there is an integer k such that $n = 2k$. It follows that $(g \circ f)(n) = g(f(n)) = g\left(\frac{n}{2}\right) = g\left(\frac{2k}{2}\right) = g(k) = 0$ because $k \in \mathbb{Z}$. If n is odd, then there is an integer k such that $n = 2k + 1$. So, $(g \circ f)(n) = g(f(n)) = g\left(\frac{n}{2}\right) = g\left(\frac{2k+1}{2}\right) = 1$ because $\frac{2k+1}{2} \in \mathbb{Q} \setminus \mathbb{Z}$. To see that $\frac{2k+1}{2} \in \mathbb{Q}$, simply observe that $2k + 1 \in \mathbb{Z}$, $2 \in \mathbb{Z}$, and $2 \neq 0$. To see that $\frac{2k+1}{2} \notin \mathbb{Z}$, first note that, $\frac{2k+1}{2} = \frac{2k}{2} + \frac{1}{2} = k + \frac{1}{2}$ (Check this!). Now, let $m = k + \frac{1}{2}$. If $m \in \mathbb{Z}$, then we would have $m - k \in \mathbb{Z}$ because when we subtract one integer from another, we always get an integer. It would then follow that $\frac{1}{2} \in \mathbb{Z}$, which we know it is not. Since assuming that $\frac{2k+1}{2} \in \mathbb{Z}$ would lead to the false statement $\frac{1}{2} \in \mathbb{Z}$, we know that the statement $\frac{2k+1}{2} \in \mathbb{Z}$ must be false.

Note: In part 2 of Example 3.4 above, to show that $m \notin \mathbb{Z}$, we began with the assumption that $m \in \mathbb{Z}$, and then used a logically valid argument to derive a false statement. This is known as a proof by contradiction. See the proof of Theorem 4.6 for a formal proof by contradiction and a detailed explanation as to how it works. Theorem 5.2 provides another example.

It will be important to know that when we take the composition of bijective functions, we always get a bijective function. We will prove this in two steps. We will first show that the composition of injective functions is injective. We will then show that the composition of surjective functions is surjective.

Theorem 3.5: If $f: A \hookrightarrow B$ and $g: B \hookrightarrow C$, then $g \circ f: A \hookrightarrow C$.

Note: We are given that f and g are injections, and we want to show that $g \circ f$ is an injection. We can show this directly using the definition of injectivity, or we can use the contrapositive of the definition of injectivity. Let's do it both ways.

Direct proof of Theorem 3.5: Suppose that $f: A \hookrightarrow B$ and $g: B \hookrightarrow C$, and let $x, y \in A$ with $x \neq y$. Since f is injective, $f(x) \neq f(y)$. Since g is injective, $g(f(x)) \neq g(f(y))$. So, $(g \circ f)(x) \neq (g \circ f)(y)$. Since $x, y \in A$ were arbitrary, $g \circ f: A \hookrightarrow C$. □

Contrapositive proof of Theorem 3.5: Suppose that $f: A \hookrightarrow B$ and $g: B \hookrightarrow C$, let $x, y \in A$ and suppose that $(g \circ f)(x) = (g \circ f)(y)$. Then $g(f(x)) = g(f(y))$. Since g is injective, $f(x) = f(y)$. Since f is injective, $x = y$. Since $x, y \in A$ were arbitrary, $g \circ f: A \hookrightarrow C$. □

Theorem 3.6: If $f: A \mapsto B$ and $g: B \mapsto C$, then $g \circ f: A \mapsto C$.

Proof: Suppose that $f: A \mapsto B$ and $g: B \mapsto C$, and let $c \in C$. Since g surjective, there is $b \in B$ with $g(b) = c$. Since f is surjective, there is $a \in A$ with $f(a) = b$. So, $(g \circ f)(a) = g(f(a)) = g(b) = c$. Since $c \in C$ was arbitrary, $g \circ f$ is surjective. □

Corollary 3.7: If $f: A \cong B$ and $g: B \cong C$, then $g \circ f: A \cong C$.

Proof: Suppose that $f: A \cong B$ and $g: B \cong C$. Then f and g are injective. By Theorem 3.5, $g \circ f$ is injective. Also, f and g are surjective. By Theorem 3.6, $g \circ f$ is surjective. Since $g \circ f$ is both injective and surjective, $g \circ f$ is bijective. □

Note: A **corollary** is a theorem that follows easily from a theorem or theorems that have already been proved.

Identity Functions

If A is any set, then we define the **identity function** on A, written $i_A: A \to A$ by $i_A(a) = a$ for all $a \in A$. Note that the identity function on A is a bijection from A to itself.

Theorem 3.8: If $f: A \cong B$, then $f^{-1} \circ f = i_A$ and $f \circ f^{-1} = i_B$.

Proof: Let $a \in A$ with $f(a) = b$. Then $f^{-1}(b) = a$, and so, $(f^{-1} \circ f)(a) = f^{-1}(f(a)) = f^{-1}(b) = a$. Since $i_A(a) = a$, we see that $(f^{-1} \circ f)(a) = i_A(a)$. Since $a \in A$ was arbitrary, $f^{-1} \circ f = i_A$.

Now, let $b \in B$. Since $f: A \cong B$, there is a unique $a \in A$ with $f(a) = b$. Equivalently, $f^{-1}(b) = a$. We have $(f \circ f^{-1})(b) = f(f^{-1}(b)) = f(a) = b$ Since $i_B(b) = b$, we see that $(f \circ f^{-1})(b) = i_B(b)$. Since $b \in B$ was arbitrary, $f \circ f^{-1} = i_B$. □

Images and and Inverse Images

If $f: X \to Y$ and $A \subseteq X$, then the **image of A under f** is the set $f[A] = \{f(x) \mid x \in A\}$. Similarly, if $B \subseteq Y$, then the **inverse image of B under f** is the set $f^{-1}[B] = \{x \in X \mid f(x) \in B\}$.

Example 3.9:

1. Let $f: \{a, b, c, d\} \to \{0, 1, 2\}$ be defined by $f = \{(a, 0), (b, 0), (c, 1), (d, 2)\}$. Let $A = \{a\}$, $B = \{a, b\}$, $C = \{a, c\}$, and $D = \{b, c, d\}$. Then $f[A] = \{0\}$, $f[B] = \{0\}$, $f[C] = \{0, 1\}$, and $f[D] = \{0, 1, 2\}$. Now, let $X = \{0\}$, $Y = \{0, 1\}$, $Z = \{0, 2\}$, and $W = \{0, 1, 2\}$. Then we have $f^{-1}[X] = \{a, b\}$, $f^{-1}[Y] = \{a, b, c\}$, $f^{-1}[Z] = \{a, b, d\}$, and $f^{-1}[W] = \{a, b, c, d\}$.

2. Define $f: \mathbb{R} \to \mathbb{R}$ by $f(x) = x^4$. Then we have $f[\mathbb{R}] = [0, \infty)$, $f(\{-2, 0, 3\}) = \{0, 16, 81\}$, $f[(-3, 2]] = [0, 81)$, $f^{-1}[\mathbb{R}] = \mathbb{R}$, $f^{-1}[\{16\}] = \{-2, 2\}$, $f^{-1}[[0, \infty)] = \mathbb{R}$, $f^{-1}[(-\infty, 0)] = \emptyset$, and $f^{-1}[(0, \infty)] = (-\infty, 0) \cup (0, \infty) = \mathbb{R} \setminus \{0\}$.

Sequences

A function with domain $\mathbb{N}$ is called an **infinite sequence**. For example, let $f: \mathbb{N} \to \{0, 1\}$ be defined by $f(n) = \begin{cases} 0 & \text{if } n \text{ is even} \\ 1 & \text{if } n \text{ is odd} \end{cases}$. A nice way to visualize an infinite sequence is to list the "outputs" of the sequence in order in parentheses. So, we may write f as $(0, 1, 0, 1, 0, 1, \ldots)$. In general, if A is a nonempty set and $f: \mathbb{N} \to A$ is a sequence, then we can write f as $(f(0), f(1), f(2), \ldots)$.

Similarly, a **finite sequence** is a function with domain $\{0, 1, \ldots, n - 1\}$ for some n. For example, the sequence $(0, 2, 4, 6, 8, 10)$ is the function $g: \{0, 1, 2, 3, 4, 5\} \to \mathbb{N}$ defined by $g(k) = 2k$. If the domain of a finite sequence is $\{0, 1, \ldots, n - 1\}$, we say that n is the **length** of the sequence.

Observe how a finite sequence with domain $\{0, 1, \ldots, n - 1\}$ and range A looks just like an n-tuple in A^n. In fact, it's completely natural to identify a finite sequence of length n with the corresponding n-tuple. So, $(0, 2, 4, 6, 8, 10)$ can be thought of as a 6-tuple from $\mathbb{N}^6$, or as the function $g: \{0, 1, 2, 3, 4, 5\} \to \mathbb{N}$ defined by $g(k) = 2k$.

Informally, we can think of an infinite sequence as an infinite length tuple. As one more example, $(1, 2, 4, 8, 16, 32, \ldots)$ represents the sequence $h: \mathbb{N} \to \mathbb{N}$ defined by $h(n) = 2^n$. If $f: \mathbb{N} \to X$ is defined by $f(n) = x_n$, then we may represent the sequence f using the notation $(x_n)_{n \in \mathbb{N}}$ or simply (x_n). For example, the sequence h can be represented as $(2^n)_{n \in \mathbb{N}}$ or (2^n). x_n is called the **nth term** of the sequence. The 0th term of the sequence h is 1, the first term of the sequence h is 2, and so on. In general, the nth term of the sequence h is 2^n.

Notes: (1) Using the definition $n = \{0, 1, 2, \ldots, n - 1\}$ (see Note 2 following Example 3.3), we can say that a finite sequence of length n is a function $f: n \to A$ for some set A. For example, the function g above has domain 6, so that $g: 6 \to \mathbb{N}$.

(2) In some real analysis books (and math books in general), 0 is excluded from the set of natural numbers. In this case, sequences would start from a 1st term instead of a 0th term.

Example 3.10:

1. There is exactly one finite sequence of length 0, namely the empty sequence, ∅. The empty sequence can be described using function notation as $f: 0 \to X$, where X is any set (remember from Note 2 following Example 3.3 that $0 = \emptyset$).

2. The infinite sequence $(0, -1, 2, -3, 4, -5, \ldots)$ is a function from $\mathbb{N}$ to $\mathbb{Z}$. If we name this function g, then we have that $g: \mathbb{N} \to \mathbb{Z}$ and g is defined by $g(n) = (-1)^n n$. So, the nth term of the sequence g is $(-1)^n n$ and we can represent the sequence as $((-1)^n n)$. This function g is an example of an *integer-valued function* (or $\mathbb{Z}$-*valued function* or $\mathbb{Z}$-*valued sequence*) because the codomain of g consists of only integers.

3. The infinite sequence $\left(\frac{1}{n+1}\right)$ is a function from $\mathbb{N}$ to $\mathbb{Q}$. The nth term of this sequence is $\frac{1}{n+1}$. If we name this function h, then we have that $h: \mathbb{N} \to \mathbb{Q}$ and h is defined by $h(n) = \frac{1}{n+1}$. This function h is an example of a *rational-valued function* (or $\mathbb{Q}$-*valued function* or $\mathbb{Q}$-*valued sequence*) because the codomain of h consists of only rational numbers. Since the "outputs" of h take on only positive values, we can "shrink" the codomain of h to $\mathbb{Q}^+$, the set of positive rational numbers. So, we can write $h: \mathbb{N} \to \mathbb{Q}^+$. We can visualize this sequence as follows:

$$\left(1, \frac{1}{2}, \frac{1}{3}, \frac{1}{4}, \frac{1}{5}, \ldots\right)$$

4. The infinite sequence $\left(n^2 + \sqrt{n}i\right)$ is a function from $\mathbb{N}$ to $\mathbb{C}$. The nth term of this sequence is $n^2 + \sqrt{n}i$. If we name this function k, then we have that $k: \mathbb{N} \to \mathbb{C}$ and k is defined by $k(n) = n^2 + \sqrt{n}i$. This function k is an example of a *complex-valued function* (or $\mathbb{C}$-*valued function* or $\mathbb{C}$-*valued sequence*) because the codomain of k consists of only complex numbers. We can visualize this sequence as follows: $\left(0, 1 + i, 4 + \sqrt{2}i, 9 + \sqrt{3}i, 16 + 2i, \ldots\right)$

In this book, we will mostly be interested in infinite **sequences of real numbers**. These are *real-valued functions* (or $\mathbb{R}$-*valued functions*) with domain $\mathbb{N}$. In other words, the sequences we look at will be functions of the form $f: \mathbb{N} \to \mathbb{R}$. Observe that every rational-valued function is also a real-valued function because $\mathbb{Q} \subseteq \mathbb{R}$.

Example 3.11:

1. Let r be a real number and define $f: \mathbb{N} \to \mathbb{R}$ by $f(n) = r$. Then f is called a **constant sequence** because it always takes on the same value. The nth term of the sequence f is r and we can represent the sequence as (r) or $(r, r, r, r, r, \ldots)$.

    As a specific example, if we let $r = \sqrt{2}$, then the sequence $(\sqrt{2})$ is the constant function whose only output is $\sqrt{2}$.

2. Let r be a real number and define $g: \mathbb{N} \to \mathbb{R}$ by $g(n) = r^n$. The nth term of the sequence g is r^n and we can represent the sequence as (r^n) or $(1, r, r^2, r^3, r^4, \ldots)$.

More generally, if a and r are real numbers, then the sequence $g: \mathbb{N} \to \mathbb{R}$ given by $g(n) = ar^n$ is known as a **geometric sequence**. The real number a is the first term of the sequence and r is called the **common ratio** of the geometric sequence.

As a specific example of a geometric sequence, if we let $a = 1$ and $r = \frac{1}{2}$, then the sequence $\left(\left(\frac{1}{2}\right)^n\right) = \left(\frac{1}{2^n}\right)$ can be visualized as $\left(1, \frac{1}{2}, \frac{1}{4}, \frac{1}{8}, \frac{1}{16}, \ldots\right)$. Note that this particular example is also an infinite sequence of rational numbers (so, it is a $\mathbb{Q}$-valued sequence).

As one more example, if we let $a = 5$ and $r = \sqrt{2}$, then the sequence $\left(5(\sqrt{2})^n\right)$ can be visualized as $\left(5, 5\sqrt{2}, 10, 10\sqrt{2}, 20, \ldots\right)$

Convention: From now on, whenever we say "sequence," we will always mean an infinite real-valued sequence. If we wish to speak of a different kind of sequence, we will be more specific (for example, we might say that $\left(1, \frac{1}{2}, \frac{1}{3}, \frac{1}{4}, \frac{1}{5}, \frac{1}{6}, \frac{1}{7}\right)$ is a finite rational-valued sequence of length 7).

Subsequences

The formal definition of a subsequence may seem a bit confusing at first. So, let's begin by looking at a simple example.

Consider the sequence $(s_n) = (\sqrt{n}) = (0, 1, \sqrt{2}, \sqrt{3}, 2, \sqrt{5}, \sqrt{6}, \sqrt{7}, \ldots)$.

An example of a subsequence of this sequence is $(0, \sqrt{2}, 2, \sqrt{6}, \ldots)$. Notice how we're forming this new sequence by "picking out" every other member of the range of the original sequence.

Using function notation, the original sequence can be described as the function $f: \mathbb{N} \to \mathbb{R}$ defined by $f(n) = \sqrt{n}$.

The subsequence can be described as the function $h: \mathbb{N} \to \mathbb{R}$ defined by $h(k) = \sqrt{2k}$ (I am using the variable k instead of n for the function h only to make the following explanation easier to follow).

Observe that we get from the function f to the function h by replacing n by $2k$. Formally, h is the composition of the functions f and g ($h = f \circ g$), where $g: \mathbb{N} \to \mathbb{N}$ is defined by $g(k) = 2k$. Indeed, we have $(f \circ g)(k) = f(g(k)) = f(2k) = \sqrt{2k} = h(k)$.

Getting back to our original notation, if we let $s_n = f(n) = \sqrt{n}$ and we let $n_k = g(k) = 2k$, then,
$$(s_{n_k}) = (s_{2k}) = \sqrt{2k}.$$

Let S be a set and let (s_n) be a sequence in S. In other words, (s_n) is a function $f: \mathbb{N} \to S$, where $f(n) = s_n$ for each $n \in \mathbb{N}$. Let $g: \mathbb{N} \to \mathbb{N}$ be a **strictly increasing function** ($n < m \to g(n) < g(m)$). Then $f \circ g: \mathbb{N} \to S$ is called a **subsequence** of f.

Note: If we let $n_k = g(k)$ for each $k \in \mathbb{N}$, we see that $(f \circ g)(k) = f(g(k)) = f(n_k) = s_{n_k}$. So, we may use the notation (s_{n_k}) to represent a subsequence of (s_n). Since g is a strictly increasing function, it follows that $n_0 < n_1 < n_2 < \cdots$.

Example 3.12:

1. Let $(s_n) = ((-1)^n) = (1, -1, 1, -1, 1, -1, ...)$. Then (s_n) is a $\mathbb{Z}$-valued sequence. That is, (s_n) is the function $f: \mathbb{N} \to \mathbb{Z}$ given by $f(n) = (-1)^n$. Now, $(s_{n_k}) = (s_{2k}) = (1) = (1, 1, 1, ...)$ is a subsequence of (s_n). Here, we have $n_k = 2k$. In function notation, n_k is the function $g: \mathbb{N} \to \mathbb{N}$ given by $g(k) = 2k$, and s_{n_k} is the composition $f \circ g: \mathbb{N} \to \mathbb{Z}$, where

$$(f \circ g)(k) = f(g(k)) = f(2k) = (-1)^{2k} = 1.$$

Another subsequence is given by $(s_{n_j}) = (s_{2j+1}) = (-1) = (-1, -1, -1, ...)$. Here, we have $n_j = 2j + 1$. In function notation, n_j is the function $g: \mathbb{N} \to \mathbb{N}$ given by $g(j) = 2j + 1$, and s_{n_j} is the composition $f \circ g: \mathbb{N} \to \mathbb{Z}$, where

$$(f \circ g)(j) = f(g(j)) = f(2j + 1) = (-1)^{2j+1} = -1.$$

There are many other subsequences of (s_n). For example, (s_{n_t}), where $n_t = \frac{t(t+1)}{2}$ is the subsequence $(1, -1, -1, 1, 1, -1, -1, 1, 1, ...)$. (Convince yourself of this!)

2. Let $(s_n) = (1, 1.4, 1.41, 1.414, 1.4142, ...)$ (this is a $\mathbb{Q}$-valued sequence that looks like it's trying to give better and better approximations to $\sqrt{2}$). Here, (s_n) is a function $f: \mathbb{N} \to \mathbb{Q}$. An example of a subsequence of (s_n) is $(s_{n_k}) = (s_{k^2}) = (1, 1.4, 1.4142, ...)$. Here, we have $n_k = k^2$. In function notation, n_k is the function $g: \mathbb{N} \to \mathbb{N}$ given by $g(k) = k^2$, and s_{n_k} is the composition $f \circ g: \mathbb{N} \to \mathbb{Z}$, where

$$(f \circ g)(k) = f(g(k)) = f(k^2) = s_{k^2}.$$

Convergent and Cauchy Sequences

Informally, we say that a sequence (s_n) of real numbers **converges** to a real number s if the real numbers s_n get "closer and closer" to the real number s, as n gets larger and larger. For example, it looks like the sequence $(s_n) = \left(\frac{1}{n+1}\right) = \left(1, \frac{1}{2}, \frac{1}{3}, \frac{1}{4}, \frac{1}{5}, ...\right)$ converges to 0. Below is a picture of the graph of this sequence. Notice how the values of the sequence seem to get closer to 0 as n gets larger.

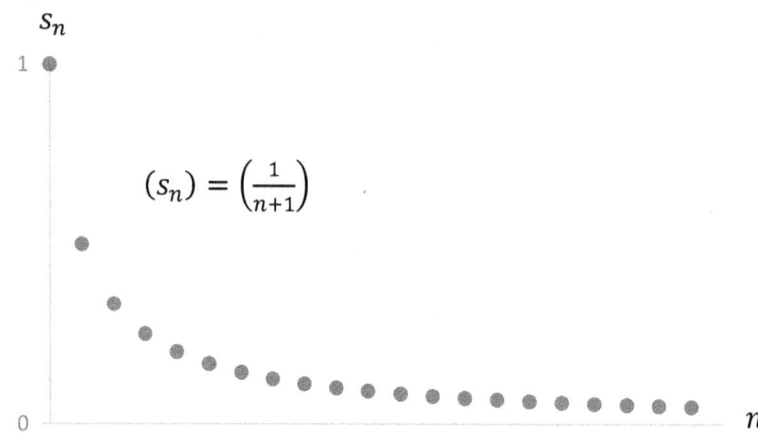

The sequence $(t_n) = (n) = (0, 1, 2, 3, 4, 5, ...)$ does not seem to converge to a fixed number. In this case, we say that the sequence **diverges**. Below is a picture of the graph of this sequence. Notice how the values of the sequence seem to increase without bound as n gets larger.

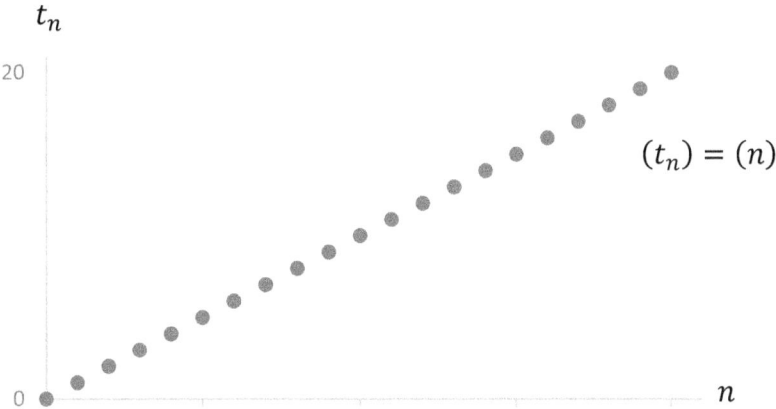

Convergent sequences will be defined more formally in Lesson 8.

Informally, we say that a sequence (s_n) of real numbers is a **Cauchy sequence** if the real numbers s_n get "closer and closer to each other" as we go further out into the sequence.

For example, it looks like the sequence $(s_n) = \left(\frac{1}{n+1}\right) = \left(1, \frac{1}{2}, \frac{1}{3}, \frac{1}{4}, \frac{1}{5}, ...\right)$ is a Cauchy sequence. We will see in Lesson 8 that for real-valued sequences, the notions of convergent sequence and Cauchy sequence coincide. In other words, every convergent real-valued sequence is a Cauchy sequence and every Cauchy sequence of real numbers is convergent. This is one of the main characteristics that sets the real numbers apart from the rational numbers.

A Cauchy sequence always seems to be converging to a fixed value. However, whether this value actually exists depends on the type of sequence. As stated at the end of the last paragraph, real-valued Cauchy sequences will always converge, whereas rational-valued Cauchy sequences may or may not converge (they will always converge to a *real number*, but they may or may not converge to a rational number).

The sequence $(t_n) = (n) = (0, 1, 2, 3, 4, 5, ...)$ is not a Cauchy sequence, as the distance between any two consecutive terms of the sequence is always 1. So, the real numbers are not getting closer to each other as we go further out into the sequence.

Cauchy sequences will be defined more formally in Lessons 4 and 8.

Example 3.13:
1. The constant sequence $(x_n) = (r)$ is a Cauchy sequence for any real number r. Indeed, the distance between any two terms of the sequence is 0. Furthermore, the sequence converges to the real number r.

2. The sequence $(s_n) = (1, 1.4, 1.41, 1.414, 1.4142, ...)$ defined in part 2 of Example 3.12 is a Cauchy sequence. The terms are getting closer and closer together because as we go further out into the sequence, the numbers in the sequence agree to more and more decimal places. If we are thinking of (s_n) as a real-valued sequence, then the sequence converges to the real number $\sqrt{2}$. On the other hand, if we are thinking of (s_n) as a rational-valued sequence, then the sequence **does not** converge. Indeed, we will see in Lesson 6 that $\sqrt{2}$ is **not** a rational number.

3. If a and r are real numbers with $a \neq 0$ and $r > 1$, then the geometric sequence $(g_n) = (ar^n)$ is **not** a Cauchy sequence. The sequence also does **not** converge.

Real-valued Functions

In this book, we will mostly be interested in functions $f: A \to \mathbb{R}$, where $A \subseteq \mathbb{R}$. For example, if $A = \mathbb{N}$, then $f: \mathbb{N} \to \mathbb{R}$ is a real-valued sequence. Let's take a look at some other common real-valued functions and their graphs.

Let $r \in \mathbb{R}$ and define $f: \mathbb{R} \to \mathbb{R}$ by $f(x) = r$. Then f is called a **constant function**. The graph of a constant function is a horizontal line in the xy-plane.

Example 3.14: Define $f: \mathbb{R} \to \mathbb{R}$ by $f(x) = 2$. Then f is a constant function. The graph of f looks as follows:

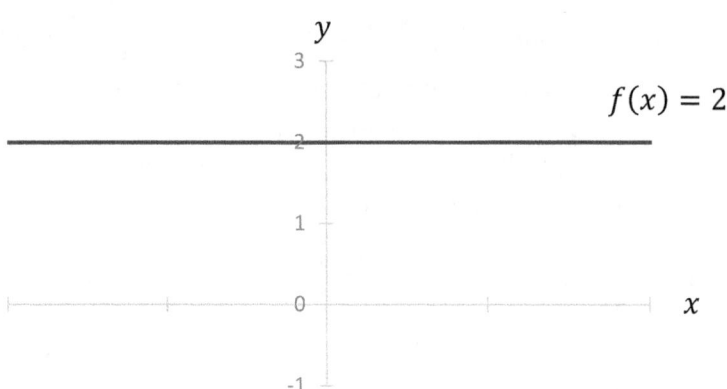

Let $m, b \in \mathbb{R}$ and define $f: \mathbb{R} \to \mathbb{R}$ by $f(x) = mx + b$. Then f is called a **linear function**. The graph of this function is a line with **slope** m and **y-intercept** $(0, b)$.

Observe that a linear function with slope $m = 0$ is a constant function. Example 3.14 above provides an example of such a function.

Linear functions will be discussed in more detail in Lesson 10. For now, let's look at a simple example of a linear function that is **not** a constant function.

Example 3.15: Define $g: \mathbb{R} \to \mathbb{R}$ by $g(x) = \frac{3}{2}x + 1$. Then g is a linear function with slope $m = \frac{3}{2}$ and y-intercept $(0, 1)$. The graph of g looks as follows:

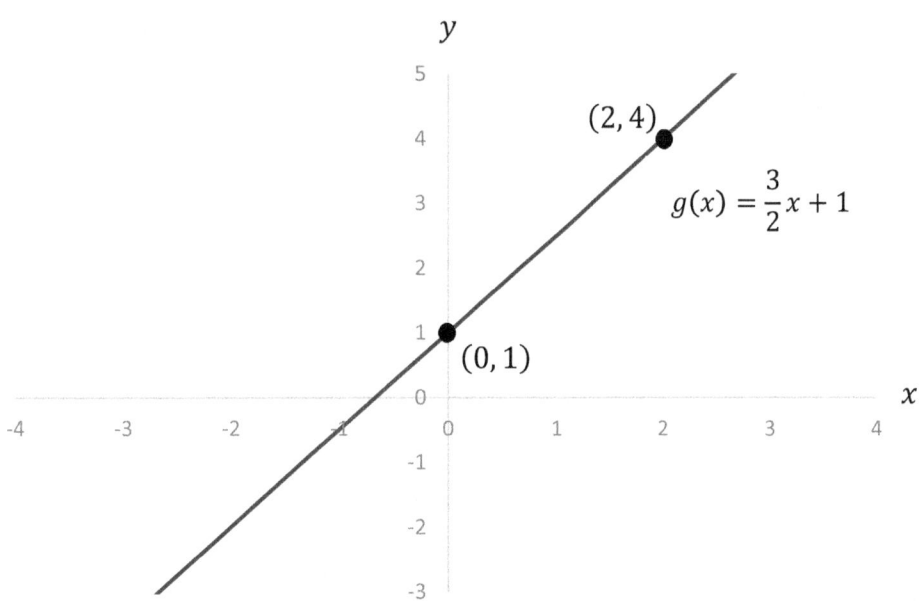

Note: The point $(0, 1)$ is on the line because $b = 1$. We can then use the fact that the slope of the line is $m = \frac{3}{2}$ to find another point on the line. From the point $(0, 1)$, we move up 3 units and right 2 units to get to the point $(2, 4)$. Since a line is completely determined by two points, once we plot these two points, we can draw the whole line.

Let $a, b, c \in \mathbb{R}$ with $a \neq 0$ and define $f: \mathbb{R} \to \mathbb{R}$ by $f(x) = ax^2 + bx + c$. Then f is called a **quadratic function**. The graph of a quadratic function is called a **parabola**. The point $\left(-\frac{b}{2a}, f\left(-\frac{b}{2a}\right)\right)$ is called the **vertex** of the parabola. If $a > 0$, then the parabola opens upwards and if $a < 0$, then the parabola opens downwards.

Example 3.16: Define $h: \mathbb{R} \to \mathbb{R}$ by $h(x) = x^2$. Then h is a quadratic function with $a = 1, b = 0$, and $c = 0$. The vertex of this parabola is $\left(-\frac{b}{2a}, f\left(-\frac{b}{2a}\right)\right) = (0, 0)$. Since $a = 1 > 0$, the parabola opens upwards. The graph of h looks as follows:

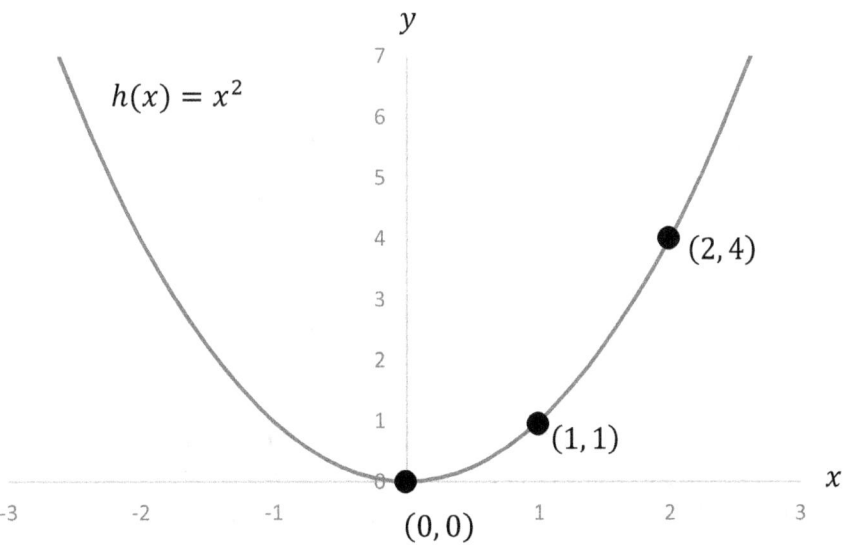

Let $a_0, a_1, \ldots, a_{k-1}, a_k \in \mathbb{R}$ and define $f: \mathbb{R} \to \mathbb{R}$ by $f(x) = a_k x^k + a_{k-1} x^{k-1} + \cdots + a_1 x + a_0$. Then f is called a **polynomial function**. If $a_k \neq 0$, then the natural number k is called the **degree** of the polynomial. The **zero** polynomial, 0, has no degree. The integers $a_0, a_1, \ldots, a_{k-1}, a_k$ are called the **coefficients** of the polynomial and if $a_k \neq 0$, then a_k is the **leading coefficient**. The expressions $a_k x^k, a_{k-1} x^{k-1}, \ldots, a_1 x, a_0$ are called the **terms** of the polynomial and if $a_k \neq 0$, then $a_k x^k$ is the **leading term**. a_0 is called the **constant term**. In general, for each $i = 0, 1, \ldots, k$, a_i is called the ith term. Constant functions are polynomials of degree 0 (with the exception of the 0 polynomial, which has no degree). Linear functions are polynomials of degree 1. Quadratic functions are polynomials of degree 2. Polynomials of degree 3 are called **cubic functions**.

Note: We define the **powers** of x as follows:
$$x^0 = 1; \quad x^1 = x; \quad x^2 = x \cdot x; \quad x^3 = x^2 \cdot x; \quad \ldots \text{ and so on.}$$

In general, for $n \in \mathbb{N}$, x^k is the product of x with itself k times. We can write $x^{k+1} = x^k \cdot x$.

Example 3.17: Define $j: \mathbb{R} \to \mathbb{R}$ by $j(x) = x^3$. Then j is a polynomial of degree 3 (or a cubic polynomial) with $a_3 = 1$ and $a_2 = a_1 = a_0 = 0$. The graph of j looks as follows:

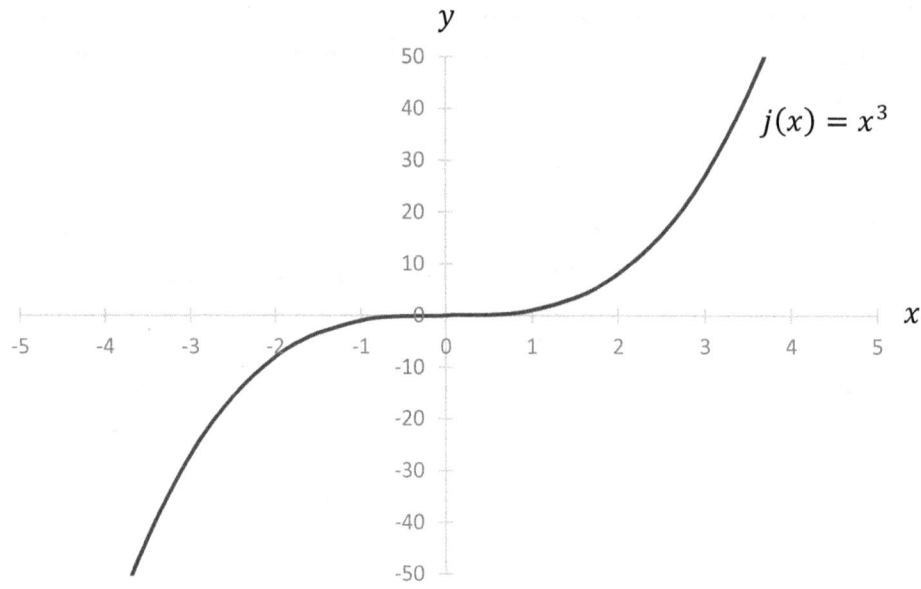

Let $f, g: \mathbb{R} \to \mathbb{R}$ be polynomial functions and let $A = \{x \in \mathbb{R} \mid g(x) \neq 0\}$. The function $h: A \to \mathbb{R}$ defined by $h(x) = \frac{f(x)}{g(x)}$ is called a **rational function**. Note that every polynomial function is a rational function (set $g(x) = 1$ in the definition).

Example 3.18: Define $k: \mathbb{R}^* \to \mathbb{R}$ by $k(x) = \frac{1}{x}$. Then k is a rational function. In the graph of k displayed below the line $y = 0$ is called a **horizontal asymptote** and the line $x = 0$ is called a **vertical asymptote**.

Notes: Informally speaking, a **vertical asymptote** is a vertical line that a curve approaches as the input values get closer to a fixed real number. A **horizontal asymptote** is a horizontal line that a curve approaches as the input values approach infinity or negative infinity.

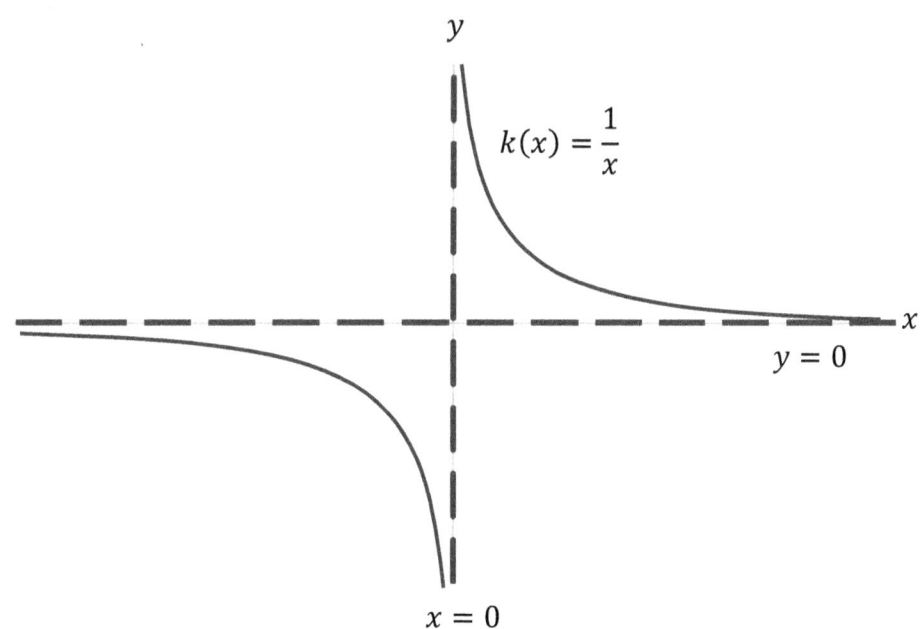

Informally, an **algebraic function** is one that can be formed by using the operations of addition, multiplication, division, and taking roots. Polynomials are formed using only addition and multiplication. Therefore, all polynomials are algebraic functions. Rational functions are formed using addition, multiplication, and division. Therefore, they are also algebraic functions.

Notes: (1) A formal definition of algebraic function is quite complicated and outside the scope of this book. For the purpose of this book, the informal description will suffice.

(2) If x and y are real numbers such that $y = x^2$, then we call x a **square root** of y. If x is a positive real number, then we say that x is the **positive square root** of y and we write $x = \sqrt{y}$ or $x = y^{\frac{1}{2}}$.

(3) If $n \in \mathbb{Z}^+, n > 2$, and $y = x^n$, we say that x is an **nth root** of y. If n is even, we insist that x be a positive real number. Once again, if x is a positive real number, then we write $x = \sqrt[n]{y}$ or $x = y^{\frac{1}{n}}$.

Example 3.19: Define $a: [0, \infty) \to \mathbb{R}$ by $a(x) = \sqrt{x}$. Then a is an algebraic function that is **not** a rational function. The domain and range of a are both $[0, \infty)$. The graph of a looks as follows:

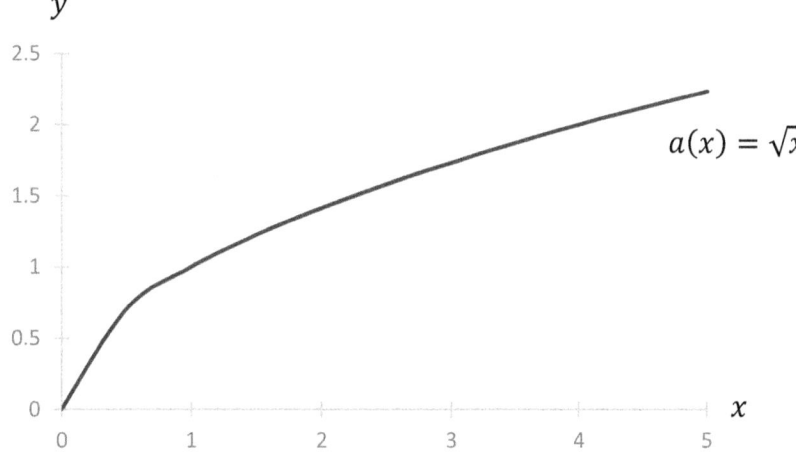

Let $r \in \mathbb{R}^+$ with $r \neq 1$ and define $f: \mathbb{R} \to \mathbb{R}$ by $f(x) = r^x$. Then f is called an **exponential function**. The domain of an exponential function is $\mathbb{R}$ and the range of an exponential function is $(0, \infty)$. Every exponential function has a horizontal asymptote of $y = 0$.

Example 3.20: Define $f: \mathbb{R} \to \mathbb{R}$ by $f(x) = 2^x$. Then f is an exponential function. The graph of f looks as follows:

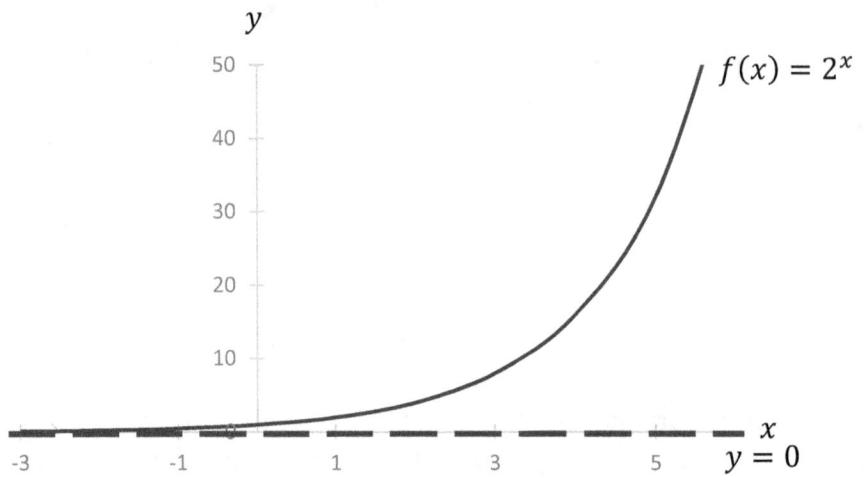

Since the exponential function with base r, $f(x) = r^x$, is injective, it has an inverse. If we let g be the inverse of this exponential function, then $g: \mathbb{R} \to \mathbb{R}$ is called a **logarithmic function**. The base of this logarithmic function is also r and we write $g(x) = \log_r x$. The domain of a logarithmic function is $(0, \infty)$ and the range of a logarithmic function is $\mathbb{R}$. Every logarithmic function has a vertical asymptote of $x = 0$.

Example 3.21: Define $g: (0, \infty) \to \mathbb{R}$ by $g(x) = \log_2 x$. Then g is a logarithmic function. The graph of g looks as follows:

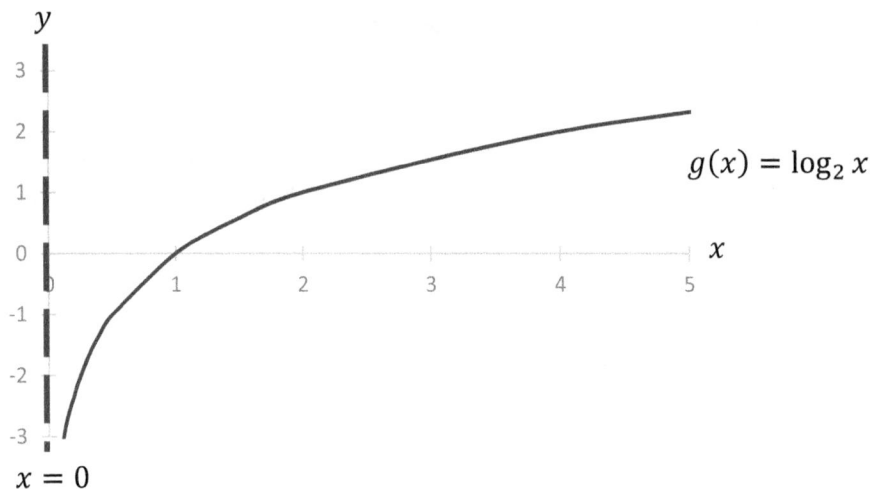

We will finish this lesson with definitions of the six **trigonometric functions**. We begin with a brief discussion of circles.

A **circle** in the Cartesian plane is the set of all points that are at a fixed distance (called the **radius** of the circle) from a fixed point (called the **center** of the circle).

The **circumference** of a circle is the distance around the circle.

If C and C' are the circumferences of two circles with radii r and r', respectively, then it turns out that $\frac{C}{2r} = \frac{C'}{2r'}$. In other words, the value of the ratio $\frac{\text{Circumference}}{2(\text{radius})}$ is independent of the circle that we use to form this ratio. We leave the proof of this fact for the interested reader to investigate themselves. We call the common value of this ratio π (pronounced "pi"). So, we have $\frac{C}{2r} = \pi$, or equivalently, $C = 2\pi r$.

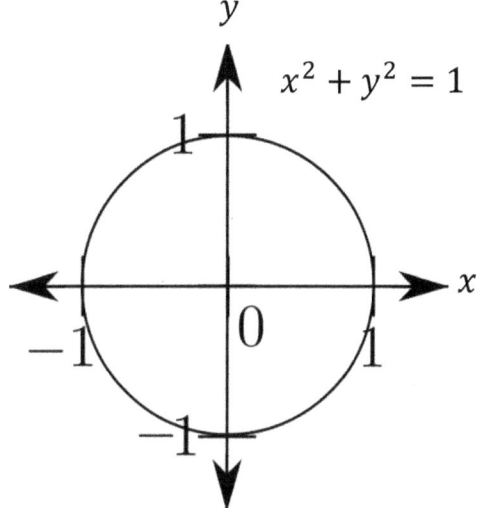

Example 3.22: The **unit circle** is the circle with radius 1 and center $(0,0)$. The equation of this circle is $x^2 + y^2 = 1$. To the right is a picture of the unit circle in the Cartesian plane.

The circumference of the unit circle is $2\pi \cdot 1 = 2\pi$.

An **angle in standard position** consists of two **rays**, both of which have their initial point at the origin, and one of which is the positive x-axis. We call the positive x-axis the **initial ray** and we call the second ray the **terminal ray**. The **radian measure** of the angle is the part of the circumference of the unit circle beginning at the point $(1, 0)$ on the positive x-axis and *eventually* ending at the point on the unit circle intercepted by the second ray. If the motion is in the counterclockwise direction, the radian measure is positive and if the motion is in the clockwise direction, the radian measure is negative.

Example 3.23: Let's draw a few angles where the terminal ray lies along the line $y = x$.

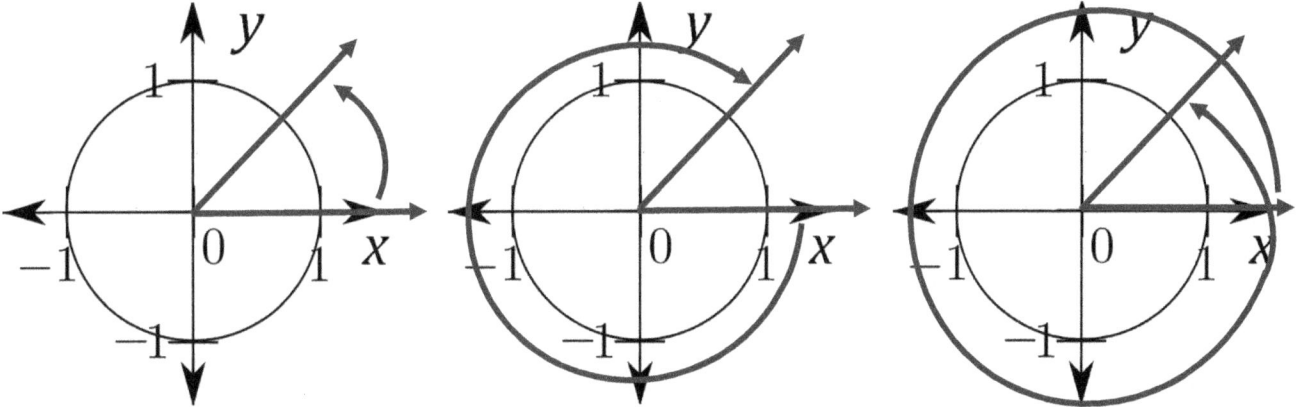

Observe that in the leftmost picture, the arc intercepted by the angle has a length that is one-eighth of the circumference of the circle. Since the circumference of the unit circle is 2π and the motion is in the counterclockwise direction, the angle has a radian measure of $\frac{2\pi}{8} = \frac{\pi}{4}$.

Similarly, in the center picture, the arc intercepted by the angle has a length that is seven-eighths of the circumference of the circle. This time the motion is in the clockwise direction, and so, the radian measure of the angle is $-\frac{7}{8} \cdot 2\pi = -\frac{7\pi}{4}$.

In the rightmost picture, the angle consists of a complete rotation, tracing out the entire circumference of the circle, followed by tracing out an additional length that is one-eighth the circumference of the circle. Since the motion is in the counterclockwise direction, the radian measure of the angle is $2\pi + \frac{2\pi}{8} = \frac{8\pi}{4} + \frac{\pi}{4} = \frac{9\pi}{4}$.

Let's find the point of intersection of the unit circle with the terminal ray of the angle $\frac{\pi}{4}$ that lies along the line with equation $y = x$ (as shown in the leftmost figure from Example 3.23 above). If we call this point (a, b), then we have $b = a$ (because (a, b) is on the line $y = x$) and $a^2 + b^2 = 1$ (because (a, b) is on the unit circle). Replacing b by a in the second equation gives us $a^2 + a^2 = 1$, or equivalently, $2a^2 = 1$. So, $a^2 = \frac{1}{2}$. The two solutions to this equation are $a = \pm\sqrt{\frac{1}{2}} = \pm\frac{\sqrt{1}}{\sqrt{2}} = \pm\frac{1}{\sqrt{2}}$. From the picture, it should be clear that we are looking for the positive solution, so that $a = \frac{1}{\sqrt{2}}$. Since $b = a$, we also have $b = \frac{1}{\sqrt{2}}$. Therefore, the point of intersection is $\left(\frac{1}{\sqrt{2}}, \frac{1}{\sqrt{2}}\right)$.

Notes: (1) The number $\frac{1}{\sqrt{2}}$ can also be written in the form $\frac{\sqrt{2}}{2}$. To see that these two numbers are equal, observe that we have

$$\frac{1}{\sqrt{2}} = \frac{1}{\sqrt{2}} \cdot 1 = \frac{1}{\sqrt{2}} \cdot \frac{\sqrt{2}}{\sqrt{2}} = \frac{1 \cdot \sqrt{2}}{\sqrt{2} \cdot \sqrt{2}} = \frac{\sqrt{2}}{2}.$$

(2) In the figure below on the left, we see a visual representation of the circle, the given angle, and the desired point of intersection.

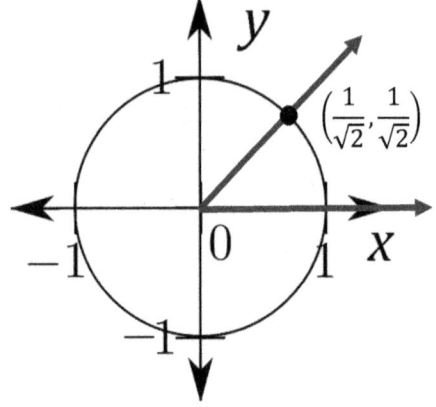 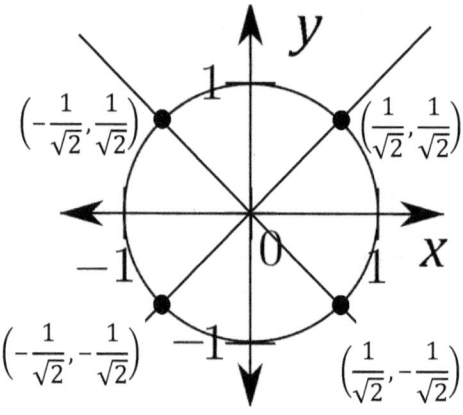

(3) In the figure above on the right, we have divided the Cartesian plane into eight regions using the lines with equations $y = x$ and $y = -x$ (together with the x- and y-axes). We then used the symmetry of the circle to label the four points of intersection of the unit circle with each of these two lines.

If θ (pronounced "theta") is the radian measure of an angle in standard position such that the terminal ray intersects the unit circle at the point (x, y), then we will say that $W(\theta) = (x, y)$. This expression defines a function $W: \mathbb{R} \to \mathbb{R} \times \mathbb{R}$ called the **wrapping function**. Observe that the inputs of the wrapping function are real numbers, which we think of as the radian measure of angles in standard position. The outputs of the wrapping function are pairs of real numbers, which we think of as points in the Cartesian plane. Also, observe that the range of the wrapping function is the unit circle.

We now define the cosine and sine of the angle θ by $\cos\theta = x$ and $\sin\theta = y$, where $W(\theta) = (x, y)$. For convenience, we also define the tangent, secant, cosecant, and cotangent of the angle as follows:

$$\tan\theta = \frac{\sin\theta}{\cos\theta} = \frac{y}{x} \qquad \sec\theta = \frac{1}{\cos\theta} = \frac{1}{x} \qquad \csc\theta = \frac{1}{\sin\theta} = \frac{1}{y} \qquad \cot\theta = \frac{\cos\theta}{\sin\theta} = \frac{x}{y}$$

Notes: (1) The wrapping function is **not** injective. For example, $W\left(\frac{\pi}{2}\right) = (0, 1)$ and $W\left(\frac{5\pi}{2}\right) = (0, 1)$. However, $\frac{\pi}{2} \neq \frac{5\pi}{2}$. There are actually infinitely many real numbers that map to $(0, 1)$ under the wrapping function. Specifically, $W\left(\frac{\pi}{2} + 2k\pi\right) = (0, 1)$ for every $k \in \mathbb{Z}$.

In general, each point on the unit circle is the image of infinitely many real numbers. Indeed, if $W(\theta) = (a, b)$, then $W(\theta + 2k\pi) = (a, b)$ for all $k \in \mathbb{Z}$.

(2) The six functions $\cos\theta$, $\sin\theta$, $\tan\theta$, $\sec\theta$, $\csc\theta$, and $\cot\theta$ are collectively known as **trigonometric functions**.

(3) The wrapping function gives us a convenient way to associate an angle θ in standard position with the corresponding point (x, y) on the unit circle. It is mostly used only as a notational convenience. We will usually be more interested in the expressions $\cos\theta = x$ and $\sin\theta = y$ (as well as the other four trigonometric functions).

Example 3.24: Using the rightmost figure above, we can make the following computations:

$$W\left(\frac{\pi}{4}\right) = \left(\frac{1}{\sqrt{2}}, \frac{1}{\sqrt{2}}\right) \quad W\left(\frac{3\pi}{4}\right) = \left(-\frac{1}{\sqrt{2}}, \frac{1}{\sqrt{2}}\right) \quad W\left(\frac{5\pi}{4}\right) = \left(-\frac{1}{\sqrt{2}}, -\frac{1}{\sqrt{2}}\right) \quad W\left(\frac{7\pi}{4}\right) = \left(\frac{1}{\sqrt{2}}, -\frac{1}{\sqrt{2}}\right)$$

$$\cos\frac{\pi}{4} = \frac{1}{\sqrt{2}} \qquad \sin\frac{\pi}{4} = \frac{1}{\sqrt{2}} \qquad \cos\frac{3\pi}{4} = -\frac{1}{\sqrt{2}} \qquad \sin\frac{3\pi}{4} = \frac{1}{\sqrt{2}}$$

$$\cos\frac{5\pi}{4} = -\frac{1}{\sqrt{2}} \qquad \sin\frac{5\pi}{4} = -\frac{1}{\sqrt{2}} \qquad \cos\frac{7\pi}{4} = \frac{1}{\sqrt{2}} \qquad \sin\frac{7\pi}{4} = -\frac{1}{\sqrt{2}}$$

It's also easy to compute the cosine and sine of the four **quadrantal angles** $0, \frac{\pi}{2}, \pi$, and $\frac{3\pi}{2}$. Here we use the fact that the points $(1, 0), (0, 1), (-1, 0)$, and $(0, -1)$ lie on the unit circle.

$$W(0) = (1, 0) \quad W\left(\frac{\pi}{2}\right) = (0, 1) \quad W(\pi) = (-1, 0) \quad W\left(\frac{3\pi}{2}\right) = (0, -1)$$

$$\cos 0 = 1 \qquad \sin 0 = 0 \qquad \cos\frac{\pi}{2} = 0 \qquad \sin\frac{\pi}{2} = 1$$

$$\cos\pi = -1 \qquad \sin\pi = 0 \qquad \cos\frac{3\pi}{2} = 0 \qquad \sin\frac{3\pi}{2} = -1$$

Also, if we add any integer multiple of 2π to an angle, the cosine and sine of the new angle have the same values as the old angle. For example, $\cos\frac{9\pi}{4} = \cos\left(\frac{\pi}{4} + \frac{8\pi}{4}\right) = \cos\left(\frac{\pi}{4} + 2\pi\right) = \cos\frac{\pi}{4} = \frac{1}{\sqrt{2}}$. This is a direct consequence of the fact that $W(\theta + 2k\pi) = W(\theta)$ for all $k \in \mathbb{Z}$.

We can also compute the tangent of each angle by dividing the sine of the angle by the cosine of the angle. For example, we have

$$\tan\frac{\pi}{4} = \frac{\sin\frac{\pi}{4}}{\cos\frac{\pi}{4}} = \frac{\frac{1}{\sqrt{2}}}{\frac{1}{\sqrt{2}}} = 1.$$

Similarly, we have

$$\tan\frac{3\pi}{4} = -1 \qquad \tan\frac{5\pi}{4} = 1 \qquad \tan\frac{7\pi}{4} = -1 \qquad \tan 0 = 0 \qquad \tan \pi = 0$$

When $\theta = \frac{\pi}{2}$ or $\frac{3\pi}{2}$, $\tan \theta$ is **undefined**.

Let's also compute the secant, cosecant, and cotangent of some of these angles (these are the **reciprocals** of cosine, sine and tangent, respectively).

$$\sec\frac{\pi}{4} = \sqrt{2} \qquad \csc\frac{\pi}{4} = \sqrt{2} \qquad \cot\frac{\pi}{4} = 1$$

$$\sec\frac{3\pi}{4} = -\sqrt{2} \qquad \csc\frac{3\pi}{4} = \sqrt{2} \qquad \cot\frac{3\pi}{4} = -1$$

$$\sec\frac{5\pi}{4} = -\sqrt{2} \qquad \csc\frac{5\pi}{4} = -\sqrt{2} \qquad \cot\frac{5\pi}{4} = 1$$

$$\sec\frac{7\pi}{4} = \sqrt{2} \qquad \csc\frac{7\pi}{4} = -\sqrt{2} \qquad \cot\frac{7\pi}{4} = -1$$

$$\sec 0 = 1 \qquad \csc 0 = \text{undefined} \qquad \cot 0 = \text{undefined}$$

$$\sec\frac{\pi}{2} = \text{undefined} \qquad \csc\frac{\pi}{2} = 1 \qquad \cot\frac{\pi}{2} = 0$$

$$\sec \pi = -1 \qquad \csc \pi = \text{undefined} \qquad \cot \pi = \text{undefined}$$

$$\sec\frac{3\pi}{2} = \text{undefined} \qquad \csc\frac{3\pi}{2} = -1 \qquad \cot\frac{3\pi}{2} = 0$$

Notes: (1) It is standard to use the abbreviations $\cos^2 \theta$ and $\sin^2 \theta$ for $(\cos \theta)^2$ and $(\sin \theta)^2$, respectively.

From the definition of cosine and sine, we have the following formula called the **Pythagorean Identity**:

$$\cos^2 \theta + \sin^2 \theta = 1$$

(2) Also, from the definition of cosine and sine, we have the following two formulas called the **Negative Identities**:

$$\cos(-\theta) = \cos \theta \qquad \sin(-\theta) = -\sin \theta.$$

Theorem 3.25: Let θ and ϕ be the radian measures of angles A and B, respectively. Then we have

$$\cos(\theta + \phi) = \cos \theta \cos \phi - \sin \theta \sin \phi$$
$$\sin(\theta + \phi) = \sin \theta \cos \phi + \cos \theta \sin \phi.$$

Notes: (1) The two formulas appearing in Theorem 3.25 are called the **Sum Identities**. You will be asked to prove Theorem 3.25 in Problem 13 below (parts (i) and (v)).

(2) θ and ϕ are Greek letters pronounced "theta" and "phi," respectively. These letters are often used to represent angle measures.

Example 3.26: Define $f: \mathbb{R} \to \mathbb{R}$ by $f(x) = \sin x$. Then f is a trigonometric function. The graph of f looks as follows:

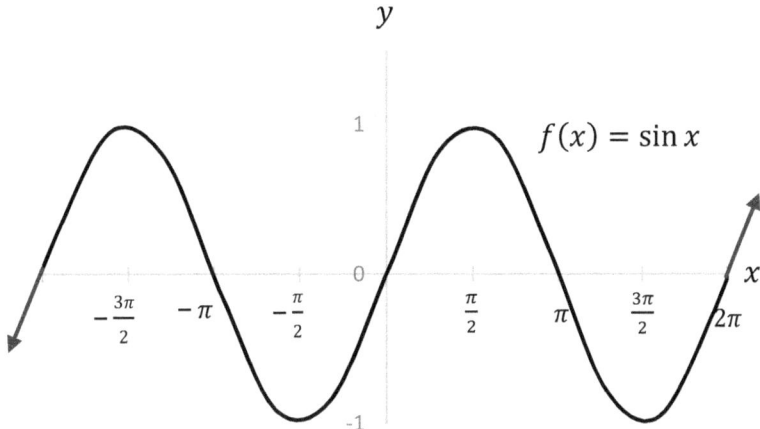

The reader is encouraged to sketch graphs of the other five trigonometric functions. Try using the definitions to sketch these graphs, rather than just looking up the answers.

Problem Set 3

Full solutions to these problems are available for free download here:
www.SATPrepGet800.com/RABQXZ

LEVEL 1

1. Determine if each of the following relations are functions. For each such function, determine if it is injective. State the domain and range of each function.

    (i) $R = \{(a,b), (b,b), (c,d), (e,a)\}$

    (ii) $S = \{(a,a), (a,b), (b,a)\}$

    (iii) $T = \{(a,b) \mid a, b \in \mathbb{R} \land b < 0 \land a^2 + b^2 = 9\}$

2. Define $f: \mathbb{Z} \to \mathbb{Z}$ by $f(n) = n^2$. Let $A = \{0, 1, 2, 3, 4\}$, $B = \mathbb{N}$, and $C = \{-2n \mid n \in \mathbb{N}\}$. Evaluate each of the following:

    (i) $f[A]$

    (ii) $f^{-1}[A]$

    (iii) $f^{-1}[B]$

    (iv) $f^{-1}[B \cup C]$

3. In Problems 10 and 11 below, you will be asked to show that $W\left(\frac{\pi}{3}\right) = \left(\frac{1}{2}, \frac{\sqrt{3}}{2}\right)$ and $W\left(\frac{\pi}{6}\right) = \left(\frac{\sqrt{3}}{2}, \frac{1}{2}\right)$. Use this information to compute the sine, cosine, and tangent of each of the following angles:

    (i) $\frac{\pi}{6}$

    (ii) $\frac{\pi}{3}$

    (iii) $\frac{2\pi}{3}$

    (iv) $\frac{5\pi}{6}$

    (v) $\frac{7\pi}{6}$

    (vi) $\frac{4\pi}{3}$

    (vii) $\frac{5\pi}{3}$

    (viii) $\frac{11\pi}{6}$

LEVEL 2

4. Find sets A and B and a function f such that $f[A \cap B] \neq f[A] \cap f[B]$.

5. Let $f: A \to B$ and let $V \subseteq B$. Prove that $f[f^{-1}[V]] \subseteq V$.

6. Use the sum identities (Theorem 3.25) to compute the cosine, sine, and tangent of each of the following angles:

    (i) $\frac{5\pi}{12}$

    (ii) $\frac{\pi}{12}$

    (iii) $\frac{11\pi}{12}$

    (iv) $\frac{19\pi}{12}$

LEVEL 3

7. For $f, g \in {}^{\mathbb{R}}\mathbb{R}$, define $f \preccurlyeq g$ if and only if for all $x \in \mathbb{R}$, $f(x) \leq g(x)$. Is $({}^{\mathbb{R}}\mathbb{R}, \preccurlyeq)$ a poset? Is it a linearly ordered set? What if we replace $\preccurlyeq$ by $\preccurlyeq^*$, where $f \preccurlyeq^* g$ if and only if there is an $x \in \mathbb{R}$ such that $f(x) \leq g(x)$?

8. Prove that the function $f: \mathbb{N} \to \mathbb{Z}$ defined by $f(n) = \begin{cases} \frac{n}{2} & \text{if } n \text{ is even} \\ -\frac{n+1}{2} & \text{if } n \text{ is odd} \end{cases}$ is a bijection.

9. Define $f: \mathbb{R} \setminus \{2\} \to \mathbb{R}$ by $f(x) = \frac{3x}{x-2}$.

    (i) Prove that f is injective.

    (ii) Prove that f is **not** surjective.

    (iii) Find a set X such that $f: \mathbb{R} \setminus \{2\} \to X$ is a bijection.

    (iv) What is f^{-1}?

LEVEL 4

10. Consider triangle AOP, where $O = (0,0)$, $A = (1,0)$, and P is the point on the unit circle so that angle POA has radian measure $\frac{\pi}{3}$. Prove that triangle AOP is equilateral, and then use this to prove that $W\left(\frac{\pi}{3}\right) = \left(\frac{1}{2}, \frac{\sqrt{3}}{2}\right)$. You may use the following facts about triangles: (i) The interior angle measures of a triangle sum to π radians; (ii) Two sides of a triangle have the same length if and only if the interior angles of the triangle opposite these sides have the same measure; (iii) If two sides of a triangle have the same length, then the line segment beginning at the point of intersection of those two sides and terminating on the opposite base midway between the endpoints of that base is perpendicular to that base.

11. Prove that $W\left(\frac{\pi}{6}\right) = \left(\frac{\sqrt{3}}{2}, \frac{1}{2}\right)$. You can use facts (i), (ii), and (iii) described in Problem 10.

12. Let θ and ϕ be the radian measure of angles A and B, respectively. Prove the following identity:
$$\cos(\theta - \phi) = \cos\theta\cos\phi + \sin\theta\sin\phi$$

13. Let θ and ϕ be the radian measure of angles A and B, respectively. Prove the following identities:
    (i) $\cos(\theta + \phi) = \cos\theta\cos\phi - \sin\theta\sin\phi$
    (ii) $\cos(\pi - \theta) = -\cos\theta$
    (iii) $\cos\left(\frac{\pi}{2} - \theta\right) = \sin\theta$
    (iv) $\sin\left(\frac{\pi}{2} - \theta\right) = \cos\theta$
    (v) $\sin(\theta + \phi) = \sin\theta\cos\phi + \cos\theta\sin\phi$
    (vi) $\sin(\pi - \theta) = \sin\theta$

LEVEL 5

14. Let X be a nonempty set of sets and let f be a function such that $\bigcup X \subseteq \text{dom } f$. Prove each of the following:
    (i) $f[\bigcup X] = \bigcup\{f[A] \mid A \in X\}$
    (ii) $f[\bigcap X] \subseteq \bigcap\{f[A] \mid A \in X\}$
    (iii) $f^{-1}[\bigcup X] = \bigcup\{f^{-1}[A] \mid A \in X\}$
    (iv) $f^{-1}[\bigcap X] = \bigcap\{f^{-1}[A] \mid A \in X\}$

15. Given ordered pairs $(x_1, y_1), (x_2, y_2)$, with x_1 and x_2 distinct, prove that there is a unique linear function f such that $f(x_1) = y_1$ and $f(x_2) = y_2$.

CHALLENGE PROBLEMS

16. Given ordered pairs $(x_1, y_1), (x_2, y_2), (x_3, y_3)$ with x_1, x_2, and x_3 distinct, prove that there is a unique linear or quadratic function f such that $f(x_1) = y_1, f(x_2) = y_2$, and $f(x_3) = y_3$.

17. Prove that for $x \in \left(-\frac{\pi}{2}, 0\right) \cup \left(0, \frac{\pi}{2}\right)$, $\cos x \leq \frac{\sin x}{x} \leq 1$.

LESSON 4
NUMBER SYSTEMS AND INDUCTION

The Natural Numbers

At this point, let's provide more formal definitions of the number systems we use most frequently, beginning with the set of natural numbers.

We define the following:

$$0 = \emptyset$$
$$1 = \{\emptyset\} = \{0\}$$
$$2 = \{\emptyset, \{\emptyset\}\} = \{0, 1\}$$
$$3 = \{\emptyset, \{\emptyset\}, \{\emptyset, \{\emptyset\}\}\} = \{0, 1, 2\}$$

In general, we let $n = \{0, 1, 2, \ldots, n-1\}$ and we define the **natural numbers** to be the set

$$\mathbb{N} = \{0, 1, 2, 3, 4, \ldots\}$$

If n is a natural number, we define the **successor** of n, written n^+, to be the natural number $n^+ = n \cup \{n\}$. Note that $n^+ = \{0, 1, 2, \ldots, n-1, n\}$.

Example 4.1:

1. $0^+ = 0 \cup \{0\} = \{0\} = 1$.
2. $1^+ = 1 \cup \{1\} = \{0\} \cup \{1\} = \{0, 1\} = 2$.
3. $2^+ = 2 \cup \{2\} = \{0, 1\} \cup \{2\} = \{0, 1, 2\} = 3$.

If n is a natural number such that $n \neq 0$, we define the **predecessor** of n, written n^-, to be the natural number k such that $n = k^+$ (by Problem 4, part (vii) below, k is unique). Thus, $n = n^- \cup \{n^-\}$.

Note that for all $n \in \mathbb{N}$, $(n^+)^- = n$ and for all $n \in \mathbb{N}$ with $n \neq 0$, $(n^-)^+ = n$.

Example 4.2:

1. Since $1 = 0^+$, $1^- = 0$.
2. Since $2 = 1^+$, $2^- = 1$.
3. Since $3 = 2^+$, $3^- = 2$.

We define the ordering $<_\mathbb{N}$ on $\mathbb{N}$ by $n <_\mathbb{N} m$ if and only if $n \in m$. We will usually abbreviate $<_\mathbb{N}$ simply by $<$, especially if it is already clear that we are working with the natural numbers.

With this definition, $<_\mathbb{N}$ is a strict linear ordering on $\mathbb{N}$. You will be asked to prove this as part of Problem 20 below.

Example 4.3:

1. $0 \in \{0\}$. Therefore, $0 \in 0 \cup \{0\} = 1$. So, $0 < 1$.
2. $1 = \{0\} \notin \emptyset = 0$. So, $1 \not< 0$.
3. $1 = \{0\} \notin 0$ and $1 = \{0\} \notin \{0\}$. Therefore, $1 = \{0\} \notin 0 \cup \{0\} = 1$. So, $1 \not< 1$.
4. If $n \neq 0$, then $n^- \in \{n^-\}$. Therefore, $n^- \in n^- \cup \{n^-\} = n$. So, $n^- < n$.

We add the natural numbers m and n as follows:

(i) If $n = 0$, then $m + n = m$.
(ii) If $n = k^+$, then $m + n = (m + k)^+$.

We multiply the natural numbers m and n as follows:

(i) If $n = 0$, then $mn = 0$.
(ii) If $n = k^+$, then $mn = (mk) + m$.

Notes: (1) With the definition of addition given above, we see that for all natural numbers m, we have $m + 1 = (m + 0)^+ = m^+$. So, from now on, we can write $m + 1$ instead of m^+. In particular, clause (ii) in the definition of the sum of two natural numbers can be written as follows:

(ii) If $n = k + 1$, then $m + n = (m + k) + 1$.

(2) Whenever it will not cause confusion, we will use the usual order of operations that we learned in elementary school. So, for example, we can abbreviate $(mk) + m$ as $mk + m$.

(3) Addition and multiplication of natural numbers are examples of **recursive definitions**. A full exploration of recursive definitions lies outside the scope of this book. Here we will just accept the existence of recursively defined functions without further mention.

Example 4.4:

1. $0 + 0 = 0$ by the first part of the definition of addition of natural numbers.
2. $0 + 1 = 0 + 0^+ = (0 + 0)^+ = 0^+ = 0 \cup \{0\} = \emptyset \cup \{\emptyset\} = \{\emptyset\} = 1$.
3. $0 + 2 = 0 + 1^+ = (0 + 1)^+ = 1^+ = 1 \cup \{1\} = \{0\} \cup \{1\} = \{0, 1\} = 2$.
4. $2 + 0 = 2$ by the first part of the definition of addition of natural numbers.
5. $2 + 1 = 2 + 0^+ = (2 + 0)^+ = 2^+ = 2 \cup \{2\} = \{0, 1\} \cup \{2\} = \{0, 1, 2\} = 3$.
6. $2 + 2 = 2 + 1^+ = (2 + 1)^+ = 3^+ = 3 \cup \{3\} = \{0, 1, 2\} \cup \{3\} = \{0, 1, 2, 3\} = 4$.
7. $2 \cdot 0 = 0$ by the first part of the definition of multiplication of natural numbers.
8. $2 \cdot 1 = 2 \cdot 0^+ = 2 \cdot 0 + 2 = 0 + 2 = 2$ (by part 3 above).
9. $2 \cdot 2 = 2 \cdot 1^+ = 2 \cdot 1 + 2 = 2 + 2 = 4$ (by part 6 above).

Well Ordering and the Principle of Mathematical Induction

If A is a subset of natural numbers, then we say that a is the **least** element of A if a is less than every other element of A. Symbolically, $\forall b \in A(b \neq a \to a < b)$.

Example 4.5: The least element of $\mathbb{N}$ itself is 0.

We now describe two important principles that are equivalent to each other.

The **Well Ordering Principle** (abbreviated **WOP**) is the following statement: Every nonempty subset of natural numbers has a least element.

The **Principle of Mathematical Induction** (abbreviated **POMI**) is the following statement: Let S be a set of natural numbers such that (i) $0 \in S$ and (ii) for all $k \in \mathbb{N}$, $k \in S \to k^+ \in S$. Then $S = \mathbb{N}$.

Notes: (1) The Principle of Mathematical Induction works like a chain reaction. We know that $0 \in S$ (this is condition (i)). Substituting 0 in for k in the expression "$k \in S \to k^+ \in S$" (condition (ii)) gives us $0 \in S \to 1 \in S$. So, we have that 0 is in the set S, and "if 0 is in the set S, then 1 is in the set S." So, $1 \in S$ must also be true.

(2) In terms of logic, if we let p be the statement $0 \in S$ and q the statement $1 \in S$, then we are given that $p \wedge (p \to q)$ is true. Observe that the only way that this statement can be true is if q is also true. Indeed, we must have both $p \equiv \text{T}$ and $p \to q \equiv \text{T}$. If q were false, then we would have $p \to q \equiv \text{T} \to \text{F} \equiv \text{F}$. So, we must have $q \equiv \text{T}$.

(3) Now that we showed $1 \in S$ is true (from Note 1 above), we can substitute 1 for k in the expression "$k \in S \to k^+ \in S$" (condition (ii)) to get $1 \in S \to 2 \in S$. So, we have $1 \in S \wedge (1 \in S \to 2 \in S)$ is true. So, $2 \in S$ must also be true.

(4) In general, we get the following chain reaction:
$$0 \in S \to 1 \in S \to 2 \in S \to 3 \in S \to \cdots$$
I hope that the "argument" presented in Notes 1 through 4 above convinces you that the Principle of Mathematical Induction is a reasonable one.

Now let's show that the Principle of Mathematical Induction follows from the Well Ordering Principle (In Problem 8 below, you will be asked to show that the Well Ordering Principle follows from the Principle of Mathematical Induction, thus showing that the two statements are equivalent). Proofs involving the Well Ordering Principle are generally done by contradiction.

Theorem 4.6: WOP $\to$ POMI.

Proof: Assume WOP and let S be a set of natural numbers such that $0 \in S$ (condition (i)), and such that whenever $k \in S$, $k^+ \in S$ (condition (ii)). Assume toward contradiction that $S \neq \mathbb{N}$. Let $A = \{k \in \mathbb{N} \mid k \notin S\}$ (so, A is the set of natural numbers **not** in S). Since $S \neq \mathbb{N}$, A is nonempty. So, by the Well Ordering Principle, A has a least element, let's call it a. $a \neq 0$ because $0 \in S$ and $a \notin S$. So, $a^- \in \mathbb{N}$. Letting $k = a^-$, we have $a^- \in S \to k \in S \to k^+ \in S \to (a^-)^+ \in S \to a \in S$. But $a \in A$, which means that $a \notin S$. This is a contradiction, and so, $S = \mathbb{N}$. $\square$

Note: A proof by contradiction works as follows:

1. We assume the negation of what we are trying to prove.
2. We use a logically valid argument to derive a statement which is false.
3. Since the argument was logically valid, the only possible error is our original assumption. Therefore, the negation of our original assumption must be true.

In this problem we are trying to prove that $S = \mathbb{N}$. The negation of this statement is that $S \neq \mathbb{N}$, and so that is what we assume.

We then define a set A which contains elements of $\mathbb{N}$ that are not in S. In reality, this set is empty (because the conclusion of the theorem is $S = \mathbb{N}$). However, our (wrong!) assumption that $S \neq \mathbb{N}$ tells us that this set A actually has something in it. Saying that A has something in it is an example of a false statement that was derived from a logically valid argument. This false statement occurred not because of an error in our logic, but because we started with an incorrect assumption ($S \neq \mathbb{N}$).

The Well Ordering Principle then allows us to pick out the least element of this set A. Note that we can do this because A is a subset of $\mathbb{N}$. This wouldn't work if we knew only that A was a subset of $\mathbb{Z}$, as $\mathbb{Z}$ does **not** satisfy the Well Ordering Principle (for example, $\mathbb{Z}$ itself has no least element).

Again, although the argument that A has a least element is logically valid, A does not actually have any elements at all. We are working from the (wrong!) assumption that $S \neq \mathbb{N}$.

Once we have our hands on this least element a, we can get our contradiction. What can this least element a be? Well a was chosen to **not** be in S, so a cannot be 0 (because 0 **is** in S). Also, we know that $a^- \in S$ (because a is the **least** element not in S and $a^- < a$). But condition (ii) then forces a to be in S (because $a = (a^-)^+$).

So, we wind up with $a \in S$, contradicting the fact that a is the least element **not** in S.

The Principle of Mathematical Induction is often written in the following way:

($\star$) Let $P(n)$ be a statement and suppose that (i) $P(0)$ is true and (ii) for all $k \in \mathbb{N}$, $P(k) \to P(k^+)$. Then $P(n)$ is true for all $n \in \mathbb{N}$.

In Problem 11 below, you will be asked to show that statement ($\star$) is equivalent to POMI.

There are essentially two steps involved in a proof by mathematical induction. The first step is to prove that $P(0)$ is true (this is called the **base case**), and the second step is to assume that $P(k)$ is true, and use this to show that $P(k^+)$ is true (this is called the **inductive step**). While doing the inductive step, the statement "$P(k)$ is true" is often referred to as the **inductive hypothesis**.

Theorem 4.7: The sum of two natural numbers is a natural number.

Note: Consider the sum $m + n$ of the natural numbers m and n. We will prove Theorem 4.7 by induction on n (and **not** by induction on m). We will start by letting m be an arbitrary natural number.

The base case will be to show that $m + 0$ is a natural number.

The inductive step will be to assume that $m + k$ is a natural number and then to prove that $m + k^+$ (or equivalently, $m + (k + 1)$) is a natural number.

Let's begin:

Proof: Assume that m is a natural number.

Base Case ($k = 0$): $m + 0 = m$ (by clause (i) in the definition of the sum of two natural numbers), which we assumed was a natural number. Thus, we have shown that $m + 0$ is a natural number.

Inductive Step: Let k be a natural number and assume that $m + k$ is also a natural number. Then $m + k^+ = (m + k)^+$ (or equivalently, $m + (k + 1) = (m + k) + 1$), which is also a natural number.

Here we used the fact that the successor of a natural number is a natural number.

By the Principle of Mathematical Induction, $m + n$ is a natural number for all natural numbers n.

Since m was an arbitrary natural number, we have shown that the sum of any two natural numbers is a natural number. □

Note: Since the sum of two natural numbers is always a natural number (by Theorem 4.7), we say that $\mathbb{N}$ is **closed** under addition. We may also say that $+$ is a **binary operation** on $\mathbb{N}$.

Theorem 4.8: For all natural numbers n, $0 + n = n$.

Proof: Base Case ($k = 0$): $0 + 0 = 0$ by the definition of addition of natural numbers.

Inductive Step: Let $k \in \mathbb{N}$ and assume that $0 + k = k$. Then $0 + (k + 1) = (0 + k) + 1 = k + 1$, as desired.

For the first equality, we used the definition of addition of natural numbers. For the second equality, we used the inductive hypothesis.

By the Principle of Mathematical Induction, for all natural numbers n, $0 + n = n$. □

Notes: (1) It's also true that for all natural numbers n, $n + 0 = n$. This follows right from the definition of addition of natural numbers.

(2) Since for all natural numbers n, we have $0 + n = n + 0 = 0$, we say that 0 is an **identity** with respect to addition, or that 0 is an **additive identity**.

Theorem 4.9: For all natural numbers m, n, and t, $(m + n) + t = m + (n + t)$.

Proof: Let m and n be natural numbers.

Base Case ($k = 0$): $(m + n) + 0 = m + n = m + (n + 0)$ by the definition of addition in $\mathbb{N}$.

Inductive Step: Let $k \in \mathbb{N}$ and assume that $(m + n) + k = m + (n + k)$. Then we have

$$(m+n) + (k+1) = \big((m+n) + k\big) + 1 = \big(m + (n+k)\big) + 1$$
$$= m + \big((n+k) + 1\big) = m + \big(n + (k+1)\big).$$

For the first, third, and fourth equalities, we used the definition of addition of natural numbers. For the second equality, we used the inductive hypothesis.

By the Principle of Mathematical Induction, for all natural numbers t, $(m+n) + t = m + (n+t)$.

Since m and n were arbitrary natural numbers, we have shown that for all natural numbers m, n, and t, $(m+n) + t = m + (n+t)$. □

Notes: (i) Since for all natural numbers m, n, and t, we have $(m+n) + t = m + (n+t)$, we say that addition is **associative** in $\mathbb{N}$.

(ii) A **monoid** is a pair $(S, \star)$, where S is a set, $\star$ is an associative binary operation on S, and there is an identity $e \in S$ with respect to the operation $\star$. Theorems 4.7, 4.8, and 4.9 prove that $(\mathbb{N}, +)$ is a monoid.

(iii) Addition is also **commutative** in $\mathbb{N}$. That is, for all natural numbers m and n, $m + n = n + m$. You will be asked to prove this in part (i) of Problem 4 below. We say that $(\mathbb{N}, +)$ is a **commutative monoid**.

Just like the sum operation, the product operation on the set of natural numbers has many nice algebraic properties such as

1. **Closure:** For all natural numbers m and n, mn is a natural number.
2. **Identity:** For all natural numbers n, $1 \cdot n = n \cdot 1 = n$.
3. **Associativity:** For all natural numbers m, n, and t, $(mn)t = m(nt)$.
4. **Commutativity:** For all natural numbers m and n, $mn = nm$.
5. **Distributivity:** For all natural numbers m, n, and t, $m(n+t) = mn + mt$.

The proofs that these properties hold are very similar to the proofs already given for addition. You will be asked to provide detailed proofs in Problem 4 below.

If $n \in \mathbb{N}$, we define n^2 to be the product of n with itself. So, $n^2 = n \cdot n$.

We define the **difference** between two natural numbers as follows: we say that $m - n = t$ if and only if $m = t + n$. For example, $7 - 2 = 5$ because $7 = 5 + 2$.

A natural number n is called **even** if there is another natural number b such that $n = 2b$.

Example 4.10:
1. 6 is even because $6 = 2 \cdot 3$.
2. 14 is even because $14 = 2 \cdot 7$.
3. We can write $1 = 2 \cdot \frac{1}{2}$, but this does **not** show that 1 is even (and as we all know, it is not). In the definition of even, it is very important that b is a natural number. The problem here is that $\frac{1}{2}$ is not a natural number, and so, it cannot be used as a value for b in the definition of even.

Let's use the Principle of Mathematical Induction to prove a theorem about natural numbers using these definitions.

Theorem 4.11: For all natural numbers n, $n^2 - n$ is an even natural number.

The proof of this result will require two very simple technical theorems about differences between natural numbers. A **lemma** is a theorem whose primary purpose is to prove a more important theorem. Let's prove these two preliminary lemmas first before we prove the main theorem.

Lemma 4.12: For all natural numbers m, n, and t, if $n - t \in \mathbb{N}$, then $m(n - t) = mn - mt$.

Proof: Suppose that $n - t \in \mathbb{N}$, say $n - t = a$. Then $n = a + t$. So, $mn = m(a + t) = ma + mt$ (using the distributivity of multiplication over addition in $\mathbb{N}$). Therefore, $mn - mt = ma = m(n - t)$. □

Lemma 4.13: For all natural numbers m, n, and t, $(m + n) - n = m$.

Proof: The equation $m + n = m + n$ is equivalent to $(m + n) - n = m$. □

Proof of Theorem 4.11: We will prove this theorem by induction on n.

Base Case ($k = 0$): $0^2 - 0 = 2 \cdot 0$ because $0^2 = 2 \cdot 0 + 0$. So, $0^2 - 0$ is even.

Inductive Step: Let $k \in \mathbb{N}$ and assume that $k^2 - k$ is an even natural number. Then $k^2 - k = 2b$ for some natural number b. Now,

$$(k+1)^2 - (k+1) = (k+1)(k+1) - (k+1) \cdot 1 = (k+1)[(k+1) - 1] = (k+1) \cdot k$$
$$= k(k+1) = k \cdot k + k \cdot 1 = k^2 + k = (k^2 - k) + 2k = 2b + 2k = 2(b + k).$$

Since $\mathbb{N}$ is closed under addition, $b + k \in \mathbb{N}$. Therefore, $(k+1)^2 - (k+1)$ is an even natural number.

By the Principle of Mathematical Induction, $n^2 - n$ is an even natural number for all $n \in \mathbb{N}$. □

Notes: (1) For the first equality, we used the definition of $(k+1)^2$ and the multiplicative identity property in $\mathbb{N}$.

(2) For the second equality, we used Lemma 4.12.

(3) For the third equality, we used Lemma 4.13.

(4) For the fourth equality, we used the commutativity of multiplication in $\mathbb{N}$.

(5) For the fifth and ninth equalities, we used the distributivity of multiplication over addition in $\mathbb{N}$.

(6) For the sixth equality, we used the definition of k^2 and the multiplicative identity property in $\mathbb{N}$.

(7) For the seventh equality, we used what I call the "**Standard Advanced Calculus Trick**." I sometimes abbreviate this as **SACT**. The trick is simple. If you need something to appear, just put it in. Then correct it by performing the opposite of what you just did.

In this case, in order to use the inductive hypothesis, we need $k^2 - k$ to appear, but unfortunately, we have $k^2 + k$ instead. Using SACT, I do the following:

- I simply put in what I need (and exactly where I need it): $k^2 - \boldsymbol{k} + k$
- Now, I undo the damage by performing the reverse operation: $k^2 - k + \boldsymbol{k} + k$
- Finally, I leave the part I need as is, and simplify the rest: $(k^2 - k) + 2k$

(8) For the eighth equality, we simply replaced $k^2 - k$ by $2b$. We established that these two quantities were equal in the second sentence of the inductive step.

Sometimes a statement involving the natural numbers may be false for 0, but true from some natural number on. In this case, we can still use induction. We just need to adjust the base case.

Theorem 4.14: $n^2 > 2n + 1$ for all natural numbers $n \geq 3$.

Proof: Base Case ($k = 3$): $3^2 = 9$ and $2 \cdot 3 + 1 = 6 + 1 = 7$. So, $3^2 > 2 \cdot 3 + 1$.

Inductive Step: Let $k \in \mathbb{N}$ with $k \geq 3$ and assume that $k^2 > 2k + 1$. Then we have

$$(k+1)^2 = (k+1)(k+1) = (k+1)k + (k+1)(1) = k^2 + k + k + 1 > (2k+1) + k + k + 1$$
$$= 2k + 2 + k + k = 2(k+1) + k + k \geq 2(k+1) + 1 \text{ (because } k + k \geq 3 + 3 = 6 \geq 1\text{)}.$$

By the Principle of Mathematical Induction, $n^2 > 2n + 1$ for all $n \in \mathbb{N}$ with $n \geq 3$. □

Notes: (1) If we have a sequence of equations and inequalities of the form $=$, $\geq$, and $>$ (with at least one inequality symbol appearing), beginning with a and ending with b, then the final result is $a > b$ if $>$ appears at least once and $a \geq b$ otherwise.

For example, if $a = j = h = m > n = p = q \geq b$, then $a > b$. The sequence that appears in the solution above has this form.

(2) By definition, $x^2 = x \cdot x$. We used this in the first equality in the inductive step to write $(k+1)^2$ as $(k+1)(k+1)$.

(3) For the second equality in the inductive step, we used distributivity to write $(k+1)(k+1)$ as $(k+1)k + (k+1)(1)$.

(4) For the third equality in the inductive step, we used commutativity, distributivity, and the multiplicative identity property to write $(k+1)k$ as $k(k+1) = k \cdot k + k \cdot 1 = k^2 + k$. We also used the multiplicative identity property to write $(k+1)(1) = k+1$.

(5) Associativity of addition is being used when we write the expression $k^2 + k + k + 1$. Notice the lack of parentheses. Technically speaking, we should have written $(k^2 + k) + (k + 1)$ and then taken another step to rewrite this as $k^2 + \bigl(k + (k+1)\bigr)$. However, since we have associativity, we can simply drop all those parentheses.

(6) The inequality "$k^2 + k + k + 1 > (2k + 1) + k + k + 1$" was attained by using the inductive hypothesis "$k^2 > 2k + 1$" together with part (x) from Problem 4 below.

(7) The dedicated reader should verify that the remaining equalities and inequalities in the proof are valid by determining which properties were used at each step.

Theorem 4.15: For every natural number n, there is a natural number j such that $n = 2j$ or $n = 2j + 1$.

Proof: Base Case ($k = 0$): $0 = 2 \cdot 0$ by definition of mutiplication in $\mathbb{N}$.

Inductive Step: Suppose that $k \in \mathbb{N}$ and there is $j \in \mathbb{N}$ such that $k = 2j$ or $k = 2j + 1$. If $k = 2j$, then $k + 1 = 2j + 1$. If $k = 2j + 1$, then $k + 1 = (2j + 1) + 1 = 2j + (1 + 1) = 2j + 2 = 2(j + 1)$. Here we used associativity of addition in $\mathbb{N}$ and distributivity of multiplication over addition in $\mathbb{N}$. Since $\mathbb{N}$ is closed under addition, $j + 1 \in \mathbb{N}$.

By the Principle of Mathematical Induction, for every natural number n, there is a natural number j such that $n = 2j$ or $n = 2j + 1$. □

The Integers

To motivate the definition of the integers, note that we can think of every integer as a difference of two natural numbers. For example, the integer -3 can be thought of as $1 - 4$. However, -3 can also be thought of as $2 - 5$. So, we must insist that $1 - 4 = 2 - 5$, or equivalently, $1 + 5 = 2 + 4$.

We define a relation R on $\mathbb{N} \times \mathbb{N}$ by $R = \{((a,b),(c,d)) \in (\mathbb{N} \times \mathbb{N})^2 \mid a + d = b + c\}$. In part 5 of Example 2.12, we showed that this relation is an equivalence relation. We can now define the set of integers to be the set of equivalence classes for this equivalence relation. That is, we define the set of integers to be $\mathbb{Z} = \{[(a,b)] \mid (a,b) \in \mathbb{N} \times \mathbb{N}\}$.

We identify the integer $[(n, 0)]$ with the natural number n. In this way, we have $\mathbb{N} \subseteq \mathbb{Z}$.

We define the ordering $<_\mathbb{Z}$ on $\mathbb{Z}$ by $[(a,b)] <_\mathbb{Z} [(c,d)]$ if and only if $a + d <_\mathbb{N} b + c$, where $<_\mathbb{N}$ is the usual ordering on $\mathbb{N}$ ($n <_\mathbb{N} m$ if and only if $n \in m$).

In Problem 9 below, you will be asked to show that $<_\mathbb{Z}$ is a well-defined strict linear ordering on $\mathbb{Z}$.

We add and multiply two integers using the following rules:
$$[(a,b)] + [(c,d)] = [(a+c, b+d)]$$
$$[(a,b)] \cdot [(c,d)] = [(ac + bd, ad + bc)]$$

In Problems 12 and 18 below, you will be asked to show that these two operations are well-defined.

Notes: (1) We will usually abbreviate $<_\mathbb{Z}$ simply by $<$, especially if it is already clear that we are working with the integers.

(2) If $a, b \in \mathbb{N}$ with $a \geq b$, then $[(a,b)] = [(a - b, 0)]$. If $a < b$, then $[(a,b)] = [(0, b - a)]$. In this way, we see that every integer can be written in the form $[(n, 0)]$ or $[(0, n)]$ for some $n \in \mathbb{N}$. We abbreviate $[(n, 0)]$ by n and we abbreviate $[(0, n)]$ by $-n$. For example, $[(2,7)] = [(0,5)] = -5$.

Example 4.16:

1. $[(k, k)] = [(0,0)]$ for all $k \in \mathbb{N}$ because $k + 0 = k + 0$.
2. $[(5,0)] = [(6,1)]$ because $5 + 1 = 0 + 6$. Similarly, we have $[(5,0)] = [(7,2)] = [(8,3)]$, and in general $[(5,0)] = [(5+k,k)]$ for any natural number k. $[(5,0)]$ is the most "natural" way to express the natural number 5 as an integer. More generally, the natural number n can be expressed as an integer as $[(n,0)]$.
3. We usually abbreviate the integer $[(0,k)]$ as $-k$. For example, -3 is an abbreviation for $[(0,3)]$. We can also write -3 as $[(1,4)]$ because $1 + 3 = 4 + 0$.
4. $[(0,0)] < [(4,0)]$ because $0 + 0 < 0 + 4$. More generally, for any natural number $k \neq 0$, we have $[(0,0)] < [(k,0)]$ because $0 + 0 < 0 + k$. This shows that for any natural number $k \neq 0$, the natural number k satisfies $0 < k$.
5. $[(0,4)] < [(0,0)]$ because $0 + 0 < 4 + 0$. More generally, for any natural number $k \neq 0$, we have $[(0,k)] < [(0,0)]$ because $0 + 0 < k + 0$. This shows that for any natural number $k \neq 0$, the integer $-k$ satisfies $-k < 0$.
6. $7 + (-2) = [(7,0)] + [(0,2)] = [(7,2)] = [(5,0)] = 5$.
7. $-3 \cdot 5 = [(0,3)] \cdot [(5,0)] = [(0 \cdot 5 + 3 \cdot 0, 0 \cdot 0 + 3 \cdot 5)] = [(0, 15)] = -15$.

The Rational Numbers

In part 6 of Example 2.12, we showed that $R = \{((a,b),(c,d)) \in (\mathbb{Z} \times \mathbb{Z}^*)^2 \mid ad = bc\}$ is an equivalence relation on $\mathbb{Z} \times \mathbb{Z}^*$. For each $a \in \mathbb{Z}$ and $b \in \mathbb{Z}^*$, we define the **rational number** $\frac{a}{b}$ to be the equivalence class of (a, b). So, $\frac{a}{b} = [(a,b)]$, and we have $\frac{a}{b} = \frac{c}{d}$ if and only if $(a,b)R(c,d)$ if and only if $ad = bc$. The set of rational numbers is $\mathbb{Q} = \{\frac{a}{b} \mid a \in \mathbb{Z} \land b \in \mathbb{Z}^*\}$. In words, $\mathbb{Q}$ is "the set of quotients a over b such that a and b are integers and b is not zero." (Recall that $\mathbb{Z}^* = \mathbb{Z} \setminus \{0\}$.)

We identify the rational number $\frac{a}{1}$ with the integer a. In this way, we have $\mathbb{Z} \subseteq \mathbb{Q}$.

We define $<_\mathbb{Q}$ on $\mathbb{Q}$ by $\frac{a}{b} <_\mathbb{Q} \frac{c}{d}$ if and only if $ad <_\mathbb{Z} bc$, where $<_\mathbb{Z}$ is the usual ordering on $\mathbb{Z}$.

In Problem 20 below, you will be asked to show that $<_\mathbb{Q}$ is a well-defined strict linear ordering on $\mathbb{Q}$.

We add and multiply two rational numbers using the following rules:

$$\frac{a}{b} + \frac{c}{d} = \frac{a \cdot d + b \cdot c}{b \cdot d} \qquad \frac{a}{b} \cdot \frac{c}{d} = \frac{a \cdot c}{b \cdot d}$$

In Problem 13 below, you will be asked to show that these two operations are well-defined.

Example 4.17:

1. $\frac{2}{3} < \frac{5}{4}$ because $2 \cdot 4 < 3 \cdot 5$. Also, $\frac{2}{3} + \frac{5}{4} = \frac{2 \cdot 4 + 3 \cdot 5}{3 \cdot 4} = \frac{23}{12}$ and $\frac{2}{3} \cdot \frac{5}{4} = \frac{2 \cdot 5}{3 \cdot 4} = \frac{10}{12} = \frac{5}{6}$ (because $10 \cdot 6 = 12 \cdot 5$).
2. $\frac{-3}{4} < \frac{-5}{7}$ because $-3 \cdot 7 < 4(-5)$, $\frac{-3}{4} + \frac{-5}{7} = \frac{-3 \cdot 7 + 4(-5)}{4 \cdot 7} = \frac{-41}{28}$, and $\frac{-3}{4} \cdot \frac{-5}{7} = \frac{-3(-5)}{4 \cdot 7} = \frac{15}{28}$.

The Real Numbers

There are several equivalent ways to define the set of real numbers. We will define this set here as equivalence classes of rational-valued **Cauchy sequences** (see Problem 24 in Problem Set 6 for another approach). We introduced Cauchy sequences informally in Lesson 3. Here we will define them more formally. We first make a few preliminary definitions.

If $n = [(k, m)]$ is an integer, we define $-n$ to be the integer $[(m, k)]$. With this definition, we have $-(-n) = n$. Indeed, if $n = [(k, m)]$, then $-(-n) = -(-[(k, m)]) = -[(m, k)] = [(k, m)] = n$.

If $x = \frac{a}{b}$ is a rational number, then we define $-x$ to be $\frac{-a}{b}$. We then define subtraction as follows: $x - y = x + (-y)$. The expression $x - y$ is called a **difference**.

The **absolute value** of the rational number x is then defined by $|x| = \begin{cases} x & \text{if } x \geq 0. \\ -x & \text{if } x < 0. \end{cases}$

For example, we have $\left|\frac{2}{3}\right| = \frac{2}{3}$, $|0| = 0$, and $\left|\frac{-5}{7}\right| = -\left(\frac{-5}{7}\right) = \frac{-(-5)}{7} = \frac{5}{7}$.

The **distance** between rational numbers x and y is $|x - y|$. For example, the distance between 3 and 5 is $|3 - 5| = |-2| = 2$ and the distance between $\frac{1}{5}$ and $\frac{1}{7}$ is $\left|\frac{1}{5} - \frac{1}{7}\right| = \left|\frac{7-5}{5 \cdot 7}\right| = \left|\frac{2}{35}\right| = \frac{2}{35}$.

Notice that the distance between 5 and 3 is the same as the distance between 3 and 5. Indeed, we also have $|5 - 3| = |2| = 2$. In general, for rational numbers x and y, we have $|x - y| = |y - x|$.

Let $f = (x_n)$ be a rational-valued sequence. We say that f is a **Cauchy sequence** if

for every $k \in \mathbb{N}^+$, there is $K \in \mathbb{N}$ such that $m \geq n > K$ implies $|x_m - x_n| < \frac{1}{k}$.

The idea is that we can make the distance between any two terms of the sequence as small as we choose by deleting a finite portion of the beginning of the sequence. If we wish to make the distance between any two terms less than $\frac{1}{k}$, we delete the first $K + 1$ terms of the sequence.

Note: We will be using rational-valued Cauchy sequences to define the real numbers. In Lesson 8, we will learn about real-valued Cauchy sequences.

Example 4.18:

1. The sequence $(x_n) = \left(\frac{1}{n+1}\right)$ is a Cauchy sequence. To see this, let $k \in \mathbb{N}$, let $K = k$, and let $m \geq n > K$. Then
$$|x_m - x_n| = \left|\frac{1}{m+1} - \frac{1}{n+1}\right| = \left|\frac{n-m}{(m+1)(n+1)}\right| \leq \left|\frac{m}{(m+1)(n+1)}\right| = \left|\frac{m}{mn+m+n+1}\right| \leq \left|\frac{m}{mn}\right| = \frac{1}{n} < \frac{1}{K} = \frac{1}{k}.$$

2. The sequence $(x_n) = ((-1)^n n)$ is **not** a Cauchy sequence. To see this, let $k = 1$ and let $K \in \mathbb{N}$. Then $K + 2 \geq K + 1 > K$, and
$$|x_{K+2} - x_{K+1}| = |(-1)^{K+2}(K+2) - (-1)^{K+1}(K+1)| = |(K+2) + (K+1)| = 2K + 3 \geq 3$$
(See the Note below). However, $\frac{1}{k} = \frac{1}{1} = 1$ and $3 \not< 1$.

3. For each $q \in \mathbb{Q}$, the **constant sequence** $(x_n) = (q)$ is a Cauchy sequence. To see this, let $k \in \mathbb{N}$, let $K = 0$, and let $m \geq n > 0$. Then
$$|x_m - x_n| = |q - q| = |0| = 0 < \frac{1}{k}.$$

Note: If K is even, then $|(-1)^{K+2}(K+2) - (-1)^{K+1}(K+1)| = (K+2) + (K+1) = 2K+3$, whereas, if K is odd, then $(-1)^{K+2}(K+2) - (-1)^{K+1}(K+1) = -(K+2) - (K+1) = -(2K+3)$.

Next, we would like to identify Cauchy sequences that seem to be converging to the same value. For example, we will identify the Cauchy sequences $(x_n) = \left(\frac{1}{n+1}\right)$ and $(y_n) = (0)$.

An equivalent way of saying that (x_n) and (y_n) converge to the same value is to say that $(x_n - y_n)$ converges to 0.

Let $A = \{(x_n) \mid (x_n) \text{ is a Cauchy sequence of rational numbers}\}$. We define a relation R on A as follows:

$(x_n)R(y_n)$ if and only if for every $k \in \mathbb{N}^+$, there is $K \in \mathbb{N}$ such that $n > K$ implies $|x_n - y_n| < \frac{1}{k}$.

In Problem 14 below, you will be asked to show that R is an equivalence relation on A.

Note: To show that R is transitive, you will need to use the **Triangle Inequality**. The Triangle Inequality says that if $a, b \in \mathbb{Q}$, then $|a + b| \leq |a| + |b|$. We will come back to the Triangle Inequality again in Theorem 7.4. You will be asked to prove this theorem in Problem 9 in Problem Set 7.

We can now define the set of real numbers as follows:

$$\mathbb{R} = \{[(x_n)] \mid (x_n) \text{ is a Cauchy sequence of rational numbers}\}.$$

We identify the real number $[(q)]$ with the rational number q. In this way, we have $\mathbb{Q} \subseteq \mathbb{R}$.

We define the ordering $\leq_\mathbb{R}$ on $\mathbb{R}$ by

$[(x_n)] \leq_\mathbb{R} [(y_n)]$ if and only if there is $K \in \mathbb{N}$ such that $n > K$ implies $x_n \leq y_n$.

We can then define $<_\mathbb{R}$ on $\mathbb{R}$ by $[(x_n)] <_\mathbb{R} [(y_n)]$ if and only if $[(x_n)] \leq_\mathbb{R} [(y_n)]$ and $[(x_n)] \neq [(y_n)]$.

In Problem 20 below, you will be asked to show that $<_\mathbb{R}$ is a well-defined strict linear ordering on $\mathbb{R}$.

We add and multiply two real numbers using the following rules:
$$[(x_n)] + [(y_n)] = [(x_n + y_n)]$$
$$[(x_n)] \cdot [(y_n)] = [(x_n \cdot y_n)]$$

In Problems 17 and 19 below, you will be asked to show that $+$ and $\cdot$ are well-defined operations on the real numbers and that the sum and product of two real numbers are real numbers.

Note: We will generally drop the subscript from $<_\mathbb{R}$ and simply write $<$.

The Complex Numbers

The set of complex numbers is $\mathbb{C} = \{a + bi \mid a, b \in \mathbb{R}\}$.

In addition to visualizing complex numbers as points in the Complex Plane, as we did in part 3 of Example 1.6, we can also visualize the complex number $a + bi$ as a directed line segment (or **vector**) starting at the origin and ending at the point (a, b). Three examples are shown to the right.

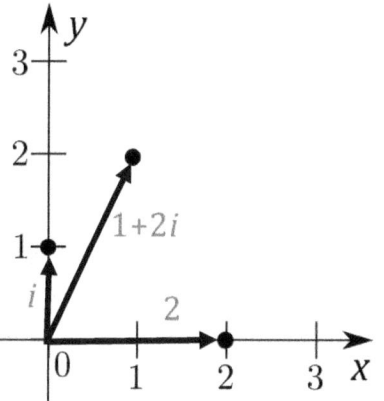

If $z = a + bi$ is a complex number, we call a the **real part** of z and b the **imaginary part** of z, and we write $a = \text{Re } z$ and $b = \text{Im } z$.

Two complex numbers are **equal** if and only if they have the same real part and the same imaginary part. In other words,

$$a + bi = c + di \text{ if and only if } a = c \text{ and } b = d.$$

We add two complex numbers by simply adding their real parts and adding their imaginary parts. So,

$$(a + bi) + (c + di) = (a + c) + (b + d)i.$$

As a point, this sum is $(a + c, b + d)$. We can visualize this sum as the vector starting at the origin that is the diagonal of the parallelogram formed from the vectors $a + bi$ and $c + di$. Here is an example showing that $(1 + 2i) + (-3 + i) = -2 + 3i$.

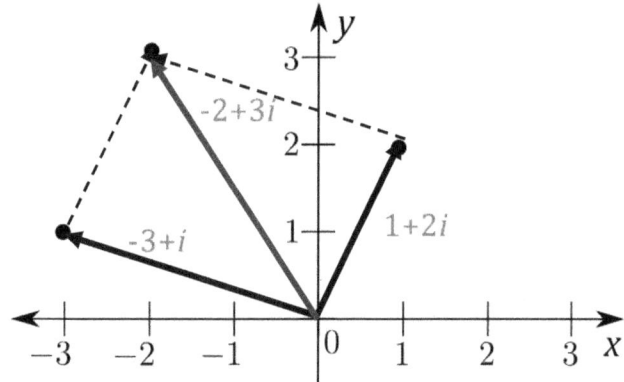

The definition for multiplying two complex numbers is a bit more complicated:

$$(a + bi)(c + di) = (ac - bd) + (ad + bc)i.$$

Notes: (1) If $b = 0$, then we call $a + bi = a + 0i = a$ a **real number**. Note that when we add or multiply two real numbers, we always get another real number.

$$(a + 0i) + (b + 0i) = (a + b) + (0 + 0)i = (a + b) + 0i = a + b.$$
$$(a + 0i)(b + 0i) = (ab - 0 \cdot 0) + (a \cdot 0 + 0b)i = (ab - 0) + (0 + 0)i = ab + 0i = ab.$$

In this way, we have $\mathbb{R} \subseteq \mathbb{C}$.

(2) If $a = 0$, then we call $a + bi = 0 + bi = bi$ a **pure imaginary number**.

(3) $i^2 = -1$. To see this, note that $i^2 = i \cdot i = (0 + 1i)(0 + 1i)$, and we have

$$(0 + 1i)(0 + 1i) = (0 \cdot 0 - 1 \cdot 1) + (0 \cdot 1 + 1 \cdot 0)i = (0 - 1) + (0 + 0)i = -1 + 0i = -1.$$

(4) The definition of the product of two complex numbers is motivated by how we expect multiplication should behave, together with replacing i^2 by -1. If we were to naïvely multiply the two complex numbers, we would have

$$(a + bi)(c + di) = (a + bi)c + (a + bi)(di) = ac + bci + adi + bdi^2$$
$$= ac + bci + adi + bd(-1) = ac + (bc + ad)i - bd = (ac - bd) + (ad + bc)i.$$

Those familiar with the mnemonic FOIL may notice that "FOILing" will always work to produce the product of two complex numbers, provided we replace i^2 by -1 and simplify.

Example 4.19: Let $z = 2 - 3i$ and $w = -1 + 5i$. Then

$$z + w = (2 - 3i) + (-1 + 5i) = (2 + (-1)) + (-3 + 5)i = \mathbf{1 + 2i}.$$
$$zw = (2 - 3i)(-1 + 5i) = \big(2(-1) - (-3)(5)\big) + \big(2 \cdot 5 + (-3)(-1)\big)i$$
$$= (-2 + 15) + (10 + 3)i = \mathbf{13 + 13i}.$$

Problem Set 4

Full solutions to these problems are available for free download here:
www.SATPrepGet800.com/RABQXZ

LEVEL 1

1. Use the Principle of Mathematical Induction to prove each of the following:

    (i) $2^n > n$ for all natural numbers $n \geq 1$.

    (ii) $0 + 1 + 2 + \cdots + n = \frac{n(n+1)}{2}$ for all natural numbers.

    (iii) $0^2 + 1^2 + 2^2 + \cdots + n^2 = \frac{n(n+1)(2n+1)}{6}$ for all natural numbers.

    (iv) $n! > 2^n$ for all natural numbers $n \geq 4$ (where $n! = 1 \cdot 2 \cdots n$ for all natural numbers $n \geq 1$).

    (v) $2^n \geq n^2$ for all natural numbers $n \geq 4$.

2. A natural number n is **divisible** by a natural number k, written $k|n$, if there is another natural number b such that $n = kb$. Prove that $n^3 - n$ is divisible by 3 for all natural numbers n.

3. Let $z = -4 - i$ and $w = 3 - 5i$. Compute each of the following:

    (i) $z + w$

    (ii) zw

    (iii) Im w

LEVEL 2

4. Prove each of the following. (You may assume that $<$ is a strict linear ordering of $\mathbb{N}$.)

    (i) Addition is commutative in $\mathbb{N}$.

    (ii) The set of natural numbers is closed under multiplication.

    (iii) 1 is a multiplicative identity in $\mathbb{N}$.

    (iv) Multiplication is distributive over addition in $\mathbb{N}$.

    (v) Multiplication is associative in $\mathbb{N}$.

    (vi) Multiplication is commutative in $\mathbb{N}$.

    (vii) For all natural numbers m, n, and k, if $m + k = n + k$, then $m = n$.

    (viii) For all natural numbers m, n, and k with $k \neq 0$, if $mk = nk$, then $m = n$.

    (ix) For all natural numbers m and n, $m < n$ if and only if there is a natural number $k > 0$ such that $n = m + k$.

    (x) For all natural numbers m, n, and k, $m < n$ if and only if $m + k < n + k$.

    (xi) For all natural numbers m and n, if $m > 0$ and $n > 0$, then $mn > 0$.

89

5. A set A is **transitive** if $\forall x(x \in A \to x \subseteq A)$ (in words, every element of A is also a subset of A). Prove that every natural number is transitive.

6. Determine if each of the following sequences are Cauchy sequences. Are any of the Cauchy sequences equivalent?

   (i) $(x_n) = \left(1 + \frac{1}{n+1}\right)$

   (ii) $(y_n) = (2^n)$

   (iii) $(z_n) = \left(1 - \frac{1}{2n+1}\right)$

LEVEL 3

7. Prove that if $n \in \mathbb{N}$ and A is a nonempty subset of n, then A has a least element.

8. Prove POMI $\to$ WOP.

9. Prove that $<_{\mathbb{Z}}$ is a well-defined strict linear ordering on $\mathbb{Z}$. You may use the fact that $<_{\mathbb{N}}$ is a well-defined strict linear ordering on $\mathbb{N}$.

LEVEL 4

10. Prove that $3^n - 1$ is even for all natural numbers n.

11. Show that the Principle of Mathematical Induction is equivalent to the following statement:

    ($\star$) Let $P(n)$ be a statement and suppose that (i) $P(0)$ is true and (ii) for all $k \in \mathbb{N}$, $P(k) \to P(k+1)$. Then $P(n)$ is true for all $n \in \mathbb{N}$.

12. Prove that addition of integers is well-defined.

13. Prove that addition and multiplication of rational numbers are well-defined.

14. Let $A = \{(x_n) \mid (x_n) \text{ is a Cauchy sequence of rational numbers}\}$ and define the relation R on A by $(x_n) R (y_n)$ if and only if for every $k \in \mathbb{N}^+$, there is $K \in \mathbb{N}$ such that $n > K$ implies $|x_n - y_n| < \frac{1}{k}$. Prove that R is an equivalence relation on A.

LEVEL 5

15. The Principle of Strong Induction is the following statement:

    ($\star\star$) Let $P(n)$ be a statement and suppose that (i) $P(0)$ is true and (ii) for all $k \in \mathbb{N}$, $\forall j \leq k \, (P(j)) \to P(k+1)$. Then $P(n)$ is true for all $n \in \mathbb{N}$.

    Use the Principle of Mathematical Induction to prove the Principle of Strong Induction.

16. Use the Principle of Mathematical Induction to prove that for every $n \in \mathbb{N}$, if S is a set with $|S| = n$, then S has 2^n subsets. (Hint: Use Problem 25 from Problem Set 1.)

17. Prove that addition of real numbers is well-defined and that the sum of two real numbers is a real number.

CHALLENGE PROBLEMS

18. Prove that multiplication of integers is well-defined.

19. Prove that multiplication of real numbers is well-defined and that the product of two real numbers is a real number.

20. Prove that $<_\mathbb{N}$ is a strict linear ordering on $\mathbb{N}$, that $<_\mathbb{Q}$ is a well-defined strict linear ordering on $\mathbb{Q}$, and that $<_\mathbb{R}$ is a well-defined strict linear ordering on $\mathbb{R}$.

21. Define a set to be **selfish** if the number of elements it has is in the set. For example, the set $K_5 = \{1, 2, 3, 4, 5\}$ is selfish because it has 5 elements and 5 is in the set. A selfish set is **minimal** if none of its proper subsets is also selfish. For example, the set K_5 is not a minimal selfish set because $\{1\}$ is a selfish subset. Let $K_n = \{1, 2, 3, \ldots n\}$. Determine with proof how many minimal selfish subsets K_n has in terms of n.

Lesson 5
Equinumerosity

Basic Definitions and Examples

We say that two sets A and B are **equinumerous**, written $A \sim B$ if there is a bijection $f: A \cong B$. In this case, we may also say that A and B have the same **cardinality**, and we can write $|A| = |B|$.

It turns out that $\sim$ **is an equivalence relation**. For any set A, the identity function $i_A: A \to A$ is a bijection, showing that $\sim$ is reflexive. For sets A and B, if $f: A \cong B$, then $f^{-1}: B \cong A$, showing that $\sim$ is symmetric. For sets A, B, and C, if $f: A \cong B$ and $g: B \cong C$, then $g \circ f: A \cong C$ by Corollary 3.7, showing that $\sim$ is transitive.

Example 5.1:

1. Let $A = \{\text{anteater}, \text{elephant}, \text{giraffe}\}$ and $B = \{\text{apple}, \text{banana}, \text{orange}\}$. Then $A \sim B$. We can define a bijection $f: A \cong B$ by $f(\text{anteater}) = \text{apple}$, $f(\text{elephant}) = \text{banana}$, and $f(\text{giraffe}) = \text{orange}$. This is not the only bijection from A to B (there are 6 distinct bijections from A to B), but we need only find one (or prove one exists) to show that the sets are equinumerous.

2. At this point it should be easy to see that two finite sets are equinumerous if and only if they have the same number of elements. It should also be easy to see that a finite set can never be equinumerous with an infinite set.

3. Let $\mathbb{N} = \{0, 1, 2, 3, 4 \ldots\}$ be the set of natural numbers and $2\mathbb{N} = \{0, 2, 4, 6, 8 \ldots\}$ the set of even natural numbers. Then $\mathbb{N} \sim 2\mathbb{N}$. We can actually see a bijection between these two sets just by looking at the sets themselves.

$$\begin{array}{ccccccc} 0 & 1 & 2 & 3 & 4 & 5 & 6 \ldots \\ 0 & 2 & 4 & 6 & 8 & 10 & 12 \ldots \end{array}$$

The function $f: \mathbb{N} \to 2\mathbb{N}$ defined by $f(n) = 2n$ is an explicit bijection. To see that f maps $\mathbb{N}$ into $2\mathbb{N}$, just observe that if $n \in \mathbb{N}$, then $2n \in 2\mathbb{N}$ by the definition of an even integer ($a \in \mathbb{N}$ is even if there is $b \in \mathbb{N}$ with $a = 2b$). f is injective because if $f(n) = f(m)$, then $2n = 2m$, and so, $n = m$ (see part (viii) of Problem 4 from Problem Set 4 for formal details). Finally, f is surjective because if $n \in 2\mathbb{N}$, then there is $k \in \mathbb{N}$ such that $n = 2k$. So, $f(k) = 2k = n$.

4. $\mathbb{N} \sim \mathbb{Z}$ via the bijection $f: \mathbb{N} \cong \mathbb{Z}$ defined by $f(n) = \begin{cases} \frac{n}{2} & \text{if } n \text{ is even.} \\ -\frac{n+1}{2} & \text{if } n \text{ is odd.} \end{cases}$

    You were asked to show that f is a bijection in Problem 8 in Problem Set 3. Let's look at this correspondence visually:

$$\begin{array}{ccccccc} 0 & 1 & 2 & 3 & 4 & 5 & 6 \ldots \\ 0 & -1 & 1 & -2 & 2 & -3 & 3 \ldots \end{array}$$

Many students get confused here because they are under the misconception that the integers should be written "in order." However, when checking to see if two sets are equinumerous, we **do not** include any other structure. In other words, we are just trying to "pair up" elements—it does not matter how we do so.

5. For A any nonempty set, $^A2 \sim \mathcal{P}(A)$. We showed this in part 4 of Example 3.3.
6. $[0, 1] \sim [0, 5]$ via the bijection $f: [0, 1] \to [0, 5]$ defined by $f(x) = 5x$. To see that f maps into $[0, 5]$, note that if $0 \leq x \leq 1$, then $0 \leq 5x \leq 5$. To see that f is injective, note that if $x \neq y$, then $5x \neq 5y$. Finally, to see that f is surjective, let $y \in [0, 5]$. Then $0 \leq y \leq 5$ and so, it follows that $0 \leq \frac{y}{5} \leq 1$ and $f\left(\frac{y}{5}\right) = 5 \cdot \frac{y}{5} = y$.

Countable and Uncountable Sets

We say that a set is **countable** if it is equinumerous with a subset of $\mathbb{N}$. It's easy to visualize a countable set because a bijection from a subset of $\mathbb{N}$ to a set A generates a list. For example, the set $2\mathbb{N}$ can be listed as 0, 2, 4, 6, ... and the set $\mathbb{Z}$ can be listed as 0, – 1, 1, – 2, 2, ... (see Example 5.1 above).

There are two kinds of countable sets: **finite** sets and **denumerable** (or **countably infinite**) sets. A **finite** set is a countable set that is equinumerous with $n = \{0, 1, 2, ..., n - 1\}$ for some $n \in \mathbb{N}$. An **infinite** set is a set that is not finite and a denumerable set is a countable set that is not finite. The dedicated reader may want to prove that a countable set is denumerable if and only if it is equinumerous with $\mathbb{N}$.

At this point, you may be asking yourself if all infinite sets are denumerable. If this were the case, then we would simply have finite sets and infinite sets, and that would be the end of it. However, there are in fact infinite sets that are **not** denumerable. An infinite set that is not denumerable is **uncountable**.

Theorem 5.2 (Cantor's Theorem): If A is any set, then A is **not** equinumerous with $\mathcal{P}(A)$.

Analysis: How can we prove that A is not equinumerous with $\mathcal{P}(A)$? Well, we need to show that there **does not** exist a bijection from A to $\mathcal{P}(A)$. Recall that a bijection is a function that is both an injection and a surjection. So, we will attempt to show that there do not exist any surjections from A to $\mathcal{P}(A)$. To do this, we will take an arbitrary function $f: A \to \mathcal{P}(A)$, and then argue that f is not surjective. We will show that $\operatorname{ran} f \neq \mathcal{P}(A)$ by finding a set $B \in \mathcal{P}(A) \setminus \operatorname{ran} f$. In words, we will find a subset of A that is **not** in the range of f.

Let's begin by looking at $\mathbb{N}$, the set of natural numbers. Given a specific function $f: \mathbb{N} \to \mathcal{P}(\mathbb{N})$, it's not too hard to come up with a set $B \in \mathcal{P}(\mathbb{N}) \setminus \operatorname{ran} f$. Let's choose a specific such f and use this example to try to come up with a procedure for describing the set B.

$$f(0) = \{\mathbf{0}, 1, 2, 3, 4, 5, 6, 7, 8, 9, 10, ...\}$$
$$f(1) = \{0, \mathbf{1}, 3, 4, 5, 6, 7, 8, 9, 10, ...\}$$
$$f(2) = \{0, 1, 4, 5, 6, 7, 8, 9, 10, ...\}$$
$$f(3) = \{0, 1, 4, 6, 7, 8, 9, 10, ...\}$$
$$f(4) = \{0, 1, \mathbf{4}, 6, 8, 9, 10, ...\}$$
$$...$$

Technical note: Recall that a **prime number** is a natural number with **exactly** two factors, 1 and itself. The set of prime numbers looks like this: $\{2, 3, 5, 7, 11, 13, 17, \ldots\}$. The function $f: \mathbb{N} \to \mathcal{P}(\mathbb{N})$ that we chose to use here is defined by $f(n) = \{k \in \mathbb{N} \mid k \text{ is not equal to one of the first } n \text{ prime numbers}\}$. Notice how $f(0)$ is just the set $\mathbb{N}$ of all natural numbers, $f(1)$ is the set of all natural numbers except 2 (we left out the first prime), $f(2)$ is the set of all natural numbers except 2 and 3 (we left out the first two primes), and so on.

Observe that the "inputs" of our function are natural numbers, and the "outputs" are sets of natural numbers. So, it's perfectly natural to ask the question "Is n in $f(n)$?"

For example, we see that $0 \in f(0)$, $1 \in f(1)$, and $4 \in f(4)$ (indicated in bold in the definition of the function above). However, we also see that $2 \notin f(2)$ and $3 \notin f(3)$.

Let's let B be the set of natural numbers n that are **not** inside their images. Symbolically, we have

$$B = \{n \in \mathbb{N} \mid n \notin f(n)\}.$$

Which natural numbers are in the set B? Well, we already said that $0 \in f(0)$. It follows that $0 \notin B$. Similarly, $1 \notin B$ and $4 \notin B$, but $2 \in B$ and $3 \in B$.

Why did we choose to define B this way? The reason is because we are trying to make sure that B cannot be equal to $f(n)$ for every n. Since $0 \in f(0)$, but $0 \notin B$, it follows that $f(0)$ and B are different sets because they differ by at least one element, namely 0. Similarly, since $1 \in f(1)$, but $1 \notin B$, B cannot be equal to $f(1)$. What about 2? Well $2 \notin f(2)$, but $2 \in B$. Therefore, $B \neq f(2)$ as well... and so on down the line. We intentionally chose to make B disagree with $f(n)$ for every natural number n, ensuring that B will not be in the range of f.

I think we are now ready to prove the theorem.

Proof of Theorem 5.2: Let $f: A \to \mathcal{P}(A)$, and let $B = \{a \in A \mid a \notin f(a)\}$. Suppose toward contradiction that $B \in \operatorname{ran} f$. Then there is $a \in A$ with $f(a) = B$. But then we have $a \in B$ if and only if $a \notin f(A)$ if and only if $a \notin B$. This contradiction tells us that $B \notin \operatorname{ran} f$, and so, f is not surjective. Since $f: A \to \mathcal{P}(A)$ was arbitrary, there does not exist a surjection from A to $\mathcal{P}(A)$, and therefore, there is no bijection from A to $\mathcal{P}(A)$. So, A is not equinumerous with $\mathcal{P}(A)$. □

Notes: (1) The proof given here is a **proof by contradiction**. A proof by contradiction works as follows:

1. We assume the negation of what we are trying to prove.

2. We use a logically valid argument to derive a statement which is false.

3. Since the argument was logically valid, the only possible error is our original assumption. Therefore, the negation of our original assumption must be true.

(2) In this problem we are trying to prove that A is **not** equinumerous with $\mathcal{P}(A)$. The negation of this statement is that A **is** equinumerous with $\mathcal{P}(A)$, and so that is what we assume. Since $A \sim \mathcal{P}(A)$, there is a bijection $f: A \to \mathcal{P}(A)$. So, f is a surjection, which means that every subset of $\mathcal{P}(A)$ is in the range of f. In particular, the set B described in the proof is a subset of $\mathcal{P}(A)$, and therefore it is in the range of f. We then use a logically valid argument to derive the obviously false statement "$a \in B$ if and only if $a \notin B$."

By Theorem 5.2, $\mathbb{N}$ is not equinumerous with $\mathcal{P}(\mathbb{N})$. Which of these two sets is the "bigger" one? Let's consider the function $f: \mathbb{N} \to \mathcal{P}(\mathbb{N})$ defined by $f(n) = \{n\}$. This function looks like this:

$$0 \quad 1 \quad 2 \quad 3 \quad 4 \ldots$$
$$\{0\} \; \{1\} \; \{2\} \; \{3\} \; \{4\} \ldots$$

Observe that we are matching up each natural number with a subset of natural numbers (a very simple subset consisting of just one natural number) in a way so that different natural numbers get matched with different subsets. In other words, we defined an injective function from $\mathbb{N}$ to $\mathcal{P}(\mathbb{N})$. It seems like there are lots of subsets of $\mathbb{N}$ that didn't get mapped to (for example, all infinite subsets of $\mathbb{N}$). So, it seems that $\mathbb{N}$ is a "smaller" set than $\mathcal{P}(\mathbb{N})$.

We use the notation $A \preccurlyeq B$ if there is an injective function from A to B.

$$A \preccurlyeq B \text{ if and only if } \exists f (f: A \hookrightarrow B)$$

Recall: The symbol $\exists$ is called an **existential quantifier**, and it is pronounced "There exists" or "There is." The expression $\exists f(f: A \hookrightarrow B)$ can be translated into English as "There exists an f such that f is an injective function from A to B."

We write $A \prec B$ if $A \preccurlyeq B$ and $A \nsim B$.

So, for example, $\mathbb{N} \prec \mathcal{P}(\mathbb{N})$.

Theorem 5.3: If A is any set, then $A \prec \mathcal{P}(A)$.

Proof: The function $f: A \to \mathcal{P}(A)$ defined by $f(a) = \{a\}$ is injective. So, $A \preccurlyeq \mathcal{P}(A)$. By Theorem 5.2, $A \nsim \mathcal{P}(A)$. It follows that $A \prec \mathcal{P}(A)$. □

Example 5.4: If we let $A = \mathcal{P}(\mathbb{N})$, we can apply Theorem 5.3 to this set A to see that $\mathcal{P}(\mathbb{N}) \prec \mathcal{P}(\mathcal{P}(\mathbb{N}))$. Continuing in this fashion, we get a sequence of increasingly larger sets.

$$\mathbb{N} \prec \mathcal{P}(\mathbb{N}) \prec \mathcal{P}(\mathcal{P}(\mathbb{N})) \prec \mathcal{P}(\mathcal{P}(\mathcal{P}(\mathbb{N}))) \prec \cdots$$

If A and B are arbitrary sets, in general it can be difficult to determine if A and B are equinumerous by producing a bijection. Luckily, the next theorem provides an easier way.

Theorem 5.5 (Cantor-Schroeder-Bernstein Theorem): If A and B are sets such that $A \preccurlyeq B$ and $B \preccurlyeq A$, then $A \sim B$.

Note: At first glance, many students think that Theorem 5.5 is obvious and that the proof must be trivial. This is not true. The theorem says that if there is an injective function from A to B and another injective function from B to A, then there is a bijective function from A to B. This is a deep result, which is far from obvious. Constructing a bijection from two arbitrary injections is not an easy thing to do. I suggest that the reader takes a few minutes to try to do it, if for no other reason than to convince themselves that the proof is difficult. I leave the proof itself as an exercise (see Problem 11 below).

Example 5.6: Let's use Theorem 5.5 to prove that the open interval of real numbers $(0, 1)$ is equinumerous with the closed interval of real numbers $[0, 1]$.

Analysis: Since $(0, 1) \subseteq [0, 1]$, there is an obvious injective function $f: (0, 1) \to [0, 1]$ (just send each element to itself).

The harder direction is finding an injective function g from $[0, 1]$ into $(0, 1)$. We will do this by drawing a line segment with endpoints $\left(0, \frac{1}{4}\right)$ and $\left(1, \frac{3}{4}\right)$. This will give us a bijection from $[0, 1]$ to $\left[\frac{1}{4}, \frac{3}{4}\right]$. We can visualize this bijection using the graph to the right. We will write an equation for this line segment in the slope-intercept form $y = mx + b$. Here m is the slope of the line and b is the y-intercept of the line. We can use the graph to see that $b = \frac{1}{4}$ and $m = \frac{\text{rise}}{\text{run}} = \frac{\frac{3}{4} - \frac{1}{4}}{1 - 0} = \frac{2}{4} = \frac{1}{2}$. So, we define $g: [0, 1] \to (0, 1)$ by $g(x) = \frac{1}{2}x + \frac{1}{4}$.

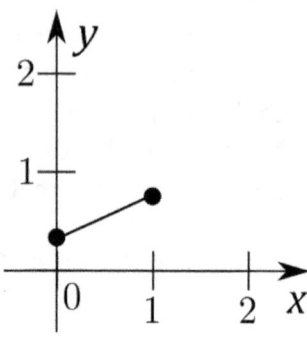

Let's write out the details of the proof.

Proof: Let $f: (0, 1) \to [0, 1]$ be defined by $f(x) = x$. Clearly, f is injective, so that $(0, 1) \preccurlyeq [0, 1]$.

Next, we define $g: [0, 1] \to \mathbb{R}$ by $g(x) = \frac{1}{2}x + \frac{1}{4}$. If $0 \leq x \leq 1$, then $0 \leq \frac{1}{2}x \leq \frac{1}{2}$, and therefore, $\frac{1}{4} \leq \frac{1}{2}x + \frac{1}{4} \leq \frac{3}{4}$. Since $0 < \frac{1}{4}$ and $\frac{3}{4} < 1$, we have $0 < g(x) < 1$. Therefore, $g: [0, 1] \to (0, 1)$. If $x \neq x'$, then $\frac{1}{2}x \neq \frac{1}{2}x'$, and so, $g(x) = \frac{1}{2}x + \frac{1}{4} \neq \frac{1}{2}x' + \frac{1}{4} = g(x')$. This shows that g is injective. It follows that $[0, 1] \preccurlyeq (0, 1)$.

Since $(0, 1) \preccurlyeq [0, 1]$ and $[0, 1] \preccurlyeq (0, 1)$, it follows from the Cantor-Schroeder-Bernstein Theorem that $(0, 1) \sim [0, 1]$. □

Notes: (1) If $A \subseteq B$, then the function $f: A \to B$ defined by $f(a) = a$ for all $a \in A$ is always injective. It is called the **inclusion map**.

(2) It is unfortunate that the same notation is used for points (ordered pairs of real numbers) and open intervals. Normally this isn't an issue, but in this particular example both usages of this notation appear. Take another look at the analysis above and make sure you can see when the notation (a, b) is being used for a point and when it is being used for an open interval.

(3) We could have used any closed interval $[a, b]$ with $0 < a < b < 1$ in place of $\left[\frac{1}{4}, \frac{3}{4}\right]$.

Example 5.7: Let's use Theorem 5.5 to prove that the half-open interval of real numbers $[0, 1)$ is equinumerous with $\mathcal{P}(\mathbb{N})$.

Proof: Let $f: \mathcal{P}(\mathbb{N}) \to [0, 1)$ be defined by $f(A) = 0.a_0 a_1 \ldots$, where $a_n = \begin{cases} 0 & \text{if } n \notin A. \\ 1 & \text{if } n \in A. \end{cases}$

If $A, B \in \mathcal{P}(\mathbb{N})$ with $A \neq B$, then there is $n \in \mathbb{N}$ with $n \in A \setminus B$ or $n \in B \setminus A$. **Without loss of generality**, assume that $n \in A \setminus B$. Assuming that $f(A) = 0.a_0 a_1 \ldots$ and $f(B) = 0.b_0 b_1 \ldots$, we have $a_n = 1$ and $b_n = 0$. Therefore, $f(A) \neq f(B)$. This shows that f is injective. It follows that $\mathcal{P}(\mathbb{N}) \preccurlyeq [0, 1)$.

Next, we define $g: [0, 1) \to \mathcal{P}(\mathbb{N})$. Let $a \in [0, 1)$. We define $A_0 = \left[0, \frac{1}{2}\right)$ and $A_1 = \left[\frac{1}{2}, 1\right)$. Then $\{A_0, A_1\}$ is a partition of $[0, 1)$, and so, it follows that $a \in A_0$ or $a \in A_1$, but not both. Similarly, we define $A_{00} = \left[0, \frac{1}{4}\right)$, $A_{01} = \left[\frac{1}{4}, \frac{1}{2}\right)$, $A_{10} = \left[\frac{1}{2}, \frac{3}{4}\right)$, and $A_{11} = \left[\frac{3}{4}, 1\right)$. Then $\{A_{00}, A_{01}, A_{10}, A_{11}\}$ is a partition of $[0, 1)$, and so, it follows that a is in exactly one of the four sets A_{00}, A_{01}, A_{10}, or A_{11}. Note that $\{A_{00}, A_{01}\}$ is a partition of A_0 and $\{A_{10}, A_{11}\}$ is a partition of A_1. It follows that if $a \in A_0$, then a must be in either A_{00} or A_{01}, and similarly if $a \in A_1$. Continuing in this fashion, we get a sequence of subscripts $a_0, a_0 a_1, a_0 a_1 a_2, \ldots$, where each subscript a_n is either 0 or 1. We let $g(a) = \{n \in \mathbb{N} \mid a_n = 1\}$. Now, suppose that $a \neq b$. Without loss of generality assume that $a < b$. If $a \in A_0$ and $b \in A_1$, then $0 \notin g(a)$, $0 \in g(b)$. Otherwise, there must be n such that $a, b \in A_{a_0 a_1 a_2 \cdots a_n}$, $a \in A_{a_0 a_1 a_2 \cdots a_n 0}$, and $b \in A_{a_0 a_1 a_2 \cdots a_n 1}$. It follows that $n + 1 \notin g(a)$ and $n + 1 \in g(b)$. Therefore, $g(a) \neq g(b)$. This shows that g is injective. It follows that $[0, 1) \preccurlyeq \mathcal{P}(\mathbb{N})$.

Since $\mathcal{P}(\mathbb{N}) \preccurlyeq [0, 1)$ and $[0, 1) \preccurlyeq \mathcal{P}(\mathbb{N})$, it follows from the Cantor-Schroeder-Bernstein Theorem that $\mathcal{P}(\mathbb{N}) \sim [0, 1)$. □

Notes: (1) In the proof above, we used the expression "Without loss of generality" twice. Recall from Note 1 following Example 3.3 that this expression can be used when an argument can be split up into two or more cases, and the proof of each of the cases is nearly identical.

The first time it appears in the proof above, the two cases are (i) $n \in A \setminus B$ and (ii) $n \in B \setminus A$. The argument for case (ii) is the same as the argument for case (i) except that the roles of A and B (and the roles of a_n and b_n for each n) are reversed.

The second time it appears in the proof, the two cases are (i) $a < b$ and (ii) $b < a$. Once again, the argument for case (ii) is the same as the argument for case (i) except that the roles of a and b are reversed.

(2) Given $a \in [0, 1)$, the number $.a_0 a_1 a_2 \ldots$ is called the **binary expansion** of a.

As an example, let's find the binary expansion of $\frac{1}{8}$.

Since $0 \leq \frac{1}{8} < \frac{1}{2}$, we have $\frac{1}{8} \in \left[0, \frac{1}{2}\right) = A_0$. So, $a_0 = 0$. (Here we are in the "leftmost" set in the partition of $[0, 1) = \left[0, \frac{1}{2}\right) \cup \left[\frac{1}{2}, 1\right)$.)

Next, since $0 \leq \frac{1}{8} < \frac{1}{4}$, we have $\frac{1}{8} \in \left[0, \frac{1}{4}\right) = A_{00}$. So, $a_1 = 0$. (Here we are once again in the "leftmost" set in the partition of $\left[0, \frac{1}{2}\right) = \left[0, \frac{1}{4}\right) \cup \left[\frac{1}{4}, \frac{1}{2}\right)$. It may help to rewrite this using only denominators of 4 as follows: $\left[0, \frac{2}{4}\right) = \left[0, \frac{1}{4}\right) \cup \left[\frac{1}{4}, \frac{2}{4}\right)$.)

Now, since $\frac{1}{8} \leq \frac{1}{8} < \frac{1}{4}$, we have $\frac{1}{8} \in \left[\frac{1}{8}, \frac{1}{4}\right) = A_{001}$. So, $a_2 = 1$. (This time we are in the "rightmost" set in the partition of $\left[0, \frac{1}{4}\right) = \left[0, \frac{1}{8}\right) \cup \left[\frac{1}{8}, \frac{1}{4}\right)$. Once again, it may help to rewrite this using only denominators of 8 as follows: $\left[0, \frac{2}{8}\right) = \left[0, \frac{1}{8}\right) \cup \left[\frac{1}{8}, \frac{2}{8}\right)$.)

Since $\frac{1}{8} \leq \frac{1}{8} < \frac{3}{16}$, we have $\frac{1}{8} \in \left[\frac{1}{8}, \frac{3}{16}\right) = A_{0010}$. So, $a_3 = 0$. (This time we are back in the "leftmost" set in the partition of $\left[\frac{1}{8}, \frac{1}{4}\right) = \left[\frac{1}{8}, \frac{3}{16}\right) \cup \left[\frac{3}{16}, \frac{1}{4}\right)$. Again, it may help to rewrite this using only denominators of 16 as follows: $\left[\frac{2}{16}, \frac{4}{16}\right) = \left[\frac{2}{16}, \frac{3}{16}\right) \cup \left[\frac{3}{16}, \frac{4}{16}\right)$.)

From this point on, it should be easy to see that we will always be choosing the "leftmost" set in the partition. So, the remaining digits will all be 0. It follows that the binary expansion of $\frac{1}{8}$ is 0.001.

(3) Using this method of writing a binary expansion, we will never get a tail of 1's. Therefore, using this method, each real number corresponds to a unique binary expansion without a tail of 1's. See part 2 of Example 1.6 for the definition of a "tail."

(4) We can use this same method to form the **decimal expansion** of a real number as well. In this case, we partition $[0, 1)$ into $A_0 = \left[0, \frac{1}{10}\right)$, $A_1 = \left[\frac{1}{10}, \frac{2}{10}\right)$, $A_2 = \left[\frac{2}{10}, \frac{3}{10}\right)$,..., and $A_9 = \left[\frac{9}{10}, 1\right)$. Using $\frac{1}{8}$ as an example again, we have $\frac{1}{8} \in \left[\frac{1}{10}, \frac{2}{10}\right) = A_1$. So, $a_0 = 1$. We then partition $\left[\frac{1}{10}, \frac{2}{10}\right) = \left[\frac{10}{100}, \frac{20}{100}\right)$ into $A_{10} = \left[\frac{10}{100}, \frac{11}{100}\right)$, $A_{11} = \left[\frac{11}{100}, \frac{12}{100}\right)$, $A_{12} = \left[\frac{12}{100}, \frac{13}{100}\right)$,..., and $A_{19} = \left[\frac{19}{100}, \frac{20}{100}\right)$. Now, we have $\frac{1}{8} \in \left[\frac{12}{100}, \frac{13}{100}\right) = A_{12}$, and so, $a_1 = 2$. You should now check that $\frac{1}{8} \in \left[\frac{125}{1000}, \frac{126}{1000}\right) = A_{125}$, and so, $a_2 = 5$. It is then not too hard to check that the remaining digits will be 0. It follows that the decimal expansion of $\frac{1}{8}$ is 0.125.

(5) We can use this same method to form n-ary expansions for any integer $n \geq 2$. If $n = 2$, we get the binary expansion, as defined in Note 2. If $n = 10$, we get the decimal expansion, as defined in Note 4. If $n = 3$, we get the **ternary expansion**. The reader may want to check that the ternary expansion of $\frac{1}{8}$ is 0.010101010101 ...

Problem Set 5

Full solutions to these problems are available for free download here:
www.SATPrepGet800.com/RABQXZ

LEVEL 1

1. Let A and B be sets such that $A \subseteq B$. Prove that $\mathcal{P}(A) \preccurlyeq \mathcal{P}(B)$.

2. Let A, B, and C be sets. Prove the following:
   - (i) $\preccurlyeq$ is transitive.
   - (ii) $\prec$ is transitive.
   - (iii) If $A \preccurlyeq B$ and $B \prec C$, then $A \prec C$.
   - (iv) If $A \prec B$ and $B \preccurlyeq C$, then $A \prec C$.

LEVEL 2

3. Define $\mathcal{P}_k(\mathbb{N})$ for each $k \in \mathbb{N}$ by $\mathcal{P}_0(\mathbb{N}) = \mathbb{N}$ and $\mathcal{P}_{k+1}(\mathbb{N}) = \mathcal{P}(\mathcal{P}_k(\mathbb{N}))$ for $k > 0$. Find a set B such that for all $k \in \mathbb{N}$, $\mathcal{P}_k(\mathbb{N}) \prec B$.

4. Prove that if $A \sim B$ and $C \sim D$, then $A \times C \sim B \times D$.

LEVEL 3

5. Prove the following:
   - (i) There is a partition $\boldsymbol{P}$ of $\mathbb{N}$ such that $\boldsymbol{P} \sim \mathbb{N}$ and for each $X \in \boldsymbol{P}$, $X \sim \mathbb{N}$
   - (ii) A countable union of countable sets is countable.

6. Let A and B be sets such that $A \sim B$. Prove that $\mathcal{P}(A) \sim \mathcal{P}(B)$.

LEVEL 4

7. Prove the following:
   - (i) $\mathbb{N} \times \mathbb{N} \sim \mathbb{N}$.
   - (ii) $\mathbb{Q} \sim \mathbb{N}$.
   - (iii) Any two intervals of real numbers are equinumerous (including $\mathbb{R}$ itself).
   - (iv) $^{\mathbb{N}}\mathbb{N} \sim \mathcal{P}(\mathbb{N})$.

8. Prove that if $A \sim B$ and $C \sim D$, then $^A C \sim {}^B D$.

LEVEL 5

9. Prove that for any sets A, B, and C, $^{B\times C}A \sim {}^C(^BA)$.

10. Prove the following:
    (i) $\mathcal{P}(\mathbb{N}) \sim \{f \in {}^\mathbb{N}\mathbb{N} \mid f \text{ is a bijection}\}$.
    (ii) $\mathbb{R} \sim \mathcal{P}(\mathbb{N})$.
    (iii) $^\mathbb{N}\mathbb{R} \nsim {}^\mathbb{R}\mathbb{N}$.

CHALLENGE PROBLEM

11. Prove the Cantor-Schroeder-Bernstein Theorem.

LESSON 6
ALGEBRAIC STRUCTURES AND COMPLETENESS

Binary Operations and Closure

A **binary operation** on a set is a rule that combines two elements of the set to produce another element of the set.

Example 6.1: Let $S = \{0, 1\}$. Multiplication on S is a binary operation, whereas addition on S is **not** a binary operation (here we are thinking of multiplication and addition in the "usual" sense, meaning the way we would think of them in elementary school or middle school).

To see that multiplication is a binary operation on S, observe that $0 \cdot 0 = 0, 0 \cdot 1 = 0, 1 \cdot 0 = 0$, and $1 \cdot 1 = 1$. Each of the four computations produces 0 or 1, both of which are in the set S.

To see that addition is not a binary operation on S, just note that $1 + 1 = 2$, and $2 \notin S$.

Formally, a **binary operation** $\star$ on a set S is a **function** $\star : S \times S \to S$. So, if $a, b \in S$, then we have $\star (a, b) \in S$. For easier readability, we usually write $\star (a, b)$ as $a \star b$.

When $\star$ is a binary operation on S, we say that S is **closed** under $\star$.

Example 6.2:

1. Let $S = \{u, v, w\}$ and define $\star$ using the following table:

$\star$	u	v	w
u	v	w	w
v	w	u	u
w	u	v	v

   The table given above is called a **multiplication table**. For $a, b \in S$, we evaluate $a \star b$ by taking the entry in the row given by a and the column given by b. For example, $v \star w = u$.

$\star$	u	v	w
u	v	w	w
v	w	u	u
w	u	v	v

   $\star$ is a binary operation on S because the only possible "outputs" are u, v, and w.

2. The operation of addition on the set of natural numbers is a binary operation because whenever we add two natural numbers we get another natural number (we proved this in Theorem 4.7). Here, the set S is $\mathbb{N}$ and the operation $\star$ is $+$.

3. Similarly, the operation of multiplication on the set of natural numbers is a binary operation because whenever we multiply two natural numbers we get another natural number (you were asked to prove this in part (ii) of Problem 4 from Problem Set 4). Here, the set S is $\mathbb{N}$ and the operation $\star$ is $\cdot$.

4. The operation of addition on the set of integers is a binary operation because whenever we add two integers we get another integer. Here, the set S is $\mathbb{Z} = \{[(a,b)] \mid (a,b) \in \mathbb{N} \times \mathbb{N}\}$, where $(a,b) \sim (c,d)$ if and only if $a+d = b+c$. The operation $\star$ is $+$, which is defined by $[(a,b)] + [(c,d)] = [(a+c, b+d)]$. You were asked to prove that this operation is well-defined in Problem 12 from Problem Set 4. By 2 above, if $a,b \in \mathbb{N}$, then $a+c, b+d \in \mathbb{N}$. So, $[(a,b)], [(c,d)] \in \mathbb{Z}$ implies that $[(a,b)] + [(c,d)] = [(a+c, b+d)] \in \mathbb{Z}$, showing that $\mathbb{Z}$ is closed under addition.

5. Similarly, the operation of multiplication on the set of integers is also a binary operation. The operation $\star$ is $\cdot$, which is defined by $[(a,b)] \cdot [(c,d)] = [(ac+bd, ad+bc)]$. You were asked to prove that this operation is well-defined in Problem 18 from Problem Set 4. By 2 and 3 above, if $a, b, c, d \in \mathbb{N}$, then $ac + bd, ad + bc \in \mathbb{N}$. So, $[(a,b)], [(c,d)] \in \mathbb{Z}$ implies that $[(a,b)] \cdot [(c,d)] = [(ac+bd, ad+bc)] \in \mathbb{Z}$, showing that $\mathbb{Z}$ is closed under multiplication.

6. The operation of addition on the set of rational numbers is a binary operation because whenever we add two rational numbers we get another rational number. Here, the set S is $\mathbb{Q} = \{\frac{a}{b} \mid a \in \mathbb{Z} \wedge b \in \mathbb{Z}^*\}$, where $\frac{a}{b} = \frac{c}{d}$ if and only if $ad = bc$. The operation $\star$ is $+$, which is defined by $\frac{a}{b} + \frac{c}{d} = \frac{ad+bc}{bd}$. You were asked to prove that this operation is well-defined in Problem 13 from Problem Set 4. By 4 and 5 above, if $a, b, c, d \in \mathbb{Z}$, then $ad + bc, bd \in \mathbb{Z}$. We need to show that for integers $b, d \neq 0$, we have $bd \neq 0$. Let $b = [(m,n)]$ and $d = [(s,t)]$ and assume that $b \cdot d = [(m,n)] \cdot [(s,t)] = [(0,0)]$ and $b = [(m,n)] \neq [(0,0)]$. We must show that $d = [(s,t)] = [(0,0)]$. Since $b \cdot d = [(0,0)]$, we have $[(ms+nt, mt+ns)] = [(0,0)]$. So, $ms + nt = mt + ns$. Since $b \neq [(0,0)]$, we have $m \neq n$. Without loss of generality, assume that $m < n$, then by part (ix) of Problem 4 from Problem Set 4, there is a natural number $k > 0$ such that $n = m + k$. Using associativity of addition and distributivity of multiplication over addition in $\mathbb{N}$, we have

$$ms + mt + kt = ms + (m+k)t = ms + nt = mt + ns = mt + (m+k)s = mt + ms + ks.$$

By part (vii) of Problem 4 from Problem Set 4 and commutativity of addition in $\mathbb{N}$, $kt = ks$. By part (viii) of Problem 4 from Problem Set 4, $t = s$. So, $d = [(s,t)] = [(0,0)]$. Thus, $\mathbb{Q}$ is closed under addition.

7. Similarly, the operation of multiplication on the set of rational numbers is also a binary operation. The operation $\star$ is $\cdot$, which is defined by $\frac{a}{b} \cdot \frac{c}{d} = \frac{ac}{bd}$. You were asked to prove that this operation is well-defined in Problem 13 from Problem Set 4. By 5 above, if $a, b, c, d \in \mathbb{Z}$, then $ac, bd \in \mathbb{Z}$. Also, if $b, d \neq 0$, then $bd \neq 0$ by the same argument given in 6 above. Thus, $\mathbb{Q}$ is closed under multiplication.

8. The operation of addition on the set of real numbers is a binary operation because whenever we add two real numbers we get another real number. Here, the set S is $\mathbb{R} = \{[(x_n)] \mid (x_n)$ is a Cauchy sequence of rational numbers$\}$, where $(x_n) \sim (y_n)$ if and only if for every $k \in \mathbb{N}^+$, there is $K \in \mathbb{N}$ such that $n > K$ implies $|x_n - y_n| < \frac{1}{k}$. The operation $\star$ is $+$, which is defined by $[(x_n)] + [(y_n)] = [(x_n + y_n)]$. You were asked to prove that this operation is well-defined and that the sum of two real numbers is a real number in Problem 17 from Problem Set 4. Thus, $\mathbb{R}$ is closed under addition.

9. Similarly, the operation of multiplication on the set of real numbers is also a binary operation. The operation $\star$ is $\cdot$, which is defined by $[(x_n)] \cdot [(y_n)] = [(x_n \cdot y_n)]$. You were asked to prove that this operation is well-defined and that the product of two real numbers is a real number in Problem 19 from Problem Set 4. Thus, $\mathbb{R}$ is closed under multiplication.

10. Subtraction on the set of natural numbers is **not** a binary operation. To see this, we just need to provide a single **counterexample**. Recall from Lesson 4 that for natural numbers m, n, and t, we say that $m - n = t$ if and only if $m = t + n$. It is straightforward to prove by induction on $t \in \mathbb{N}$ that $1 \subset t + 2$ (1 is a *proper* subset of $t + 2$) for all $t \in \mathbb{N}$. In particular, for all $t \in \mathbb{N}$, $1 \neq t + 2$. Therefore, $1 - 2$ is not defined in $\mathbb{N}$.

Some authors refer to a binary operation $\star$ on a set S even when the binary operation is not defined on all pairs of elements $a, b \in S$. We will always refer to these "false operations" as **partial binary operations**.

We say that the set S is **closed** under the partial binary operation $\star$ if whenever $a, b \in S$, we have $a \star b \in S$.

In Example 6.2, part 10 above, we saw that subtraction is a partial binary operation on $\mathbb{N}$ that is not a binary operation. In other words, $\mathbb{N}$ is not **closed** under subtraction.

Groups

A **group** is a pair $(G, \star)$ consisting of a set G together with a binary operation $\star$ on G satisfying:

(1) **(Associativity)** For all $x, y, z \in G$, $(x \star y) \star z = x \star (y \star z)$.

(2) **(Identity)** There exists an element $e \in G$ such that for all $x \in G$, $e \star x = x \star e = x$.

(3) **(Inverse)** For each $x \in G$, there is $y \in G$ such that $x \star y = y \star x = e$.

Notes: (1) If $y \in G$ is an inverse of $x \in G$, we will usually write $y = x^{-1}$.

(2) Recall that the definition of a binary operation already implies closure. However, many books on groups will mention this property explicitly:

   (Closure) For all $x, y \in G$, $x \star y \in G$.

(3) A group is **commutative** or **Abelian** if for all $x, y \in G$, $x \star y = y \star x$.

(4) The properties that define a group are called the **group axioms**. These are the statements that are **given** to be true in all groups. There are many other statements that are true in groups. However, any additional statements need to be **proved** using the axioms.

(5) If properties (1) and (2) hold, we get a **monoid**. So, a group is a monoid with the inverse property.

(6) If property (1) holds, we get a **semigroup**. So, a monoid is semigroup with an identity element.

Example 6.3:

1. Is $(\mathbb{N}, +)$ a group? By Theorem 4.7, $+$ is a binary operation on $\mathbb{N}$ (or equivalently, $\mathbb{N}$ is closed under $+$). By Theorem 4.9, $+$ is associative in $\mathbb{N}$. By Theorem 4.8, together with the definition of addition (see Note 1 following Theorem 4.8), 0 is an additive identity for $\mathbb{N}$. It follows that $(\mathbb{N}, +)$ is a monoid.

    However, $(\mathbb{N}, +)$ is **not** a group. The inverse property fails. For example, 1 has no inverse (and in fact, the only natural number with an additive inverse is 0). Indeed, suppose toward contradiction that $n \in \mathbb{N}$ with $n + 1 = 0$. Since $n + 1 = n^+ = n \cup \{n\}$, and $n \in n \cup \{n\}$, we must have $n \in 0 = \emptyset$. But the empty set has no elements. This contradiction proves that there is no natural number n such that $n + 1 = 0$. So, 1 has no additive inverse in $\mathbb{N}$. Therefore, the inverse property fails and $(\mathbb{N}, +)$ is **not** a group. It's worth mentioning that $(\mathbb{N}, +)$ is commutative. This follows from part (i) of Problem 4 from Problem Set 4. So, $(\mathbb{N}, +)$ is a commutative monoid that is **not** a group.

2. Similarly, $(\mathbb{N}, \cdot)$ is a commutative monoid with identity 1 that is not a group. I leave the details that $(\mathbb{N}, \cdot)$ is a commutative monoid to the reader. Let's prove that 2 has no multiplicative inverse in $\mathbb{N}$. We have $2 \cdot 0 = 0 \neq 1$. We now prove by induction that for all $n \geq 1$, $2n > 1$. The base case is $2 \cdot 1 = 2 > 1$. Assuming $2k > 1$, by parts (iv) and (x) of Problem 4 from Problem Set 4, we have $2(k + 1) = 2k + 2 > 1 + 2 > 1$. In particular, we showed that for all $n \in \mathbb{N}$, $2n \neq 1$. Therefore, 2 has no multiplicative inverse in $\mathbb{N}$.

3. $(\mathbb{Z}, +)$ is a commutative group with identity $0 = [(0,0)]$. We showed that $+$ is a binary operation on $\mathbb{Z}$ in part 4 of Example 6.2 above. To see that $+$ is associative in $\mathbb{Z}$, observe that for $a, b, c, d, e, f \in \mathbb{N}$, we have

    $$([(a,b)] + [(c,d)]) + [(e,f)] = [(a+c, b+d)] + [(e,f)] = [((a+c)+e, (b+d)+f)]$$
    $$= [(a+(c+e), b+(d+f))] = [(a,b)] + [(c+e, d+f)] = [(a,b)] + ([(c,d)] + [(e,f)])$$

    For the first, second, fourth and fifth equalities, we simply used the definition of addition of integers. For the third equality, we used the associativity of addition in $\mathbb{N}$. I leave it to the reader to verify that $[(0,0)]$ is an additive identity, that the inverse of $[(a,b)]$ is $[(b,a)]$, and that $+$ is commutative in $\mathbb{Z}$.

4. $(\mathbb{Z}, \cdot)$ is a commutative monoid with identity $1 = [(1,0)]$ that is not a group. I leave it to the reader to verify that $(\mathbb{Z}, \cdot)$ is a commutative monoid. Let's prove that $2 = [(2,0)]$ has no multiplicative inverse. If $[(a,b)]$ is a multiplicative inverse of 2, then we have $2a + 0b = 1$ and $2b + 0a = 0$. The first equation is equivalent to $2a = 1$. However, by 2 above, 2 has no multiplicative inverse in $\mathbb{N}$. Therefore, the equation $2a = 1$ has no solution, and so, 2 has no multiplicative inverse in $\mathbb{Z}$.

5. $(\mathbb{Q}, +)$ is a commutative group with identity $0 = \frac{0}{1}$. You will need to prove this as part of Problem 17 below.

6. $(\mathbb{R}, +)$ is a commutative group with identity $[(0)]$. We showed that $+$ is a binary operation on $\mathbb{R}$ in part 8 of Example 6.2 above. To see that $+$ is associative in $\mathbb{R}$, we use the associativity of $+$ in $\mathbb{Q}$. If $[(x_n)], [(y_n)], [(z_n)] \in \mathbb{R}$, then

$$([(x_n)] + [(y_n)]) + [(z_n)] = [(x_n + y_n)] + [(z_n)] = [((x_n + y_n) + z_n)]$$
$$= [(x_n + (y_n + z_n))] = [(x_n)] + [(y_n + z_n)] = [(x_n)] + ([(y_n)] + [(z_n)]).$$

To see that $[(0)]$ is the additive identity, using the fact that 0 is the additive identity in $\mathbb{Q}$, we have for $[(x_n)] \in \mathbb{R}$,

$$[(0)] + [(x_n)] = [(0 + x_n)] = [(x_n)] \text{ and } [(x_n)] + [(0)] = [(x_n + 0)] = [(x_n)].$$

The inverse of the real number $[(x_n)]$ is $[(-x_n)]$, where for each $n \in \mathbb{N}$, $-x_n$ is the additive inverse of x_n in $\mathbb{Q}$. To see this, simply observe that

$$[(x_n)] + [(-x_n)] = [(x_n + (-x_n))] = [(0)] \text{ and } [(-x_n)] + [(x_n)] = [(-x_n + x_n)] = [(0)].$$

Finally, to see that $+$ is commutative in $\mathbb{R}$, we use the commutativity of $+$ in $\mathbb{Q}$. If $[(x_n)], [(y_n)] \in \mathbb{R}$, then

$$[(x_n)] + [(y_n)] = [(x_n + y_n)] = [(y_n + x_n)] = [(y_n)] + [(x_n)].$$

7. $(\mathbb{Q}, \cdot)$ and $(\mathbb{R}, \cdot)$ fail to be groups, but only because 0 has no inverse (this follows immediately from part (iii) of Problem 7 below). However, $(\mathbb{Q}^*, \cdot)$ and $(\mathbb{R}^*, \cdot)$ are both commutative groups. You will need to prove this as part of Problem 17 below.

Fields

A **field** is a triple $(F, +, \cdot)$, where F is a set and $+$ and $\cdot$ are binary operations on F satisfying

(1) $(F, +)$ is a commutative group.

(2) $(F^*, \cdot)$ is a commutative group.

(3) $\cdot$ is **distributive** over $+$ in F. That is, for all $x, y, z \in F$, we have

$$x \cdot (y + z) = x \cdot y + x \cdot z \quad \text{and} \quad (y + z) \cdot x = y \cdot x + z \cdot x.$$

(4) $0 \neq 1$ (where 0 is the identity of $(F, +)$ and 1 is the identity of $(F^*, \cdot)$).

We will refer to the operation $+$ as addition, the operation $\cdot$ as multiplication, the additive identity as 0, the multiplicative identity as 1, the additive inverse of an element $x \in F$ as $-x$, and the multiplicative inverse of an element $x \in F^*$ as x^{-1}. We will often abbreviate $x \cdot y$ as xy.

Notes: (1) $(F, +)$ a commutative group means the following:

- **(Closure)** For all $x, y \in F$, $x + y \in F$.
- **(Associativity)** For all $x, y, z \in F$, $(x + y) + z = x + (y + z)$.
- **(Commutativity)** For all $x, y \in F$, $x + y = y + x$.

- **(Identity)** There exists an element $0 \in F$ such that for all $x \in F$, $0 + x = x + 0 = x$.
- **(Inverse)** For each $x \in F$, there is $-x \in F$ such that $x + (-x) = (-x) + x = 0$.

(2) Similarly, $(F^*, \cdot)$ a commutative group means the following:
- **(Closure)** For all $x, y \in F^*$, $xy \in F^*$.
- **(Associativity)** For all $x, y, z \in F^*$, $(xy)z = x(yz)$.
- **(Commutativity)** For all $x, y \in F^*$, $xy = yx$.
- **(Identity)** There exists an element $1 \in F^*$ such that for all $x \in F^*$, $1x = x \cdot 1 = x$.
- **(Inverse)** For each $x \in F^*$, there is $x^{-1} \in F^*$ such that $xx^{-1} = x^{-1}x = 1$.

(3) Recall that F^* is the set of nonzero elements of F. We can write $F^* = \{x \in F \mid x \neq 0\}$ (pronounced "the set of x in F such that x is not equal to 0") or $F^* = F \setminus \{0\}$ (pronounced "F with 0 removed").

(4) The properties that define a field are called the **field axioms**. These are the statements that are **given** to be true in all fields. There are many other statements that are true in fields. However, any additional statements need to be **proved** using the axioms.

(5) If we replace the condition that "$(F^*, \cdot)$ is a commutative group" by "$(F, \cdot)$ is a monoid," then the resulting structure is called a **ring**. The most well-known example of a ring is $\mathbb{Z}$, the ring of integers.

We also do not require 0 and 1 to be distinct in the definition of a ring. If $0 = 1$, we get the zero ring, which consists of just one element, namely 0 (Why?). The operations of addition and multiplication are defined by $0 + 0 = 0$ and $0 \cdot 0 = 0$. The reader may want to verify that the zero ring is in fact a ring.

The main difference between a ring and a field is that in a ring, there can be nonzero elements that do not have multiplicative inverses. For example, in $\mathbb{Z}$, 2 has no multiplicative inverse (see part 4 of Example 6.3 above). So, the equation $2x = 1$ has no solution.

(6) If we also replace "$(F, +)$ is a commutative group" by "$(F, +)$ is a commutative monoid," then the resulting structure is a **semiring**. The most well-known example of a semiring is $\mathbb{N}$, the semiring of natural numbers.

The main difference between a semiring and a ring is that in a semiring, there can be elements that do not have additive inverses. For example, in $\mathbb{N}$, 1 has no additive inverse (see part 1 of Example 6.3). Thus, the equation $x + 1 = 0$ has no solution.

Technical note: For a semiring, we include one additional axiom: For all $x \in F$, $0 \cdot x = x \cdot 0 = 0$. This statement follows from the ring axioms (see part (iii) of Problem 7 below), but **not** from the other semiring axioms. We need the additive inverse property to prove it.

(7) Every field is a commutative ring. Although this is not too hard to show (you will be asked to show this in Problem 12 below), it is worth observing that this is not completely obvious. For example, if $(F, +, \cdot)$ is a ring, then since $(F, \cdot)$ is a monoid with identity 1, it follows that $1 \cdot 0 = 0 \cdot 1 = 0$. However, in the definition of a field given above, this property of 0 is not given as an axiom. We **are** given that $(F^*, \cdot)$ is a commutative group, and so, it follows that 1 is an identity for F^*. But $0 \notin F^*$, and so, $1 \cdot 0 = 0 \cdot 1 = 0$ needs to be proved.

Similarly, in the definition of a field given above, 0 is excluded from associativity and commutativity. These need to be checked.

Example 6.4:

1. $(\mathbb{Z}, +, \cdot)$ is a ring that is **not** a field. In parts 3 and 4 of Example 6.3 above, we saw that $(\mathbb{Z}, +)$ is a commutative group and $(\mathbb{Z}, \cdot)$ is a monoid. All that is left to check is distributivity. We first check left distributivity. For $a, b, c, d, e, f \in \mathbb{N}$, we have

$$[(a,b)]([(c,d)] + [(e,f)]) = [(a,b)] \cdot [(c+e, d+f)]$$
$$= [(a(c+e) + b(d+f), a(d+f) + b(c+e))]$$
$$= [((ac+ae) + (bd+bf), (ad+af) + (bc+be))]$$
$$= [((ac+bd) + (ae+bf), (ad+bc) + (af+be))]$$
$$= [(ac+bd, ad+bc)] + [(ae+bf, af+be)] = [(a,b)] \cdot [(c,d)] + [(a,b)] \cdot [(e,f)].$$

For the first, second, fifth and sixth equalities, we simply used the definitions of addition and multiplication of integers. For the third equality, we used the distributivity of multiplication over addition in $\mathbb{N}$. For the fourth equality, we used the associativity and commutativity of addition in $\mathbb{N}$. Since multiplication is commutative in $\mathbb{Z}$, right distributivity follows immediately from left distributivity.

$$([(c,d)] + [(e,f)])[(a,b)] = [(a,b)]([(c,d)] + [(e,f)])$$
$$= [(a,b)] \cdot [(c,d)] + [(a,b)] \cdot [(e,f)] = [(c,d)] \cdot [(a,b)] + [(e,f)] \cdot [(a,b)].$$

Therefore, $(\mathbb{Z}, +, \cdot)$ is a ring. We already showed in part 4 of Example 6.3 that 2 has no multiplicative inverse in $\mathbb{Z}$. So, $(\mathbb{Z}, +, \cdot)$ is **not** a field.

2. $(\mathbb{Q}, +, \cdot)$ and $(\mathbb{R}, +, \cdot)$ are both fields. You will need to prove this as part of Problem 17 below.

3. $(\mathbb{C}, +, \cdot)$ is field. You will be asked to prove this in Problem 17 below. The proof is very straightforward and mostly uses the fact that $(\mathbb{R}, +, \cdot)$ is a field. For example, to verify that addition is commutative in $\mathbb{C}$, we have

$$(a+bi) + (c+di) = (a+c) + (b+d)i = (c+a) + (d+b)i = (c+di) + (a+bi).$$

We have $a + c = c + a$ because $a, c \in \mathbb{R}$ and addition is commutative in $\mathbb{R}$. For the same reason, we have $b + d = d + b$.

We note a few additional things of importance here: The identity for addition is $0 = 0 + 0i$. The identity for multiplication is $1 = 1 + 0i$. The additive inverse of $z = a + bi$ is $-z = -(a+bi) = -a - bi$. The multiplicative inverse of $z = a + bi$ is $z^{-1} = \frac{a}{a^2+b^2} - \frac{b}{a^2+b^2}i$.

Subtraction and Division: If $a, b \in F$, we define $a - b = a + (-b)$ and for $b \neq 0$, $\frac{a}{b} = ab^{-1}$.

Ordered Rings and Fields

An **ordered ring** is a quadruple $(R, +, \cdot, \leq)$, where $(R, +, \cdot)$ is a ring, and $(R, \leq)$ is a linearly ordered set such that:

(1) If $a, b, c \in R$, then $a \leq b \rightarrow a + c \leq b + c$.

(2) If $a, b \in R$ with $0 \leq a$ and $0 \leq b$, then $0 \leq ab$.

Example 6.5:

1. $(\mathbb{Z}, +, \cdot, \leq)$ is an ordered ring. We already showed that $(\mathbb{Z}, +, \cdot)$ is a ring in part 1 of Example 6.4.

   In Problem 9 from Problem Set 4, you were asked to show that $<$ is a well-defined strict linear ordering on $\mathbb{Z}$. $\leq$ is the corresponding linear ordering given by Theorem 2.8.

   Let's check that properties (1) and (2) above are satisfied.

   For (1), let $a, b, c, d, e, f \in \mathbb{N}$, and assume that $[(a, b)] \leq [(c, d)]$. Then $a + d \leq b + c$, and so, by part (x) of Problem 4 from Problem Set 4, $(a + d) + (e + f) \leq (b + c) + (e + f)$. We can then use associativity and commutativity of addition in $\mathbb{N}$, to get the inequality $(a + e) + (d + f) \leq (b + f) + (c + e)$. By definition, $[(a + e, b + f)] \leq [(c + e, d + f)]$. Therefore, $[(a, b)] + [(e, f)] \leq [(c, d)] + [(e, f)]$. For (2), let $a, b, c, d \in \mathbb{N}$, and assume that $[(0, 0)] \leq [(a, b)]$ and $[(0, 0)] \leq [(c, d)]$. Then $b \leq a$, and $d \leq c$. By part (ix) of Problem 4 from Problem Set 4, there are natural numbers $k > 0$ and $t > 0$ such that $a = b + k$ and $c = d + t$. Then we have
   $$ad + bc = (b + k)d + b(d + t) = bd + kd + bd + bt.$$
   $$ac + bd = (b + k)(d + t) + bd = (b + k)d + (b + k)t + bd = bd + kd + bt + kt + bd.$$
   It follows that $ac + bd = (ad + bc) + kt$. Therefore, again by part (ix) of Problem 4 from Problem Set 4, we have $ad + bc \leq ac + bd$. Thus, $[(0, 0)] \leq [(ac + bd, ad + bc)]$. Therefore, $[(0, 0)] \leq [(a, b)] \cdot [(c, d)]$.

2. $(\mathbb{Q}, +, \cdot, \leq)$ and $(\mathbb{R}, +, \cdot, \leq)$ are ordered fields. You will be asked to prove this in Problem 17 below.

3. $(\mathbb{C}, +, \cdot)$ **cannot** be ordered. To see this, assume toward contradiction that $\leq$ is an ordering of $(\mathbb{C}, +, \cdot)$. If $i \geq 0$, then $-1 = i^2 = i \cdot i \geq 0$ by property (2) of an ordered field. If $i \leq 0$, then by property (1) of an ordered field, $0 = -i + i \leq -i + 0 = -i$, and so, by property (2) of an ordered field, we have $-1 = i^2 = 1i^2 = (-1)(-1)i \cdot i = (-1i)(-1i) = (-i)(-i) \geq 0$ (here we have also used parts (vi) and (vii) of Problem 7 below together with commutativity and associativity of multiplication in $\mathbb{C}$, and the multiplicative identity property in $\mathbb{C}$). So, $-1 \geq 0$. By property (1) of an ordered field, we have $0 = -1 + 1 \geq 0 + 1 = 1$ and by property (2) of an ordered field, we have $1 = (-1)(-1) \geq 0$. Since $\leq$ is antisymmetric, $1 = 0$, a contradiction.

If $(R, +, \cdot, \leq)$ is an ordered ring, we will write $a < b$ if $a \leq b$ and $a \neq b$. By Theorem 2.7, $<$ is a strict linear ordering on R. We may also write $b \geq a$ in place of $a \leq b$ and $b > a$ in place of $a < b$.

In general, we may just use the name of the underlying set for a whole structure when there is no danger of confusion. For example, we may refer to the ring R or the ordered field F instead of the ring $(R, +, \cdot)$ or the ordered field $(F, +, \cdot, \leq)$.

Fields are particularly nice to work with because all the arithmetic and algebra we've learned through the years can be used in fields. For example, in the field of rational numbers, we can solve the equation $2x = 1$. The multiplicative inverse property allows us to do this. Indeed, the multiplicative inverse of 2 is $\frac{1}{2}$, and therefore, $x = \frac{1}{2}$ is a solution to the given equation. Compare this to the ring of integers. If we restrict ourselves to the integers, then the equation $2x = 1$ has no solution (see part 4 of Example 6.3).

Working with ordered fields is very nice as well. In the Problem Set below, you will be asked to derive some additional properties of fields and ordered fields that follow from the axioms. We will prove a few such properties now as examples.

Theorem 6.6: Let $(F, \leq)$ be an ordered field. Then for all $x \in F^*$, $x \cdot x > 0$.

Proof: There are two cases to consider:

Case 1: If $x > 0$, then $x \geq 0$ and $x \neq 0$. By property (2) of an ordered field, $x \cdot x \geq 0$. If $x \cdot x = 0$, then since x has an inverse, $x = 1 \cdot x = (x^{-1} \cdot x) \cdot x = x^{-1} \cdot (x \cdot x) = x^{-1} \cdot 0 = 0$ (by part (iii) of Problem 7 below), contrary to our assumption that $x \in F^*$. So, $x \cdot x \neq 0$, and therefore, $x \cdot x > 0$.

Case 2: If $x < 0$, then $x \leq 0$. By property (1) of an ordered field, $x + (-x) \leq 0 + (-x)$, and so, $0 \leq -x$. Therefore, $(-x)(-x) \geq 0$, by property (2) of an ordered field. Now, using Problem 7 (parts (vi) and (vii)) below, together with commutativity and associativity of multiplication in F, and the multiplicative identity property in F, we have

$$(-x)(-x) = (-1x)(-1x) = (-1)(-1)x \cdot x = 1(x \cdot x) = x \cdot x.$$

So, again we have $x \cdot x \geq 0$. The same argument used in case 1 can be used to rule out $x \cdot x = 0$. Therefore, $x \cdot x > 0$. □

Theorem 6.7: Every ordered field $(F, +, \cdot, \leq)$ contains a copy of the natural numbers. Specifically, F contains a subset $\overline{\mathbb{N}} = \{\overline{n} \mid n \in \mathbb{N}\}$ such that for all $n, m \in \mathbb{N}$, we have $\overline{n + m} = \overline{n} + \overline{m}$, $\overline{n \cdot m} = \overline{n} \cdot \overline{m}$, and $n < m \leftrightarrow \overline{n} < \overline{m}$.

Proof: Let $(F, +, \cdot, \leq)$ be an ordered field. By the definition of a field, $0, 1 \in F$ and $0 \neq 1$.

We let $\overline{0} = 0$ and $\overline{n} = 1 + 1 + \cdots + 1$, where 1 appears n times. Let $\overline{\mathbb{N}} = \{\overline{n} \mid n \in \mathbb{N}\}$. Then $\overline{\mathbb{N}} \subseteq F$.

We first prove by induction on m that for all $n, m \in \mathbb{N}$, $\overline{n + m} = \overline{n} + \overline{m}$.

Base case ($k = 0$): $\overline{n + 0} = \overline{n} = \overline{n} + 0 = \overline{n} + \overline{0}$.

Inductive step: Suppose that $\overline{n + k} = \overline{n} + \overline{k}$. Then we have

$$\overline{n + (k+1)} = \overline{(n+k) + 1} = \overline{n+k} + 1 = (\overline{n} + \overline{k}) + 1 = \overline{n} + (\overline{k} + 1) = \overline{n} + \overline{k+1}.$$

By the Principle of Mathematical Induction, for all natural numbers m, $\overline{n+m} = \overline{n} + \overline{m}$.

Since $n \in \mathbb{N}$ was arbitrary, we have that for all $n, m \in \mathbb{N}$, $\overline{n+m} = \overline{n} + \overline{m}$.

Similarly, we now prove by induction on m that for all $n, m \in \mathbb{N}$, $\overline{n \cdot m} = \overline{n} \cdot \overline{m}$.

Base case ($k = 0$): $\overline{n \cdot 0} = \overline{0} = \overline{n} \cdot \overline{0}$.

Inductive step: Suppose that $\overline{n \cdot k} = \overline{n} \cdot \overline{k}$. Then we have
$$\overline{n \cdot (k+1)} = \overline{nk + n} = \overline{nk} + \overline{n} = \overline{n} \cdot \overline{k} + \overline{n} = \overline{n}(\overline{k} + 1) = \overline{n}(\overline{k + 1}) = \overline{n}(\overline{k+1}).$$

By the Principle of Mathematical Induction, for all natural numbers m, $\overline{n \cdot m} = \overline{n} \cdot \overline{m}$.

Since $n \in \mathbb{N}$ was arbitrary, we have that for all $n, m \in \mathbb{N}$, $\overline{n \cdot m} = \overline{n} \cdot \overline{m}$.

We now wish to prove that for all $n, m \in \mathbb{N}$, $n < m \leftrightarrow \overline{n} < \overline{m}$.

We first note that since $1 > 0$ (because $1 = 1 \cdot 1 > 0$ by Theorem 6.6), we have for any natural number n, $\overline{n+1} = \overline{n} + 1 > \overline{n} + 0 = \overline{n}$ by property (1) of an ordered field.

We now prove by induction on $m \in \mathbb{N}$ that $n < m \to \overline{n} < \overline{m}$.

The base case $k = 0$ is vacuously true because $n < 0$ never occurs. For the inductive step, let $k \in \mathbb{N}$ and assume that $n < k \to \overline{n} < \overline{k}$. Now, suppose that $n < k + 1 = k \cup \{k\}$. Then $n < k$ or $n = k$. If $n < k$, then $\overline{n} < \overline{k}$ by the inductive hypothesis. By property (1) of an ordered field and $0 < 1$, we have $\overline{k} = \overline{k} + 0 < \overline{k} + 1 = \overline{k+1}$. By the transitivity of $<$, $\overline{n} < \overline{k+1}$. If $n = k$, then by the same reasoning in the last sentence, $\overline{k} < \overline{k+1}$. By the Principle of Mathematical Induction, we have for all $m \in \mathbb{N}$, $n < m \to \overline{n} < \overline{m}$.

Since $n \in \mathbb{N}$ was arbitrary, we have that for all $n, m \in \mathbb{N}$, $n < m \to \overline{n} < \overline{m}$.

Finally, we need to prove that $\overline{n} < \overline{m} \to n < m$. By the contrapositive, this statement is equivalent to $m \leq n \to \overline{m} \leq \overline{n}$. If $m < n$, then $\overline{m} < \overline{n}$, by the same reasoning used in the last paragraph. Otherwise, $m = n$, in which case $\overline{m} = \overline{n}$. □

Note: The function that sends $n \in \mathbb{N}$ to $\overline{n} \in \overline{\mathbb{N}}$ is called an **isomorphism**. It has the following properties: (i) $\overline{n+m} = \overline{n} + \overline{m}$; (ii) $\overline{n \cdot m} = \overline{n} \cdot \overline{m}$; (iii) $n < m$ if and only if $\overline{n} < \overline{m}$; (iv) the function provides a bijection between the elements of $\mathbb{N}$ and the elements of $\overline{\mathbb{N}}$.

So, when we say that every field contains a "copy" of the natural numbers, we mean that there is a subset $\overline{\mathbb{N}}$ of the field so that $(\overline{\mathbb{N}}, +, \cdot, \leq)$ is isomorphic to $(\mathbb{N}, +, \cdot, \leq)$.

Theorem 6.8: Let $(F, \leq)$ be an ordered field and let $x \in F$ with $x > 0$. Then $\frac{1}{x} > 0$.

Proof: Since $x \neq 0$, $\frac{1}{x} = x^{-1}$ exists and is nonzero.

Assume toward contradiction that $\frac{1}{x} < 0$. Then $\frac{1}{x} \leq 0$, and so, by property (1) of an ordered field, $\frac{1}{x} + \left(-\frac{1}{x}\right) \leq -\frac{1}{x}$, or equivalently, $0 \leq -\frac{1}{x}$. Using Problem 7 (part (vi)) below, together with commutativity and associativity of multiplication, the multiplicative inverse property, and the multiplicative identity property, $x\left(-\frac{1}{x}\right) = x(-1)x^{-1} = -1xx^{-1} = -1 \cdot 1 = -1$. Since $x > 0$ and $-\frac{1}{x} \geq 0$, by property (2) of an ordered field, $-1 = x\left(-\frac{1}{x}\right) \geq 0$. So, by property (1) of an ordered field, $0 = 1 + (-1) \geq 1$. But by Theorem 6.6, $1 = 1 \cdot 1 > 0$. This is a contradiction. Thus, $\frac{1}{x} > 0$. □

Why Isn't $\mathbb{Q}$ Enough?

At first glance, it would appear that the ordered field of rational numbers would be sufficient to solve all "real world" problems. However, a long time ago, a group of people called the Pythagoreans showed that this is not the case. The problem was first discovered when applying the now well-known Pythagorean Theorem.

Theorem 6.9 (Pythagorean Theorem): In a right triangle with legs of lengths a and b, and a hypotenuse of length c, $c^2 = a^2 + b^2$.

The picture to the right shows a right triangle. The vertical and horizontal segments (labeled a and b, respectively) are called the **legs** of the right triangle, and the side opposite the right angle (labeled c) is called the **hypotenuse** of the right triangle.

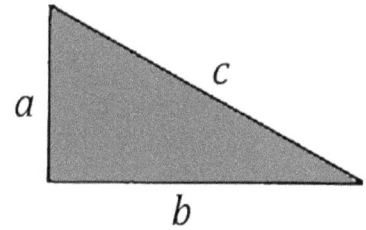

There are many ways to prove the Pythagorean Theorem. Here, we will provide a simple geometric argument. For the proof we will want to recall that the area of a square with side length s is $A = s^2$, and the area of a triangle with base b and height h is $A = \frac{1}{2}bh$. Notice that in our right triangle drawn here, the base is labeled b (how convenient), and the height is labeled a. So, the area of this right triangle is $A = \frac{1}{2}ba = \frac{1}{2}ab$.

Proof of Theorem 6.9: We draw 2 squares, each of side length $a + b$, by rearranging 4 copies of the given triangle in 2 different ways:

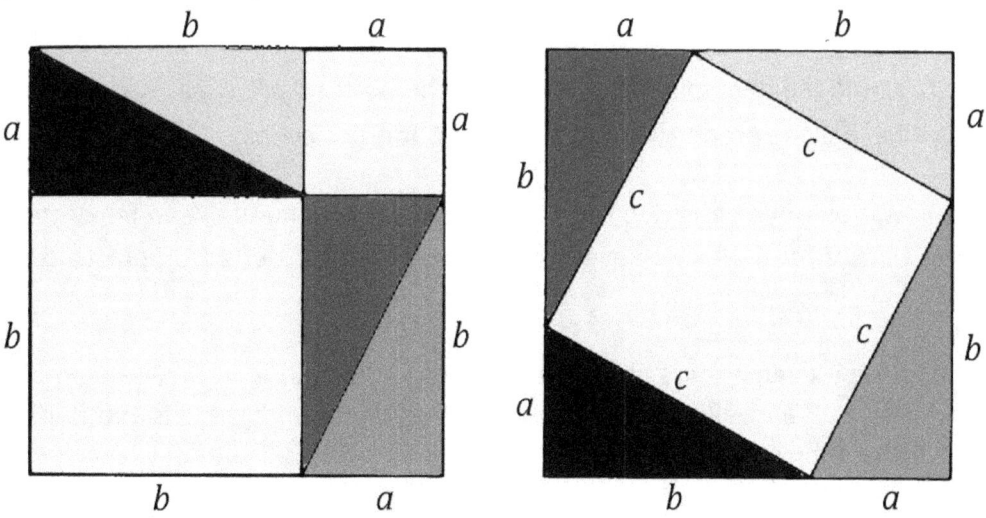

We can get the area of each of these squares by adding the areas of all the figures that comprise each square.

The square on the left consists of 4 copies of the given right triangle, a square of side length a and a square of side length b. It follows that the area of this square is $4 \cdot \frac{1}{2}ab + a^2 + b^2 = 2ab + a^2 + b^2$.

The square on the right consists of 4 copies of the given right triangle, and a square of side length c. It follows that the area of this square is $4 \cdot \frac{1}{2}ab + c^2 = 2ab + c^2$.

Since the areas of both squares of side length $a + b$ are equal (both areas are equal to $(a+b)^2$), $2ab + a^2 + b^2 = 2ab + c^2$. Cancelling $2ab$ from each side of this equation yields $a^2 + b^2 = c^2$. □

Question: In a right triangle where both legs have length 1, what is the length of the hypotenuse?

Let's try to answer this question. If we let c be the length of the hypotenuse of the triangle, then by the Pythagorean Theorem, we have $c^2 = 1^2 + 1^2 = 1 + 1 = 2$. Since $c^2 = c \cdot c$, we need to find a number with the property that when you multiply that number by itself you get 2. The Pythagoreans showed that if we use only numbers in $\mathbb{Q}$, then no such number exists.

Theorem 6.10: There does not exist a rational number a such that $a^2 = 2$.

Analysis: We will prove this theorem by assuming that there is a rational number a such that $a^2 = 2$, and arguing until we reach a contradiction. A first attempt at a proof would be to let $a = \frac{m}{n} \in \mathbb{Q}$ satisfy $\left(\frac{m}{n}\right)^2 = 2$. It follows that $m^2 = 2n^2$ ($\frac{m^2}{n^2} = \frac{m \cdot m}{n \cdot n} = \frac{m}{n} \cdot \frac{m}{n} = \left(\frac{m}{n}\right)^2$ and $2 = \frac{2}{1} \Rightarrow \frac{m^2}{n^2} = \frac{2}{1} \Rightarrow m^2 = 2n^2$), showing that m^2 **is even**. We will then use this information to show that both m and n are even (we will need a preliminary Lemma to help us with this).

Now, in our first attempt, the fact that m and n both turned out to be even did not produce a contradiction. However, we can modify the beginning of the argument to make this happen.

Remember that every rational number has infinitely many representations. For example, $\frac{6}{12}$ is the same rational number as $\frac{2}{4}$ (because $6 \cdot 4 = 12 \cdot 2$). Notice that in both representations, the numerator (number on the top) and the denominator (number on the bottom) are even. However, they are both equivalent to $\frac{1}{2}$, which has the property that the numerator is not even.

In Problem 16 below, you will be asked to show that every rational number can be written in the form $\frac{m}{n}$, where at least one of m or n is **not** even. We can now adjust our argument to get the desired contradiction.

Recall that an integer n is **even** if there is another integer b such that $n = 2b$. For example, 6 and -22 are even because $6 = 2 \cdot 3$ and $-22 = 2 \cdot (-11)$. An integer n is **odd** if there is another integer b such that $n = 2b + 1$. For example, 7 and -23 are odd because $7 = 2 \cdot 3 + 1$ and $-23 = 2 \cdot (-12) + 1$.

Lemma 6.11: Every integer is even or odd, but not both.

Proof: We already proved in Theorem 4.15 that every natural number is even or odd. If $n < 0$ is an integer, then $-n > 0$, and so, there is a natural number j such that $-n = 2j$ or $-n = 2j + 1$. If $-n = 2j$, then $n = 2(-j)$ (and since $j \in \mathbb{N}$, $-j \in \mathbb{Z}$). If $-n = 2j + 1$, then

$$n = -(2j + 1) = -2j - 1 = -2j - 1 - 1 + 1 \text{ (SACT)} = -2j - 2 + 1 = 2(-j - 1) + 1.$$

Here we used the fact that $(\mathbb{Z}, +, \cdot)$ is a ring. Since $\mathbb{Z}$ is closed under addition, it follows that $-j - 1 = -j + (-1) \in \mathbb{Z}$.

Now, if $n = 2j$ and $n = 2k + 1$, then $2j = 2k + 1$. So, we have

$$2(j - k) = 2j - 2k = (2k + 1) - 2k = 2k + (1 - 2k) = 2k + (-2k + 1)$$
$$= (2k - 2k) + 1 = 0 + 1 = 1.$$

So, $2(j - k) = 1$. But by part 4 of Example 6.3, 2 does not have a multiplicative inverse in $\mathbb{Z}$, and so, this is a contradiction. □

Lemma 6.12: The product of two odd integers is odd.

Proof: Let m and n be odd integers. Then there are integers j and k such that $m = 2j + 1$ and $n = 2k + 1$. So,

$$m \cdot n = (2j + 1) \cdot (2k + 1) = (2j + 1)(2k) + (2j + 1)(1) = (2k)(2j + 1) + (2j + 1)$$
$$= ((2k)(2j) + 2k) + (2j + 1) = (2(k(2j)) + 2k) + (2j + 1) = 2(k(2j) + k) + (2j + 1)$$
$$= (2(k(2j) + k) + 2j) + 1 = 2((k(2j) + k) + j) + 1.$$

Here we used the fact that $(\mathbb{Z}, +, \cdot)$ is a ring. (Which properties did we use?) Since $\mathbb{Z}$ is closed under addition and multiplication, we have $(k(2j) + k) + j \in \mathbb{Z}$. Therefore, mn is odd. □

Proof of Theorem 6.10: Assume, toward contradiction, that there is a rational number a such that $a^2 = 2$. Since a is a rational number, there are $m \in \mathbb{Z}$ and $n \in \mathbb{Z}^*$, **not both even**, so that $a = \frac{m}{n}$.

So, we have $\frac{m^2}{n^2} = \frac{m \cdot m}{n \cdot n} = \frac{m}{n} \cdot \frac{m}{n} = a \cdot a = a^2 = 2 = \frac{2}{1}$. Thus, $m^2 \cdot 1 = n^2 \cdot 2$. So, $m^2 = 2n^2$. Therefore, m^2 is even. If m were odd, then by Lemma 6.12, $m^2 = m \cdot m$ would be odd. So, **m is even**.

Since m is even, there is $k \in \mathbb{Z}$ such that $m = 2k$. Replacing m by $2k$ in the equation $m^2 = 2n^2$ gives us $2n^2 = m^2 = (2k)^2 = (2k)(2k) = 2(k(2k))$. So, $n^2 = k(2k)$ (we can use part (viii) of Problem 4 from Problem Set 4 to easily prove that if $a, b, c \in \mathbb{Z}$, then $ca = cb \rightarrow a = b$). Using associativity and commutativity of multiplication in $\mathbb{Z}$, we have $k(2k) = (k \cdot 2)k = (2k)k = 2(k \cdot k)$. So, $n^2 = 2(k \cdot k)$, and we see that n^2 is even. Again, by Lemma 6.12, **n is even**.

So, we have m even and n even, contrary to our original assumption that m and n are not both even. Therefore, there is no rational number a such that $a^2 = 2$. □

It turns out that $\mathbb{Q}$ fails to have an element a such that $a^2 = 2$ because $\mathbb{Q}$ is not "complete." Luckily $\mathbb{R}$ doesn't have this deficiency, as we will show shortly.

Completeness

Let F be an ordered field and let S be a nonempty subset of F. We say that S is **bounded above** if there is $M \in F$ such that for all $s \in S$, $s \leq M$. Each such number M is called an **upper bound** of S.

In words, an upper bound of a set S is simply an element from the field that is at least as big as every element in S.

Similarly, we say that S is **bounded below** if there is $K \in F$ such that for all $s \in S$, $K \leq s$. Each such number K is called a **lower bound** of S.

In words, a lower bound of a set S is simply an element from the field that is no bigger than any element in S.

We will say that S is **bounded** if it is both bounded above and bounded below. Otherwise S is **unbounded**.

A **least upper bound** of a set S is an upper bound that is smaller than any other upper bound of S, and a **greatest lower bound** of S is a lower bound that is larger than any other lower bound of S.

Example 6.13: Let F be an ordered field with $\mathbb{Q} \subseteq F$.

Note: The only two examples of F that we are interested in are $\mathbb{Q}$ (the set of rational numbers) and $\mathbb{R}$ (the set of real numbers). As you look at the set in each example below, think about what it looks like as a subset of $\mathbb{Q}$ and as a subset of $\mathbb{R}$.

1. $S = \{1, 2, 3, 4, 5\}$ is bounded.

    5 is an upper bound of S, as is any number larger than 5. The number 5 is special in the sense that there are no upper bounds smaller than it. So, 5 is the **least** upper bound of S.

    Similarly, 1 is a lower bound of S, as is any number smaller than 1. The number 1 is the **greatest** lower bound of S because there are no lower bounds larger than it.

    Notice that the least upper bound and greatest lower bound of S are inside the set S itself. This will always happen when the set S is finite.

2. $T = \{x \in F \mid -2 < x \leq 2\}$ is also bounded. Any number greater than or equal to 2 is an upper bound of T, and any number less than or equal to -2 is a lower bound of T.

    2 is the least upper bound of T and -2 is the greatest lower bound of T.

    Note that the least upper bound of T is in T, whereas the greatest lower bound of T is not in T.

3. $U = \{x \in F \mid x < -3\}$ is bounded above by any number greater than or equal to -3, and -3 is the least upper bound of U. The set U is not bounded below, and therefore, U is unbounded.

4. $V = \{x \in F \mid x^2 < 2\}$ is bounded above by 2. To see this, note that if $x > 2$, then $x^2 > 4 \geq 2$ (the reader should verify that for all $a, b \in F^+$, $a > b \rightarrow a^2 > b^2$), and therefore, $x \notin V$. Any number greater than 2 is also an upper bound. Is 2 the least upper bound of V? It's not! For example, $\frac{3}{2}$ is also an upper bound. Indeed, if $x > \frac{3}{2}$, then $x^2 > \frac{9}{4} \geq 2$.

Does V have a least upper bound? A moment's thought might lead you to suspect that a least upper bound M would satisfy $M^2 = 2$. And it turns out that you are right! (Proving this, however, is quite difficult—see Problem 22 below). Clearly, this least upper bound M is not in the set V. The big question is "Does M exist at all?"

Well, if $F = \mathbb{Q}$, then by Theorem 6.10, M **does not** exist in F. In this case, V is an example of a set which is bounded above in $\mathbb{Q}$, but has no least upper bound in $\mathbb{Q}$.

So, if we want an ordered field F containing $\mathbb{Q}$ where M does exist, we can insist that F has the property that any set which is bounded above in F has a least upper bound in F. It turns out that $\mathbb{R}$ is such an ordered field.

Many authors use the term **supremum** for "least upper bound" and **infimum** for "greatest lower bound," and they may write $\sup A$ and $\inf A$ for the supremum and infimum of a set A, respectively (if they exist).

In Example 6.13 above, we stated the least upper bound and greatest lower bound of the sets S, T, U, and V without proof. Intuitively, it seems reasonable that those numbers are correct. Let's do one of the examples carefully.

Theorem 6.14: Let $U = \{x \in F \mid x < -3\}$. Then $\sup U = -3$.

Analysis: We need to show that -3 is an upper bound of U, and that any number less than -3 is **not** an upper bound of U. That -3 is an upper bound of U follows immediately from the definition of U.

The harder part of the argument is showing that a number less than -3 is not an upper bound of U. However, conceptually it's not hard to see that this is true. If $a < -3$, we simply need to find some number x between a and -3. Here is a picture of the situation.

Notice that a can be very close to -3 and we don't know exactly what a is—we know only that it's less than -3. So, we need to be careful how we choose x. The most natural choice for x would be to go midway between a and -3. In other words, we can take the average of a and -3. So, we will let $x = \frac{1}{2}(a + (-3))$. Then we just need to verify that $a < x$ and that $x \in U$ (that is, $x < -3$).

Proof of Theorem 6.14: If $x \in U$, then $x < -3$ by definition, and so, -3 is an upper bound of U.

Suppose that $a < -3$ (or equivalently, $-a - 3 > 0$). We want to show that a is **not** an upper bound of U. To do this, we let $x = \frac{1}{2}(a - 3) = 2^{-1}(a + (-3))$. $x \in F$ because F is closed under addition and multiplication, and the multiplicative inverse property holds in F^*. We will show that $a < x < -3$.

$$x - a = \frac{1}{2}(a-3) - a = \frac{1}{2}(a-3) - \frac{1}{2}(2a) = \frac{1}{2}(a - 3 - 2a) = \frac{1}{2}(a - 2a - 3) = \frac{1}{2}(-a - 3).$$

Since $\frac{1}{2} > 0$ (by Theorem 6.8) and $-a - 3 > 0$, it follows that $x - a > 0$, and therefore, $x > a$.

$$-3-x = -3 - \frac{1}{2}(a-3) = \frac{1}{2}(-6) - \frac{1}{2}a + \frac{1}{2} \cdot 3 = \frac{1}{2}(-6 - a + 3) = \frac{1}{2}(-a - 3).$$

Again, since $\frac{1}{2} > 0$ and $-a - 3 > 0$, it follows that $-3 - x > 0$, and therefore, $x < -3$. Thus, $x \in U$.

So, we found an element $x \in U$ (because $x < -3$) with $a < x$. This shows that a is **not** an upper bound of U. It follows that $-3 = \sup U$. □

An ordered field F has the **Completeness Property** if every nonempty subset of F that is bounded above in F has a least upper bound in F. In this case, we say that F is a **complete ordered field**.

Our next goal is to provide an outline for a proof that the ordered field $\mathbb{R}$ is complete. In order to do this, we will need a few preliminary results.

Theorem 6.15: Every Cauchy sequence of rational numbers is bounded by a rational number.

Analysis: First note that any finite set A of rational numbers is bounded. If a and b are the least and greatest rational numbers in A, respectively, then for any $q \in A$, we have $a \leq q \leq b$. In this case, we can also write $|q| \leq M$, where $M = \max\{|a|, |b|\}$ (note that $|q| \leq M$ is equivalent to $-M \leq q \leq M$), and we may say that A is bounded by M.

Also, note that if A is bounded by M_1 and B is bounded by M_2, then $A \cup B$ is bounded by $\max\{M_1, M_2\}$.

By the previous remarks, we need only show that some "**tail**" of the Cauchy sequence of rational numbers is bounded (by a tail, we mean we can forget about finitely many terms of the sequence and just consider the infinitely many terms from some point on). By the definition of Cauchy sequence, we can choose a natural number K so that the distance between x_{K+1} and x_m is less than 1 for all $m > K$. That is, we find $K \in \mathbb{N}$ so that $\forall m > K(|x_m - x_{K+1}| < 1)$. Finally, we use the following little trick:

$$|x_m| = |(x_m - x_{K+1}) + x_{K+1}| \leq |x_m - x_{K+1}| + |x_{K+1}| < 1 + |x_{K+1}|$$

For the first equality we used the Standard Advanced Calculus Trick (SACT) from Note 7 following the proof of Theorem 4.11. We wanted $x_m - x_{K+1}$ to appear, so we made it appear. We then added x_{K+1} to correct the damage that we did.

We then used the Triangle Inequality, which was described in the second Note following Example 4.18.

Let's write out the details of the proof.

Proof: Let (x_n) be a Cauchy sequence. Then for every $k \in \mathbb{N}^+$, there is $K \in \mathbb{N}$ such that $m \geq n > K$ implies $|x_m - x_n| < \frac{1}{k}$. In particular, by letting $k = 1$, we see that there is $K \in \mathbb{N}$ such that $m \geq n > K$ implies $|x_m - x_n| < 1$. By the Triangle Inequality (and SACT), we have

$$|x_m| = |(x_m - x_n) + x_n| \leq |x_m - x_n| + |x_n|.$$

Therefore, $m \geq n > K$ implies $|x_m| \leq |x_m - x_n| + |x_n| < 1 + |x_n|$. Letting $n = K + 1$, we see that $m > K$ implies $|x_m| < 1 + |x_{K+1}|$.

Let $M = \max\{|x_0|, |x_1|, \ldots, |x_K|, 1 + |x_{K+1}|\}$. Then for all $m \in \mathbb{N}$, $|x_m| \leq M$. □

By Theorem 6.7, every ordered field F contains an isomorphic copy of $\mathbb{N}$. For example, $\mathbb{Q}$ contains $\left\{\frac{n}{1} \mid n \in \mathbb{N}\right\}$ and $\mathbb{R}$ contains $\left\{\left[\left(\frac{n}{1}\right)\right] \mid n \in \mathbb{N}\right\}$. To avoid using messy notation, we will usually abuse notation just a bit and use the name $\mathbb{N}$ for any isomorphic copy of $\mathbb{N}$. So, if F is an ordered field, we can write $\mathbb{N} \subseteq F$.

We say that an ordered field F has the **Archimedean Property** if $\mathbb{N}$ is unbounded in F. Symbolically, we write $\forall x \in F \, \exists n \in \mathbb{N}(n > x)$.

Theorem 6.16: $\mathbb{Q}$ and $\mathbb{R}$ both have the Archimedean Property.

Proof: We first show that $\mathbb{Q}$ has the Archimedean Property. Let $q \in \mathbb{Q}$. If $q \leq 0$, then since $0 < 1$, by the transitivity of $<$, we have $q < 1$. If $q > 0$, then there are $a, b \in \mathbb{Z}^+$ with $q = \frac{a}{b}$. We have

$$(a+1) - \frac{a}{b} = \frac{(a+1)}{1} + \left(\frac{-a}{b}\right) = \frac{(a+1)b + 1 \cdot (-a)}{1b} = \frac{ab + b - a}{b} = \frac{a(b-1) + b}{b}.$$

Since $b \in \mathbb{Z}^+$, $b - 1 \geq 0$, and so, $a(b-1) \geq 0$. Therefore, $a(b-1) + b \geq 0 + b > 0$. Thus, $(a+1) - \frac{a}{b} = \frac{a(b-1)+b}{b} > 0$. Therefore, we have $a + 1 > \frac{a}{b} = q$. Since $a, 1 \in \mathbb{N}$ and $\mathbb{N}$ is closed under addition, it follows that $a + 1 \in \mathbb{N}$. Since $q \in \mathbb{Q}$ was arbitrary, we have shown that $\mathbb{Q}$ has the Archimedean Property.

Next, we show that $\mathbb{R}$ has the Archimedean Property. Let $[(x_n)] \in \mathbb{R}$. We want to find a natural number $[(t)]$ such that $[(t)] > [(x_n)]$. Suppose toward contradiction, that for all $t \in \mathbb{N}$, $[(t)] \leq [(x_n)]$. Then for each $t \in \mathbb{N}$, there is K_t such that $n > K_t$ implies $t \leq x_n$. Since (x_n) is a Cauchy sequence, by Theorem 6.15, (x_n) is bounded by a rational number M. So, for all $n \in \mathbb{N}$, $x_n \leq M$. Let $t \in \mathbb{N}$ and choose K_t such that $n > K_t$ implies $t \leq x_n$. Since $x_n \leq M$, by the transitivity of $\leq$, we have $t \leq M$. Since $t \in \mathbb{N}$ was arbitrary, we see that $\mathbb{N}$ is bounded by the rational number M, contradicting that $\mathbb{Q}$ has the Archimedean Property. This contradiction proves that $\mathbb{R}$ has the Archimedean Property. $\square$

Theorem 6.17 (The Density Theorem): If $x, y \in \mathbb{R}$ with $x < y$, then there is $q \in \mathbb{Q}$ with $x < q < y$.

In other words, the Density Theorem says that between any two real numbers we can always find a rational number. We say that $\mathbb{Q}$ is **dense** in $\mathbb{R}$.

To help understand the proof, let's first run a simple simulation using a specific example. Let's let $x = \frac{16}{3}$ and $y = \frac{17}{3}$. We begin by subtracting to get $y - x = \frac{1}{3}$. This is the distance between x and y. We wish to find a natural number n such that $\frac{1}{n}$ is smaller than this distance. In other words, we want $\frac{1}{n} < \frac{1}{3}$, or equivalently, $n > 3$. So, we can let n be any natural number greater than 3, say $n = 4$. We now want to "shift" $\frac{1}{n} = \frac{1}{4}$ to the right to get a rational number between x and y. We can do this as follows. We multiply n times x to get $nx = 4 \cdot \frac{16}{3} = \frac{64}{3}$. We then let m be the **least** integer greater than nx. So, $m = \frac{66}{3} = 22$. Finally, we let $q = \frac{m}{n} = \frac{22}{4} = \frac{11}{2}$. And we did it! Indeed, we have $\frac{16}{3} < \frac{11}{2} < \frac{17}{3}$. The reader should confirm that these inequalities hold. Let's write out the details of the proof.

Proof: Let's first consider the case where $0 \leq x < y$. Let $z = y - x = y + (-x)$. Since $\mathbb{R}$ has the additive inverse property and is closed under addition, $z \in \mathbb{R}$. Also, $z > 0$. By the Archimedean Property, there is $n \in \mathbb{N}$ such that $n > \frac{1}{z}$. Using Problem 11 (part (i)) below, we have $\frac{1}{n} < z$.

By the Archimedean Property once again, there is $m \in \mathbb{N}$ such that $m > nx$. Therefore, $\frac{m}{n} > x$ (Check this!). So, $\left\{m \in \mathbb{N} \mid \frac{m}{n} > x\right\} \neq \emptyset$. By the Well Ordering Principle, $\left\{m \in \mathbb{N} \mid \frac{m}{n} > x\right\}$ has a least element, let's call it k. Since $k > 0$, (because $x \geq 0$ and $n > 0$) and k is the **least** natural number such that $\frac{k}{n} > x$, it follows that $k - 1 \in \mathbb{N}$ and $\frac{k-1}{n} \leq x$, or equivalently, $\frac{k}{n} - \frac{1}{n} \leq x$. Therefore, we have $\frac{k}{n} \leq x + \frac{1}{n} < x + z = x + (y - x) = y$. Thus, $x < \frac{k}{n} < y$. Since $k, n \in \mathbb{N}$, we have $\frac{k}{n} \in \mathbb{Q}$.

Now, we consider the case where $x < 0$ and $x < y$. By the Archimedean Property, there is $t \in \mathbb{N}$ such that $t > -x$. Then, we have $0 < x + t < y + t$. So, $x + t$ and $y + t$ satisfy the first case above. Thus, there is $q \in \mathbb{Q}$ with $x + t < q < y + t$. It follows that $x < q - t < y$. Since $t \in \mathbb{N}$, $-t \in \mathbb{Z}$. Since $\mathbb{Z} \subseteq \mathbb{Q}$, $-t \in \mathbb{Q}$. So, we have $q, -t \in \mathbb{Q}$. Since $\mathbb{Q}$ is closed under addition, $q - t = q + (-t) \in \mathbb{Q}$. □

Theorem 6.18: $\mathbb{R}$ is a complete ordered field.

The proof of this theorem requires several steps. I will give a brief outline of these steps and leave the details as an exercise for the reader (see Problem 23 below).

Proof outline for Theorem 6.18: We already know that $\mathbb{R}$ is an ordered field. All that's left to show is that $\mathbb{R}$ is complete. So, let $S \subseteq \mathbb{R}$ with $S \neq \emptyset$ be bounded above. Then there is $b_0 \in \mathbb{R}$ such that $\forall s \in S, s \leq b_0$ (b_0 is an upper bound of S). Since $S \neq \emptyset$, there is $a_0 \in S$. Let $c_0 = \frac{1}{2}(a_0 + b_0)$. If c_0 is an upper bound of S, let $a_1 = a_0$ and $b_1 = c_0$. If c_0 is **not** an upper bound of S, let $a_1 = c_0$ and $b_1 = b_0$. We continue in this fashion to inductively create sequences of real numbers (a_n), (b_n), and (c_n), where $c_n = \frac{1}{2}(a_n + b_n)$, if c_n is an upper bound of S, $a_{n+1} = a_n$ and $b_{n+1} = c_n$, and if c_n is **not** an upper bound of S, $a_{n+1} = c_n$ and $b_{n+1} = b_n$. Then let $c_{n+1} = \frac{1}{2}(a_n + b_n)$.

Step 1: Use induction on $\mathbb{N}$ to prove that for all $n \in \mathbb{N}$, $a_{n+1} \geq a_n$ and $b_{n+1} \leq b_n$.

Step 2: Use Step 1 to prove that (a_n) and (b_n) are **real-valued** Cauchy sequences.

Step 3: Use Theorems 6.16 and 6.17 to prove that there is $x \in \mathbb{R}$ such that the following holds:

for every $k \in \mathbb{N}^+$, there is $K \in \mathbb{N}$ such that $n > K$ implies $|b_n - x| < \frac{1}{k}$.

In this case, we say that (b_n) converges to x and we write $b_n \to x$.

Step 4: Use Theorem 6.16 to prove that $a_n \to x$.

Step 5: Use the sequence (b_n) to prove that x is an upper bound of S.

Step 6: Use the sequence (a_n) to prove that if $y < x$, then y is **not** an upper bound of S.

Completing these steps will prove the theorem. □

Problem Set 6

Full solutions to these problems are available for free download here:
www.SATPrepGet800.com/RABQXZ

LEVEL 1

1. Show that there are exactly two monoids on the set $S = \{e, a\}$, where e is the identity. Which of these monoids are groups? Which of these monoids are commutative?

2. The addition and multiplication tables below are defined on the set $S = \{0, 1\}$. Show that $(S, +, \cdot)$ does **not** define a ring.

+	0	1
0	0	1
1	1	0

$\cdot$	0	1
0	1	0
1	0	1

3. The addition and multiplication tables below are defined on the set $S = \{0, 1, 2\}$. Show that $(S, +, \cdot)$ does **not** define a field.

+	0	1	2
0	0	1	2
1	1	2	0
2	2	0	1

$\cdot$	0	1	2
0	0	0	0
1	0	1	2
2	0	2	2

4. Let $F = \{0, 1\}$, where $0 \neq 1$. Show that there is exactly one field $(F, +, \cdot)$, where 0 is the additive identity and 1 is the multiplicative identity.

LEVEL 2

5. Let $G = \{e, a, b\}$ and let $(G, \star)$ be a group with identity element e. Draw a multiplication table for $(G, \star)$.

6. Prove that in any monoid $(M, \star)$, the identity element is unique.

7. Let $(F, +, \cdot)$ be a field. Prove each of the following:

   (i) If $a, b \in F$ with $a + b = b$, then $a = 0$.

   (ii) If $a \in F$, $b \in F^*$, and $ab = b$, then $a = 1$.

   (iii) If $a \in F$, then $a \cdot 0 = 0$.

   (iv) If $a \in F^*$, $b \in F$, and $ab = 1$, then $b = \frac{1}{a}$.

   (v) If $a, b \in F$ and $ab = 0$, then $a = 0$ or $b = 0$.

   (vi) If $a \in F$, then $-a = -1a$.

   (vii) $(-1)(-1) = 1$.

8. Let $(F, +, \cdot)$ be a field with $\mathbb{N} \subseteq F$. Prove that $\mathbb{Q} \subseteq F$.

LEVEL 3

9. Assume that a group $(G, \star)$ of order 4 exists with $G = \{e, a, b, c\}$, where e is the identity, $a^2 = b$ and $b^2 = e$. Construct the multiplication table for the operation of such a group.

10. Prove that in any group $(G, \star)$, each element has a unique inverse.

11. Let $(F, +, \cdot, \leq)$ be an ordered field. Prove each of the following:
    (i) If $a, b \in F^+$ and $a > b$, then $\frac{1}{a} < \frac{1}{b}$.
    (ii) If $a, b \in F$, then $a \geq b$ if and only if $-a \leq -b$.
    (iii) If $a, b \in F^+$, then $a \leq b$ if and only if $a^2 \leq b^2$.

12. Let $(F, +, \cdot)$ be a field. Show that $(F, \cdot)$ is a commutative monoid.

LEVEL 4

13. Let $(G, \star)$ be a group with $a, b \in G$, and let a^{-1} and b^{-1} be the inverses of a and b, respectively. Prove
    (i) $(a \star b)^{-1} = b^{-1} \star a^{-1}$.
    (ii) the inverse of a^{-1} is a.

14. Prove that there is no smallest positive real number.

15. Let a be a nonnegative real number. Prove that $a = 0$ if and only if a is less than every positive real number. (Note: a nonnegative means $a \geq 0$.)

16. Prove that every rational number can be written in the form $\frac{m}{n}$, where $m \in \mathbb{Z}$, $n \in \mathbb{Z}^*$, and at least one of m or n is **not** even.

LEVEL 5

17. Prove that $(\mathbb{Q}, +, \cdot, \leq)$ and $(\mathbb{R}, +, \cdot, \leq)$ are ordered fields. Also prove that $(\mathbb{C}, +, \cdot)$ is field that cannot be ordered.

18. Prove that every nonempty set of real numbers that is bounded below has a greatest lower bound in $\mathbb{R}$.

19. Show that between any two real numbers there is a real number that is **not** rational.

20. Let $T = \{x \in F \mid -2 < x \leq 2\}$. Prove $\sup T = 2$ and $\inf T = -2$.

CHALLENGE PROBLEMS

21. Prove that every interval of real numbers is of one of the nine types described after Example 2.9.

22. Let $V = \{x \in F \mid x^2 < 2\}$ and let $a = \sup V$. Prove that $a^2 = 2$.

23. Prove that $\mathbb{R}$ is a complete ordered field.

24. Let $D = \{A, B\}$ be a partition of $\mathbb{Q}$ such that $A \neq \emptyset$, $A \neq \mathbb{Q}$, A has no greatest element, and every element of A is less than every element of B. D is called a **Dedekind cut**. Let $X = \{D \mid D \text{ is a Dedekind cut}\}$. Prove that $+$, $\cdot$, and $\leq$ can be defined on X so that $(X, +, \cdot, \leq)$ is a complete ordered field that is isomorphic to $\mathbb{R}$.

25. Prove that any two complete ordered fields are isomorphic.

LESSON 7
BASIC TOPOLOGY OF $\mathbb{R}$

Absolute Value and Distance

Recall from Lesson 3 that if x and y are real numbers such that $y = x^2$, then we call x a **square root** of y. If x is a positive real number, then we say that x is the **positive square root** of y and we write $x = \sqrt{y}$.

Example 7.1:

1. Since $2^2 = 4$, $2 \in \mathbb{R}$, and $2 > 0$, we see that 2 is the positive square root of 4 and we write $2 = \sqrt{4}$.

2. We have $(-2)^2 = 4$, but $-2 < 0$, and so we **do not** write $-2 = \sqrt{4}$. However, -2 is still a square root of 4, and we can write $-2 = -\sqrt{4}$.

The **absolute value** of the real number x is the nonnegative real number $|x| = \sqrt{x^2}$.

Note: $|x|$ is equal to x if $x \geq 0$ and $-x$ if $x < 0$.

For example, $|4| = \sqrt{4^2} = \sqrt{16} = 4$ and $|-4| = \sqrt{(-4)^2} = \sqrt{16} = 4 = -(-4)$.

The statement "$|x| = -x$ for $x < 0$" often confuses students. This confusion is understandable, as a minus sign is usually used to indicate that an expression is negative, whereas here we are negating a negative number to make it positive. Unfortunately, this is the simplest way to say, "delete the minus sign in front of the number" using basic notation.

Geometrically, the absolute value of a real number x is the distance between x and 0 (0 is sometimes called the **origin**) on the real line.

Example 7.2: Which of the following real numbers is closest to the origin, 3 or -2?

$$|3| = 3 \text{ and } |-2| = 2$$

Since $2 < 3$, we see that -2 is closest to the origin.

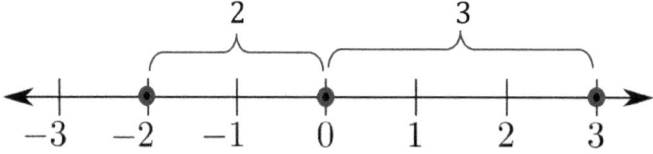

Recall that if a and b are real numbers, then we define subtraction as follows: $a - b = a + (-b)$. The expression $a - b$ is called a **difference**.

The **distance** between the real numbers x and y is

$$d(x, y) = x - y = |x - y|.$$

Example 7.3:
1. The distance between 2 and 5 is $|2 - 5| = |-3| = 3$.

    Observe that the distance between 5 and 2 is the same. Indeed, we have $|5 - 2| = |3| = 3$.
2. The distance between -4 and -1 is
$$|-4 - (-1)| = |-4 + 1| = |-3| = 3.$$
    Once again, we could also do this computation as $|-1 - (-4)| = |-1 + 4| = |3| = 3$.

Note: In general, for real numbers x and y, $|x - y| = |y - x|$. This follows immediately from the fact that $y - x = -(x - y)$.

Theorem 7.4 (The Triangle Inequality): For all $x, y \in \mathbb{R}$, $|x + y| \leq |x| + |y|$.

I leave the proof of the Triangle Inequality as an exercise (see Problem 9 below).

Example 7.5: Let's look at the sum $3 + (-2) = 1$. We have
$$|3 + (-2)| = |1| = 1 < 5 = 3 + 2 = |3| + |-2|.$$

If x and a are real numbers, then the equation $|x - a| = r$ is equivalent to the statement "$x - a = r$ or $x - a = -r$," or equivalently, "$x = a + r$ or $x = a - r$."

Example 7.6: The equation $|x - 1| = 2$ is equvalent to "$x = 1 + 2 = 3$ or $x = 1 - 2 = -1$". In other words, there are exactly two real numbers that are at a distance of 2 from 1. These two numbers are 3 and -1.

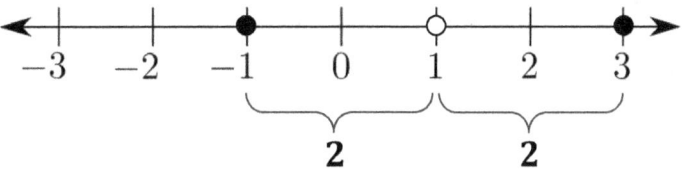

Neighborhoods

In real analysis, we will often need to work with inequalities of the form $|x - a| < r$. We call the set of real numbers x that satisfy $|x - a| < r$ the **r-neighborhood of a** and we write
$$N_r(a) = \{x \in \mathbb{R} \mid |x - a| < r\}.$$

Notes: (1) Neighborhoods are also sometimes referred to as **open disks** or **open balls**. The reason for this is because in higher dimensions neighborhoods can look like these objects.

(2) If x and a are real numbers, then we have
$$x \in N_r(a) \Leftrightarrow |x - a| < r \Leftrightarrow \sqrt{(x - a)^2} < r \Leftrightarrow 0 \leq (x - a)^2 < r^2$$
$$\Leftrightarrow -r < x - a < r \Leftrightarrow a - r < x < a + r \Leftrightarrow x \in (a - r, a + r).$$

So, in $\mathbb{R}$, the r-neighborhood of a is the **open interval** $N_r(a) = (a - r, a + r)$. Notice that the length (or **diameter**) of this interval is $2r$. Also, notice that a is at the center of this interval.

Example 7.7: Let's draw a picture of $N_2(1) = (1-2, 1+2) = (-1, 3)$. Observe that the center of this neighborhood (or open interval) in $\mathbb{R}$ is the real number 1, the radius of the neighborhood is 2, and the diameter of the neighborhood (or length of the interval) is 4.

We will also sometimes be interested in including the "boundary" of a neighborhood. In this case, we get the corresponding **closed interval** (or **closed disk** or **closed ball**). If a is the center of the closed interval and r is the radius of the closed interval, then any point x inside the closed interval satisfies $|x - a| \leq r$. For example, the closed interval with center 1 and radius 2 is $[1-2, 1+2] = [-1, 3]$

We will often want to remove the center of a neighborhood. The **deleted r-neighborhood of a** (or **punctured open interval with center a and radius r**) is the set

$$N_r^{\circ}(a) = \{x \in \mathbb{R} \mid 0 < |x - a| < r\}.$$

Note that $0 < |x - a|$ is equivalent to $0 \neq |x - a|$ because $|x - a|$ must be nonnegative. In turn, this is equivalent to $0 \neq x - a$, or better yet, $x \neq a$.

So, $N_r^{\circ}(a)$ consists of all the points in the interior of the open interval $(a - r, a + r)$ **except** for the center of the interval (namely, a).

If a is the center of the punctured open interval and r is the radius of the open interval, then any point x inside the punctured open interval satisfies $|x - a| < r$ and $x \neq a$. So,

$$N_r^{\circ}(a) = (a - r, a + r) \setminus \{a\} = (a - r, a) \cup (a, a + r).$$

This is the open interval centered at a of length (or diameter) $2r$ with a removed.

Example 7.8: Let's draw a picture of $N_2^{\circ}(1) = (-1, 3) \setminus \{1\} = (-1, 1) \cup (1, 3)$.

Open and Closed Sets in $\mathbb{R}$

A subset X of $\mathbb{R}$ is said to be **open** in $\mathbb{R}$ if for every real number $x \in X$, there is an open interval (a, b) with $x \in (a, b)$ and $(a, b) \subseteq X$.

In words, a set is open in $\mathbb{R}$ if every number in the set has "some space" on both sides of that number inside the set. If you think of each point in the set as an animal, then each animal in the set should be able to move a little to the left and a little to the right without ever leaving the set. Another way to think of this is that no number is on "the edge" or "the boundary" of the set, about to fall out of it.

Example 7.9:
1. Every bounded open interval is open in $\mathbb{R}$. To see this, let $X = (a, b)$ and let $x \in X$. Then $X = (a, b)$ itself is an open interval with $x \in (a, b)$ and $(a, b) \subseteq X$. For example, $(0, 1)$ and $\left(-\sqrt{2}, \frac{3}{5}\right)$ are open in $\mathbb{R}$.

2. We will prove in the theorems below that **all** open intervals are open in $\mathbb{R}$. For example, $(-2, \infty)$, $(-\infty, 5)$, and $(-\infty, \infty)$ are all open in $\mathbb{R}$.

3. $(0, 1]$ is **not** open in $\mathbb{R}$ because the "boundary point" 1 is included in the set. If (a, b) is any open interval containing 1, then $(a, b) \not\subseteq (0, 1]$ because there are numbers greater than 1 inside (a, b). For example, let $x = \frac{1}{2}(1 + b)$ (the average of 1 and b). Since $b > 1$, we have that $x > \frac{1}{2}(1 + 1) = \frac{1}{2} \cdot 2 = 1$. So, $x > 1$. Also, since $1 > a$, $x > a$. Now, since $1 < b$, we have that $x < \frac{1}{2}(b + b) = \frac{1}{2}(2b) = \left(\frac{1}{2} \cdot 2\right)b = 1b = b$. So, $x \in (a, b)$.

4. We can use reasoning similar to that used in 3 to see that all half-open intervals and closed intervals are **not** open in $\mathbb{R}$.

Theorem 7.10: Let $a \in \mathbb{R}$. The infinite interval (a, ∞) is open in $\mathbb{R}$.

The idea behind the proof is quite simple. If $x \in (a, \infty)$, then $(a, x + 1)$ is an open interval with x inside of it and with $(a, x + 1) \subseteq (a, \infty)$.

Proof of Theorem 7.10: Let $x \in (a, \infty)$ and let $b = x + 1$.

Since $x \in (a, \infty)$, $x > a$. Since $(x + 1) - x = 1 > 0$, we have $b = x + 1 > x$.

So, we have $a < x < b$. That is, $x \in (a, b)$. Also, $(a, b) \subseteq (a, \infty)$. Since $x \in (a, \infty)$ was arbitrary, (a, ∞) is open in $\mathbb{R}$. □

In Problem 8 below (part (i)), you will be asked to show that an interval of the form $(-\infty, b)$ is also open in $\mathbb{R}$.

Theorem 7.11: $\emptyset$ and $\mathbb{R}$ are both open in $\mathbb{R}$.

Proof: The statement that $\emptyset$ is open in $\mathbb{R}$ is vacuously true (since $\emptyset$ has no elements, there is nothing to check).

If $x \in \mathbb{R}$, then $x \in (x - 1, x + 1)$ and $(x - 1, x + 1) \subseteq \mathbb{R}$. Since x was an arbitrary element of $\mathbb{R}$, we have shown that for every $x \in \mathbb{R}$, there is an open interval (a, b) with $x \in (a, b)$ and $(a, b) \subseteq \mathbb{R}$. So, $\mathbb{R}$ is open in $\mathbb{R}$. □

Many authors define "open" in a slightly different way from the definition we've been using. This next theorem will show that the definition we have been using is equivalent to theirs.

Theorem 7.12: A subset X of $\mathbb{R}$ is open in $\mathbb{R}$ if and only if for every real number $x \in X$, there is a positive real number c such that $(x - c, x + c) \subseteq X$.

Analysis: The harder direction of the proof is showing that if X is open in $\mathbb{R}$, then for every real number $x \in X$, there is a positive real number c such that $(x - c, x + c) \subseteq X$.

To see this, suppose that X is open in $\mathbb{R}$ and let $x \in X$. Then there is an open interval (a, b) with $x \in (a, b)$ and $(a, b) \subseteq X$. We want to replace the interval (a, b) by an interval that has x right in the center.

The following picture should help us to come up with an argument.

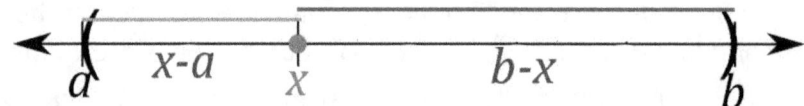

In the picture, we have an open interval (a, b), containing x. In this particular picture, x is a bit closer to a than it is to b. However, we should remember to be careful that our argument doesn't assume this (as we have no control over where x "sits" inside of (a, b)).

In the picture, we see that $x - a$ is the distance from a to x, and $b - x$ is the distance from x to b. Since the distance from a to x is smaller, let's let c be that smaller distance. In other words, we let $c = x - a$. From the picture, it looks like the interval $(x - c, x + c)$ will be inside the interval (a, b).

In general, if x is closer to a, we would let $c = x - a$, and if x is closer to b, we would let $c = b - x$. We can simply define c to be the smaller of $x - a$ and $b - x$. That is, $c = \min\{x - a, b - x\}$. From the picture, it seems like with this choice of c, the interval $(x - c, x + c)$ should give us what we want.

Proof of Theorem 7.12: Let X be an open subset of $\mathbb{R}$ and let $x \in X$. Then there is an open interval (a, b) with $x \in (a, b)$ and $(a, b) \subseteq X$. Let $c = \min\{x - a, b - x\}$. We claim that $(x - c, x + c)$ is an open interval containing x and contained in (a, b). We need to show $a \leq x - c < x < x + c \leq b$.

Since $c = \min\{x - a, b - x\}$, $c \leq x - a$. So, $-c \geq -(x - a)$. It follows that
$$(x - c) - a \geq (x - (x - a)) - a = (x - x + a) - a = a - a = 0.$$

So, $x - c \geq a$.

Since $c = \min\{x - a, b - x\}$, $c \leq b - x$. So, $-c \geq -(b - x)$. It follows that
$$b - (x + c) = b - x - c \geq b - x - (b - x) = 0.$$

So, $b \geq x + c$, or equivalently, $x + c \leq b$.

Note that $x > a$, so that $x - a > 0$, and $x < b$, so that $b - x > 0$. It follows that $c > 0$.

We have $x - (x - c) = c > 0$, so that $x > x - c$. We also have $(x + c) - x = c > 0$, so that $x + c > x$.

We have shown $a \leq x - c < x < x + c \leq b$, as desired.

Since $(x - c, x + c) \subseteq (a, b)$ and $(a, b) \subseteq X$, by the transitivity of $\subseteq$ (Theorem 1.14), we have $(x - c, x + c) \subseteq X$.

The converse is immediate since for $x \in X$, $(x - c, x + c)$ is an open interval containing x. □

Theorem 7.13: The union of two open sets in $\mathbb{R}$ is open in $\mathbb{R}$.

Proof: Let A and B be open sets in $\mathbb{R}$, and let $x \in A \cup B$. Then $x \in A$ or $x \in B$. **Without loss of generality**, we may assume that $x \in A$. Since A is open in $\mathbb{R}$, there is an interval (a, b) with $x \in (a, b)$ and $(a, b) \subseteq A$. By Theorem 1.26, $A \subseteq A \cup B$. Since $\subseteq$ is transitive (Theorem 1.14), $(a, b) \subseteq A \cup B$. Therefore, $A \cup B$ is open. □

Note: In the proof of Theorem 7.13, we used the expression "Without loss of generality." Recall from Note 1 following Example 3.3 that this expression can be used when an argument can be split up into two or more cases, and the proof of each of the cases is nearly identical.

For Theorem 7.13, the two cases are (i) $x \in A$ and (ii) $x \in B$. The argument for case (ii) is the same as the argument for case (i), essentially word for word—only the roles of A and B are interchanged.

Example 7.14: $(-5, 2)$ is open in $\mathbb{R}$ by part 1 of Example 7.9 and $(7, \infty)$ is open in $\mathbb{R}$ by Theorem 7.10. Therefore, by Theorem 7.13, $(-5, 2) \cup (7, \infty)$ is also open in $\mathbb{R}$.

If you look at the proof of Theorem 7.13 closely, you should notice that the proof would still work if we were taking a union of more than two sets. In fact, **any** union of open sets is open, as we now prove.

Theorem 7.15: Let X be a set of open subsets of $\mathbb{R}$. Then $\bigcup X$ is open in $\mathbb{R}$.

Proof: Let X be a set of open subsets of $\mathbb{R}$ and let $x \in \bigcup X$. Then $x \in A$ for some $A \in X$. Since A is open in $\mathbb{R}$, there is an interval (a, b) with $x \in (a, b)$ and $(a, b) \subseteq A$. By part 1 of Problem 26 from Problem Set 1, $A \subseteq \bigcup X$. Since $\subseteq$ is transitive (Theorem 1.14), $(a, b) \subseteq \bigcup X$. Therefore, $\bigcup X$ is open in $\mathbb{R}$. □

Example 7.16:
1. $(1,2) \cup (2,3) \cup (3,4) \cup (4, \infty)$ is open in $\mathbb{R}$.
2. $\mathbb{R} \setminus \mathbb{Z}$ is open because it is a union of open intervals. It looks like this:
$$\cdots (-2, -1) \cup (-1, 0) \cup (0, 1) \cup (1, 2) \cup \cdots$$
   $\mathbb{R} \setminus \mathbb{Z}$ can also be written as
$$\bigcup \{(n, n+1) \mid n \in \mathbb{Z}\} \quad \text{or} \quad \bigcup_{n \in \mathbb{Z}} (n, n+1)$$
3. If we take the union of all intervals of the form $\left(\frac{1}{n+1}, \frac{1}{n}\right)$ for positive integers n, we get an open set. We can visualize this open set as follows:
$$\bigcup \left\{\left(\tfrac{1}{n+1}, \tfrac{1}{n}\right) \mid n \in \mathbb{Z}^+\right\} = \cdots \cup \left(\tfrac{1}{5}, \tfrac{1}{4}\right) \cup \left(\tfrac{1}{4}, \tfrac{1}{3}\right) \cup \left(\tfrac{1}{3}, \tfrac{1}{2}\right) \cup \left(\tfrac{1}{2}, 1\right)$$

Theorem 7.17: Every nonempty open set in $\mathbb{R}$ can be expressed as a union of bounded open intervals.

The main idea of the argument will be the following. Every real number that is in an open set is inside an open interval that is a subset of the set. Just take the union of all these open intervals (one interval for each real number in the set).

Proof of Theorem 7.17: Let X be a nonempty open set in $\mathbb{R}$. Since X is open, for each $x \in X$, there is an interval (a_x, b_x) with $x \in (a_x, b_x)$ and $(a_x, b_x) \subseteq X$. We Let $Y = \{(a_x, b_x) \mid x \in X\}$. We will show that $X = \bigcup Y$.

First, let $x \in X$. Then $x \in (a_x, b_x)$. Since $(a_x, b_x) \in Y$, $x \in \bigcup Y$. Since x was arbitrary, $X \subseteq \bigcup Y$.

Now, let $x \in \bigcup Y$. Then there is $z \in X$ with $x \in (a_z, b_z)$. Since $(a_z, b_z) \subseteq X$, $x \in X$. Since $x \in X$ was arbitrary, $\bigcup Y \subseteq X$.

Since $X \subseteq \bigcup Y$ and $\bigcup Y \subseteq X$, it follows that $X = \bigcup Y$. □

We can actually get even more specific than Theorem 7.17.

Theorem 7.18: Every nonempty open set in $\mathbb{R}$ can be expressed as a countable union of pairwise disjoint open intervals.

The proof of Theorem 7.18 is more difficult than the proof of Theorem 7.17. We leave it as an exercise for the reader. See Problem 20 below.

Theorem 7.19: The intersection of two open sets in $\mathbb{R}$ is open in $\mathbb{R}$.

Proof: Let A and B be open sets in $\mathbb{R}$ and let $x \in A \cap B$. Then $x \in A$ and $x \in B$. Since A is open, there is an open interval (a, b) with $x \in (a, b)$ and $(a, b) \subseteq A$. Since B is open, there is an open interval (c, d) with $x \in (c, d)$ and $(c, d) \subseteq B$. Let $C = (a, b) \cap (c, d)$. Since $x \in (a, b)$ and $x \in (c, d)$, $x \in C$. By Problem 8 below (part (ii)), C is an open interval. By Problem 18 and part (ii) of Problem 12, both from Problem Set 1, $C \subseteq A$ and $C \subseteq B$. It follows that $C \subseteq A \cap B$ (Prove this!). Since $x \in A \cap B$ was arbitrary, $A \cap B$ is open in $\mathbb{R}$. □

In Problem 8 below (part (iii)), you will be asked to show that the intersection of **finitely** many open sets in $\mathbb{R}$ is open in $\mathbb{R}$. In problem 15 below, you will be asked to show that an **arbitrary** intersection of open sets in $\mathbb{R}$ does **not** need to be open in $\mathbb{R}$.

A subset X of $\mathbb{R}$ is said to be **closed** in $\mathbb{R}$ if $\mathbb{R} \setminus X$ is open in $\mathbb{R}$.

$\mathbb{R} \setminus X$ is called the **complement** of X in $\mathbb{R}$, or simply the complement of X. It consists of all real numbers **not** in X.

Example 7.20:

1. Every closed interval is closed in $\mathbb{R}$. For example, $[0, 1]$ is closed in $\mathbb{R}$ because its complement in $\mathbb{R}$ is $\mathbb{R} \setminus [0, 1] = (-\infty, 0) \cup (1, \infty)$. This is a union of open intervals, which is open in $\mathbb{R}$.

    Similarly, $[3, \infty)$ is closed in $\mathbb{R}$ because $\mathbb{R} \setminus [3, \infty) = (-\infty, 3)$, which is open in $\mathbb{R}$.

2. Half-open intervals are neither open nor closed in $\mathbb{R}$. For example, we saw in part 3 of Example 7.9 that $(0,1]$ is **not** open in $\mathbb{R}$. We see that $(0,1]$ is not closed in $\mathbb{R}$ by observing that $\mathbb{R} \setminus (0,1] = (-\infty, 0] \cup (1, \infty)$, which is not open in $\mathbb{R}$.

3. $\emptyset$ is closed in $\mathbb{R}$ because $\mathbb{R} \setminus \emptyset = \mathbb{R}$ is open in $\mathbb{R}$. Also, $\mathbb{R}$ is closed in $\mathbb{R}$ because $\mathbb{R} \setminus \mathbb{R} = \emptyset$ is open in $\mathbb{R}$. $\emptyset$ and $\mathbb{R}$ are the only two sets of real numbers that are both open and closed.

4. If $a \in \mathbb{R}$, then the set $\{a\}$ consisting of just one real number is closed in $\mathbb{R}$ because it is the complement of the open set $(-\infty, a) \cup (a, \infty)$.

5. By Theorem 7.18, every closed set in $\mathbb{R}$ can be expressed as $\mathbb{R} \setminus U$, where U is a countable union of pairwise disjoint open intervals. In other words, we can form a closed set by removing countably many pairwise disjoint open intervals from the real line. This procedure sounds innocent enough, but we can create some strange looking sets this way. For example, see part 2 of Example 7.22 below.

Theorem 7.21: The intersection of two closed sets in $\mathbb{R}$ is closed in $\mathbb{R}$.

Proof: Let A and B be closed in $\mathbb{R}$. Then $\mathbb{R} \setminus A$ and $\mathbb{R} \setminus B$ are open in $\mathbb{R}$. By Theorem 7.13 (or 7.15), $(\mathbb{R} \setminus A) \cup (\mathbb{R} \setminus B)$ is open in $\mathbb{R}$. Therefore, $\mathbb{R} \setminus [(\mathbb{R} \setminus A) \cup (\mathbb{R} \setminus B)]$ is closed in $\mathbb{R}$. So, it suffices to show that $A \cap B = \mathbb{R} \setminus [(\mathbb{R} \setminus A) \cup (\mathbb{R} \setminus B)]$. Well, $x \in A \cap B$ if and only if $x \in A$ and $x \in B$ if and only if $x \notin \mathbb{R} \setminus A$ and $x \notin \mathbb{R} \setminus B$ if and only if $x \notin (\mathbb{R} \setminus A) \cup (\mathbb{R} \setminus B)$ if and only if $x \in \mathbb{R} \setminus [(\mathbb{R} \setminus A) \cup (\mathbb{R} \setminus B)]$. So, $A \cap B = \mathbb{R} \setminus [(\mathbb{R} \setminus A) \cup (\mathbb{R} \setminus B)]$, completing the proof. □

A similar argument can be used to show that the union of two closed sets in $\mathbb{R}$ is a closed set in $\mathbb{R}$. This result can be extended to the union of finitely many closed sets in $\mathbb{R}$ with the help of Problem 8 below (part (iii)). The dedicated reader should prove this. In Problem 13 below, you will be asked to show that an arbitrary intersection of closed sets in $\mathbb{R}$ is closed in $\mathbb{R}$. In Problem 15 you will be asked to show that an arbitrary union of closed sets in $\mathbb{R}$ does **not** need to be closed.

Example 7.22:

1. Since the union of finitely many closed sets is closed and a set containing a single point is closed (by part 4 of Example 7.20), it follows that every finite set is closed.

2. The **Cantor set**, C, can be defined as follows: Let C_0 be the closed unit interval $[0,1]$. Remove the open interval $\left(\frac{1}{3}, \frac{2}{3}\right)$ from C_0 to form the set $C_1 = \left[0, \frac{1}{3}\right] \cup \left[\frac{2}{3}, 1\right]$. Note that we have removed the middle third from C_0. We now delete the middle thirds from each of $\left[0, \frac{1}{3}\right]$ and $\left[\frac{2}{3}, 1\right]$ to form the set $C_2 = \left[0, \frac{1}{9}\right] \cup \left[\frac{2}{9}, \frac{1}{3}\right] \cup \left[\frac{2}{3}, \frac{7}{9}\right] \cup \left[\frac{8}{9}, 1\right]$. We continue in this fashion, creating a sequence (C_n) of sets. Each C_n is a finite union of closed sets in $\mathbb{R}$, and therefore, each C_n is closed in $\mathbb{R}$. The Cantor set is defined to be $C = \cap \{C_n \mid n \in \mathbb{N}\}$. Since an arbitrary intersection of closed sets in $\mathbb{R}$ is closed in $\mathbb{R}$, the Cantor set is closed in $\mathbb{R}$.

    Which real numbers are in C? Well, C certainly contains the endpoints of each interval that make up each C_n. For example, $0, \frac{1}{3}, \frac{2}{3}, 1 \in C$ because these real numbers are endpoints of the intervals that make up C_1. Similarly, $\frac{1}{9}, \frac{2}{9}, \frac{7}{9}, \frac{8}{9} \in C$.

Are there any other points that are in C besides these endpoints? Well, there are only countably many such endpoints, and it turns out that C is uncountable. So, there are "many" other points in C. You will be asked to prove that C is uncountable in Problem 19 below.

We would now like to give an alternate definition of a closed set in $\mathbb{R}$ in terms of "accumulation points."

If $S \subseteq \mathbb{R}$, then $x \in \mathbb{R}$ is called an **accumulation point** of S if every open interval containing x contains at least one point of S different from x. Symbolically, we have that x is an accumulation point of S if and only if

$$\forall a, b \in \mathbb{R}(a < x < b \to \exists y \in S(a < y < b \land y \neq x)).$$

In words, an accumulation point of S is really, really close to the set S. It's so close to the set S that given any positive distance, we can find an element of S that is within that distance of the element. For example, suppose that 3 is an accumulation point of S and let's choose a distance of $r = 0.001$. Since 3 is an accumulation point, we can find a point $y \in S \cap (2.999, 3.001)$ that is different from 3. Notice that the distance between y and 3 is less than 0.001. That is, $|y - 3| < 0.001$.

Example 7.23:

1. Let $S = \left\{1, \frac{1}{2}, \frac{1}{3}, \dots\right\}$. Then 0 is an accumulation point of S. To see this, let (a, b) be an open interval containing 0. By the Archimedean Property of the real numbers (Theorem 6.16), there is $n \in \mathbb{N}$ such that $n > \frac{1}{b}$, or equivalently, $\frac{1}{n} < b$. So, $\frac{1}{n} \in S$, $\frac{1}{n} \in (a, b)$, and $\frac{1}{n} \neq 0$.

    It is not too hard to check that 0 is the **only** accumulation point of S. I leave it to the reader to show that S has no other accumulation points.

2. The half-open interval $(0, 1]$ has many accumulation points. In fact, the set of accumulation points of $(0, 1]$ is the closed interval $[0, 1]$. Notice that 0 is an accumulation point of $(0, 1]$ that is **not** in the set, whereas 1 is an accumulation point of $(0, 1]$ that is in the set.

3. The set of rational numbers, $\mathbb{Q}$, has every real number as an accumulation point. To see this, let $x \in \mathbb{R}$ and let (a, b) be an open interval containing x. By the Density Theorem (Theorem 6.17), there is a rational number q with $a < q < x$. So, $q \in \mathbb{Q}$, $q \in (a, b)$ and $q \neq x$.

4. The set of integers, $\mathbb{Z}$, has **no** accumulation points. Indeed, if $x \in \mathbb{R}$, let n be the least integer such that $x < n$. Then $n - 1 \leq x < n$. If $x = n - 1$, then $(x - 1, x + 1)$ is an interval containing x that contains no integers other than x. Otherwise, if we let $c = \min\{x - (n - 1), n - x\}$, then $c > 0$ and the interval $(x - c, x + c)$ is an interval containing x that contains no integers. See the analysis after Theorem 7.12 as well as the proof of Theorem 7.12 for details.

Theorem 7.24: A subset C of $\mathbb{R}$ is closed in $\mathbb{R}$ if and only if C contains each of its accumulation points.

Proof: First assume that C is closed in $\mathbb{R}$. Then $\mathbb{R} \setminus C$ is open in $\mathbb{R}$. Let x be any point **not** in C. Then $x \in \mathbb{R} \setminus C$. Since $\mathbb{R} \setminus C$ is open, there is an open interval (a, b) with $x \in (a, b)$ and $(a, b) \subseteq \mathbb{R} \setminus C$. So, if $y \in (a, b)$, then $y \notin C$. It follows that x is **not** an accumulation point of C. So, we have shown that $x \notin C$ implies that x is not an accumulation point of C. The contrapositive of this statement is "if x is an accumulation point of C, then $x \in C$." So, C contains each of its accumulation points.

Conversely, assume that C contains each of its accumulation points. We need to show that $\mathbb{R} \setminus C$ is open in $\mathbb{R}$. Since $\emptyset$ is open in $\mathbb{R}$, we can assume that $\mathbb{R} \setminus C \neq \emptyset$. Let $x \in \mathbb{R} \setminus C$. Then x is **not** an accumulation point of C. So, there is an open interval (a,b) containing x and no points of C. So, $(a,b) \subseteq \mathbb{R} \setminus C$. Since $x \in \mathbb{R} \setminus C$ was arbitrary, we have shown that $\mathbb{R} \setminus C$ is open in $\mathbb{R}$. □

If $S \subseteq \mathbb{R}$, then the **closure** of S in $\mathbb{R}$, written $\overline{S}$, is the intersection of all closed sets containing S. Symbolically, we have

$$\overline{S} = \bigcap \{C \mid S \subseteq C \wedge C \text{ is closed in } \mathbb{R}\}.$$

Since $\overline{S}$ is an intersection of closed sets in $\mathbb{R}$, $\overline{S}$ is closed in $\mathbb{R}$. We now prove a few basic facts about the closure of a set S.

Theorem 7.25: Let $S \subseteq \mathbb{R}$.

1. $S \subseteq \overline{S}$.
2. If C is closed in $\mathbb{R}$ with $S \subseteq C$, then $\overline{S} \subseteq C$.
3. $\overline{S} = S \cup \{x \in \mathbb{R} \mid x \text{ is an accumulation point of } S\}$.
4. S is closed in $\mathbb{R}$ if and only if $S = \overline{S}$.
5. $x \in \overline{S}$ if and only if every open interval containing x contains at least one point of S.

Proof:

1. Let $x \in S$. If $S \subseteq C$ with C closed, then $x \in C$. So, $x \in \bigcap\{C \mid S \subseteq C \wedge C \text{ is closed}\} = \overline{S}$. Since $x \in S$ was arbitrary, $\forall x (x \in S \to x \in \overline{S})$. Therefore, $S \subseteq \overline{S}$.

2. Let C be closed in $\mathbb{R}$ with $S \subseteq C$ and let $x \in \overline{S}$. By the definition of $\overline{S}$, $x \in C$. Since $x \in \overline{S}$ was arbitrary, $\forall x (x \in \overline{S} \to x \in C)$. Therefore, $\overline{S} \subseteq C$.

3. Let $C = S \cup \{x \in \mathbb{R} \mid x \text{ is an accumulation point of } S\}$. We first prove that C is closed in $\mathbb{R}$. Let x be an accumulation point of C and let (a,b) be an open interval containing x. Let $y \in C$ with $y \in (a,b)$ and $y \neq x$. Without loss of generality, assume that $x < y$. Since $y \in C$, $y \in S$ or y is an accumulation point of S. If y is an accumulation point of S, then there is $z \in S$ with $z \in (x, b)$. In either case, we have found an element of S in (a,b) different from x. So, x is an accumulation point of S, and therefore, $x \in C$. Since x was an arbitrary accumulation point of C, it follows that C contains each of its accumulation points. By Theorem 7.24, C is closed in $\mathbb{R}$.

   Since $S \subseteq C$ and C is closed in $\mathbb{R}$, it follows from 2 that $\overline{S} \subseteq C$.

   Now suppose $x \notin \overline{S}$. Since $S \subseteq \overline{S}$ (from 1), $x \notin S$. Since $x \notin \overline{S}$, x is in the open set $\mathbb{R} \setminus \overline{S}$. Therefore, there is an interval (a,b) with $x \in (a,b)$ and $(a,b) \subseteq \mathbb{R} \setminus \overline{S}$. Since $S \subseteq \overline{S}$, we have $\mathbb{R} \setminus \overline{S} \subseteq \mathbb{R} \setminus S$. By Theorem 1.14, $(a,b) \subseteq \mathbb{R} \setminus S$. So, x is **not** an accumulation point of S. Since $x \notin S$ and x is not an accumulation point of S, $x \notin C$. Since $x \notin \overline{S}$ was arbitrary, we have $\forall x (x \notin \overline{S} \to x \notin C)$. By the contrapositive, $\forall x (x \in C \to x \in \overline{S})$. So, $C \subseteq \overline{S}$.

   Since $\overline{S} \subseteq C$ and $C \subseteq \overline{S}$, we have $\overline{S} = C$

4. First assume that S is closed in $\mathbb{R}$. By 1 above, $S \subseteq \overline{S}$. Since S is closed and $S \subseteq S$, by 2 above, we have $\overline{S} \subseteq S$. So, $S = \overline{S}$.

   Conversely, assume that $S = \overline{S}$. Since $\overline{S}$ is closed in $\mathbb{R}$ and $S = \overline{S}$, S is closed in $\mathbb{R}$.

5. First let $x \in \overline{S}$ and let (a, b) be an open interval containing x. By 3, $x \in S$ or x is an accumulation point of S. If $x \in S$, then we have found a point in (a, b) that is in S, namely x. If x is an accumulation point of S, then there is $y \in (a, b)$ with $y \in S$ and $y \neq x$.

   Conversely, assume that every open interval containing x contains at least one point of S. If $x \in S$, then by 1, $x \in \overline{S}$. So, assume that $x \notin S$. Then every open interval containing x contains a point of S different from x. Thus, x is an accumulation point of S. By 3, $x \in \overline{S}$. □

Example 7.26:

1. Let $A = (a, b)$. Then $\overline{A} = [a, b]$.

2. Let C be the Cantor set from part 2 of Example 7.22. Since C is closed in $\mathbb{R}$, by part 4 of Theorem 7.25, we have $\overline{C} = C$.

3. Let $S = \left\{1, \frac{1}{2}, \frac{1}{3}, \ldots\right\}$. In part 1 of Example 7.23, we saw that the only accumulation point of S is 0. It follows from part 3 of Theorem 7.25 that $\overline{S} = S \cup \{0\}$.

4. $\overline{\mathbb{Q}} = \mathbb{R}$. This follows immediately from part 3 of Example 7.23 and part 3 of Theorem 7.25.

5. $\mathbb{Z}$ has no accumulation points by part 4 of Example 7.23. It follows from part 3 of Theorem 7.25 that $\overline{\mathbb{Z}} = \mathbb{Z} \cup \emptyset = \mathbb{Z}$.

Compactness

Let $S \subseteq \mathbb{R}$. A collection $\mathcal{C}$ of subsets of $\mathbb{R}$ is a **covering** of S (or we can say that $\mathcal{C}$ **covers** S) if $S \subseteq \bigcup \mathcal{C}$. If $\mathcal{C}$ consists of only open sets, then we will say that $\mathcal{C}$ is an **open covering** of S.

$S \subseteq \mathbb{R}$ is **compact** if every open covering of S contains a finite subcollection that still covers S (we will call such a finite subcollection a finite **subcover** of the covering).

Example 7.27:

1. If S is a finite subset of $\mathbb{R}$, then S is compact. Indeed, if $\mathcal{C}$ is an open covering of S, then for each $x \in S$, we can choose $U_x \in \mathcal{C}$ with $x \in U_x$. Then $\mathcal{D} = \{U_x \mid x \in S\}$ is a finite subcover of S.

2. $\mathbb{R}$ is **not** compact. Indeed, $\{(n, n+2) \mid n \in \mathbb{Z}\}$ is an open covering of $\mathbb{R}$ with no finite subcollection covering $\mathbb{R}$.

3. Let $A \subseteq \mathbb{R}$ be defined by $A = \left\{\frac{1}{n} \mid n \in \mathbb{Z}^+\right\} \cup \{0\}$. Then A is compact. To see this, let $\mathcal{C}$ be an open covering of A. Since $0 \in A$, there is $U \in \mathcal{C}$ with $0 \in U$. Then U contains an open interval (a, b) with $0 \in (a, b)$. By the Archimedean Property of the real numbers, there is $n \in \mathbb{N}$ such that $n > \frac{1}{b}$, or equivalently, $\frac{1}{n} < b$. So, $\left\{\frac{1}{k} \mid k \geq n\right\} \cup \{0\} \subseteq U$. For each $k \in \mathbb{Z}^+$ with $k < n$, let U_k be an open set in $\mathcal{C}$ such that $\frac{1}{k} \in U_k$. Then $\{U, U_1, U_2, \ldots, U_{k-1}\}$ is a finite subcover of $\mathcal{C}$.

Note that if we remove 0 from the set A to get the set $B = \left\{\frac{1}{n} \mid n \in \mathbb{Z}^+\right\}$, then we wind up with a subset of $\mathbb{R}$ that is **not** compact. Indeed, $\left\{\left(\frac{1}{n+1}, \frac{1}{n-1}\right) \mid n \in \mathbb{N} \setminus \{0, 1\}\right\} \cup \{(0, 2)\}$ is an open cover of B with no finite subcollection covering B.

4. The open interval $(0, 1)$ is **not** compact. The collection of open sets $\mathcal{C} = \left\{\left(\frac{1}{n}, 1\right) \mid n \in \mathbb{Z}^+\right\}$ is an open covering of $(0, 1)$ with no finite subcover. Furthermore, it is not hard to show that any open or half-open interval is not compact. See Problem 2 below.

The previous examples might be a bit misleading. Although those examples were quite simple, in general, it can be difficult to determine if a given subset of $\mathbb{R}$ is compact. For example, consider the closed interval $[0, 1]$. As it turns out, this set is compact. However, a direct proof using the definition of compactness is not so easy. Luckily, the compactness of $[0, 1]$ follows immediately from the Heine-Borel Theorem, which describes all compact subsets of $\mathbb{R}$.

Recall from Lesson 6 that $A \subseteq \mathbb{R}$ is **bounded** in $\mathbb{R}$ if there exist $a, b \in \mathbb{R}$ such that for all $x \in A$, $a \leq x \leq b$. If we let $M = \max\{|a|, |b|\}$, then we have $-M \leq x \leq M$, or equivalently, $|x| \leq M$. Conversely, if $-M \leq x \leq M$ for all $x \in A$, then clearly A is bounded.

Theorem 7.28 (Heine-Borel Theorem): $A \subseteq \mathbb{R}$ is compact if and only if A is closed and bounded in $\mathbb{R}$.

We will prove the Heine-Borel Theorem though a series of lemmas.

Lemma 7.29: Let $A \subseteq \mathbb{R}$ be compact. Then A is bounded in $\mathbb{R}$.

Proof: Let $A \subseteq \mathbb{R}$ be compact. To see that A is bounded in $\mathbb{R}$, consider the following open cover of A: $\mathcal{C} = \{(-k, k) \mid k \in \mathbb{Z}^+\}$. By the compactness of A, $\mathcal{C}$ has a finite subcover of A, let's say $\{(-k_0, k_0), (-k_1, k_1), \ldots, (-k_n, k_n)\}$. Let $N = \max\{k_0, k_1, \ldots k_n\}$. Then $A \subseteq (-N, N)$. So, for all $x \in A$, we have $|x| \leq N$. Therefore, A is bounded in $\mathbb{R}$. □

Lemma 7.30: Let $A \subseteq \mathbb{R}$ be compact. Then A is closed in $\mathbb{R}$.

Proof: Let $A \subseteq \mathbb{R}$ be compact. Suppose toward contradiction that A is **not** closed in $\mathbb{R}$. By part 4 of Theorem 7.25, $\overline{A} \setminus A \neq \emptyset$. Let $x \in \overline{A} \setminus A$. For each $k \in \mathbb{Z}^+$, let $U_k = \left(-\infty, x - \frac{1}{k}\right) \cup \left(x + \frac{1}{k}, \infty\right)$. Then $U_1 \subseteq U_2 \subseteq \cdots$. Using the Archimedean Property of the real numbers, we can see that $\bigcup\{U_k \mid k \in \mathbb{Z}^+\} = \mathbb{R} \setminus \{x\}$ (Prove this!). Since $x \notin A$, we have $A \subseteq \mathbb{R} \setminus \{x\} = \bigcup\{U_k \mid k \in \mathbb{Z}^+\}$. So, $\mathcal{C} = \{U_k \mid k \in \mathbb{Z}^+\}$ is an open cover of A. By compactness, $\mathcal{C}$ has a finite subcover of A, say $\{U_{k_0}, U_{k_1}, \ldots, U_{k_n}\}$. If we let $N = \max\{k_0, k_1, \ldots, k_n\}$, then $A \subseteq U_N = \left(-\infty, x - \frac{1}{N}\right) \cup \left(x + \frac{1}{N}, \infty\right)$. So, $\left(x - \frac{1}{N}, x + \frac{1}{N}\right) \cap A = \emptyset$. By part 5 of Theorem 7.25, $x \notin \overline{A}$, contrary to our assumption. □

Lemma 7.31: Let A be a nonempty bounded subset of $\mathbb{R}$. Then $\sup A$ and $\inf A$ are both in $\overline{A}$.

Proof: Let A be a nonempty bounded subset of $\mathbb{R}$. Since A is bounded, by the Completeness Property of $\mathbb{R}$ (see Lesson 6), $\sup A$ and $\inf A$ both exist. Suppose that $\inf A = a$ and $\sup A = b$. Then $A \subseteq [a, b]$. Let (c, d) be an open interval containing b. Then $(c, d) \cap A \neq \emptyset$ (otherwise, $c < b$ and c is an upper bound of A, contradicting that $b = \sup A$). By part 5 of Theorem 7.25, $b \in \overline{A}$. Similarly, if (e, f) is an open interval containing a, then $(e, f) \cap A \neq \emptyset$, and so, by part 5 of Theorem 7.25, $a \in \overline{A}$. □

Lemma 7.32: Let A be a closed and bounded subset of $\mathbb{R}$. Then A is compact.

Proof: Let A be a closed and bounded subset of $\mathbb{R}$ and let $\mathcal{C}$ be an open covering of A by open sets in $\mathbb{R}$. For each $x \in \mathbb{R}$, let $A_x = (-\infty, x] \cap A$ and let

$$Z = \{x \in \mathbb{R} \mid \text{there exists a finite subcover } \mathcal{D} \text{ of } \mathcal{C} \text{ such that } \mathcal{D} \text{ covers } A_x\}.$$

Since A is closed, by part 4 of Theorem 7.25, $A = \overline{A}$. By Lemma 7.31, $\sup A$ and $\inf A$ are both in $\overline{A} = A$. Let $a = \inf A$ and $b = \sup A$.

$A_a = (-\infty, a] \cap A = \{a\}$. Since $\{a\}$ can be covered by one element of $\mathcal{C}$, $a \in Z$. This shows that $Z \neq \emptyset$.

Suppose toward contradiction that Z is bounded above and let $c = \sup Z$. If $c \in A$, then there is $U \in \mathcal{C}$ such that $c \in U$ (because $\mathcal{C}$ covers A). Since U is open, there is an interval $[e, f] \subseteq U$ with $c \in (e, f)$. Since $e < c$ and $c = \sup Z$, there is a finite subcover $\mathcal{D}$ of $\mathcal{C}$ so that $\mathcal{D}$ covers A_e. But then $\mathcal{D} \cup \{U\}$ covers A_f, so that $f \in Z$. This contradicts $c = \sup Z$. So, $c \notin A$. Now, since A is closed, there is $r > 0$ such that $(c - r, c + r) \cap A = \emptyset$. But then $A_{c-r} = A_{c+r}$. Since $c - r \in Z$, we have $c + r \in Z$. This again contradicts $c = \sup Z$. It follows that Z is not bounded from above.

Let $g \in Z$ with $g > b$. Then $A_g = (-\infty, g] \cap A = A$. Therefore, there is a finite subcover of $\mathcal{C}$ that covers $A_g = A$. It follows that A is compact. □

Together, Lemmas 7.29 through 7.32 provide a proof of the Heine-Borel Theorem (Theorem 7.28).

Connectedness

Let $S \subseteq \mathbb{R}$. A **disconnection** of S is a pair (U, V), where U and V are nonempty open sets in $\mathbb{R}$ such that $U \cap S \neq \emptyset$, $V \cap S \neq \emptyset$, $(U \cap V) \cap S = \emptyset$ and $(U \cup V) \cap S = S$.

Note: Some authors use the expression "**separation**" instead of "**disconnection**."

$S \subseteq \mathbb{R}$ is **disconnected** if there exists a disconnection of S. Otherwise S is **connected**.

Example 7.33:

1. $\mathbb{R}$ is connected. This will follow immediately from Theorem 7.34 below. By the same theorem, every interval is connected.

2. $\mathbb{Q}$ is disconnected. Indeed, we have $(-\infty, \sqrt{2}) \cap \mathbb{Q} \neq \emptyset$, $(\sqrt{2}, \infty) \cap \mathbb{Q} \neq \emptyset$, $\left((-\infty, \sqrt{2}) \cap (\sqrt{2}, \infty)\right) \cap \mathbb{Q} = \emptyset$, and $\left((-\infty, \sqrt{2}) \cup (\sqrt{2}, \infty)\right) \cap \mathbb{Q} = \mathbb{Q}$.

So, $\big((-\infty, \sqrt{2}), (\sqrt{2}, \infty)\big)$ is a disconnection of $\mathbb{Q}$.

3. $S = [0, 1] \cup [9, 10]$ is disconnected. Indeed, we have $(-1, 2) \cap S \neq \emptyset$, $(8, 11) \cap S \neq \emptyset$, $\big((-1, 2) \cap (8, 11)\big) \cap S = \emptyset$ and $\big((-1, 2) \cup (8, 11)\big) \cap S = S$.

So, $\big((-1, 2), (8, 11)\big)$ is a disconnection of S.

We will now show that the connected subsets of the real line are precisely the intervals (see Lesson 2 for the nine types of intervals in $\mathbb{R}$). Let's begin by showing that every interval (including $\mathbb{R}$ itself) is connected.

Theorem 7.34: Let $I \subseteq \mathbb{R}$ be an interval. Then I is connected.

Proof: Let $I \subseteq \mathbb{R}$ be an interval and suppose toward contradiction that (U, V) is a disconnection of I. Since $U \cap I$ and $V \cap I$ are nonempty, we can choose $a \in U \cap I$ and $b \in V \cap I$. Since $(U \cap V) \cap I = \emptyset$, $a \neq b$. Without loss of generality, assume that $a < b$. Since I is an interval, $[a, b] \subseteq I$. Since $(U \cup V) \cap I = I$, each element of I is in either U or V. Let $c = \sup(U \cap [a, b])$. Since $a \leq c \leq b$, $c \in I$. Since V is open in $\mathbb{R}$, $\mathbb{R} \setminus V$ is closed in $\mathbb{R}$. Also, $U \cap [a, b] \subseteq U \cap I \subseteq \mathbb{R} \setminus V$. So, by Lemma 7.31, $c \in \mathbb{R} \setminus V$. Since $(U \cup V) \cap I = I$, $c \in U$. Now, $b \in V \cap I$ and $(U \cap V) \cap I = \emptyset$. Therefore, we must have $c < b$. For every $r \in \mathbb{R}^+$ with $c + r \leq b$, by the definition of c, we must have $c + r \in V \cap I$. Therefore, c is an accumulation point of $V \cap I$. Since U is open in $\mathbb{R}$, $\mathbb{R} \setminus U$ is closed in $\mathbb{R}$. Also, we have $V \cap I \subseteq \mathbb{R} \setminus U$. By Theorem 7.24, $c \in \mathbb{R} \setminus U$. Since $(U \cup V) \cap I = I$, $c \in V$. So, $c \in (U \cap V) \cap I$, contradicting $(U \cap V) \cap I = \emptyset$. Therefore, there does not exist a disconnection of I, and so, I is connected. $\square$

We now show that there are no other connected subsets of $\mathbb{R}$.

Theorem 7.35: Let $A \subseteq \mathbb{R}$ be connected. Then A is an interval.

Proof: Let $A \subseteq \mathbb{R}$ and assume that A is **not** an interval. Then there are $a, b, c \in \mathbb{R}$ with $a < c < b$, $a, b \in A$, and $c \notin A$. Let $U = (-\infty, c)$ and $V = (c, \infty)$. We will show that (U, V) is a disconnection of A. Since $a < c$, $a \in U$. So, $a \in U \cap A$. Since $c < b$, $b \in V$. So, $b \in V \cap A$. It follows that $U \cap A$ and $V \cap A$ are both nonempty. U and V are both open because they are open intervals. If $x \in (U \cap V) \cap A$, then $x < c$ and $c < x$, from which it follows that $x < x$. Since $<$ is a strict linear ordering on $\mathbb{R}$, this is impossible. So, $(U \cap V) \cap A = \emptyset$. Finally, let $x \in A$. If $x < c$, then $x \in U$. If $x > c$, then $x \in V$. So, $x \in U$ or $x \in V$. Therefore, $x \in U \cup V$, and so, $(U \cup V) \cap A = A$.

Since (U, V) is a disconnection of A, A is disconnected. $\square$

Problem Set 7

Full solutions to these problems are available for free download here:
www.SATPrepGet800.com/RABQXZ

LEVEL 1

1. Define a set of real numbers with exactly two accumulation points.

2. Let I be an open or half-open interval of real numbers. Prove that I is **not** compact.

3. Prove that a finite union of compact subsets of $\mathbb{R}$ is compact.

4. Prove that the Cantor set is a compact subset of $\mathbb{R}$.

LEVEL 2

5. Let S be a set of real numbers and let S' be the set of accumulation points of S. Prove that S' is closed in $\mathbb{R}$.

6. Determine the accumulation points of each of the following subsets of $\mathbb{R}$:

    (i) $\{(-1)^n \mid n \in \mathbb{Z}^+\}$

    (ii) $\{x \mid |x| < 1\}$

    (iii) $\{x \mid 0 < |x - 2| \leq 3\}$

7. Must the union of countably many compact subsets of $\mathbb{R}$ be compact? If so, provide a proof. If not, provide a counterexample.

LEVEL 3

8. Prove the following:

    (i) For all $b \in \mathbb{R}$, the infinite interval $(-\infty, b)$ is open in $\mathbb{R}$.

    (ii) The intersection of two open intervals in $\mathbb{R}$ is either empty or an open interval in $\mathbb{R}$.

    (iii) The intersection of finitely many open sets in $\mathbb{R}$ is an open set in $\mathbb{R}$.

9. Prove the Triangle Inequality (Theorem 7.4).

10. Let x and y be real numbers. Prove $||x| - |y|| \leq |x \pm y| \leq |x| + |y|$.

11. Let $x \in \mathbb{R}$ be an accumulation point of a set S. Prove that every open interval containing x contains infinitely many points of S.

12. Let X, Y, and Z be sets of real numbers with $Z = X \cup Y$. Prove that $\overline{Z} = \overline{X} \cup \overline{Y}$.

13. Prove that if X is a nonempty set of closed subsets of $\mathbb{R}$, then $\cap X$ is closed.

14. Prove that a closed subset of a compact set is compact.

LEVEL 4

15. Give an example of an infinite collection of open sets in $\mathbb{R}$ whose intersection is not open in $\mathbb{R}$. Also, give an example of an infinite collection of closed sets in $\mathbb{R}$ whose union is not closed in $\mathbb{R}$. Provide a proof for each example.

16. Determine if each of the following subsets of $\mathbb{R}$ is open, closed, both, or neither. Give a proof in each case.

    (i) $\{x \in \mathbb{R} \mid |x| > 1\}$

    (ii) $\{(-1)^n \mid n \in \mathbb{Z}^+\}$

    (iii) $\{x \in \mathbb{R} \mid 2 < |x - 2| < 4\}$

LEVEL 5

17. Prove that every closed set in $\mathbb{R}$ can be written as an intersection $\cap X$, where each element of X is a union of at most 2 closed intervals.

18. A real number x is an **interior point** of a set S of real numbers if there is a neighborhood of x that contains only points in S, whereas x is a **boundary point** of S if each neighborhood of x contains at least one point in S and one point not in S. Prove the following:

    (i)   A set of real numbers is open in $\mathbb{R}$ if and only if each point in the set is an interior point of the set.

    (ii) A set of real numbers is open in $\mathbb{R}$ if and only if it contains none of its boundary points.

    (iii) A set of real numbers is closed in $\mathbb{R}$ if and only if it contains all its boundary points.

19. Let C be the Cantor set. Prove the following:

    (i) $\frac{1}{4} \in C$.

    (ii) C is uncountable.

CHALLENGE PROBLEMS

20. Prove that every nonempty open set of real numbers can be expressed as a countable union of pairwise disjoint open intervals.

21. Let X and Y be closed subsets of $\mathbb{R}$ with $X \cap Y = \emptyset$. Prove that there are disjoint open sets of real numbers U and V such that $X \subseteq U$ and $Y \subseteq V$.

Lesson 8
Limits and Continuity

Strips and Rectangles in $\mathbb{R}$

A **horizontal strip** in $\mathbb{R} \times \mathbb{R}$ is a set of the form $\mathbb{R} \times (c, d) = \{(x, y) \mid c < y < d\}$.

Example 8.1: The horizontal strips $\mathbb{R} \times (-2, 1)$ and $\mathbb{R} \times (2.25, 2.75)$ can be visualized in the xy-plane as follows:

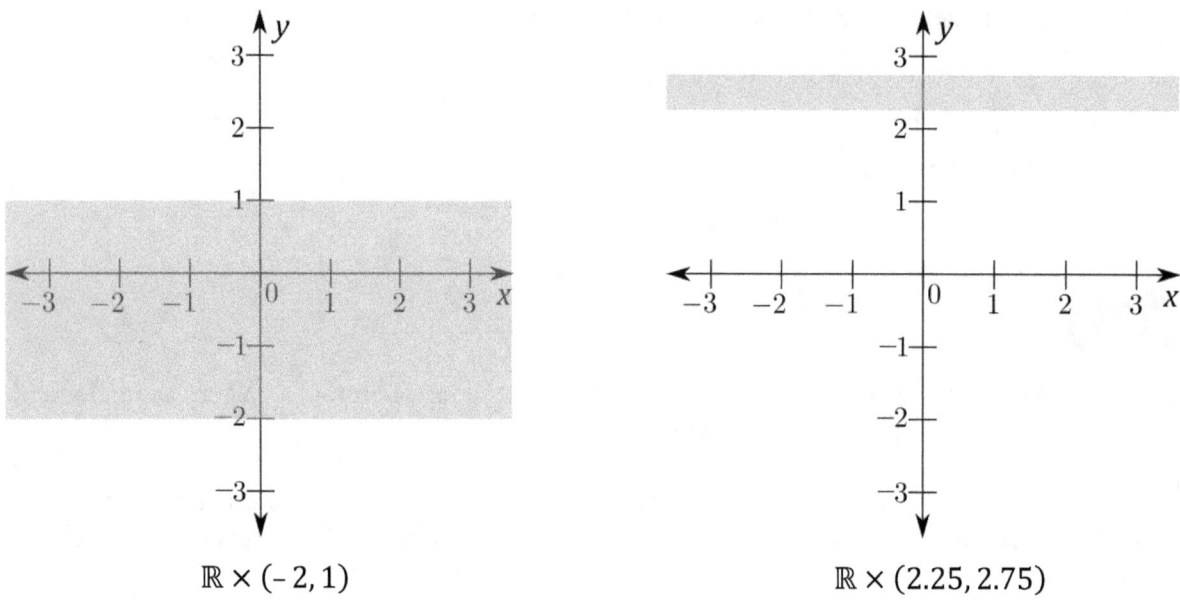

$\mathbb{R} \times (-2, 1)$ $\qquad\qquad\qquad\qquad\qquad$ $\mathbb{R} \times (2.25, 2.75)$

Similarly, a **vertical strip** in $\mathbb{R} \times \mathbb{R}$ is a set of the form $(a, b) \times \mathbb{R} = \{(x, y) \mid a < x < b\}$.

Example 8.2: The vertical strips $(-3, 0) \times \mathbb{R}$ and $(0.8, 1) \times \mathbb{R}$ can be visualized in the xy-plane as follows:

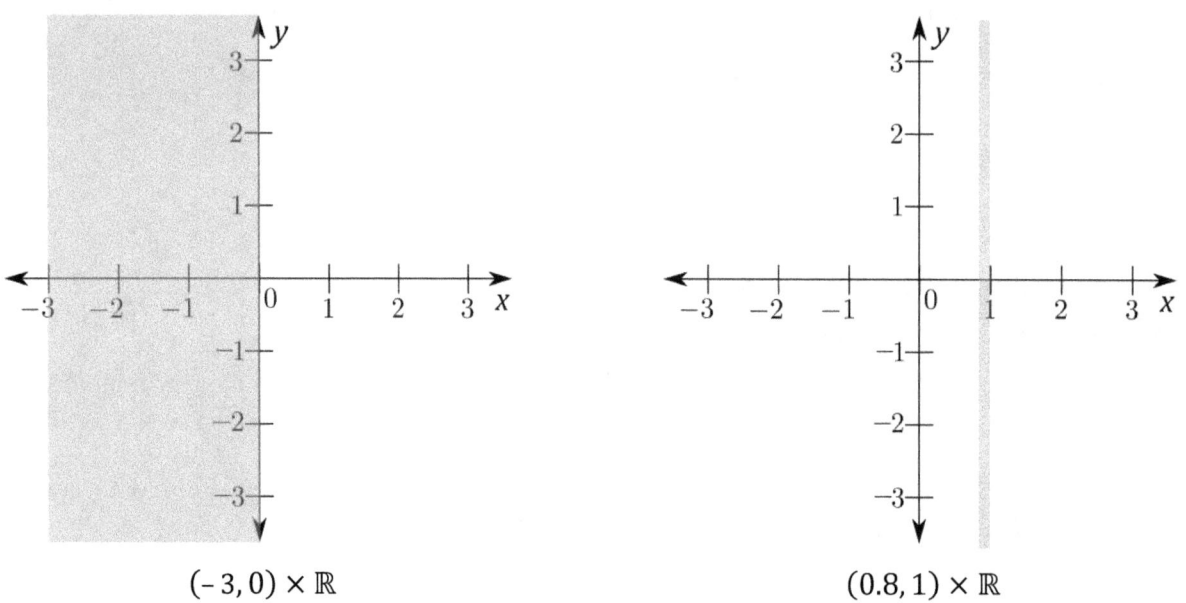

$(-3, 0) \times \mathbb{R}$ $\qquad\qquad\qquad\qquad\qquad$ $(0.8, 1) \times \mathbb{R}$

We will say that the horizontal strip $\mathbb{R} \times (c, d)$ **contains** $y \in \mathbb{R}$ if $c < y < d$. Otherwise, we will say that the horizontal strip **excludes** y.

Similarly, we will say that the vertical strip $(a, b) \times \mathbb{R}$ **contains** $x \in \mathbb{R}$ if $a < x < b$. Otherwise, we will say that the vertical strip **excludes** x.

Example 8.3: The horizontal strip $\mathbb{R} \times (2.25, 2.75)$ contains 2.5 and excludes 3. One way to visualize this is to draw the horizontal lines $y = 2.5$ and $y = 3$. Below in the figure on the left, we used a solid line for the line $y = 2.5$ because it is contained in the horizontal strip and we used a dashed line for the line $y = 3$ because it is not contained in the horizontal strip.

Similarly, the vertical strip $(0.8, 1) \times \mathbb{R}$ contains 0.9 and excludes 2. Again, we can visualize this by drawing the vertical lines $x = 0.9$ and $x = 2$. These vertical lines are shown below in the figure on the right.

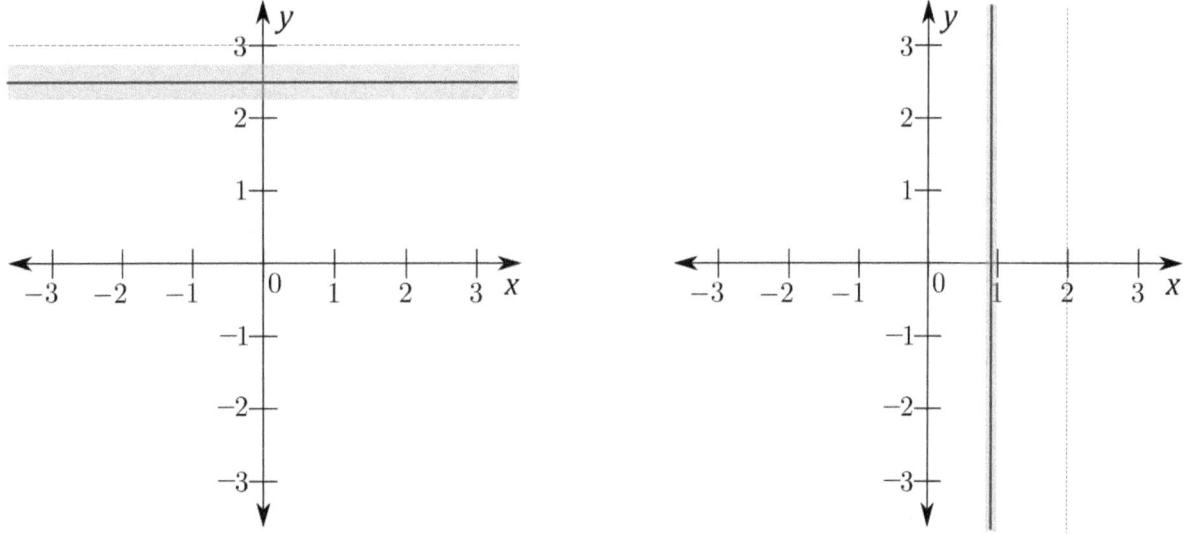

An **open rectangle** is a set of the form $(a, b) \times (c, d) = \{(x, y) \mid a < x < b \wedge c < y < d\}$. Note that the open rectangle $(a, b) \times (c, d)$ is the intersection of the horizontal strip $\mathbb{R} \times (c, d)$ and the vertical strip $(a, b) \times \mathbb{R}$. We will say that an open rectangle **traps** the point (x, y) if $x, y \in \mathbb{R}$ and (x, y) is in the open rectangle. Otherwise, we will say that (x, y) **escapes** from the open rectangle.

Example 8.4: The open rectangle $R = (-3, 0) \times (-2, 1)$ is the intersection of the horizontal strip $H = \mathbb{R} \times (-2, 1)$ and the vertical strip $V = (-3, 0) \times \mathbb{R}$. So, $R = H \cap V$. The rectangle R traps $(-1, 0)$, whereas $(-2, 3)$ escapes from R. This can be seen in the figure below on the left.

The open rectangle $R = (0.8, 1) \times (2.25, 2.75)$ is the intersection of the horizontal strip $H = \mathbb{R} \times (2.25, 2.75)$ and the vertical strip $V = (0.8, 1) \times \mathbb{R}$. So, $R = H \cap V$. The rectangle R traps $(0.9, 2.5)$, whereas $(0.9, 2)$ escapes from R. This can be seen in the figure below on the right.

Observe that in this example, I chose points that escape the given rectangles in the vertical direction. They fall outside the rectangle because they're too high or too low. This is the only type of escape that we will be interested in here. We do not care about points that escape to the left or right of a rectangle.

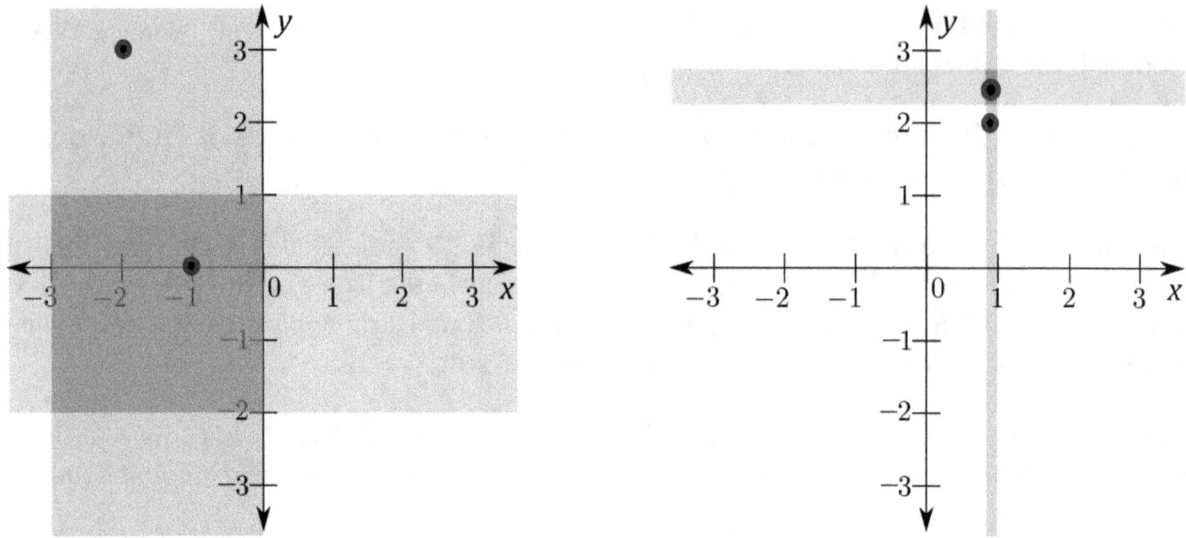

Let $A \subseteq \mathbb{R}$, let $f: A \to \mathbb{R}$, and let $R = (a,b) \times (c,d)$ be an open rectangle. We say that R **traps** f if for all $x \in (a,b)$, R traps $(x, f(x))$. Otherwise we say that f **escapes** from R.

Example 8.5: Let $f: \mathbb{R} \to \mathbb{R}$ be defined by $f(x) = x + 1$. Consider the open rectangles $R = (0,2) \times (1,3)$ and $S = (0,2) \times (0,2)$. Then R traps f, as can be seen in the figure below on the left, whereas f escapes from S, as can be seen in the figure below on the right. I put a box around the points of the form $(x, f(x))$ that escape from S. For example, the point $(1.2, f(1.2)) = (1.2, 2.2)$ escapes from S because $0 < 1.2 < 2$, but $f(1.2) = 2.2 \geq 2$.

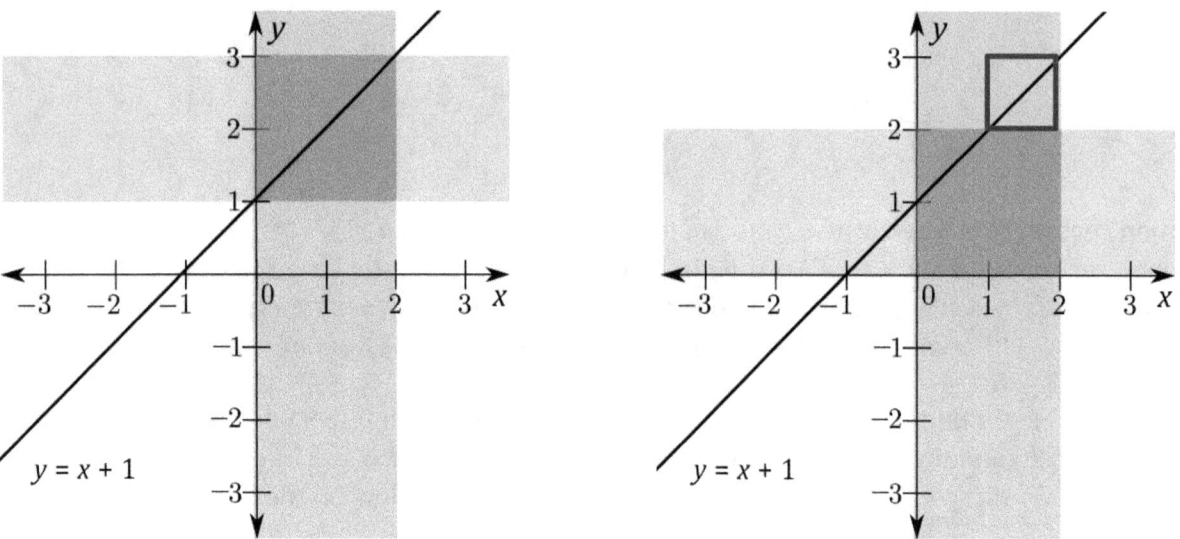

When we are checking the limiting behavior near a real number r, we don't care if the point $(r, f(r))$ escapes. Therefore, before we define a limit, we need to modify our definitions of "traps" and "escapes" slightly to account for this.

Let $A \subseteq \mathbb{R}$, let $f: A \to \mathbb{R}$, and let $R = (a,b) \times (c,d)$ be an open rectangle. We say that R **traps** f **around** r if for all $x \in (a,b) \setminus \{r\}$, R traps $(x, f(x))$. Otherwise, we say f **escapes from** R **around** r.

Limits and Continuity in $\mathbb{R}$

Let $A \subseteq \mathbb{R}$, let $f: A \to \mathbb{R}$, and let $r, L \in \mathbb{R}$. We say that the **limit of f as x approaches r is L**, written $\lim_{x \to r} f(x) = L$, if for every horizontal strip H that contains L there is a vertical strip V that contains r such that the rectangle $H \cap V$ traps f around r.

Technical note: According to the definition of limit just given, in order for $\lim_{x \to r} f(x)$ to exist, the set A needs to contain a deleted neighborhood of r, say $N_\epsilon^\odot(r) = (r - \epsilon, r) \cup (r, r + \epsilon)$. As an example, suppose that $A = \{0\}$ and $f: A \to \mathbb{R}$ is defined by $f(0) = 1$. What is the value of $\lim_{x \to 0} f(x)$? Well, any rectangle of the form $H \cap V$ does not trap any points of the form $(x, f(x))$ with $x \neq 0$ simply because $f(x)$ is not defined when $x \neq 0$. Therefore, given a horizontal strip H, there is no vertical strip V such that $H \cap V$ traps f around r, and so, $\lim_{x \to r} f(x)$ does not exist. This agrees with our intuition.

As a less extreme example, suppose that $A = \mathbb{Q}$ and $g: \mathbb{Q} \to \mathbb{R}$ is the constant function where $g(x) = 1$ for all $x \in \mathbb{Q}$. Then for any $r \in \mathbb{Q}$, we should probably have $\lim_{x \to r} g(x) = 1$. But if we use our current definition of limit, then $\lim_{x \to r} g(x)$ does not exist. A more general definition of limit would yield finite values for limits defined on certain sets (like $\mathbb{Q}$) that do not contain a neighborhood of r.

Specifically, we really should insist only that for each $j \in \mathbb{R}^+$, $((r - j, r + j) \cap A) \setminus \{r\} \neq \emptyset$. The definition of limit given above could be modified slightly to accommodate this more general situation. For example, we could replace "R traps f around r" by "R traps f around r **relative to A**," the meaning of which would be "for all $x \in ((a, b) \cap A) \setminus \{r\}$, R traps $(x, f(x))$." If we were to use this more general definition, it is very important that we also insist that the set A has the property given at the beginning of this paragraph. Otherwise, we would have an issue with the function f defined at the beginning of this note. The interested reader may want to investigate this.

In this lesson, we will avoid these more complicated domains and stick with the simpler definition of limit. To keep things simple, we will always assume that if $\lim_{x \to r} f(x)$ exists, then f is defined on some deleted neighborhood of r.

Example 8.6: Let $f: \mathbb{R} \to \mathbb{R}$ be defined by $f(x) = x + 1$, let $r = 1.5$, and let $L = 1$. Let's show that $\lim_{x \to 1.5} f(x) \neq 1$. If $H = \mathbb{R} \times (0, 2)$ and V is any vertical strip that contains 1.5, then $H \cap V$ does **not** trap f around 1.5. Indeed, if $V = (a, b) \times \mathbb{R}$, then if we let $x = \frac{1}{2}(1.5 + b)$, we will show that $x \in (a, b)$ and $f(x) = \frac{1}{2}(1.5 + b) + 1 > 2$ (see the figure to the right).

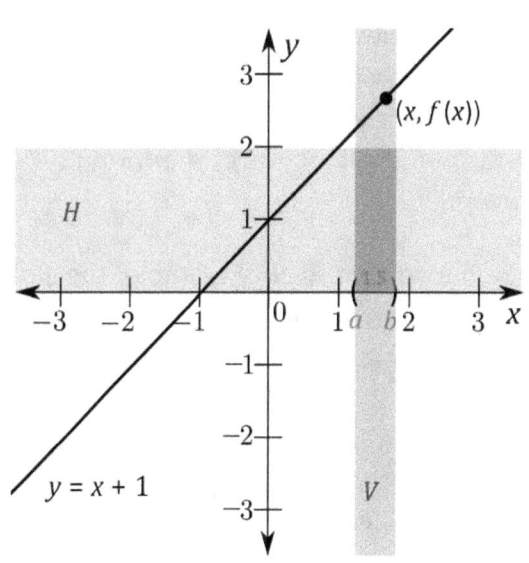

To see that $x \in (a, b)$, note that since $b > 1.5$, we have $x = \frac{1}{2}(1.5 + b) > \frac{1}{2}(1.5 + 1.5) = \frac{1}{2} \cdot 3 = 1.5 > a$, and we have $x = \frac{1}{2}(1.5 + b) < \frac{1}{2}(b + b) = \frac{1}{2} \cdot 2b = b$.

To see that $f(x) > 2$, note that

$$f(x) = \tfrac{1}{2}(1.5 + b) + 1 > \tfrac{1}{2}(1.5 + 1.5) + 1 = \tfrac{1}{2} \cdot 3 + 1 = 1.5 + 1 = 2.5 > 2.$$

So, what is $\lim_{x \to 1.5} f(x)$ equal to? From the picture above, a good guess would be 2.5. To verify that this is true, let $H = \mathbb{R} \times (c, d)$ be a horizontal strip that contains 2.5. Next, let $V = (c - 1, d - 1) \times \mathbb{R}$. We will show that $H \cap V = (c - 1, d - 1) \times (c, d)$ traps f around 1.5. Let $x \in (c - 1, d - 1) \setminus \{1.5\}$, so that $c - 1 < x < d - 1$ and $x \neq 1.5$. Adding 1 to each part of this sequence of inequalities gives $c < x + 1 < d$, so that $c < f(x) < d$, or equivalently, $f(x) \in (c, d)$. Since $x \in (c - 1, d - 1) \setminus \{1.5\}$ and $f(x) \in (c, d)$, it follows that $(x, f(x)) \in (c - 1, d - 1) \times (c, d) = H \cap V$. Therefore, $H \cap V$ traps f around 1.5.

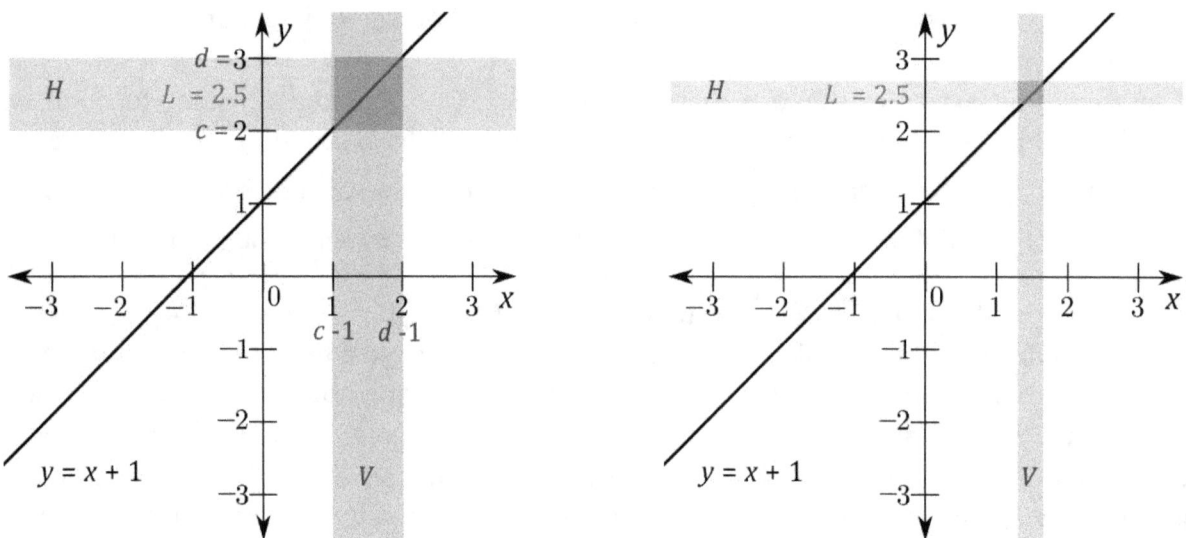

Notes: (1) The figures above give a visual representation of the argument just presented. In the figure on the left, we let $c = 2$ and $d = 3$, so that $H = \mathbb{R} \times (2, 3)$. Our choice of V is then $(1, 2) \times \mathbb{R}$, and therefore, $H \cap V = (1, 2) \times (2, 3)$. Now, if $1 < x < 2$, then $2 < x + 1 < 3$. So, $(x, f(x)) \in H \cap V$.

In the figure on the right, we started with a thinner horizontal strip without being specific about its exact definition. Notice that we then need to use a thinner vertical strip to prevent f from escaping. If the vertical strip were just a little wider on the right, then some points of the form $(x, f(x))$ would escape the rectangle because they would be too high. If the vertical strip were just a little wider on the left, then some points of the form $(x, f(x))$ would escape the rectangle because they would be too low.

(2) Notice that in this example, the point $(1.5, f(1.5))$ itself always stays in the rectangle. In the argument given above, we excluded this point from consideration. Even if $(1.5, f(1.5))$ were to escape the rectangle, it would not change the result here. We would still have $\lim_{x \to 1.5} f(x) = 2.5$. I indicated the parts of the argument where $(1.5, f(1.5))$ was being excluded from consideration in Example 8.6 above by placing rectangles around that part of the text. If we delete all the parts of the argument inside those rectangles, the resulting argument would still be correct. We will examine this situation more carefully in the next example.

Let $A \subseteq \mathbb{R}$, let $f: A \to \mathbb{R}$, and let $R = (a,b) \times (c,d)$ be an open rectangle. We say that R **traps** f **relative to** A if for all $x \in (a,b) \cap A$, R traps $(x, f(x))$.

A few small modifications to the definition of a limit leads to the definition of continuity. Specifically, we have the following:

Let $A \subseteq \mathbb{R}$, let $f: A \to \mathbb{R}$, and let $r \in A$. We say that the function f is **continuous** at r if for every horizontal strip H that contains $f(r)$, there is a vertical strip V that contains r such that the rectangle $H \cap V$ traps f relative to A.

Notes: (1) Quite often, the statements "$H \cap V$ traps f" and "$H \cap V$ traps f relative to A" lead to the same definition of continuity. For example, if A is an open set, then there is no difference. More generally, if A contains a neighborhood of r, then we can replace "$H \cap V$ traps f relative to A" by "$H \cap V$ traps f."

(2) There will be instances where the distinction is important. For example, define $f: [0,1] \to \mathbb{R}$ by $f(x) = x$. Is f continuous at 0?

It is as long as we use the correct definition. To see this, let $H = \mathbb{R} \times (c,d)$ be a horizontal strip containing 0 and let $V = (c,d) \times \mathbb{R}$. Then V contains 0 and $H \cap V = (c,d) \times (c,d)$ traps f relative to $[0,1]$. To see this, note that if $x \in (c,d) \cap [0,1]$, then $f(x) = x$, and so, $f(x) \in (c,d)$. Thus, $(x, f(x)) = (x,x) \in (c,d) \times (c,d) = H \cap V$.

On the other hand, there is no vertical strip V containing 0 such that $H \cap V$ traps f. Indeed, if $V = (a,b) \times \mathbb{R}$ contains 0, then $\frac{a}{2} \in (a,b)$ and since $\frac{a}{2} < 0$, $\frac{a}{2} \notin \text{dom } f$. So, R does **not** trap $\left(\frac{a}{2}, f\left(\frac{a}{2}\right)\right)$ simply because $f\left(\frac{a}{2}\right)$ does not exist.

(3) In this lesson, all the examples we look at will have functions whose domains are open sets. Therefore, we will **not** need to write "relative to A." However, in later lessons, we will sometimes be exploring the continuity of functions whose domains are not open sets. For these functions it will be important that we use the technically correct definition. Otherwise, we can run into the issue described in Note 2 above.

Example 8.7:

1. If we delete all the text that I placed in rectangles in Example 8.6 above, then the resulting argument shows that the function f defined by $f(x) = x + 1$ is continuous at $x = 1.5$.

   To summarize, given a horizontal strip H containing $f(1.5) = 2.5$, we found a vertical strip V containing 1.5 such that $H \cap V$ traps f. Notice once again that in this example we do not exclude $x = 1.5$ from consideration, and when we mention trapping f, we do not say "around 1.5." We need to trap $(1.5, f(1.5)) = (1.5, 2.5)$ as well.

2. Let's consider the function $g: \mathbb{R} \to \mathbb{R}$ defined by $g(x) = \begin{cases} x+1 & \text{if } x \neq 1.5 \\ -2 & \text{if } x = 1.5 \end{cases}$. This function is nearly identical to the function f we have been discussing. It differs from the previous function only at $x = 1.5$. It should follow that $\lim_{x \to 1.5} g(x) = \lim_{x \to 1.5} f(x)$. And, in fact it does. The same exact argument that we gave in Example 8.6 shows that $\lim_{x \to 1.5} g(x) = 2.5$. The figures below illustrate the situation.

This time however, we cannot delete the text inside the rectangles in Example 8.6. $x = 1.5$ needs to be excluded from consideration for the argument to go through. In the leftmost figure below, we see that if H is the horizontal strip $H = \mathbb{R} \times (2, 3)$, then for any vertical strip $V = (a, b) \times \mathbb{R}$ that contains 1.5, the point $(1.5, -2)$ will escape the rectangle $H \cap V$. Indeed, $H \cap V = (a, b) \times (2, 3)$, and $(1.5, g(1.5)) = (1.5, -2) \notin (a, b) \times (2, 3)$ because $-2 < 2$. This shows that g is **not** continuous at $x = 1.5$.

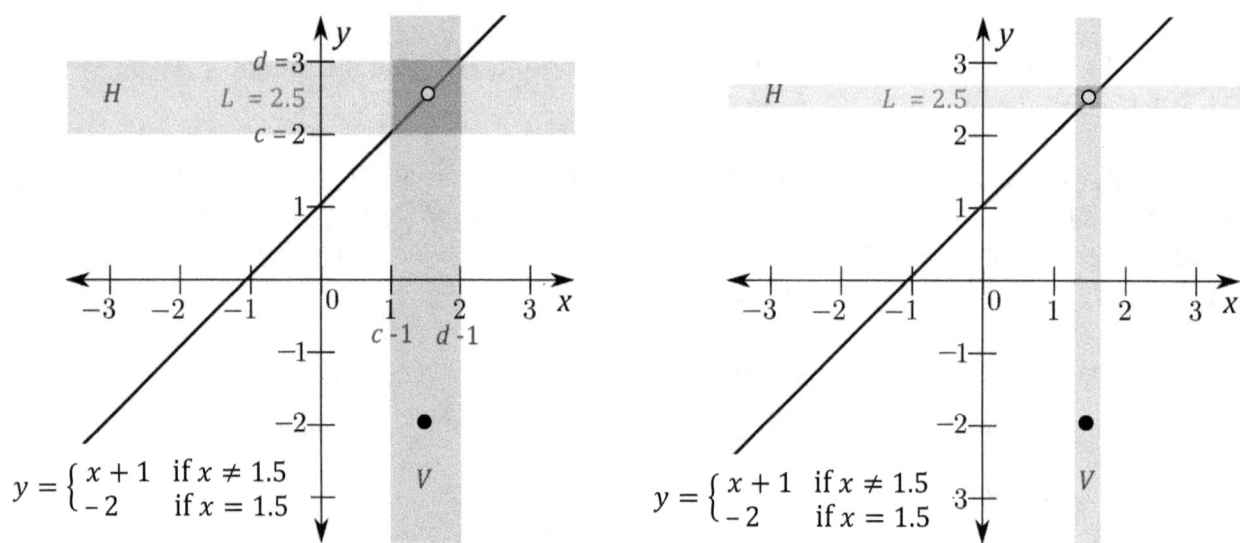

The strip game: Suppose we want to determine if $\lim_{x \to r} f(x) = L$. Consider the following game between two players: Player 1 "attacks" by choosing a horizontal strip H_0 containing L. Player 2 then tries to "defend" by choosing a vertical strip V_0 containing r such that $H_0 \cap V_0$ traps f around r. If Player 2 cannot find such a vertical strip, then Player 1 wins and $\lim_{x \to r} f(x) \neq L$. If Player 2 defends successfully, then Player 1 chooses a new horizontal strip H_1 containing L. If Player 1 is smart, then he/she will choose a "much thinner" horizontal strip that is contained in H_0 (compare the two figures above). The thinner the strip, the harder it will be for Player 2 to defend. Player 2 once again tries to choose a vertical strip V_1 such that $H_1 \cap V_1$ traps f around r. This process continues indefinitely. Player 1 wins the strip game if at some stage, Player 2 cannot defend successfully. Player 2 wins the strip game if he or she defends successfully at every stage.

Player 1 has a winning strategy for the strip game if and only if $\lim_{x \to r} f(x) \neq L$, while Player 2 has a winning strategy for the strip game if and only if $\lim_{x \to r} f(x) = L$.

Note that if it's possible for Player 1 to win the strip game, then Player 1 can win with a single move—just choose the horizontal strip immediately that Player 2 cannot defend against.

For example, if $f(x) = x + 1$, then $\lim_{x \to 1.5} f(x) \neq 1$. Player 1 can win the appropriate strip game immediately by choosing the horizontal strip $H = \mathbb{R} \times (0, 2)$. Indeed, if Player 2 chooses any vertical strip $V = (a, b) \times \mathbb{R}$ that contains 1.5, let $x \in (a, b)$ with $x > 1.5$. Then we have

$$f(x) = x + 1 > 1.5 + 1 = 2.5 > 2.$$

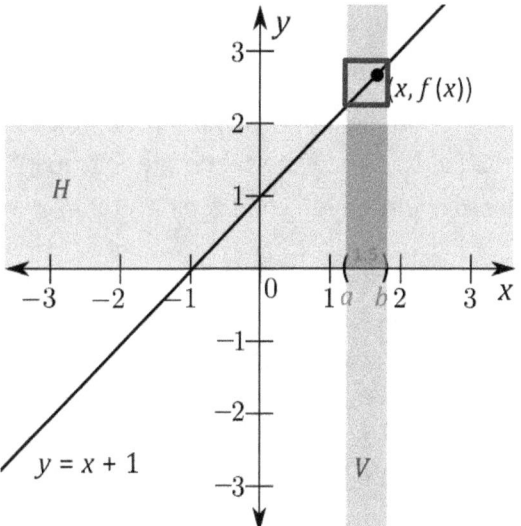

So, $(x, f(x))$ escapes $H \cap V$. In the figure to the right, we see that Player 1 has chosen $H = \mathbb{R} \times (0, 2)$ and Player 2 chose $V = (a, b) \times \mathbb{R}$ for some $a, b \in \mathbb{R}$ with $a < 1.5 < b$. The part of the line inside the square is an illustration of where f escapes $H \cap V$ between a and b. Observe that no matter how much thinner we try to make that vertical strip, if it contains 1.5, then it will contain a portion of the line that is inside the square.

Now, if it's possible for Player 2 to win the game, then we need to describe how Player 2 defends against an arbitrary attack from Player 1. Suppose again that $f(x) = x + 1$ and we are trying to show that $\lim_{x \to 1.5} f(x) = 2.5$. We have already seen how Player 2 can defend against an arbitrary attack from Player 1 in Example 8.6. If at stage n, Player 1 attacks with the horizontal strip $H_n = \mathbb{R} \times (a, b)$, then Player 2 can successfully defend with the vertical strip $V_n = (a - 1, b - 1) \times \mathbb{R}$.

Equivalent Definitions of Limits and Continuity in $\mathbb{R}$

The definitions of limit and continuity can be written using open intervals instead of strips. Specifically, we have the following:

Theorem 8.8: Let $A \subseteq \mathbb{R}$, let $f: A \to \mathbb{R}$, and let $r, L \in \mathbb{R}$. The following are equivalent:

1. $\lim_{x \to r} f(x) = L$.

2. For every open interval (c, d) with $L \in (c, d)$, there is an open interval (a, b) with $r \in (a, b)$ such that whenever $x \in (a, b)$ and $x \neq r$, $f(x) \in (c, d)$.

3. For every positive real number ϵ, there is a positive real number δ such that whenever $x \in (r - \delta, r + \delta)$ and $x \neq r$, $f(x) \in (L - \epsilon, L + \epsilon)$.

As in the proof of Theorem 2.14, we will prove that these three statements are equivalent with the following chain: $1 \to 2 \to 3 \to 1$. In other words, we will assume statement 1 and use it to prove statement 2. We will then assume statement 2 and use it to prove statement 3. Finally, we will assume statement 3 and use it to prove statement 1.

Proof of Theorem 8.8: (1→2) Suppose that $\lim_{x \to r} f(x) = L$ and let $L \in (c, d)$. Then the horizontal strip $\mathbb{R} \times (c, d)$ contains L. Since $\lim_{x \to r} f(x) = L$, there is a vertical strip $(a, b) \times \mathbb{R}$ that contains r such that the rectangle $R = (a, b) \times (c, d)$ traps f around r. Since the vertical strip $(a, b) \times \mathbb{R}$ contains r, $r \in (a, b)$. Since the rectangle R traps f around r, for all $x \in (a, b) \setminus \{r\}$, R traps $(x, f(x))$. In other words, whenever $x \in (a, b)$ and $x \neq r$, we have $(x, f(x)) \in (a, b) \times (c, d)$, and thus, $f(x) \in (c, d)$.

(2→3) Suppose 2 holds and let ϵ be a positive real number. Then $L - \epsilon < L < L + \epsilon$, or equivalently, $L \in (L - \epsilon, L + \epsilon)$. By 2, there is an open interval (a, b) with $r \in (a, b)$ such that whenever $x \in (a, b)$ and $x \neq r$, we have $f(x) \in (L - \epsilon, L + \epsilon)$. Let $\delta = \min\{r - a, b - r\}$. Since $\delta \leq r - a$, we have $-\delta \geq -(r - a) = -r + a$. Therefore, $r - \delta \geq r + (-r + a) = a$. Furthermore, since $\delta \leq b - r$, we have $r + \delta \leq r + (b - r) = b$. So, $(r - \delta, r + \delta) \subseteq (a, b)$. If $x \in (r - \delta, r + \delta)$ and $x \neq r$, then since $(r - \delta, r + \delta) \subseteq (a, b)$, $x \in (a, b)$. Therefore, $f(x) \in (L - \epsilon, L + \epsilon)$.

(3→1) Suppose 3 holds and $H = \mathbb{R} \times (c, d)$ is a horizontal strip that contains L. Since $c < L < d$, we have $L - c > 0$ and $d - L > 0$. Therefore, $\epsilon = \min\{L - c, d - L\} > 0$. So, there is $\delta > 0$ such that whenever $x \in (r - \delta, r + \delta)$ and $x \neq r$, then $f(x) \in (L - \epsilon, L + \epsilon)$. Let $V = (r - \delta, r + \delta) \times \mathbb{R}$. Then V contains r. We now show that $H \cap V = (r - \delta, r + \delta) \times (c, d)$ traps f around r. Let $x \in (r - \delta, r + \delta)$ with $x \neq r$. Then $f(x) \in (L - \epsilon, L + \epsilon)$. So, $f(x) > L - \epsilon \geq L - (L - c) = c$ and $f(x) < L + \epsilon \leq L + (d - L) = d$. Therefore, $f(x) \in (c, d)$, and so, $H \cap V$ traps f around r. □

Notes: (1) ϵ and δ are Greek letters pronounced "epsilon" and "delta," respectively. Mathematicians tend to use these two symbols to represent arbitrarily small numbers.

(2) Recall from lesson 7 that if $a \in \mathbb{R}$ and $\epsilon > 0$, then the ϵ-neighborhood of a is the interval $N_\epsilon(a) = (a - \epsilon, a + \epsilon)$ and the deleted ϵ-neighborhood of a is the "punctured" interval $N_\epsilon^\odot(a) = (a - \epsilon, a) \cup (a, a + \epsilon)$. We can visualize the deleted ϵ-neighborhood $N_\epsilon^\odot(a)$ as follows:

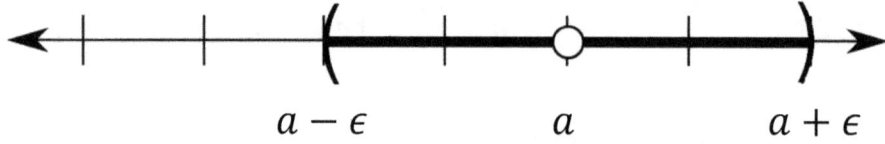

For a specific example, let's look at $N_2^\odot(1) = (1 - 2, 1) \cup (1, 1 + 2) = (-1, 1) \cup (1, 3)$.

(3) The third part of Theorem 8.8 can be written in terms of neighborhoods as follows:

"For every positive real number ϵ, there is a positive real number δ such that whenever $x \in N_\delta^\odot(r)$, $f(x) \in N_\epsilon(L)$."

(4) $x \in (a - \epsilon, a + \epsilon)$ is equivalent to $a - \epsilon < x < a + \epsilon$. If we subtract a from each part of this inequality, we get $-\epsilon < x - a < \epsilon$. This last expression is equivalent to $|x - a| < \epsilon$. So, we have the following sequence of equivalences:

$$x \in N_\epsilon(a) \Leftrightarrow x \in (a - \epsilon, a + \epsilon) \Leftrightarrow a - \epsilon < x < a + \epsilon \Leftrightarrow |x - a| < \epsilon.$$

(5) $x \neq a$ is equivalent to $x - a \neq 0$. Since the absolute value of a real number can never be negative, $x - a \neq 0$ is equivalent to $|x - a| > 0$. This can also be written $0 < |x - a|$. So, we have the following sequence of equivalences:

$$x \in N_\epsilon^{\circ}(a) \Leftrightarrow x \in (a - \epsilon, a) \cup (a, a + \epsilon) \Leftrightarrow 0 < |x - a| < \epsilon.$$

(6) The third part of Theorem 8.8 can be written using absolute values as follows:

"For every positive real number ϵ, there is a positive real number δ such that whenever $0 < |x - r| < \delta$, $|f(x) - L| < \epsilon$."

(7) We can abbreviate the expression from Note 6 using quantifiers as follows:

$$\forall \epsilon > 0 \, \exists \delta > 0 \, (0 < |x - r| < \delta \rightarrow |f(x) - L| < \epsilon)$$

We will refer to this expression as the $\epsilon - \delta$ **definition of a limit**.

For each equivalent formulation of a limit, we have a corresponding formulation for the definition of continuity.

Theorem 8.9: Let $A \subseteq \mathbb{R}$, let $f: A \rightarrow \mathbb{R}$, and let $r \in A$. The following are equivalent:

1. f is continuous at r.
2. For every open interval (c, d) with $f(r) \in (c, d)$, there is an open interval (a, b) with $r \in (a, b)$ such that whenever $x \in (a, b) \cap A$, $f(x) \in (c, d)$.
3. For every positive real number ϵ, there is a positive real number δ such that whenever $x \in (r - \delta, r + \delta) \cap A$, $f(x) \in (f(r) - \epsilon, f(r) + \epsilon)$.
4. $\forall \epsilon > 0 \, \exists \delta > 0 \, \forall x \in A \, (|x - r| < \delta \rightarrow |f(x) - f(r)| < \epsilon)$.

The proof of Theorem 8.9 is left to the reader. It is very similar to the proof of Theorem 8.8.

Basic Examples in $\mathbb{R}$

Example 8.10: Let's use the $\epsilon - \delta$ definition of a limit to prove that $\lim_{x \to 1}(2x + 1) = 3$.

Analysis: Given $\epsilon > 0$, we need to find $\delta > 0$ so that $0 < |x - 1| < \delta$ implies $|(2x + 1) - 3| < \epsilon$. First note that $|(2x + 1) - 3| = |2x - 2| = |2(x - 1)| = |2||x - 1| = 2|x - 1|$. So, $|(2x + 1) - 3| < \epsilon$ is equivalent to $|x - 1| < \frac{\epsilon}{2}$. Therefore, $\delta = \frac{\epsilon}{2}$ should work.

Proof: Let $\epsilon > 0$ and let $\delta = \frac{\epsilon}{2}$. Suppose that $0 < |x - 1| < \delta$. Then we have

$$|(2x + 1) - 3| = |2x - 2| = |2(x - 1)| = |2||x - 1| = 2|x - 1| < 2\delta = 2 \cdot \frac{\epsilon}{2} = \epsilon.$$

Since $\epsilon > 0$ was arbitrary, we have $\forall \epsilon > 0 \, \exists \delta > 0 \, (0 < |x - 1| < \delta \to |(2x + 1) - 3| < \epsilon)$.

Therefore, $\lim_{x \to 1}(2x + 1) = 3$. □

Notes: (1) Even though we're using the "$\epsilon - \delta$ definition" instead of the "strip definition," we can still visualize the situation in terms of the strip game. When we say "Let $\epsilon > 0$," we can think of this as Player 1 "attacking" with the horizontal strip $H = \mathbb{R} \times (3 - \epsilon, 3 + \epsilon)$. In the proof above, Player 2 is then "defending" with the vertical strip $V = \left(1 - \frac{\epsilon}{2}, 1 + \frac{\epsilon}{2}\right) \times \mathbb{R}$. This defense is successful because when $1 - \frac{\epsilon}{2} < x < 1 + \frac{\epsilon}{2}$, we have $2 - \epsilon < 2x < 2 + \epsilon$, and so, $3 - \epsilon < 2x + 1 < 3 + \epsilon$, or equivalently, $2x + 1 \in (3 - \epsilon, 3 + \epsilon)$. In other words, for $x \in \left(1 - \frac{\epsilon}{2}, 1 + \frac{\epsilon}{2}\right)$, $H \cap V$ traps f.

(2) Instead of playing the strip game, we can play the $\epsilon - \delta$ game instead. The idea is the same. Suppose we are trying to figure out if $\lim_{x \to r} f(x) = L$. Player 1 "attacks" by choosing a positive number ϵ. This is equivalent to Player 1 choosing the horizontal strip $H = \mathbb{R} \times (L - \epsilon, L + \epsilon)$. Player 2 then tries to "defend" by finding a positive number δ. This is equivalent to Player 2 choosing the vertical strip $V = (r - \delta, r + \delta) \times \mathbb{R}$. The defense is successful if whenever $x \in (r - \delta, r + \delta)$, $x \neq r$, we have $f(x) \in (L - \epsilon, L + \epsilon)$. This is equivalent to $H \cap V$ trapping f around r.

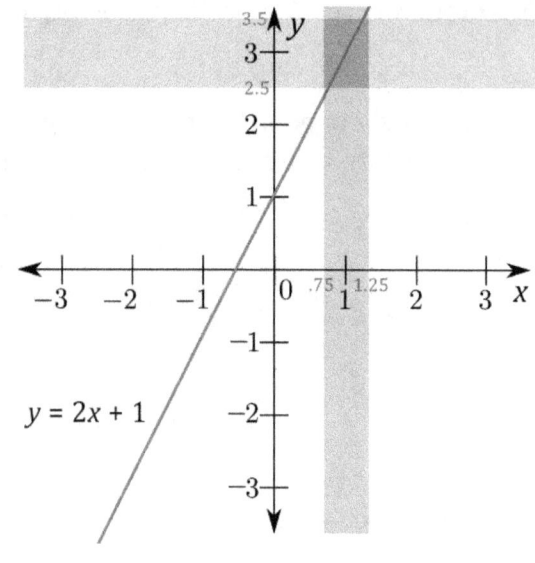

The figure to the right shows what happens during one round of the $\epsilon - \delta$ game corresponding to checking if $\lim_{x \to 1}(2x + 1) = 3$. In the figure, Player 1 chooses $\epsilon = 0.5$, so that $L - \epsilon = 3 - 0.5 = 2.5$ and $L + \epsilon = 3 + 0.5 = 3.5$. Notice how we drew the corresponding horizontal strip $H = \mathbb{R} \times (2.5, 3.5)$. According to our proof, Player 1 chooses $\delta = \frac{\epsilon}{2} = \frac{0.5}{2} = 0.25$. So $r - \delta = 1 - 0.25 = 0.75$ and $r + \delta = 1 + 0.25 = 1.25$. Notice how we drew the corresponding vertical strip $V = (0.75, 1.25) \times \mathbb{R}$. Also notice how the rectangle $H \cap V$ traps f.

(3) Observe that the value for δ that Player 2 chose here is the largest value of δ that would result in a successful defense. If we widen the vertical strip at all on either side, then f would escape from the resulting rectangle. However, any smaller value of δ will still work. If we shrink the vertical strip, then f is still trapped. After all, we have less that we need to trap.

(4) In the next round, Player 1 will want to choose a smaller value for ϵ. If Player 1 chooses a larger value for ϵ, then the same δ that was already played will work to defend against that larger ϵ. But for this problem, it doesn't matter how small a value for ϵ Player 1 chooses—Player 1 simply cannot win. All Player 2 needs to do is defend with $\delta = \frac{\epsilon}{2}$ (or any smaller positive number).

(5) Essentially the same argument can be used to show that the function f defined by $f(x) = 2x + 1$ is continuous at $x = 1$. Simply replace the expression $0 < |x - 1| < \delta$ by the expression $|x - 1| < \delta$ everywhere it appears in the proof. The point is that $f(1) = 2 \cdot 1 + 1 = 3$. Since this value is equal to $\lim_{x \to 1}(2x + 1)$, we don't need to exclude $x = 1$ from consideration when trying to trap f.

Example 8.11: Let's use the $\epsilon - \delta$ definition of a limit to prove that $\lim_{x \to 3}(x^2 - 2x + 1) = 4$.

Analysis: This is quite a bit more difficult than Example 8.10.

Given $\epsilon > 0$, we need to find $\delta > 0$ so that $0 < |x - 3| < \delta$ implies $|(x^2 - 2x + 1) - 4| < \epsilon$. First note that $|(x^2 - 2x + 1) - 4| = |x^2 - 2x - 3| = |(x - 3)(x + 1)| = |x - 3||x + 1|$. Therefore, $|(x^2 - 2x + 1) - 4| < \epsilon$ is equivalent to $|x - 3||x + 1| < \epsilon$.

There is a small complication here. The $|x - 3|$ is not an issue because we're going to be choosing δ so that this expression is small enough. But to make the argument work we need to make $|x + 1|$ small too. Remember from Note 3 after Example 8.10 that if we find a value for δ that works, then any smaller positive number will work too. This allows us to start by assuming that δ is smaller than any positive number we choose. So, let's just assume that $\delta \leq 1$ and see what effect that has on $|x + 1|$.

Well, if $\delta \leq 1$ and $0 < |x - 3| < \delta$, then $|x - 3| < 1$. Therefore, $-1 < x - 3 < 1$. We now add 4 to each part of this inequality to get $3 < x + 1 < 5$. Since $-5 < 3$, this implies that $-5 < x + 1 < 5$, which is equivalent to $|x + 1| < 5$.

So, if we assume that $\delta \leq 1$, then $|(x^2 - 2x + 1) - 4| = |x - 3||x + 1| < \delta \cdot 5 = 5\delta$. Therefore, if we want to make sure that $|(x^2 - 2x + 1) - 4| < \epsilon$, then is suffices to choose δ so that $5\delta \leq \epsilon$, as long as we also have $\delta \leq 1$. So, we will let $\delta = \min\left\{1, \frac{\epsilon}{5}\right\}$.

Proof: Let $\epsilon > 0$ and let $\delta = \min\left\{1, \frac{\epsilon}{5}\right\}$. Suppose that $0 < |x - 3| < \delta$. Then since $\delta \leq 1$, we have $|x - 3| < 1$, and so, $|x + 1| < 5$ (see the algebra in the analysis above). Also, since $\delta \leq \frac{\epsilon}{5}$, we have $|x - 3| < \frac{\epsilon}{5}$. It follows that $|(x^2 - 2x + 1) - 4| = |x^2 - 2x - 3| = |x - 3||x + 1| < \frac{\epsilon}{5} \cdot 5 = \epsilon$.

Since $\epsilon > 0$ was arbitrary, we have $\forall \epsilon > 0 \,\exists \delta > 0 \,(0 < |x - 3| < \delta \to |(x^2 - 2x + 1) - 4| < \epsilon)$. Therefore, $\lim_{x \to 1}(x^2 - 2x + 1) = 4$. □

Example 8.12: Let $m, b \in \mathbb{R}$ with $m \neq 0$. Let's use the $\epsilon - \delta$ definition of continuity to prove that the function $f: \mathbb{R} \to \mathbb{R}$ defined by $f(x) = mx + b$ is continuous everywhere.

Recall from Lesson 3 that a function of the form $f(x) = mx + b$, where $m, b \in \mathbb{R}$ is called a **linear function**. So, we will now show that every nonconstant linear function is continuous everywhere ($m \neq 0$ implies that the linear function is not a constant function).

Analysis: Given $a \in \mathbb{R}$ and $\epsilon > 0$, we will find $\delta > 0$ so that $|x - a| < \delta$ implies $|f(x) - f(a)| < \epsilon$. First note that $|f(x) - f(a)| = |(mx + b) - (ma + b)| = |mx - ma| = |m||x - a|$. Therefore, $|f(x) - f(a)| < \epsilon$ is equivalent to $|x - a| < \frac{\epsilon}{|m|}$. So, $\delta = \frac{\epsilon}{|m|}$ should work.

Proof: Let $a \in \mathbb{R}$, let $\epsilon > 0$, and let $\delta = \frac{\epsilon}{|m|}$. Suppose that $|x - a| < \delta$. Then we have

$$|f(x) - f(a)| = |(mx + b) - (ma + b)| = |mx - ma| = |m||x - a| < |m|\delta = |m| \cdot \frac{\epsilon}{|m|} = \epsilon.$$

Since $\epsilon > 0$ was arbitrary, we have $\forall \epsilon > 0 \, \exists \delta > 0 \, \forall x \in \mathbb{R} \, (|x - a| < \delta \to |f(x) - f(a)| < \epsilon)$. Therefore, f is continuous at $x = a$. Since $a \in \mathbb{R}$ was arbitrary, f is continuous everywhere. □

Notes: (1) We proved $\forall a \in \mathbb{R} \, \forall \epsilon > 0 \, \exists \delta > 0 \, \forall x \in \mathbb{R}(|x - a| < \delta \to |f(x) - f(a)| < \epsilon)$. In words, we proved that for every real number a, given a positive real number ϵ, we can find a positive real number δ such that whenever the distance between x and a is less than δ, the distance between $f(x)$ and $f(a)$ is less than ϵ. And of course, a simpler way to say this is "for every real number a, f is continuous at a," or $\forall a \in \mathbb{R} \, (f$ is continuous at $a)$."

(2) If we move the expression $\forall a \in \mathbb{R}$ next to $\forall x \in \mathbb{R}$, we get a concept that is stronger than continuity. We say that a function $f: A \to \mathbb{R}$ is **uniformly continuous** on A if

$$\forall \epsilon > 0 \, \exists \delta > 0 \, \forall a, x \in A \, (|x - a| < \delta \to |f(x) - f(a)| < \epsilon).$$

(3) As a quick example of uniform continuity, every linear function is uniformly continuous on $\mathbb{R}$. We can see this by modifying the proof above just slightly:

New proof: Let $\epsilon > 0$ and let $\delta = \frac{\epsilon}{|m|}$. Let $a, x \in \mathbb{R}$ and suppose that $|x - a| < \delta$. Then we have

$$|f(x) - f(a)| = |(mx + b) - (ma + b)| = |mx - ma| = |m||x - a| < |m|\delta = |m| \cdot \frac{\epsilon}{|m|} = \epsilon.$$

Since $\epsilon > 0$ was arbitrary, we have $\forall \epsilon > 0 \, \exists \delta > 0 \, \forall a, x \in \mathbb{R} \, (|x - a| < \delta \to |f(x) - f(a)| < \epsilon)$. Therefore, f is uniformly continuous on $\mathbb{R}$. □

(4) The difference between continuity and uniform continuity on a set A can be described as follows: In both cases, an ϵ is given and then a δ is chosen. For continuity, for each value of x, we are allowed to choose a different δ. For uniform continuity, once we choose a δ for some value of x, we need to be able to use the **same** δ for every other value of x in A.

In terms of strips, once a horizontal strip is given, we need to be more careful how we choose a vertical strip. As we check different x-values, we can move the vertical strip left and right. However, we are not allowed to decrease the width of the vertical strip.

Try to come up with a function that is continuous on a set A, but not uniformly continuous on A. This will be explored a little more in the Problem Set below.

(5) The function $f: \mathbb{R} \to \mathbb{R}$ defined by $f(x) = mx + b$ with $m \neq 0$ is a bijection.

To see that f is injective, note that if $x \neq y$, then since $m \neq 0$, $mx \neq my$. It follows that $f(x) = mx + b \neq my + b = f(y)$.

To see that f is surjective, observe that if $y \in \mathbb{R}$, then since $m \neq 0$, $\frac{y-b}{m} \in \mathbb{R}$ and $f\left(\frac{y-b}{m}\right) = y$. In fact, the inverse of f is the function $f^{-1}: \mathbb{R} \to \mathbb{R}$ defined by $f(x) = \frac{x-b}{m} = \frac{1}{m}x - \frac{b}{m}$.

Notice that this function has the same form as the original function f and therefore, f^{-1} is also continuous everywhere. In this case, we say that f is a **homeomorphism**. In general, a homeomorphism is a bijective continuous function with a continuous inverse.

Limit Theorems in $\mathbb{R}$

Theorem 8.13: Let $A, B \subseteq \mathbb{R}$, let $f: A \to \mathbb{R}$, $g: B \to \mathbb{R}$, let $r \in \mathbb{R}$, and suppose that $\lim_{x \to r}[f(x)]$ and $\lim_{x \to r}[g(x)]$ are both finite real numbers. Then $\lim_{x \to r}[f(x) + g(x)] = \lim_{x \to r}[f(x)] + \lim_{x \to r}[g(x)]$.

Analysis: If $\lim_{x \to r}[f(x)] = L$, then given $\epsilon > 0$, there is $\delta > 0$ such that $0 < |x - r| < \delta$ implies $|f(x) - L| < \epsilon$. If $\lim_{x \to r}[g(x)] = K$, then given $\epsilon > 0$, there is $\delta > 0$ such that $0 < |x - r| < \delta$ implies $|g(x) - K| < \epsilon$. We should acknowledge something here. If we are given a single positive real value for ϵ, there is no reason that we would necessarily choose the same δ for both f and g. However, using the fact that once we find a δ that works, any smaller δ will also work, it is easy to see that we **could** choose a single value for δ that would work for both f and g. This should be acknowledged in some way in the proof. There are several ways to work this into the argument. The way we will handle this is to use δ_1 for f and δ_2 for g, and then let δ be the smaller of δ_1 and δ_2.

Recall from Theorem 7.4 that the Triangle Inequality says that for all $x, y \in \mathbb{R}$, $|x + y| \leq |x| + |y|$. After assuming $0 < |x - r| < \delta$, we will use the Triangle Inequality to write

$$|f(x) + g(x) - (L + K)| = |(f(x) - L) + (g(x) - K)| \leq |f(x) - L| + |g(x) - K| < \epsilon + \epsilon = 2\epsilon.$$

It seems that we wound up with 2ϵ on the right-hand side instead of ϵ. Now, if ϵ is an arbitrarily small positive real number, then so is 2ϵ, and vice versa. So, getting 2ϵ on the right-hand side instead of ϵ really isn't too big of a deal. However, to be rigorous, we should prove that it is okay. There are at least two ways we can handle this. One possibility is to prove a theorem that says 2ϵ works just as well as ϵ. A second possibility (and the way I usually teach it in basic analysis courses) is to edit the original ϵ's, so it all works out to ϵ in the end. The idea is simple. If ϵ is a positive real number, then so is $\frac{\epsilon}{2}$. So, after we are given ϵ, we can pretend that Player 1 (in the $\epsilon - \delta$ game) is "attacking" with $\frac{\epsilon}{2}$ instead. Let's see how this all plays out in the proof.

Proof: Suppose that $\lim_{x \to r}[f(x)] = L$ and $\lim_{x \to r}[g(x)] = K$, and let $\epsilon > 0$. Since $\lim_{x \to r}[f(x)] = L$, there is $\delta_1 > 0$ such that $0 < |x - r| < \delta_1$ implies $|f(x) - L| < \frac{\epsilon}{2}$. Since $\lim_{x \to r}[g(x)] = K$, there is $\delta_2 > 0$ such that $0 < |x - r| < \delta_2$ implies $|g(x) - L| < \frac{\epsilon}{2}$. Let $\delta = \min\{\delta_1, \delta_2\}$ and suppose that $0 < |x - r| < \delta$. Then since $\delta \leq \delta_1$, $|f(x) - L| < \frac{\epsilon}{2}$. Since $\delta \leq \delta_2$, $|g(x) - K| < \frac{\epsilon}{2}$. By the Triangle Inequality, we have

$$|f(x) + g(x) - (L + K)| = |(f(x) - L) + (g(x) - K)| \leq |f(x) - L| + |g(x) - K| < \frac{\epsilon}{2} + \frac{\epsilon}{2} = \epsilon.$$

So, $\lim_{x \to r}[f(x) + g(x)] = L + K = \lim_{x \to r}[f(x)] + \lim_{x \to r}[g(x)]$. □

Theorem 8.14: Let $A, B \subseteq \mathbb{R}$, let $f: A \to \mathbb{R}$, $g: B \to \mathbb{R}$, let $r \in \mathbb{R}$, and suppose that $\lim_{x \to r}[f(x)]$ and $\lim_{x \to r}[g(x)]$ are both finite real numbers. Then $\lim_{x \to r}[f(x)g(x)] = \lim_{x \to r}[f(x)] \cdot \lim_{x \to r}[g(x)]$.

Analysis: As in Theorem 8.13, we let $\lim_{x \to r}[f(x)] = L$ and $\lim_{x \to r}[g(x)] = K$. If $\epsilon > 0$ is given, we will find a single $\delta > 0$ such that $0 < |x - r| < \delta$ implies $|f(x) - L| < \epsilon$ and $|g(x) - K| < \epsilon$ (like we did for Theorem 8.13). Now, we want to show that whenever $0 < |x - r| < \delta$, $|f(x)g(x) - LK| < \epsilon$. This is quite a bit more challenging than anything we had to do in Theorem 8.13.

To show that $|f(x)g(x) - LK| < \epsilon$ we will apply the Standard Advanced Calculus Trick (SACT – see Note 7 following the proof of Theorem 4.11). We would like for $|f(x) - L|$ and $|g(x) - K|$ to appear as factors in our expression. To make this happen, we subtract $Lg(x)$ from $f(x)g(x)$ to get $f(x)g(x) - Lg(x) = (f(x) - L)g(x)$. To "undo the damage," we then add back $Lg(x)$. The application of SACT together with the Triangle Inequality looks like this:

$$|f(x)g(x) - LK| = |(f(x)g(x) - Lg(x)) + (Lg(x) - LK)|$$
$$\leq |f(x)g(x) - Lg(x)| + |Lg(x) - LK| = |f(x) - L||g(x)| + |L||g(x) - K|$$
$$< \epsilon|g(x)| + |L|\epsilon = \epsilon(|g(x)| + |L|).$$

Uh oh! How can we possibly get rid of $|g(x)| + |L|$? We have seen how to handle a constant multiple of ϵ in the proof of Theorem 8.13. But this time we are multiplying ϵ by a function of x. We will resolve this issue by making sure we choose δ small enough so that $g(x)$ is sufficiently **bounded**.

We do this by taking a specific value for ϵ, and then using the fact that $\lim_{x \to r}[g(x)] = K$ to come up with a $\delta > 0$ and a bound M for g on the deleted δ-neighborhood of r. For simplicity, let's choose $\epsilon = 1$. Then since $\lim_{x \to r}[g(x)] = K$, we can find $\delta > 0$ such that $0 < |x - r| < \delta$ implies $|g(x) - K| < 1$. Now, $|g(x) - K| < 1 \Leftrightarrow -1 < g(x) - K < 1 \Leftrightarrow K - 1 < g(x) < K + 1$. For example, if $K = 5$, we would have $4 < g(x) < 6$. Since this implies $-6 < g(x) < 6$, or equivalently, $|g(x)| < 6$, we could choose $M = 6$. If, on the other hand, $K = -3$, we would have $-4 < g(x) < -2$. Since this implies $-4 < g(x) < 4$, or equivalently, $|g(x)| < 4$, we could choose $M = 4$. In general, we will let $M = \max\{|K - 1|, |K + 1|\}$.

We will now be able to get $|f(x)g(x) - (LK)| < \epsilon(|g(x)| + |L|) < \epsilon(M + |L|)$. Great! Now it looks just like the situation we had in Theorem 8.13. The number $M + |L|$ looks messier, but it is just a number, and so we can finish cleaning up the argument by replacing Player 1's ϵ-attacks by $\frac{\epsilon}{M+|L|}$.

Proof: Suppose that $\lim_{x \to r}[f(x)] = L$ and $\lim_{x \to r}[g(x)] = K$, and let $\epsilon > 0$. Since $\lim_{x \to r}[g(x)] = K$, there is $\delta_1 > 0$ such that $0 < |x - r| < \delta_1$ implies $|g(x) - K| < 1$. Now, $|g(x) - K| < 1$ is equivalent to $-1 < g(x) - K < 1$, or by adding K, $K - 1 < g(x) < K + 1$. Let $M = \max\{|K - 1|, |K + 1|\}$. Then, $0 < |x - r| < \delta_1$ implies $-M < g(x) < M$, or equivalently, $|g(x)| < M$. Note also that $M > 0$. Therefore, $M + |L| > 0$.

Now, since $\lim_{x \to r}[f(x)] = L$, there is $\delta_2 > 0$ such that $0 < |x - r| < \delta_2$ implies $|f(x) - L| < \frac{\epsilon}{M+|L|}$. Since $\lim_{x \to r}[g(x)] = K$, there is $\delta_3 > 0$ such that $0 < |x - r| < \delta_3$ implies $|g(x) - L| < \frac{\epsilon}{M+|L|}$. Let $\delta = \min\{\delta_1, \delta_2, \delta_3\}$ and suppose that $0 < |x - r| < \delta$. Then since $\delta \leq \delta_1$, $|g(x)| < M$. Since $\delta \leq \delta_2$, $|f(x) - L| < \frac{\epsilon}{M+|L|}$. Since $\delta \leq \delta_3$, $|g(x) - K| < \frac{\epsilon}{M+|L|}$. By the Triangle Inequality (and SACT), we have

$$|f(x)g(x) - LK| = |(f(x)g(x) - Lg(x)) + (Lg(x) - LK)|$$
$$\leq |f(x)g(x) - Lg(x)| + |Lg(x) - LK| = |f(x) - L||g(x)| + |L||g(x) - K|$$
$$< \frac{\epsilon}{M+|L|} \cdot M + |L|\frac{\epsilon}{M+|L|} = \frac{\epsilon}{M+|L|}(M + |L|) = \epsilon.$$

So, $\lim_{x \to r}[f(x)g(x)] = LK = \lim_{x \to r}[f(x)] \cdot \lim_{x \to r}[g(x)]$. □

Theorem 8.15: If $\lim_{x \to r} f(x)$ exists, then it is unique.

Proof: Suppose that $\lim_{x \to r} f(x) = L$ and $\lim_{x \to r} f(x) = K$. Let $\epsilon > 0$. Since $\lim_{x \to r} f(x) = L$, we can find $\delta_1 > 0$ such that $0 < |x - r| < \delta_1 \to |f(x) - L| < \frac{\epsilon}{2}$. Since $\lim_{x \to r} f(x) = K$, we can find $\delta_2 > 0$ such that $0 < |x - r| < \delta_2 \to |f(x) - K| < \frac{\epsilon}{2}$. Let $\delta = \min\{\delta_1, \delta_2\}$. Suppose that $0 < |x - r| < \delta$. Then $|L - K| = |(f(x) - K) - (f(x) - L)|$ (**SACT**) $\leq |f(x) - K| + |f(x) - L|$ (**TI**) $< \frac{\epsilon}{2} + \frac{\epsilon}{2} = \epsilon$. Since ϵ was an arbitrary positive real number, by Problem 15 from Problem Set 6, we have $|L - K| = 0$. So, $L - K = 0$, and therefore, $L = K$. □

Note: SACT stands for the Standard Advanced Calculus Trick and **TI** stands for the Triangle Inequality.

Limits in $\mathbb{R}$ Involving Infinity

Recall that a **horizontal strip** in $\mathbb{R} \times \mathbb{R}$ is a set of the form $\mathbb{R} \times (c, d) = \{(x, y) \mid c < y < d\}$ and a **vertical strip** is a set of the form $(a, b) \times \mathbb{R} = \{(x, y) \mid a < x < b\}$. If we allow a and/or c to take on the value $-\infty$ (in which case we say that the strip **contains** $-\infty$) and we allow b and/or d to take on the value $+\infty$ (in which case we say that the strip **contains** $+\infty$), we can extend our definition of limit to handle various situations involving infinity.

Note: Strips that contain $+\infty$ or $-\infty$ are usually called **half planes**. Here, we will continue to use the expression "strip" because it allows us to handle all types of limits (finite and infinite) without having to discuss every case individually.

Example 8.16: Consider the horizontal strip $\mathbb{R} \times (1, +\infty)$ and the vertical strip $(-\infty, -2) \times \mathbb{R}$ in the xy-plane.

The horizontal strip $\mathbb{R} \times (1, +\infty)$ contains $+\infty$ and the vertical strip $(-\infty, -2) \times \mathbb{R}$ contains $-\infty$.

These strips can be visualized as follows:

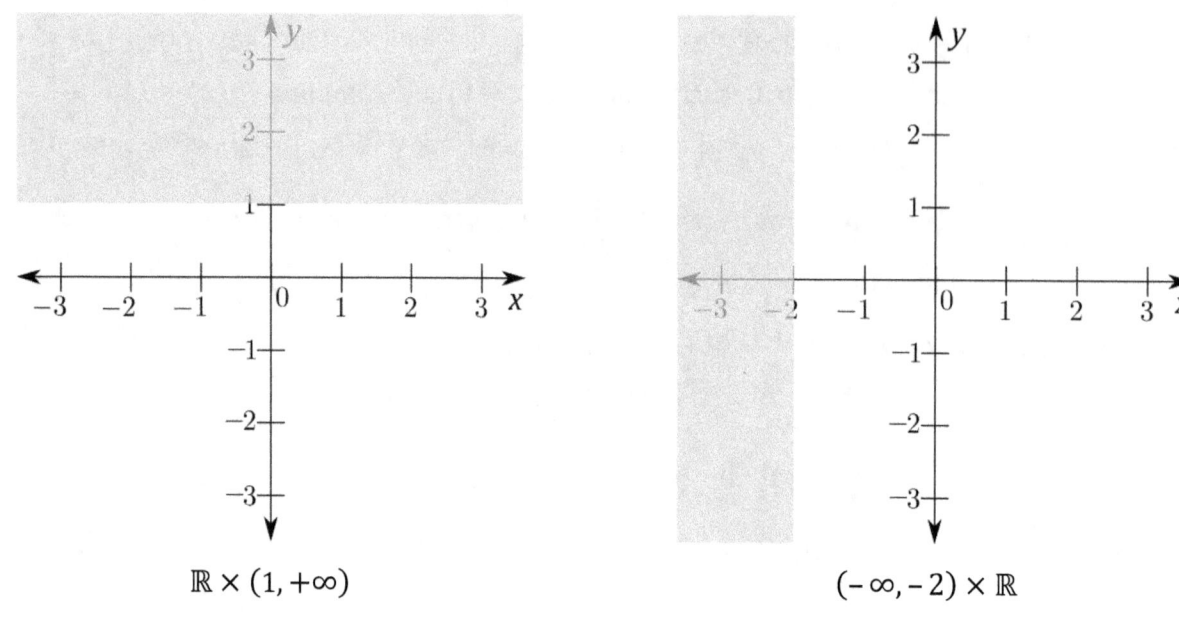

By allowing strips to contain $+\infty$ or $-\infty$, intersections of horizontal and vertical strips can now be **unbounded**. The resulting open rectangles $(a, b) \times (c, d)$ can have a and/or c taking on the value $-\infty$ and b and/or d taking on the value $+\infty$.

Example 8.17: Consider the horizontal strip $H = \mathbb{R} \times (1, +\infty)$ and the vertical strip $V = (-2, -1) \times \mathbb{R}$. The intersection of these strips is the open rectangle $R = (-2, -1) \times (1, +\infty)$. The rectangle R traps $(-1.5, 3)$, whereas $(-1.5, 0)$ escapes from R. This can be seen in the figure below on the left.

Also, consider the horizontal strip $H = \mathbb{R} \times (1, +\infty)$ and the vertical strip $V = (-\infty, -2) \times \mathbb{R}$. The intersection of these strips is the open rectangle $S = (-\infty, -2) \times (1, +\infty)$. The rectangle S traps $(-3, 2)$, whereas $(-3, -1)$ escapes from S. This can be seen in the figure below on the right.

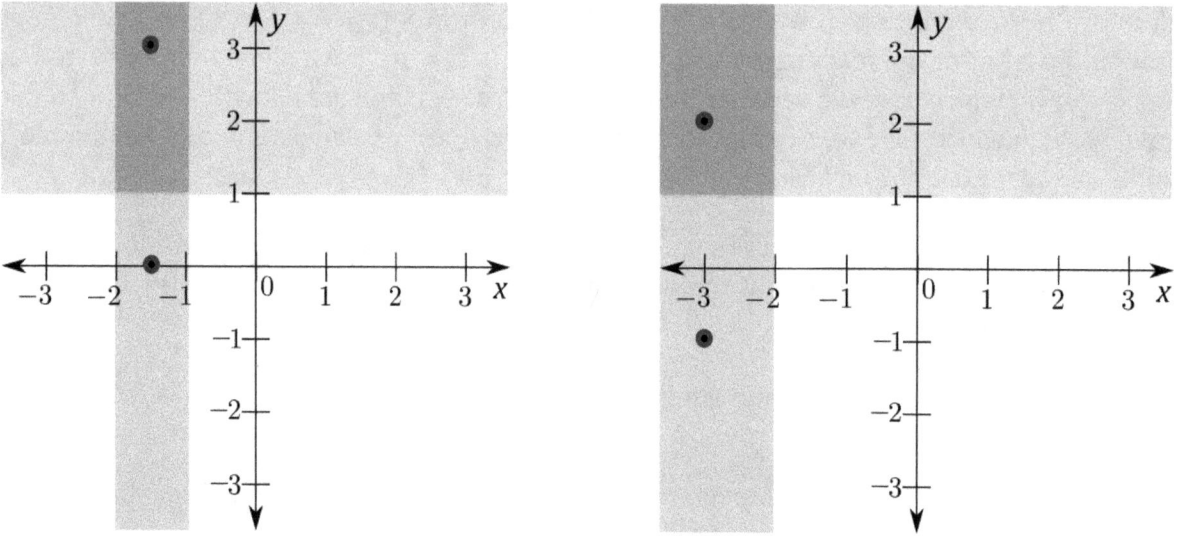

When we allow $+\infty$ and $-\infty$, the definitions of "trap" and "escape" are just about the same. We just need to make the following minor adjustment.

Small technicality: If $r = +\infty$ or $r = -\infty$, then we define **R traps f around r** to simply mean that **R traps f**. In other words, when checking a limit that is approaching $+\infty$ or $-\infty$, we do not exclude any point from consideration as we would do if r were a finite real number.

Example 8.18: Let $f: \mathbb{R} \to \mathbb{R}$ be defined by $f(x) = \frac{1}{x^2}$, let $r = 0$, and let $L = +\infty$. Let's show that $\lim_{x \to 0} f(x) = +\infty$. Let $H = \mathbb{R} \times (c, +\infty)$ be a horizontal strip that contains $+\infty$. Next, let $V = \left(-\frac{1}{\sqrt{c}}, \frac{1}{\sqrt{c}}\right) \times \mathbb{R}$. We will show that $H \cap V = \left(-\frac{1}{\sqrt{c}}, \frac{1}{\sqrt{c}}\right) \times (c, +\infty)$ traps f around 0. Let $x \in \left(-\frac{1}{\sqrt{c}}, \frac{1}{\sqrt{c}}\right) \setminus \{0\}$, so that $-\frac{1}{\sqrt{c}} < x < \frac{1}{\sqrt{c}}$ and $x \neq 0$. Then $-\frac{1}{\sqrt{c}} < x < 0$ or $0 < x < \frac{1}{\sqrt{c}}$. In either case, $x^2 < \frac{1}{c}$, and therefore, $c < \frac{1}{x^2} = f(x)$. Since $x \in \left(-\frac{1}{\sqrt{c}}, \frac{1}{\sqrt{c}}\right) \setminus \{0\}$ and $f(x) \in (c, +\infty)$, it follows that $(x, f(x)) \in \left(-\frac{1}{\sqrt{c}}, \frac{1}{\sqrt{c}}\right) \times (c, +\infty) = H \cap V$. So, $H \cap V$ traps f around 0. Thus, $\lim_{x \to 0} f(x) = +\infty$.

Example 8.19: Let $f: \mathbb{R} \to \mathbb{R}$ be defined by $f(x) = -x + 3$, let $r = +\infty$, and let $L = -\infty$. Let's show that $\lim_{x \to +\infty} f(x) = -\infty$. Let $H = \mathbb{R} \times (-\infty, d)$ be a horizontal strip that contains $-\infty$. Next, let $V = (3 - d, +\infty) \times \mathbb{R}$. We will show that $H \cap V = (3 - d, +\infty) \times (-\infty, d)$ traps f around $+\infty$ (or more simply, that $(3 - d, +\infty) \times (-\infty, d)$ traps f). Let $x \in (3 - d, +\infty)$, so that $x > 3 - d$. Then we have $-x < d - 3$, and so, $f(x) = -x + 3 < d$. Since $x \in (3 - d, +\infty)$ and $f(x) \in (-\infty, d)$, it follows that $(x, f(x)) \in (3 - d, +\infty) \times (-\infty, d) = H \cap V$. Therefore, $H \cap V$ traps f. So, $\lim_{x \to +\infty} f(x) = -\infty$.

We can find equivalent definitions for limits involving infinity on a case-by-case basis. We will do one example here and you will look at others in Problem 16 below.

Theorem 8.20: $\lim_{x \to r} f(x) = +\infty$ if and only if $\forall M > 0 \, \exists \delta > 0 \, (0 < |x - r| < \delta \to f(x) > M)$.

Proof: Suppose that $\lim_{x \to r} f(x) = +\infty$ and let $M > 0$. Let $H = \mathbb{R} \times (M, +\infty)$. Since $\lim_{x \to r} f(x) = +\infty$, there is a vertical strip $V = (a, b) \times \mathbb{R}$ that contains r such that the rectangle $(a, b) \times (M, +\infty)$ traps f around r. Let $\delta = \min\{r - a, b - r\}$ and let $0 < |x - r| < \delta$. Then $x \neq r$ and $-\delta < x - r < \delta$. So, $r - \delta < x < r + \delta$. Since $\delta \leq r - a$, we have $a \leq r - \delta$. Since $\delta \leq b - r$, we have $b \geq r + \delta$. Therefore, $a < x < b$, and so, $x \in (a, b)$. Since $x \neq r$, $x \in (a, b)$, and $(a, b) \times (M, +\infty)$ traps f around r, we have $f(x) \in (M, +\infty)$. Thus, $f(x) > M$.

Conversely, suppose that $\forall M > 0 \, \exists \delta > 0 \, (0 < |x - r| < \delta \to f(x) > M)$. Let $H = \mathbb{R} \times (c, +\infty)$ be a horizontal strip containing $+\infty$ and let $M = \max\{c, 1\}$. Then there is $\delta > 0$ such that $0 < |x - r| < \delta$ implies $f(x) > M$. Let $V = (r - \delta, r + \delta) \times \mathbb{R}$ and let $R = H \cap V = (r - \delta, r + \delta) \times (c, +\infty)$. We show that R traps f around r. Indeed, if $x \in (r - \delta, r + \delta)$ and $x \neq r$, then $0 < |x - r| < \delta$ and so, $f(x) > M$. So, $(x, f(x)) \in (r - \delta, r + \delta) \times (M, +\infty) \subseteq (r - \delta, r + \delta) \times (c, +\infty)$ (because $c \leq M$). So, R traps f around r. □

One-sided Limits in $\mathbb{R}$

Let $A \subseteq \mathbb{R}$, let $f: A \to \mathbb{R}$, and let $r \in \mathbb{R}$ and $L \in \mathbb{R} \cup \{-\infty, +\infty\}$. We say that the **limit of f as x approaches r from the right is L**, written $\lim_{x \to r^+} f(x) = L$, if for every horizontal strip H that contains L there is a vertical strip V of the form $(r, b) \times \mathbb{R}$ such that the rectangle $H \cap V$ traps f.

Example 8.21: Let $f: \mathbb{R} \setminus \{1\} \to \mathbb{R}$ be defined by $f(x) = \frac{1}{x-1}$, let $r = 1$, and let $L = +\infty$. Let's show that $\lim_{x \to 1^+} f(x) = +\infty$. Let $H = \mathbb{R} \times (c, +\infty)$ be a horizontal strip that contains $+\infty$ and let $M = \max\{1, c\}$. Let $V = \left(1, \frac{1}{M} + 1\right) \times \mathbb{R}$. We will show that $H \cap V = \left(1, \frac{1}{M} + 1\right) \times (c, +\infty)$ traps f. Let $x \in \left(1, \frac{1}{M} + 1\right)$, so that $1 < x < \frac{1}{M} + 1$. Then we have $0 < x - 1 < \frac{1}{M}$, and so, $\frac{1}{x-1} > M \geq c$. So, $f(x) > c$. Since $x \in \left(1, \frac{1}{M} + 1\right)$ and $f(x) \in (c, +\infty)$, $(x, f(x)) \in \left(1, \frac{1}{M} + 1\right) \times (c, +\infty) = H \cap V$. Therefore, $H \cap V$ traps f. So, $\lim_{x \to 1^+} f(x) = +\infty$.

Theorem 8.22: Let L be a finite real number. Then we have $\lim_{x \to r^+} f(x) = L$ if and only if $\forall \epsilon > 0 \, \exists \delta > 0 \, (0 < x - r < \delta \to |f(x) - L| < \epsilon)$.

Proof: Suppose that $\lim_{x \to r^+} f(x) = L$ and let $\epsilon > 0$. Let $H = \mathbb{R} \times (L - \epsilon, L + \epsilon)$. Since $\lim_{x \to r^+} f(x) = L$, there is a vertical strip $V = (r, b) \times \mathbb{R}$ such that the rectangle $H \cap V = (r, b) \times (L - \epsilon, L + \epsilon)$ traps f. Let $\delta = b - r$ and let $0 < x - r < \delta$. Then $r < x < b$ and so, $x \in (r, b)$. Since $(r, b) \times (L - \epsilon, L + \epsilon)$ traps f, we have $f(x) \in (L - \epsilon, L + \epsilon)$. Thus, $L - \epsilon < f(x) < L + \epsilon$, or equivalently, $|f(x) - L| < \epsilon$.

Conversely, suppose that $\forall \epsilon > 0 \, \exists \delta > 0 \, (0 < x - r < \delta \to |f(x) - L| < \epsilon)$. Let $H = \mathbb{R} \times (c, d)$ be a horizontal strip containing L and let $\epsilon = \min\{L - c, d - L\}$. Then there is $\delta > 0$ such that $0 < x - r < \delta$ implies $|f(x) - L| < \epsilon$. Let $V = (r, r + \delta) \times \mathbb{R}$ and $R = H \cap V = (r, r + \delta) \times (c, d)$. We show that R traps f. If $x \in (r, r + \delta)$, then $r < x < r + \delta$, or equivalently, $0 < x - r < \delta$. So, $|f(x) - L| < \epsilon$. Therefore, $-\epsilon < f(x) - L < \epsilon$, or equivalently, $L - \epsilon < f(x) < L + \epsilon$. Thus, $(x, f(x)) \in (r, r + \delta) \times (L - \epsilon, L + \epsilon) \subseteq (r, r + \delta) \times (c, d)$ (Check this!). So, R traps f. □

Limits of Sequences

Recall the following from Lesson 3:

- A real-valued sequence is a function $f: \mathbb{N} \to \mathbb{R}$. If $f(n) = s_n$, then we use the notation (s_n) to represent the sequence.

- Informally, we say that a real-valued sequence (s_n) **converges** to a real number s if the real numbers s_n get "closer and closer" to the real number s, as n gets larger and larger.

Let's now define convergent sequences more formally.

Let $f = (s_n)$ be a real-valued sequence. We say that f **converges** to $s \in \mathbb{R}$, written $\lim_{n \to \infty} s_n = s$ (sometimes abbreviated as $s_n \to s$) if

$$\forall \epsilon > 0 \, \exists K \in \mathbb{N} (n > K \to |s_n - s| < \epsilon).$$

If $f = (s_n)$ does not converge to any real number, then we say that f **diverges**.

The idea is that if $f = (s_n)$ converges, then we can make the distance between any term of the sequence and s as small as we choose by deleting a finite portion of the beginning of the sequence. If we wish to make the distance between any term of the sequence and s less than ϵ, we delete the first $K + 1$ terms of the sequence.

We may write $(s_n) \downarrow$ or $(s_n) \uparrow$ to indicate that (s_n) converges or diverges, respectively. If $(s_n) \downarrow$, then we say that the sequence is **convergent** If $(s_n) \uparrow$, then we say that the sequence is **divergent**.

Example 8.23:

1. The sequence $(s_n) = \left(\frac{1}{n+1}\right)$ converges to 0. To see this, let $\epsilon > 0$, let K be a positive integer such that $K > \frac{1}{\epsilon}$ (we can find such a K by the Archimedean Property of $\mathbb{R}$), and let $n > K$. Then

$$|s_n - 0| = \left|\frac{1}{n+1} - 0\right| = \left|\frac{1}{n+1}\right| = \frac{1}{n+1} < \frac{1}{n} < \frac{1}{K} < \epsilon.$$

2. For each $r \in \mathbb{R}$, the **constant sequence** $(t_n) = (r)$ converges to r. To see this, let $\epsilon > 0$, let $K = 0$, and let $n > 0$. Then

$$|t_n - r| = |r - r| = |0| = 0 < \epsilon.$$

3. The sequence $(u_n) = (n)$ diverges. To see this, assume toward contradiction that there is $u \in \mathbb{R}$ such that $u_n \to u$. By the Archimedian Property of $\mathbb{R}$, we can find $m \in \mathbb{N}$ with $m > u$. Let $\epsilon = 1$, let $K \in \mathbb{N}$, and let $L = \max\{K+1, m+1\}$. Then $L > K$ and $L \geq m+1$. Since $L > m$, we have $|u_L - u| = |L - u| = L - u \geq m + 1 - u > m + 1 - m = 1 = \epsilon$. This contradiction shows that there can be no $u \in \mathbb{R}$ such that $u_n \to u$. So, $(u_n) \uparrow$.

Note: If (s_n) converges, then $\lim_{n \to \infty} s_n$ is unique. The proof of this is very similar to the proof of Theorem 8.15. You will be asked to write in proof in Problem 1 in Problem Set 14.

We now provide a few equivalent ways to recognize that a sequence converges.

Theorem 8.24: Let (s_n) be a real-valued sequence. The following are equivalent.

1. $s_n \to s$.
2. For every open set U containing s, there is $K \in \mathbb{N}$ such that $n > K$ implies $s_n \in U$.
3. For every neighborhood $N_\epsilon(s)$ with center s, there is $K \in \mathbb{N}$ such that $n > K$ implies $s_n \in N_\epsilon(s)$.

Proof: (1→2) Suppose that $s_n \to s$ and let U be an open set containing s. Since U is open, by Theorem 7.12, there is a positive real number c such that $(s-c, s+c) \subseteq U$. By 1, there is $K \in \mathbb{N}$ such that $n > K$ implies $|s_n - s| < c$. But, if $|s_n - s| < c$, then $-c < s_n - s < c$, or $s - c < s_n < s + c$. So, $n > K$ implies that $s_n \in (s-c, s+c)$. Since $(s-c, s+c) \subseteq U$, we see that $n > K$ implies that $s_n \in U$, as desired.

(2→3) Suppose that 2 holds and let $N_\epsilon(s)$ be a neighborhood with center s. Since $N_\epsilon(s)$ is open, by 2, there is $K \in \mathbb{N}$ such that $n > K$ implies $s_n \in N_\epsilon(s)$.

(3→1) Suppose 3 holds and let $\epsilon > 0$. By 3, there is $K \in \mathbb{N}$ such that $n > K$ implies $s_n \in N_\epsilon(s)$. But $s_n \in N_\epsilon(s)$ if and only if $|s_n - s| < \epsilon$. So, $n > K \to |s_n - s| < \epsilon$. Therefore, $s_n \to s$. □

A real-valued sequence (s_n) is called a **Cauchy sequence** if

$$\forall \epsilon > 0 \, \exists K \in \mathbb{N} (m \geq n > K \to |s_m - s_n| < \epsilon).$$

Note: This definition is analogous to the definition of a rational-valued Cauchy sequence that was used to define the real numbers (see Lesson 4).

Theorem 8.25: Let (s_n) be a convergent sequence. Then (s_n) is a Cauchy sequence.

Proof: Suppose that $s_n \to s$ and let $\epsilon > 0$. Then there is $K \in \mathbb{N}$ such that $n > K$ implies $|s_n - s| < \frac{\epsilon}{2}$. Let $m \geq n > K$. By the Triangle Inequality, we have

$$|s_m - s_n| = |s_m - s + s - s_n| \leq |s_m - s| + |s - s_n| < \frac{\epsilon}{2} + \frac{\epsilon}{2} = \epsilon.$$

Therefore, (s_n) is a Cauchy sequence. □

For a real-valued sequence (s_n), we will prove that the converse of Theorem 8.25 is also true. This will require several steps.

We will say that the real-valued sequence (s_n) is **bounded** if the set $\{s_n\}$ is bounded. In other words, (s_n) is bounded if there is $M \in \mathbb{R}^+$ such that for all $n \in \mathbb{N}$, $|s_n| \leq M$ (see the discussion before the statement of Theorem 7.28). We will now show that all Cauchy sequences are bounded.

Note: In Problem 4 below, you will be asked to prove that (s_n) is bounded in $\mathbb{R}$ if and only if there is $M \in \mathbb{R}^+$ such that for all $m, n \in \mathbb{N}$, $|s_m - s_n| \leq M$. This will be used in the proof of Theorem 8.26 below.

Theorem 8.26: Let (s_n) be a real-valued Cauchy sequence. Then (s_n) is bounded.

Proof: Let (s_n) be a real-valued Cauchy sequence. Then for every $\epsilon > 0$, there is $K \in \mathbb{N}$ such that $m \geq n > K$ implies $|s_m - s_n| < \epsilon$. In particular, by letting $\epsilon = 1$, we see that there is $K \in \mathbb{N}$ such that $m \geq n > K$ implies $|s_m - s_n| < 1$. Let

$$M = \max\{|s_0 - s_{K+1}|, |s_1 - s_{K+1}|, \ldots, |s_K - s_{K+1}|, 1\}.$$

Then if $m, n \in \mathbb{N}$, by the Triangle Inequality (and SACT), we have

$$|s_m - s_n| = |(s_m - s_{K+1}) + (s_{K+1} - s_n)| \leq |s_m - s_{K+1}| + |s_{K+1} - s_n| \leq M + M = 2M.$$

Therefore, by Problem 4 below, (s_n) is bounded in S. □

Theorem 8.27 (The Bolzano-Weierstrass Theorem): Let (s_n) be a bounded real-valued sequence. Then (s_n) has a convergent subsequence.

You will be asked to prove Theorem 8.27 in Problem 24 below. An alternative proof will be suggested by Problem 13 in Problem Set 14.

Theorem 8.28 (Cauchy Criterion for Convergence): Let (s_n) be a real-valued Cauchy sequence. Then there is $s \in \mathbb{R}$ such that $s_n \to s$.

Proof: Let (s_n) be a real-valued Cauchy sequence.

By Theorem 8.26, (s_n) is bounded.

By Theorem 8.27, there is a subsequence (s_{n_k}) of (s_n) such that $s_{n_k} \to s$.

Let $\epsilon > 0$. Since (s_n) is a Cauchy sequence, there is $K_1 \in \mathbb{N}$ such that $m \geq n > K_1$ implies $|s_m - s_n| < \frac{\epsilon}{2}$. Since $s_{n_k} \to s$, there is $K_2 \in \mathbb{N}$ so that $k > K_2$ implies $|s_{n_k} - s| < \frac{\epsilon}{2}$. Let $K = \max\{K_1, K_2\} + 1$ and let $m > K$. Since $K > K_2$, $|s_{n_K} - s| < \frac{\epsilon}{2}$. Since $K > K_1$, we have $m > K_1$. We also have $n_K \geq K > K_1$ (see Note 1 below). Therefore, we have either $m \geq n_K > K_1$ or $n_K \geq m > K_1$. In either case, $|s_m - s_{n_K}| < \frac{\epsilon}{2}$. It follows from the Triangle Inequality (and SACT) that

$$|s_m - s| = |(s_m - s_{n_K}) + (s_{n_K} - s)| \leq |s_m - s_{n_K}| + |s_{n_K} - s| < \frac{\epsilon}{2} + \frac{\epsilon}{2} = \epsilon.$$

So, $s_n \to s$. $\square$

Notes: (1) In the proof of Theorem 8.28 above, we used the fact that if (s_{n_k}) is a subsequence of (s_n), then $n_k \geq k$ for all $k \in \mathbb{N}$. To see this, first recall that we must have $n_0 < n_1 < n_2 < \cdots$ (see the Note just before Example 3.12). We can now prove by induction on k that $n_k \geq k$ for all $k \in \mathbb{N}$. The base case $n_0 \geq 0$ is true because 0 is the least natural number. For the inductive step, if we assume that $n_k \geq k$, then we have $n_{k+1} > n_k \geq k$, and therefore, $n_{k+1} \geq k + 1$. By the Principle of Mathematical Induction, we have now shown that $n_k \geq k$ for all $k \in \mathbb{N}$.

(2) The usefulness of the Cauchy Criterion for Convergence lies in the fact that to show that a sequence is a Cauchy sequence does not require knowing the limit of the sequence ahead of time. It is often very difficult to determine the limit of a convergent sequence—if we are interested only in whether the sequence converges or diverges, it is often much easier to prove that the sequence is a Cauchy sequence.

We summarize the main result regarding Cauchy sequences in the following Corollary.

Corollary 8.29: A sequence (s_n) is convergent if and only if it is a Cauchy sequence.

Proof: This follows immediately from Theorems 8.25 and 8.28. $\square$

Common misconception: When first learning about Cauchy sequences, it is a common misconception that a Cauchy sequence is a sequence whose **adjacent** terms are getting "closer and closer" to each other as we go further and further out into the sequence. One counterexample to this claim is the sequence $(\sqrt{n})$. We have $\sqrt{n+1} - \sqrt{n} = (\sqrt{n+1} - \sqrt{n}) \cdot \frac{\sqrt{n+1}+\sqrt{n}}{\sqrt{n+1}+\sqrt{n}} = \frac{n+1-n}{\sqrt{n+1}+\sqrt{n}} = \frac{1}{\sqrt{n+1}+\sqrt{n}} < \frac{1}{2\sqrt{n}}$. So, given $\epsilon > 0$, if we choose $K > \frac{1}{4\epsilon^2}$ (which we can do by the Archimedean Property of $\mathbb{R}$), then $4\epsilon^2 > \frac{1}{K}$, or equivalently, $\epsilon > \frac{1}{2\sqrt{K}}$ (Check this carefully!), and so, we have

$$n > K \to |\sqrt{n+1} - \sqrt{n}| < \frac{1}{2\sqrt{n}} < \frac{1}{2\sqrt{K}} < \epsilon.$$

This shows that adjacent terms of the sequence $(\sqrt{n})$ are getting arbitrarily close to each other as n gets larger. However, the sequence itself diverges. Proving that the sequence diverges is straightforward, and so, I leave it to the reader.

Problem Set 8

Full solutions to these problems are available for free download here:
www.SATPrepGet800.com/RABQXZ

LEVEL 1

1. Let $f: \mathbb{R} \to \mathbb{R}$ be defined by $f(x) = 5x - 1$.
   (i) Prove that $\lim_{x \to 3} f(x) = 14$.
   (ii) Prove that f is continuous on $\mathbb{R}$.

2. Let $r, c \in \mathbb{R}$ and let $f: \mathbb{R} \to \mathbb{R}$ be defined by $f(x) = c$. Prove that $\lim_{x \to r}[f(x)] = c$.

3. Let $A \subseteq \mathbb{R}$, let $f: A \to \mathbb{R}$, let $r, k \in \mathbb{R}$, and suppose that $\lim_{x \to r}[f(x)]$ is a finite real number. Prove that $\lim_{x \to r}[kf(x)] = k \lim_{x \to r}[f(x)]$.

4. Prove that a real-valued sequence (s_n) is bounded if and only if there is $M \in \mathbb{R}^+$ such that for all $m, n \in \mathbb{N}$, $|s_m - s_n| \leq M$.

LEVEL 2

5. Let $A \subseteq \mathbb{R}$, let $f: A \to \mathbb{R}$, and let r be an interior point of A (see Problem 18 from Problem Set 7). Prove that f is continuous at r if and only if $\lim_{x \to r}[f(x)] = f(r)$.

6. Prove that every polynomial function $p: \mathbb{R} \to \mathbb{R}$ is continuous on $\mathbb{R}$.

LEVEL 3

7. Let $g: \mathbb{R} \to \mathbb{R}$ be defined by $g(x) = 2x^2 - 3x + 7$.
   (i) Prove that $\lim_{x \to 1} g(x) = 6$.
   (ii) Prove that g is continuous on $\mathbb{R}$.

8. Suppose that $f, g: \mathbb{R} \to \mathbb{R}$, $a \in \mathbb{R}$, f is continuous at a, and g is continuous at $f(a)$. Prove that $g \circ f$ is continuous at a.

LEVEL 4

9. Let $h: \mathbb{R} \to \mathbb{R}$ be defined by $h(x) = \frac{x^3 - 4}{x^2 + 1}$. Prove that $\lim_{x \to 2} h(x) = \frac{4}{5}$.

10. Let $k: (0, \infty) \to \mathbb{R}$ be defined by $k(x) = \sqrt{x}$.

   (i) Prove that $\lim_{x \to 25} k(x) = 5$.

   (ii) Prove that f is continuous on $(0, \infty)$.

   (iii) Is f uniformly continuous on $(0, \infty)$?

11. Let $f: \mathbb{R} \to \mathbb{R}$ be defined by $f(x) = x^2$. Prove that f is continuous on $\mathbb{R}$, but not uniformly continuous on $\mathbb{R}$.

12. Let $A \subseteq \mathbb{R}$, let $f: A \to \mathbb{R}$, let $r \in \mathbb{R}$, and suppose that $\lim_{x \to r}[f(x)] > 0$. Prove that there is a deleted neighborhood N of r such that $f(x) > 0$ for all $x \in N$.

13. Let $A \subseteq \mathbb{R}$, let $f: A \to \mathbb{R}$, let $r \in \mathbb{R}$, and suppose that $\lim_{x \to r}[f(x)]$ is a finite real number. Prove that there is $M \in \mathbb{R}$ and an open interval (a, b) containing r such that $|f(x)| \leq M$ for all $x \in (a, b) \setminus \{r\}$.

14. Let $A \subseteq \mathbb{R}$, let $f, g, h: A \to \mathbb{R}$, let $r \in \mathbb{R}$, let $f(x) \leq g(x) \leq h(x)$ for all $x \in A \setminus \{r\}$, and suppose that $\lim_{x \to r}[f(x)] = \lim_{x \to r}[h(x)] = L$. Prove that $\lim_{x \to r}[g(x)] = L$. This result is known as the **Squeeze Theorem**.

LEVEL 5

15. Let $A \subseteq \mathbb{R}$, let $f, g: A \to \mathbb{R}$ such that $g(x) \neq 0$ for all $x \in A$, let $r \in \mathbb{R}$, and suppose that $\lim_{x \to r}[f(x)]$ and $\lim_{x \to r}[g(x)]$ are both finite real numbers such that $\lim_{x \to r}[g(x)] \neq 0$. Prove that
$$\lim_{x \to r}\left[\frac{f(x)}{g(x)}\right] = \frac{\lim_{x \to r} f(x)}{\lim_{x \to r} g(x)}.$$

16. Give a reasonable equivalent definition for each of the following limits (like what was done in Theorem 8.20). r and L are finite real numbers.

   (i) $\lim_{x \to r} f(x) = -\infty$

   (ii) $\lim_{x \to +\infty} f(x) = L$

   (iii) $\lim_{x \to -\infty} f(x) = L$

   (iv) $\lim_{x \to +\infty} f(x) = +\infty$

   (v) $\lim_{x \to +\infty} f(x) = -\infty$

   (vi) $\lim_{x \to -\infty} f(x) = +\infty$

   (vii) $\lim_{x \to -\infty} f(x) = -\infty$

17. Let $f(x) = -x^2 + x + 1$. Use the $M - K$ definition of an infinite limit (that you came up with in Problem 16) to prove $\lim_{x \to +\infty} f(x) = -\infty$.

18. Give a reasonable definition for each of the following limits (like what was done in Theorem 8.22). r and L are finite real numbers.

    (i) $\lim_{x \to r^-} f(x) = L$

    (ii) $\lim_{x \to r^+} f(x) = +\infty$

    (iii) $\lim_{x \to r^+} f(x) = -\infty$

    (iv) $\lim_{x \to r^-} f(x) = +\infty$

    (v) $\lim_{x \to r^-} f(x) = -\infty$

19. Use the $M - \delta$ definition of a one-sided limit (that you came up with in Problem 18) to prove that $\lim_{x \to 3^-} \frac{1}{x-3} = -\infty$.

20. Let $f(x) = \frac{x+1}{(x-1)^2}$. Prove that

    (i) $\lim_{x \to +\infty} f(x) = 0$.

    (ii) $\lim_{x \to 1^+} f(x) = +\infty$.

21. Let $f: \mathbb{R} \to \mathbb{R}$ be defined by $f(x) = \begin{cases} 0 & \text{if } x \text{ is rational.} \\ 1 & \text{if } x \text{ is irrational.} \end{cases}$ Prove that for all $r \in \mathbb{R}$, $\lim_{x \to r}[f(x)]$ does not exist.

22. Let $f, g: \mathbb{R} \to \mathbb{R}$ be defined by $f(x) = \cos x$ and $g(x) = \sin x$. Prove that f and g are uniformly continuous on $\mathbb{R}$. **Hint:** Use the fact that the least distance between two points is a straight line.

23. Prove each of the following:

    (i) $\lim_{x \to 0} \frac{\sin x}{x} = 1$

    (ii) $\lim_{x \to 0} \frac{\cos x - 1}{x} = 0$

CHALLENGE PROBLEM

24. Prove the Bolzano-Weierstrass Theorem (Theorem 8.27). **Hint:** You should first prove that if $\mathcal{C} = \{[a_k, b_k] \mid k \in \mathbb{N}\}$ is a sequence of closed intervals in $\mathbb{R}$, such that $j < k \to [a_k, b_k] \subseteq [a_j, b_j]$, then $\cap \mathcal{C} \neq \emptyset$.

LESSON 9
TOPOLOGICAL CONTINUITY

Topological Treatment of Continuous Functions

Recall from Lesson 3 that if $f: X \to Y$ and $A \subseteq X$, then the **image of A under f** is the set $f[A] = \{f(x) \mid x \in A\}$. Similarly, if $B \subseteq Y$, then the **inverse image of B under f** is the set $f^{-1}[B] = \{x \in X \mid f(x) \in B\}$.

Also, recall from Lesson 8 (Theorem 8.9) that for $A \subseteq \mathbb{R}$, $f: A \to \mathbb{R}$ is **continuous at $a \in A$** iff

$$\forall \epsilon > 0 \; \exists \delta > 0 \; \forall x \in A(|x - a| < \delta \to |f(x) - f(a)| < \epsilon).$$

Furthermore, f is **continuous on A** if f is continuous at each $a \in A$.

Notes: (1) Continuity of a function depends just as much on the set A as it does on the function. For example, *every* function with domain $\mathbb{N}$ is continuous on $\mathbb{N}$. To see this, let $f: \mathbb{N} \to \mathbb{R}$, let $n \in \mathbb{N}$, and let $\epsilon > 0$ be given. If we simply let $\delta = 1$, then if $x \in \mathbb{N}$ and $|x - n| < 1$, then $x = n$. It follows that $|f(x) - f(n)| = |f(n) - f(n)| = 0 < \epsilon$. Similarly, every function with domain $\mathbb{Z}$ is continuous on $\mathbb{Z}$.

(2) Constant functions are always continuous. To see this, let $f: A \to \mathbb{R}$ be a constant function, say $f(x) = r$ for all $x \in A$. Let $a \in A$ and let $\epsilon > 0$ be given. Then it doesn't even matter what we choose for δ. Even if we let $\delta = 1000$, if $x \in A$ with $|x - a| < 1000$, then $|f(x) - f(a)| = |r - r| = 0 < \epsilon$. So, for example constant functions with domain $\mathbb{Q}$ are continuous.

(3) The examples given in Notes 1 and 2 above might seem to go against our intuition as well as what we learned in elementary Calculus courses. After all, the graphs of these functions in the Cartesian plane have lots of holes in them. However, if you have been exposed to basic Calculus, remember that when learning that subject, we almost always restrict our attention to functions whose domains are open or closed intervals. In those two special cases, the definition given here agrees with the definitions given in most Calculus texts. For example, see Problem 5 from Problem Set 8.

Theorem 9.1: Let $A \subseteq \mathbb{R}$, let $f: A \to \mathbb{R}$, and let $a \in A$. The following are equivalent:

1. f is continuous at a.
2. For every neighborhood $N_\epsilon(f(a))$ with center $f(a)$ and radius $\epsilon > 0$, there is a neighborhood $N_\delta(a)$ with center a and radius $\delta > 0$ such that $f[N_\delta(a) \cap A] \subseteq N_\epsilon(f(a))$.
3. If $\forall n \in \mathbb{N}(a_n \in A)$ and $a_n \to a$, then $f(a_n) \to f(a)$.
4. For each open set $V \subseteq \mathbb{R}$ with $f(a) \in V$, there is an open set $U \subseteq \mathbb{R}$ with $a \in U$ such that $f[U \cap A] \subseteq V$.

Proof: Let $A \subseteq \mathbb{R}$, let $f: A \to \mathbb{R}$, and let $a \in A$.

(1→2) Suppose that f is continuous at a and let $N_\epsilon(f(a))$ be a neighborhood with center $f(a)$ and radius $\epsilon > 0$. Since f is continuous at a, by Theorem 8.9 (part 4), there is $\delta > 0$ such that for all $x \in A$, $|x - a| < \delta$ implies $|f(x) - f(a)| < \epsilon$. It then follows that

$$f(x) \in f[N_\delta(a) \cap A] \to x \in N_\delta(a) \cap A \to |x - a| < \delta \text{ and } x \in A$$
$$\to |f(x) - f(a)| < \epsilon \to f(x) \in N_\epsilon(f(a)).$$

Therefore, $f[N_\delta(a) \cap A] \subseteq N_\epsilon(f(a))$, as desired.

(2→3) Now, suppose that 2 holds and that $a_n \to a$. Let $N_\epsilon(f(a))$ be a neighborhood with center $f(a)$ and radius ϵ. By 2, there is a neighborhood $N_\delta(a)$ with center a and radius $\delta > 0$ such that $f[N_\delta(a) \cap A] \subseteq N_\epsilon(f(a))$. Since $a_n \to a$, by Theorem 8.24 (1→3), there is $K \in \mathbb{N}$ such that $n > K$ implies $a_n \in N_\delta(a)$. Since each $a_n \in A$, we see that $n > K$ implies $f(a_n) \in f[N_\delta(a) \cap A]$. Since $f[N_\delta(a) \cap A] \subseteq N_\epsilon(f(x))$, it follows that $n > K$ implies $f(a_n) \in N_\epsilon(f(a))$. Again, by Theorem 8.24 (3→ 1), $f(a_n) \to f(a)$.

(3→4) Suppose that 4 does not hold. Then there is an open set $V \subseteq \mathbb{R}$ with $f(a) \in V$ such that for each open $U \subseteq \mathbb{R}$ with $a \in U$, there is $y \in f[U \cap A] \setminus V$. For each $n \in \mathbb{N}^+$, let $U_n = \left(a - \frac{1}{n}, a + \frac{1}{n}\right)$, let $y_n \in f[U_n \cap A] \setminus V$ and let $a_n \in U_n \cap A$ with $f(a_n) = y_n$. Let's show that $a_n \to a$, but $f(a_n) \not\to f(a)$.

To see that $a_n \to a$, let $\epsilon > 0$ and choose $K \in \mathbb{N}$ such that $K > \frac{1}{\epsilon}$ (by the Archimedean Property of $\mathbb{R}$). Then $n > K$ implies that $\frac{1}{n} < \frac{1}{K}$, and so, $a_n \in \left(a - \frac{1}{n}, a + \frac{1}{n}\right) \subseteq \left(a - \frac{1}{K}, a + \frac{1}{K}\right)$. It follows that $a - \frac{1}{K} < a_n < a + \frac{1}{K}$, or equivalently, $|a_n - a| < \frac{1}{K} < \epsilon$.

To see that $f(a_n) \not\to f(a)$, simply observe that V is an open set containing $f(a)$ and for all $n \in \mathbb{N}$, $f(a_n) \notin V$. This violates part 2 of Theorem 8.24.

(4→1) Suppose that 4 holds and let $\epsilon > 0$. Let $V = N_\epsilon(f(a)) = (f(a) - \epsilon, f(a) + \epsilon)$. Since V is open, by 4, there is an open set $U \subseteq \mathbb{R}$ with $a \in U$ such that $f[U \cap A] \subseteq V$. By Theorem 7.12, there is $\delta > 0$ such that $(a - \delta, a + \delta) \subseteq U$. Let $x \in A$. Then we have

$$|x - a| < \delta \text{ and } x \in A \to x \in (a - \delta, a + \delta) \cap A$$
$$\to x \in U \cap A \to f(x) \in f[U \cap A] \subseteq V = N_\epsilon(f(a)).$$

It follows that $|f(x) - f(a)| < \epsilon$. Since $\epsilon > 0$ was arbitrary, f is continuous at a. □

In the next Example, we will use part 2 of Theorem 9.1.

Example 9.2: Define $f: \mathbb{R} \to \mathbb{R}$ by $f(x) = \begin{cases} x & \text{if } x < 0 \\ x + 1 & \text{if } x \geq 0 \end{cases}$. Then f is **not** continuous at 0. To see this, note that $f(0) = 1 \in (0,2)$ and if $0 \in (a,b)$, then $f[(a,b)] = (a,0) \cup [1, b+1) \not\subseteq (0,2)$ because $\frac{a}{2} \in (a,0)$, so that $\frac{a}{2} < 0$, and therefore, $\frac{a}{2} \notin (0,2)$.

If $a > 0$, then f is continuous at a. To see this, let (c,d) be an open interval containing $f(a) = a + 1$. Then $c < a + 1 < d$, and so, $c - 1 < a < d - 1$. Let $k = \max\{0, c - 1\}$. Then $k < a < d - 1$. So, $a \in (k, d - 1)$. Since $k \geq 0$, $f[(k, d - 1)] = (k + 1, d)$. We now show that $(k + 1, d) \subseteq (c, d)$. Let $y \in (k + 1, d)$. Then $k + 1 < y < d$. Since $k \geq c - 1$, $k + 1 \geq c$. Thus, $c < y < d$, and therefore, $y \in (c, d)$. It follows that $f[(k, d - 1)] \subseteq (c, d)$.

Also, if $a < 0$, then f is continuous at a. To see this, let (c, d) be an open interval containing $f(a) = a$. Then $c < a < d$. Let $k = \min\{0, d\}$. Then we have $c < a < k$. So, $a \in (c, k)$. Finally, note that $f[(c, k)] = (c, k) \subseteq (c, d)$.

Theorem 9.3: Let $A \subseteq \mathbb{R}$. $f: A \to \mathbb{R}$ is continuous on A if and only if for each open set $V \subseteq \mathbb{R}$, there is an open set $U \subseteq \mathbb{R}$ such that $f^{-1}[V] = U \cap A$.

Proof: Assume that $f: A \to \mathbb{R}$ is continuous on A and let V be an open subset of $\mathbb{R}$. If $f^{-1}[V] = \emptyset$, then $f^{-1}[V] = \emptyset \cap A$ and $\emptyset$ is an open subset of $\mathbb{R}$. If $f^{-1}[V] \neq \emptyset$, let $x \in f^{-1}[V]$. Then $f(x) \in V$. So, by part 4 of Theorem 9.1, there is an open set U_x with $x \in U_x$ such that $f[U_x \cap A] \subseteq V$. Let $U = \bigcup \{U_x \mid x \in f^{-1}[V]\}$. Since U is a union of open sets, by Theorem 7.15, U is open in $\mathbb{R}$. We will show that $f^{-1}[V] = U \cap A$. Let $z \in U \cap A$. Then $z \in U$ and $z \in A$. Since $z \in U$, there is $x \in A$ with $z \in U_x$. So, we have $f(z) \in f[U_x \cap A]$. Since $f[U_x \cap A] \subseteq V$, $f(z) \in V$. Thus, $z \in f^{-1}[V]$. Since $z \in U \cap A$ was arbitrary, we have shown that $U \cap A \subseteq f^{-1}[V]$. Now, let $z \in f^{-1}[V]$. Then $f(z) \in V$. So, $z \in U_z$. Since $U_z \subseteq U$, we have $z \in U$. Since $\text{dom } f = A$ and $z \in f^{-1}[V]$, we must have $z \in A$. Therefore, $z \in U \cap A$. Since $z \in f^{-1}[V]$ was arbitrary, we have shown that $f^{-1}[V] \subseteq U \cap A$. Since $U \cap A \subseteq f^{-1}[V]$ and $f^{-1}[V] \subseteq U \cap A$, we have $f^{-1}[V] = U \cap A$.

Now assume that for each open set $V \subseteq \mathbb{R}$, there is an open set $U \subseteq \mathbb{R}$ such that $f^{-1}[V] = U \cap A$. Let $a \in A$ and let V be an open subset of $\mathbb{R}$ with $f(a) \in V$. By our assumption, there is an open set $U \subseteq \mathbb{R}$ such that $f^{-1}[V] = U \cap A$. Since $f(a) \in V$, $a \in f^{-1}[V] = U \cap A$. By Problem 5 from Problem Set 3, we have $f[U \cap A] = f[f^{-1}[V]] \subseteq V$. By part 4 of Theorem 9.1, f is continuous at a. Since $a \in A$ was arbitrary, f is continuous on A. □

Continuity on Compact Sets

We begin this section by showing that the continuous image of a compact set is compact.

Theorem 9.4: Let $A \subseteq \mathbb{R}$ be compact and let $f: A \to \mathbb{R}$ be continuous on A. Then $f[A]$ is compact.

Proof: Let $\mathcal{C}$ be an open covering of $f[A]$. By Theorem 9.3, for each $V \in \mathcal{C}$, there is an open set $U \subseteq \mathbb{R}$ such that $f^{-1}[V] = U \cap A$. Let $\mathcal{D} = \{U \mid U \text{ is open and } \exists V \in \mathcal{C} \, (f^{-1}[V] = U \cap A)\}$. If $x \in A$, then since $\mathcal{C}$ is an open covering of $f[A]$, there is $V \in \mathcal{C}$ with $f(x) \in V$. Let $U \subseteq \mathbb{R}$ be an open set such that $f^{-1}[V] = U \cap A$. Since $f(x) \in V$, we have $x \in f^{-1}[V]$. So, $x \in U$ and $U \in \mathcal{D}$. Therefore, $x \in \bigcup \mathcal{D}$. This shows that $A \subseteq \bigcup \mathcal{D}$, and so, $\mathcal{D}$ is an open covering of A. Since A is compact, there is a finite subset $\{U_0, U_1, \ldots, U_n\} \subseteq \mathcal{D}$ such that $A \subseteq U_0 \cup U_1 \cup \cdots \cup U_n$. Let $V_0, V_1, \ldots, V_n \in \mathcal{C}$ satisfy $f^{-1}[V_0] = U_0 \cap A$, $f^{-1}[V_1] = U_1 \cap A, \ldots, f^{-1}[V_n] = U_n \cap A$. Then we have

$$A \subseteq (U_0 \cup U_1 \cup \cdots \cup U_n) \cap A = (U_0 \cap A) \cup (U_1 \cap A) \cup \cdots \cup (U_n \cap A)$$
$$= f^{-1}[V_0] \cup f^{-1}[V_1] \cup \cdots \cup f^{-1}[V_n].$$

By Problem 5 and part (i) of Problem 14, both from Problem Set 3, we have

$$f[A] \subseteq f[f^{-1}[V_0] \cup f^{-1}[V_1] \cup \cdots \cup f^{-1}[V_n]] = f[f^{-1}[V_0]] \cup f[f^{-1}[V_1]] \cup \cdots \cup f[f^{-1}[V_n]]$$
$$\subseteq V_0 \cup V_1 \cup \cdots \cup V_n.$$

It follows that $\{V_0, V_1, \ldots V_n\}$ is a finite subcover of $f[A]$, proving that $f[A]$ is compact. □

Let $A \subseteq \mathbb{R}$. A function $f: A \to \mathbb{R}$ is **bounded** on A if ran f is bounded. In other words, f is bounded if there is $M \in \mathbb{R}^+$ such that for all $x \in A$, $|f(x)| \leq M$ (see the discussion before the statement of Theorem 7.28).

Example 9.5:
1. Define $f:(0,1) \to \mathbb{R}$ by $f(x) = 2x$. Then f is bounded on $(0,1)$. Indeed, if $x \in (0,1)$, then $0 < x < 1$, and so, $0 < 2x < 2$. It follows that $-2 \leq 2x \leq 2$, or equivalently, $|f(x)| \leq 2$.
2. Define $f:(0,5) \to \mathbb{R}$ by $f(x) = \frac{1}{x}$. Then f is **not** bounded on $(0,5)$. To see this, let $M \in \mathbb{R}^+$ and let $x = \min\left\{1, \frac{1}{M+1}\right\}$. If $M < 1$, then $f(x) = \frac{1}{x} \geq 1 > M$ and if $M \geq 1$, then $M + 1 \geq 2$, and so, $\frac{1}{M+1} \leq \frac{1}{2} < 1$. Thus, $x = \frac{1}{M+1}$, and therefore, $f(x) = M + 1 > M$.

Corollary 9.6 (Boundedness Theorem): Let $A \subseteq \mathbb{R}$ be closed and bounded and let $f: A \to \mathbb{R}$ be continuous on A. Then $f[A]$ is closed and bounded in $\mathbb{R}$. It follows that f is a bounded function.

Proof: This follows immediately from Theorem 9.4 and the Heine-Borel Theorem (Theorem 7.28).

Corollary 9.7 (Maximum-Minimum Theorem): Let $A \subseteq \mathbb{R}$ be nonempty, closed, and bounded. Also, let $f: A \to \mathbb{R}$ be continuous on A. Then $\sup f[A]$ and $\inf f[A]$ are both in $f[A]$.

Proof: By Corollary 9.6, $f[A]$ is closed and bounded. Since $f[A]$ is closed, by part 4 of Theorem 7.25, we have $f[A] = \overline{f[A]}$. Since $f[A]$ is a nonempty bounded subset of $\mathbb{R}$, by Lemma 7.31, $\sup f[A]$ and $\inf f[A]$ are both in $\overline{f[A]} = f[A]$. $\square$

Notes: (1) $\sup f[A]$ is called the **absolute maximum** for f on A and $\inf f[A]$ is called the **absolute minimum** for f on A. An **absolute extremum** is either an absolute maximum or absolute minimum.

(2) The absolute maximum and minimum for a function on a set A can occur at more than one location. For example, define $f:[-2,2] \to \mathbb{R}$ by $f(x) = x^2$. Then $f(-2) = (-2)^2 = 4$ and $f(2) = 2^2 = 4$. The absolute maximum for f on $[-2,2]$ is 4 and it occurs at both $x = -2$ and $x = 2$.

Continuity on Connected Sets

Theorem 9.8: Let $A \subseteq \mathbb{R}$ be connected and let $f: A \to \mathbb{R}$ be a continuous function. Then $f[A]$ is connected.

Proof: Let $A \subseteq \mathbb{R}$, let $f: A \to \mathbb{R}$ be a continuous function, and suppose that $f[A]$ is not connected. Let (U, V) be a disconnection of $f[A]$. Then U and V are nonempty open sets in $\mathbb{R}$ with $(U \cap f[A]) \neq \emptyset$, $(V \cap f[A]) \neq \emptyset$, $(U \cap V) \cap f[A] = \emptyset$ and $(U \cup V) \cap f[A] = f[A]$. Since f is continuous, by Theorem 9.3, there are open sets W_1 and W_2 such that $f^{-1}[U] = W_1 \cap A$ and $f^{-1}[V] = W_2 \cap A$.

Since $U \cap f[A] \neq \emptyset$, there is $y \in U \cap f[A]$. Then $y \in U$ and $y \in f[A]$. Since $y \in f[A]$, there is $x \in A$ with $y = f(x)$. Since $f(x) = y \in U$, $x \in f^{-1}[U]$. So, $x \in f^{-1}[U] \cap A$, and therefore, $f^{-1}[U] \cap A \neq \emptyset$. Similarly, $f^{-1}[V] \cap A \neq \emptyset$.

If we could find $x \in (f^{-1}[U] \cap f^{-1}[V]) \cap A$, then $f(x) \in (U \cap V) \cap f[A]$, contradicting our assumption that $(U \cap V) \cap f[A] = \emptyset$. So, $(f^{-1}[U] \cap f^{-1}[V]) \cap A = \emptyset$.

If $x \in A$ is arbitrary, then we have $f(x) \in f[A] = (U \cup V) \cap f[A]$. So, $f(x) \in U$ or $f(x) \in V$. Thus, $x \in f^{-1}[U]$ or $x \in f^{-1}[V]$. So, $x \in (f^{-1}[U] \cap f^{-1}[V]) \cap A$. Therefore, $(f^{-1}[U] \cap f^{-1}[V]) \cap A = A$. So, $(f^{-1}[U], f^{-1}[V])$ is a disconnection of A. It follows that A is not connected. □

Corollary 9.9 (Intermediate Value Theorem): Let $A \subseteq \mathbb{R}$ be connected and let $f: A \to \mathbb{R}$ be a continuous function. Then $f[A]$ is an interval.

Proof: Let $A \subseteq \mathbb{R}$ be connected and let $f: A \to \mathbb{R}$ be continuous. By Theorem 9.8, $f[A]$ is a connected subset of $\mathbb{R}$. By Theorem 7.35, $f[A]$ is an interval. □

Homeomorphisms

Let $A \subseteq \mathbb{R}$ and let $f: A \to \mathbb{R}$ be a continuous injective function. If we let $B = f[A]$, then we can also write $f: A \to B$. By restricting the codomain of f in this way, we see that $f: A \to B$ is a bijection. Therefore, f^{-1} exists. If f^{-1} is also continuous, then we say that $f: A \to B$ is a **homeomorphism**. If there exists a homeomorphism $f: A \to B$, then we say that A and B are **homeomorphic** or **topologically equivalent**.

Example 9.10:

1. Define $f: \mathbb{R} \to \mathbb{R}$ by $f(x) = 2x + 3$. Let's check that f is a homeomorphism.

   If $x \neq y$, then $2x \neq 2y$, and so, $2x + 3 \neq 2y + 3$. Thus, $\forall x, y \in \mathbb{R}(x \neq y \to f(x) \neq f(y))$. That is, f is injective.

   Next, if $y \in \mathbb{R}$, let $x = \frac{y-3}{2}$. Then $f(x) = f\left(\frac{y-3}{2}\right) = 2\left(\frac{y-3}{2}\right) + 3 = (y - 3) + 3 = y$. So, $\forall y \in \mathbb{R} \exists x \in \mathbb{R}(f(x) = y)$. That is, f is surjective.

   Now, let (a, b) be a bounded open interval. $f^{-1}[(a,b)] = \left(\frac{a-3}{2}, \frac{b-3}{2}\right)$, which is open. So, f is continuous.

   Also, $f[(a, b)] = (2a + 3, 2b + 3)$, which is open. So, f^{-1} is continuous.

   Since f is a continuous bijection with a continuous inverse, f is a homeomorphism.

2. If we restrict the function f from part 1 above to the open interval $(0, 1)$, then we see that $f: (0, 1) \to (3, 5)$. The argument that f is injective is the same. For surjectivity, note that if $3 < y < 5$, then $0 < y - 3 < 2$, and so, $0 < \frac{y-3}{2} < 1$, and as shown above, $f\left(\frac{y-3}{2}\right) = y$. I leave it to the reader to prove that f and f^{-1} are continuous. So, we see that $(0, 1)$ and $(3, 5)$ are homeomorphic.

   A similar argument can be used to show that the closed intervals $[0, 1]$ and $[3, 5]$ are homeomorphic.

3. Any two open intervals are homeomorphic (see Problem 7 below) and any two closed intervals are homeomorphic. An open interval and a closed interval are never homeomorphic (see Problem 2 below).

Problem Set 9

Full solutions to these problems are available for free download here:
www.SATPrepGet800.com/RABQXZ

LEVEL 1

1. Let $A \subseteq \mathbb{R}$ and let $f: A \to \mathbb{R}$. Provide a counterexample showing that each of the following is false.

    (i) If A is closed and f is continuous, then f is bounded.

    (ii) If A is bounded and f is continuous, then f is bounded.

    (iii) If A is closed and bounded, then f is bounded.

    (iv) If A is a closed and f is continuous then $f[A]$ is closed.

    (v) If A is open and f is continuous, then $f[A]$ is open.

2. Let I be a closed bounded interval and let J be an open or half-open interval. Prove that I and J are not homeomorphic.

LEVEL 2

3. Let $A \subseteq \overline{A} \subseteq B \subseteq \mathbb{R}$ and let $f, g: B \to \mathbb{R}$ be continuous. Suppose that $f(x) = g(x)$ for all $x \in A$. Prove that $f(x) = g(x)$ for all $x \in \overline{A}$.

LEVEL 3

4. Let $A \subseteq \mathbb{R}$ and let $f: A \to \mathbb{R}$. Prove that f is continuous on A if and only if for each closed set $C \subseteq \mathbb{R}$, there is a closed set $B \subseteq \mathbb{R}$ such that $f^{-1}[C] = B \cap A$.

5. Prove that the image of a Cauchy sequence under a uniformly continuous function is a Cauchy sequence. If we replace "uniformly continuous" by "continuous," is the result still true?

6. Let $S \subseteq \mathbb{R}$, let A and B be disjoint closed subsets of S, and let a and b be real numbers with $a < b$. Use Urysohn's Lemma (see Problem 13 below) to prove that there is a continuous function $f: S \to [a, b]$ such that $f[A] = \{a\}$ and $f[B] = \{b\}$.

7. Prove that any two open intervals are homeomorphic. (You may exclude the interval $(-\infty, \infty)$.)

LEVEL 4

8. Let $A \subseteq \mathbb{R}$ and let $f: A \to \mathbb{R}$. Prove that f is continuous on A if and only if for each $X \subseteq A$, $f[\overline{X} \cap A] \subseteq \overline{f[X]}$.

9. Let $A \subseteq \mathbb{R}$ be compact and let $f: A \to \mathbb{R}$ be continuous and injective. Prove that f is a homeomorphism.

LEVEL 5

10. Let $A \subseteq \mathbb{R}$ be compact and let $f: A \to \mathbb{R}$ be continuous on A. Prove that f is uniformly continuous on A.

11. Let $C \subseteq A \subseteq \mathbb{R}$ with C dense in A and suppose that $f: C \to \mathbb{R}$ is uniformly continuous. Prove that f can be extended uniquely to a uniformly continuous function $g: A \to \mathbb{R}$.

CHALLENGE PROBLEMS

12. If $A \subseteq \mathbb{R}$ and $f: A \to \mathbb{R}$ satisfies $\forall x, y \in A \big(x < y \to f(x) < f(y)\big)$, then we say that f is a **strictly increasing function**. Prove that a strictly increasing continuous function whose domain is a closed interval is a homeomorphism onto its range such that the inverse function is also strictly increasing. Then state and prove the analogous result for a **strictly decreasing function**.

13. Let X and Y be closed subsets of $\mathbb{R}$ with $X \cap Y = \emptyset$. Prove that there is a continuous function $f: \mathbb{R} \to [0, 1]$ such that $f[X] = \{0\}$ and $f[Y] = \{1\}$. This result is known as **Urysohn's Lemma**.

14. Let $X \subseteq \mathbb{R}$ be a closed set, let $a, b \in \mathbb{R}$ with $a < b$, and let $f: X \to [a, b]$ be continuous. Prove that f can be extended to a continuous function $g: \mathbb{R} \to [a, b]$. This result is known as the **Tietze Extension Theorem**.

15. Prove that $\{f \in {}^{\mathbb{R}}\mathbb{R} \mid f \text{ is continuous}\}$ is equinumerous with $\mathbb{R}$.

LESSON 10
DIFFERENTIATION

Linear Functions and Slope

Recall from Lesson 3 that a function $f: \mathbb{R} \to \mathbb{R}$ of the form $f(x) = mx + b$ (where $m, b \in \mathbb{R}$) is called a **linear function** (see Example 3.15). The graph of such a function is called a **line**. If (x_1, y_1) and (x_2, y_2) are any two points on the line, then the **slope** of the line is

$$\text{Slope} = m = \frac{\text{rise}}{\text{run}} = \frac{y_2 - y_1}{x_2 - x_1}$$

Note: We will sometimes write $y = mx + b$ instead of $f(x) = mx + b$. The first equation emphasizes that (x, y) is an arbitrary point on the line that is the graph of the equation, while the second equation emphasizes that $y = f(x)$ is a function of x. Either of these forms is known as the **slope-intercept form of the equation of the line**.

Example 10.1: Let's find the slope of the line passing through the points $(3, -5)$ and $(7, 2)$.

Method 1: $\frac{2-(-5)}{7-3} = \frac{2+5}{7-3} = \frac{7}{4}$.

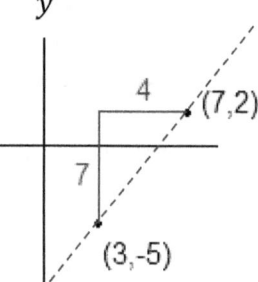

Method 2: We plot the two points as shown to the right and observe that to get from $(3, -5)$ to $(7, 2)$ we move up 7 and right 4. Therefore, the slope is $\frac{7}{4}$.

Notes: (1) If you cannot see where the 7 and 4 come from visually, then you can formally find the differences: $2 - (-5) = 7$ and $7 - 3 = 4$.

(2) Now that we know that the slope of the line is $m = \frac{7}{4}$, we see that the linear function has the form $f(x) = \frac{7}{4}x + b$, where b is some constant. We can find b by using any point on the line. For example, since we know that $(7, 2)$ is a point on the line, we have $2 = f(7) = \frac{7}{4} \cdot 7 + b = \frac{49}{4} + b$. Therefore, we have $b = 2 - \frac{49}{4} = \frac{8}{4} - \frac{49}{4} = -\frac{41}{4}$. So, the linear function is

$$f(x) = \frac{7}{4}x - \frac{41}{4}.$$

(3) Observe that $f(3) = \frac{7}{4} \cdot 3 - \frac{41}{4} = \frac{21}{4} - \frac{41}{4} = -\frac{20}{4} = -5$. So, $(3, -5)$ is on the graph of f, as it should be.

(4) Notice that $f(0) = \frac{7}{4} \cdot 0 - \frac{41}{4} = -\frac{41}{4}$. So, the point $\left(0, -\frac{41}{4}\right)$ is on the line. This point is called the **y-intercept** of the line because it lies on the y-axis. More generally, the linear function f defined by $f(x) = mx + b$ has a y-intercept of $(0, b)$.

Although the y-intercept of the line with equation $y = mx + b$ is the point $(0, b)$, we will sometimes use abbreviated language and say that the y-intercept of the line is b.

Lines with positive slope have graphs that go upwards from left to right. Lines with negative slope have graphs that go downwards from left to right. If the slope of a line is zero, it is horizontal. Vertical lines have **infinite** slope.

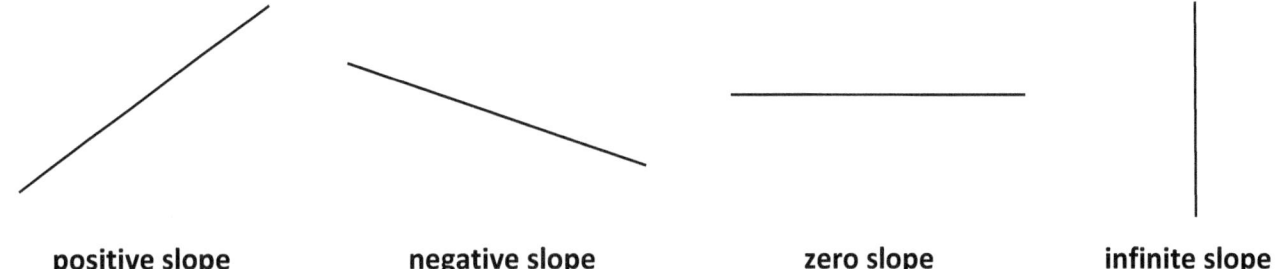

positive slope   negative slope   zero slope   infinite slope

Example 10.2:

1. $y = 2x + 7$ is an equation of a line with slope 2 and y-intercept 7.

2. $y = -x - 5$ is an equation of a line with slope -1 and y-intercept -5 (Note: $-x = (-1)x$)

3. $y = -6$ is an equation of a line with slope 0 and y-intercept -6 ($y = 0x - 6$)

4. $x = 3$ is an equation of a line with **infinite** slope and **no** y-intercept.

Example 10.3: Let's sketch graphs of the following linear equations on the same set of coordinate axes:

1. $y = -x$

2. $y = -x + 2$

3. $y = 2$

4. $x = 2$

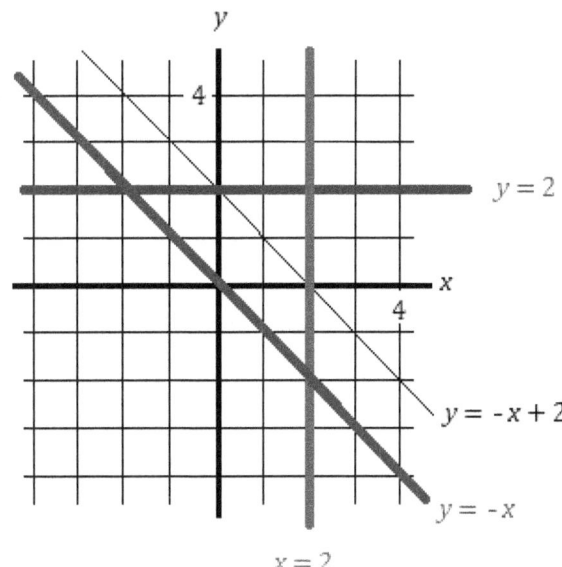

Secant Lines

Let $A \subseteq \mathbb{R}$, let $f: A \to \mathbb{R}$, and let $a, b \in A$. The line passing through the points $(a, f(a))$ and $(b, f(b))$ is called a **secant line**.

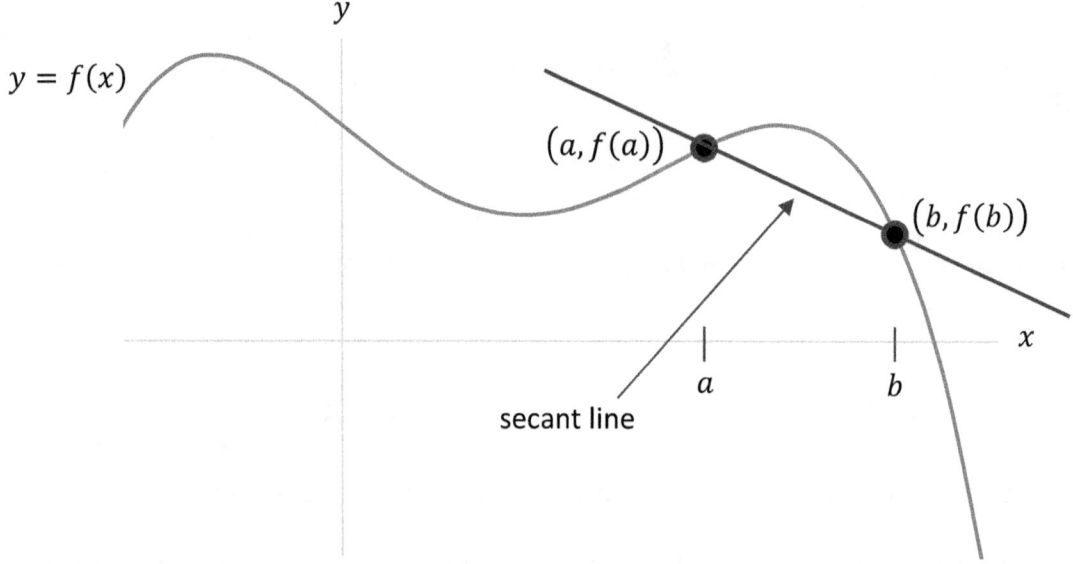

Notes: (1) The slope of the secant line shown in the picture above is $m = \frac{f(b)-f(a)}{b-a}$. This expression is sometimes called a **difference quotient**.

(2) Once we find the slope of the secant line using the formula given in Note 1 above, we can write an equation of the secant line in slope-intercept form by using the same method that we used in Note 2 following Example 10.1. See Note 2 following Example 10.4 for a specific instance of this.

Example 10.4: Define $f: \mathbb{R} \to \mathbb{R}$ by $f(x) = x^2$. Let $a = 1$ and $b = 2$. Then $f(a) = 1^2 = 1$ and $f(b) = 2^2 = 4$. Let's take a look at the graph of f together with the secant line passing through the points $(1, 1)$ and $(2, 4)$.

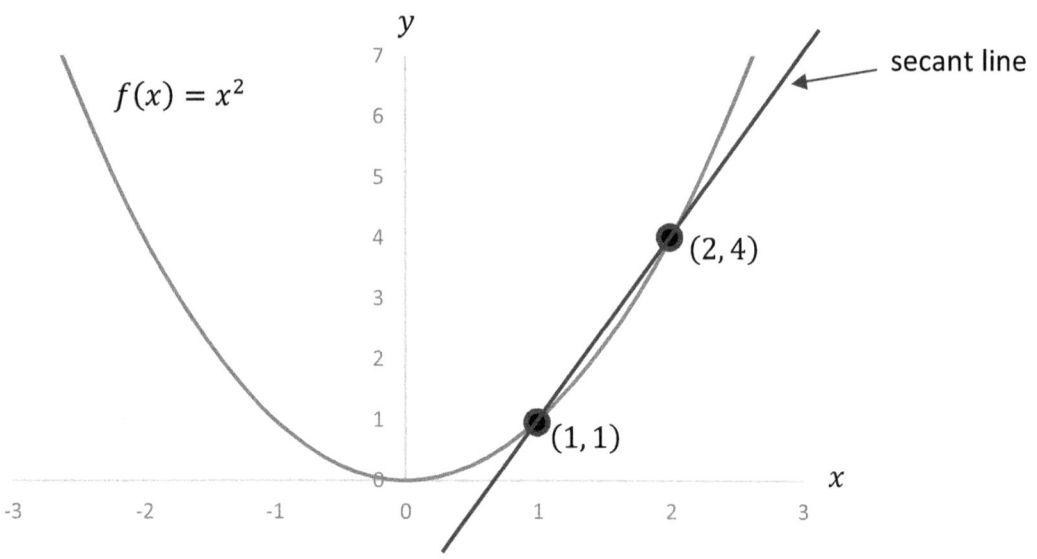

Notes: (1) The slope of the secant line shown in the picture above is $m = \frac{4-1}{2-1} = \frac{3}{1} = 3$.

(2) Let's find an equation of the secant line, as we did in Note 2 following Example 10.1. Since the slope of the line is $m = 3$, an equation of the secant line is $y = 3x + b$, where b is some constant. Since $(1, 1)$ is on the secant line, $1 = f(1) = 3 \cdot 1 + b = 3 + b$. Therefore, $b = 1 - 3 = -2$. So, the equation of the secant line in slope-intercept form is $y = 3x - 2$.

Example 10.5: As in Example 10.4, define $f : \mathbb{R} \to \mathbb{R}$ by $f(x) = x^2$. This time let a and b be arbitrary real numbers with $a \neq b$. Then the slope of the secant line passing through the points $(a, f(a)) = (a, a^2)$ and $(b, f(b)) = (b, b^2)$ is

$$m = \frac{f(b) - f(a)}{b - a} = \frac{b^2 - a^2}{b - a} = \frac{(b - a)(b + a)}{b - a} = b + a.$$

So, for example if $a = 1$ and $b = 2$, then the corresponding points are $(1, 1)$ and $(2, 4)$ and the slope of the secant line passing through these two points is $b + a = 2 + 1 = 3$, as we have seen in Note 1 following Example 10.4.

As another example, if $a = 1$ and $b = 1.5$, then the corresponding points are $(1, 1)$ and $(1.5, 2.25)$ and the slope of the secant line passing through these two points is $b + a = 1.5 + 1 = 2.5$.

In general, if $a = 1$ and b is a little more than 1, say $b = 1 + h$ for some small positive value of h, then the slope of the secant line passing through these two points is $b + a = (1 + h) + 1 = 2 + h$. Informally, we can say that if b is a little more than 1, then the slope of the secant line passing through the points $(1, 1)$ and $(b, f(b)) = (1 + h, (1 + h)^2)$ is a little more than 2.

Let's zoom in a bit on the graph of $f(x) = x^2$ and see what such a typical secant line looks like.

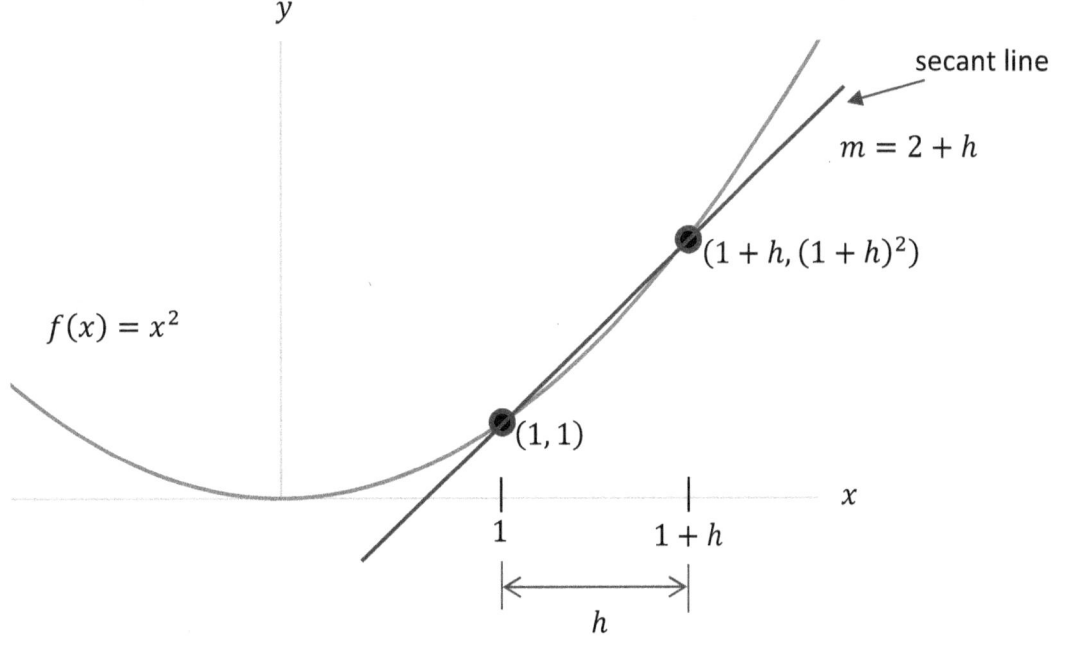

Example 10.5 suggests that we may want to look at a difference quotient (the formula for the slope of a secant line) in a slightly different way.

Given the two distinct real numbers a and b, if we let $h = b - a$, then we see that $b = a + h$ and $h \neq 0$. This observation allows us to write a difference quotient in a way that turns out to be particularly useful. We have

$$m = \frac{f(b) - f(a)}{b - a} = \frac{f(a + h) - f(a)}{(a + h) - a} = \frac{f(a + h) - f(a)}{h}.$$

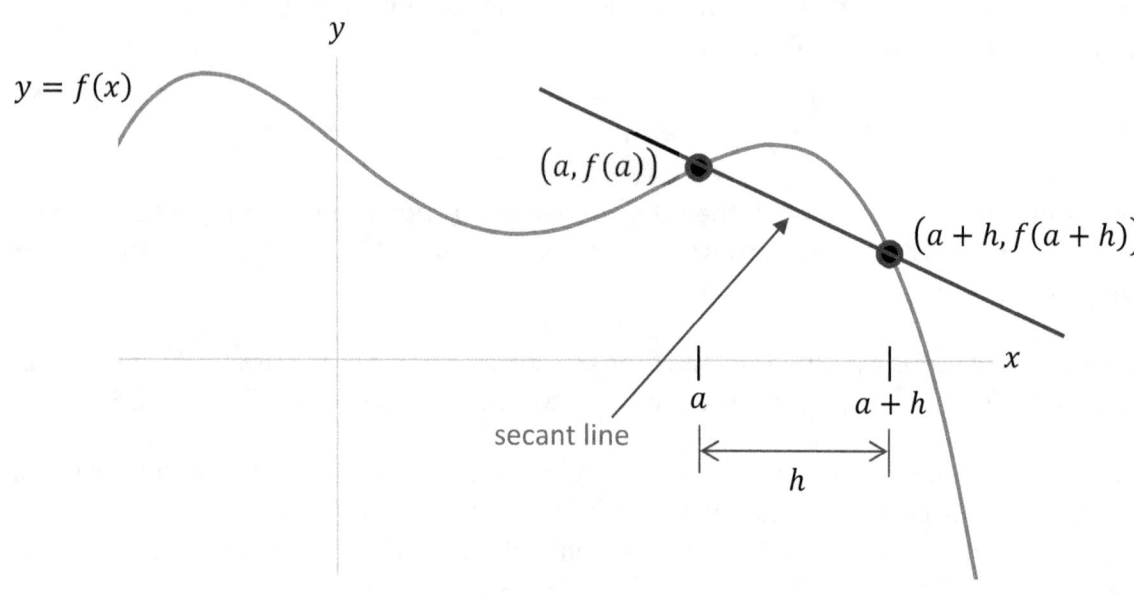

Example 10.6: As in Example 10.4, define $f \colon \mathbb{R} \to \mathbb{R}$ by $f(x) = x^2$. Let $a = 1$ and $h = 1$. Note that this is equivalent to what we did in Example 10.4 because $b = a + h = 1 + 1 = 2$. So, $f(a) = 1^2 = 1$ and $f(a + h) = f(1 + 1) = f(2) = 2^2 = 4$.

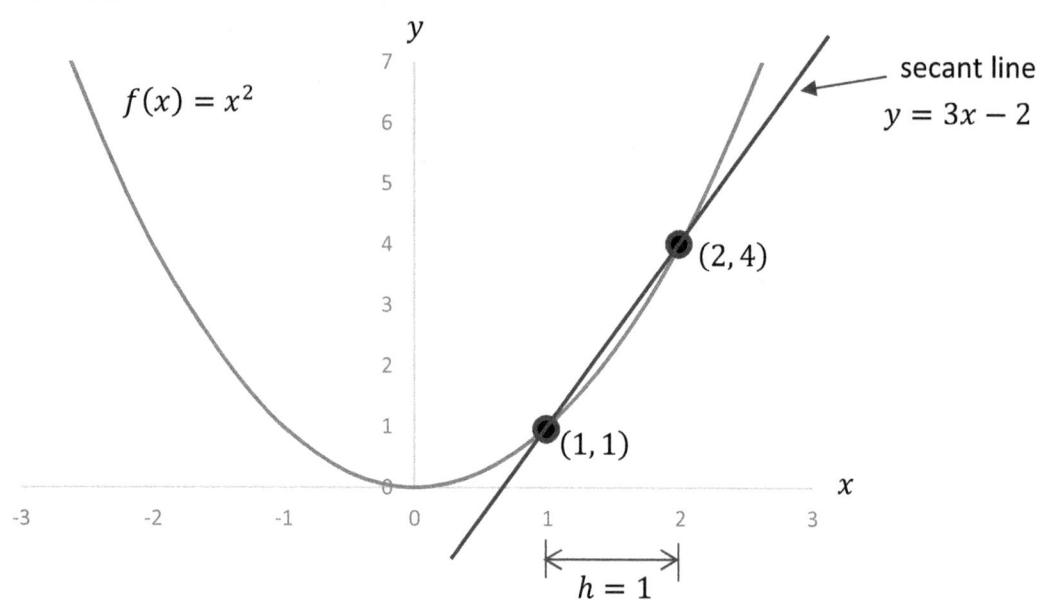

If we let $a = 1$ and $h = 0.5$, then we have $f(a + h) = f(1 + 0.5) = f(1.5) = (1.5)^2 = 2.25$. This time, the difference quotient (slope of the secant line) is $m = \frac{f(a+h)-f(a)}{h} = \frac{2.25-1}{0.5} = \frac{1.25}{0.5} = 2.5$.

Let's find an equation of the secant line. We have $y = 2.5x + b$, for some constant b. Since $(1, 1)$ is on the secant line, $1 = 2.5 \cdot 1 + b = 2.5 + b$. Therefore, $b = 1 - 2.5 = -1.5$. So, the equation of the secant line in slope-intercept form is $y = 2.5x - 1.5$.

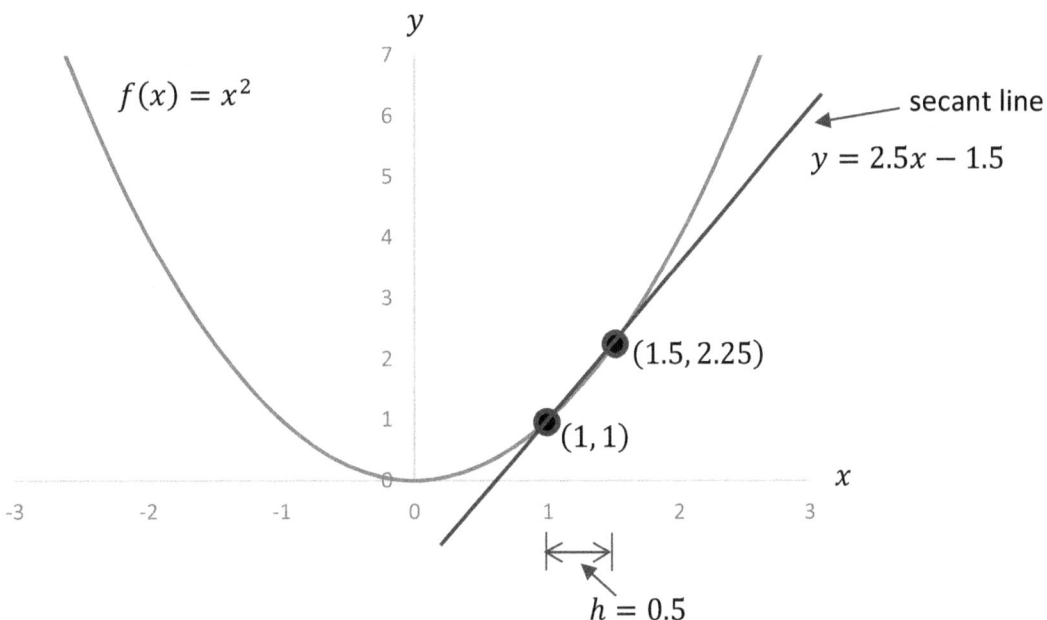

Example 10.7: As in Example 10.6, define $f: \mathbb{R} \to \mathbb{R}$ by $f(x) = x^2$. This time let a and h be arbitrary real numbers with $h \neq 0$. Then the slope of the secant line passing through the points $(a, f(a)) = (a, a^2)$ and $(a + h, f(a + h)) = (a + h, (a + h)^2)$ is

$$m = \frac{f(a+h) - f(a)}{h} = \frac{(a+h)^2 - a^2}{h} = \frac{(a+h)(a+h) - a^2}{h} = \frac{(a^2 + 2ah + h^2) - a^2}{h}$$
$$= \frac{(a^2 - a^2) + (2ah + h^2)}{h} = \frac{0 + (2ah + h^2)}{h} = \frac{h(2a + h)}{h} = 2a + h.$$

For example, if $a = 1$ and $h = 1$, then the slope of the secant line passing through $(1, 1)$ and $(1 + 1, (1 + 1)^2) = (2, 4)$ is $2a + h = 2 \cdot 1 + 1 = 2 + 1 = 3$, as we have seen in Example 10.6.

As another example, if $a = 1$ and $h = 0.5$, then the slope of the secant line passing through $(1, 1)$ and $(1 + 0.5, (1 + 0.5)^2) = (1.5, 2.25)$ is $2a + h = 2 \cdot 1 + 0.5 = 2 + 0.5 = 2.5$.

In general, if $a = 1$ and h is some small nonzero real number, then the slope of the secant line passing through $(1, 1)$ and $(1 + h, (1 + h)^2)$ is $2a + h = 2 \cdot 1 + h = 2 + h$. Informally, we can say that if h is close to 0, then the slope of the secant line passing through the points $(1, 1)$ and $(1 + h, (1 + h)^2)$ is close to 2.

The picture after Example 10.5 shows what a typical such secant line looks like when h is positive. The dedicated reader may want to draw a similar picture showing a typical secant line when h is negative.

Tangent Lines

Let $A \subseteq \mathbb{R}$, let $f: A \to \mathbb{R}$, and let $a \in A$. The **tangent line** to the graph of $y = f(x)$ at $x = a$ is the line passing through the point $(a, f(a))$ with slope

$$m = \lim_{h \to 0} \frac{f(a+h) - f(a)}{h},$$

provided that this limit is a finite real number. If the given limit is infinite, then the tangent line will be vertical. If the limit does not exist, then neither does the tangent line.

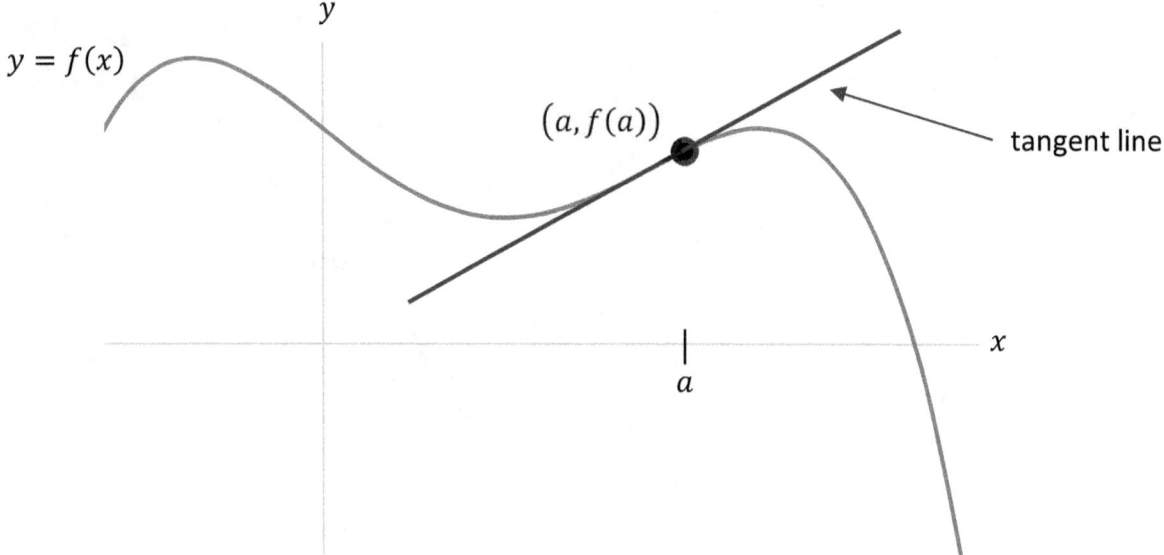

Notice that the expression inside the limit (namely, $\frac{f(a+h)-f(a)}{h}$) is the slope of the secant line passing through the points $(a, f(a))$ and $(a+h, f(a+h))$. So, if h is "near" 0, then the slope of the corresponding secant line gives a "good" approximation to the slope of the tangent line. By taking smaller and smaller values for h, we get "closer and closer" to the actual value of the slope of the tangent line.

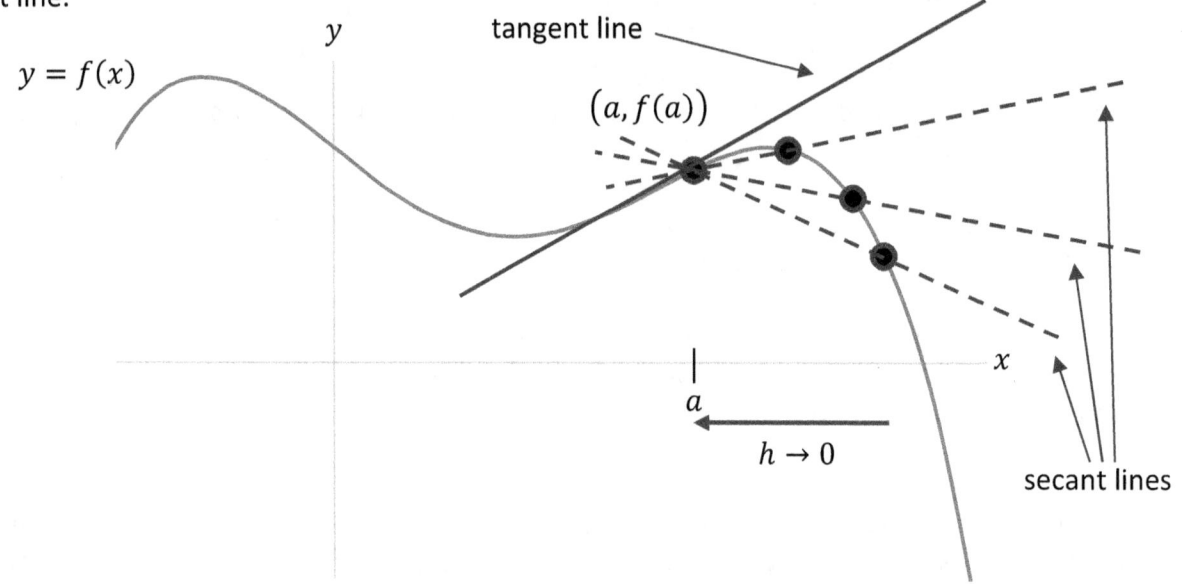

Example 10.8: Define $f:\mathbb{R} \to \mathbb{R}$ by $f(x) = x^2$ and let a and h be arbitrary real numbers with $h \neq 0$. In Example 10.7 we saw that

$$\frac{f(a+h) - f(a)}{h} = 2a + h.$$

It follows that the slope of the tangent line to the graph of $y = x^2$ at $x = a$ is

$$m = \lim_{h \to 0} \frac{f(a+h) - f(a)}{h} = \lim_{h \to 0}(2a + h) = 2a.$$

For example, at $x = 1$, the slope of the tangent line is $2 \cdot 1 = 2$.

Let's find an equation of this tangent line, as we did in Note 2 following Example 10.1. Since the slope of the line is $m = 2$, an equation of the tangent line is $y = 2x + b$, where b is some constant. Since $(1, 1)$ is on the tangent line, $1 = 2 \cdot 1 + b = 2 + b$. Therefore, $b = 1 - 2 = -1$. So, the equation of the tangent line in slope-intercept form is $y = 2x - 1$. See the figure below.

As another example, at $x = 0$, the slope of the tangent line is $2 \cdot 0 = 0$. In other words, at the point $(0, 0)$ the tangent line is horizontal. The equation of this tangent line is $y = 0$ (Check this!).

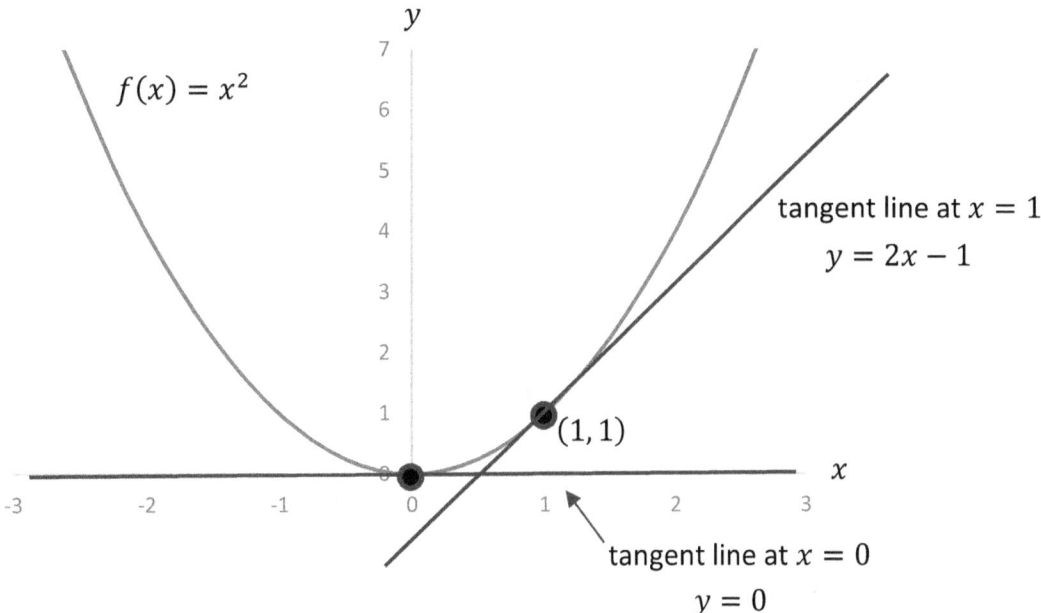

Recall that there are two equivalent ways to define a difference quotient. We have used the second definition, $\frac{f(a+h) - f(a)}{h}$, to evaluate the slope of a tangent line. We could have just as easily used the first definition of a difference quotient. In this case, the slope of the tangent line to the graph of $y = f(x)$ at $x = a$ (if it exists) can be found with the following computation:

$$m = \lim_{b \to a} \frac{f(b) - f(a)}{b - a}$$

Example 10.9: Define $f: \mathbb{R} \to \mathbb{R}$ by $f(x) = x^2$ and let a and b be arbitrary real numbers with $a \neq b$. In Example 10.5 we saw that

$$\frac{f(b) - f(a)}{b - a} = b + a.$$

It follows that the slope of the tangent line to the graph of $y = x^2$ at $x = a$ is

$$m = \lim_{b \to a} \frac{f(b) - f(a)}{b - a} = \lim_{b \to a}(b + a) = a + a = 2a.$$

Notice that this is the same result that we got in Example 10.8.

The Derivative

Let $A \subseteq \mathbb{R}$ and let $a \in A$. We say that a is an **interior point** of A if there is a neighborhood N of a (in other words, an open interval with center a) such that $N \subseteq A$.

If $f: A \to \mathbb{R}$, then we say that f is **differentiable** at an interior point $a \in A$ if $f'(a)$ exists, where

$$f'(a) = \lim_{h \to 0} \frac{f(a + h) - f(a)}{h},$$

When $f'(a)$ exists, it is called the **derivative** of f at $x = a$.

From our previous discussion, when $f'(a)$ exists, it can also be written as follows:

$$f'(a) = \lim_{x \to a} \frac{f(x) - f(a)}{x - a}$$

Example 10.10:

1. If we define $f: \mathbb{R} \to \mathbb{R}$ by $f(x) = 3x^2 - 2$, then for any $a \in \mathbb{R}$, we have

   $$f(a + h) = 3(a + h)^2 - 2 = 3(a^2 + 2ah + h^2) - 2 = 3a^2 + 6ah + 3h^2 - 2.$$

   It follows that

   $$f'(a) = \lim_{h \to 0} \frac{f(a + h) - f(a)}{h} = \lim_{h \to 0} \frac{(3a^2 + 6ah + 3h^2 - 2) - (3a^2 - 2)}{h}$$
   $$= \lim_{h \to 0} \frac{(3a^2 - 3a^2) + 6ah + 3h^2 - (2 - 2)}{h} = \lim_{h \to 0} \frac{6ah + 3h^2}{h} = \lim_{h \to 0} \frac{h(6a + 3h)}{h}$$
   $$= \lim_{h \to 0}(6a + 3h) = 6a.$$

2. If we define $g: [0, \infty) \to \mathbb{R}$ by $g(x) = \sqrt{x}$, then for any $a \in (0, \infty)$ (remember that a must be an interior point of $[0, \infty)$), we have

   $$g'(a) = \lim_{h \to 0} \frac{\sqrt{a + h} - \sqrt{a}}{h} = \lim_{h \to 0} \frac{(\sqrt{a + h} - \sqrt{a})(\sqrt{a + h} + \sqrt{a})}{h(\sqrt{a + h} + \sqrt{a})}$$
   $$= \lim_{h \to 0} \frac{a + h - a}{h(\sqrt{a + h} + \sqrt{a})} = \lim_{h \to 0} \frac{h}{h(\sqrt{a + h} + \sqrt{a})} = \lim_{h \to 0} \frac{1}{\sqrt{a + h} + \sqrt{a}} = \frac{1}{\sqrt{a} + \sqrt{a}} = \frac{1}{2\sqrt{a}}.$$

3. If we define $h: \mathbb{R} \to \mathbb{R}$ by $h(x) = \sin x$, then for any $a \in \mathbb{R}$, by Theorem 3.25, we have
$$h(a + h) = \sin(a + h) = \sin a \cos h + \cos a \sin h.$$

It follows that

$$h'(a) = \lim_{h \to 0} \frac{\sin(a+h) - \sin a}{h} = \lim_{h \to 0} \frac{\sin a \cos h + \cos a \sin h - \sin a}{h}$$

$$= \lim_{h \to 0} \frac{(\sin a \cos h - \sin a) + \cos a \sin h}{h} = \lim_{h \to 0} \frac{\sin a \, (\cos h - 1) + \cos a \sin h}{h}$$

$$= \lim_{h \to 0} \frac{\sin a \, (\cos h - 1)}{h} + \lim_{h \to 0} \frac{\cos a \sin h}{h} = (\sin a) \lim_{h \to 0} \frac{\cos h - 1}{h} + (\cos a) \lim_{h \to 0} \frac{\sin h}{h}$$

$$= (\sin a)(0) + (\cos a)(1) = \cos a \text{ (by Problem 23 in Problem Set 8).}$$

Note: In Example 10.10, we could have used the second definition of the derivative. For example, for the function $f: \mathbb{R} \to \mathbb{R}$ defined by $f(x) = 3x^2 - 2$ from part 1, we could have done the following computation instead:

$$f'(a) = \lim_{x \to a} \frac{f(x) - f(a)}{x - a} = \lim_{x \to a} \frac{(3x^2 - 2) - (3a^2 - 2)}{x - a} = \lim_{x \to a} \frac{3x^2 - 3a^2}{x - a} = \lim_{x \to a} \frac{3(x^2 - a^2)}{x - a}$$

$$= \lim_{x \to a} \frac{3(x - a)(x + a)}{x - a} = 3 \lim_{x \to a} (x + a) = 3(a + a) = 3 \cdot 2a = 6a.$$

The dedicated reader should use this alternative definition to compute the derivatives of the functions given in parts 2 and 3 as well.

Theorem 10.11: Let $A \subseteq \mathbb{R}$ and let $a \in A$. If f is differentiable at a, then f is continuous at a.

Proof: : Let $A \subseteq \mathbb{R}$ and assume that f is differentiable at $a \in A$. Then there is a real number L such that $f'(a) = L$. Therefore, we have

$$\lim_{x \to a} f(x) = \lim_{x \to a} \left[\frac{f(x) - f(a)}{x - a} (x - a) + f(a) \right] = \lim_{x \to a} \left[\frac{f(x) - f(a)}{x - a} (x - a) \right] + \lim_{x \to a} f(a)$$

$$= \lim_{x \to a} \left[\frac{f(x) - f(a)}{x - a} \right] \cdot \lim_{x \to a} (x - a) + \lim_{x \to a} f(a) = f'(a) \cdot (a - a) + f(a) = L \cdot 0 + f(a) = f(a).$$

Theorem 8.13, Theorem 8.14, and Problems 2 and 6 from Problem Set 8 can be used to justify these computations. By Problem 5 from Problem Set 8, f is continuous at a. □

In Problem 2 below, you will be asked to show that the converse of Theorem 10.11 is false. There are continuous functions that are **not** differentiable.

Note: Another notation that is often used for the derivative of a function is $\frac{d}{dx}$. For example, if $f: \mathbb{R} \to \mathbb{R}$ is defined by $f(x) = 3x^2 - 2$, we might write $\frac{d}{dx}[f(x)] = \frac{d}{dx}[3x^2 - 2] = 6x$. If we wish to evaluate this derivative at $x = a$, we could write $\frac{d}{dx}[f(x)]\big|_{x=a} = \frac{d}{dx}[3x^2 - 2]\big|_{x=a} = 6x\big|_{x=a} = 6a$.

Differentiation Rules

Theorem 10.12: Let $A, B \subseteq \mathbb{R}$ and suppose that $f: A \to \mathbb{R}$, $g: B \to \mathbb{R}$ are differentiable at $a \in A \cap B$. Then the function $f + g: A \cap B \to \mathbb{R}$ defined by $(f + g)(x) = f(x) + g(x)$ is differentiable at a and
$$(f + g)'(a) = f'(a) + g'(a).$$

Proof:
$$(f + g)'(a) = \lim_{x \to a} \frac{(f + g)(x) - (f + g)(a)}{x - a} = \lim_{x \to a} \frac{(f(x) + g(x)) - (f(a) + g(a))}{x - a}$$
$$= \lim_{x \to a} \frac{(f(x) - f(a)) + (g(x) - g(a))}{x - a} = \lim_{x \to a} \left[\frac{(f(x) - f(a))}{x - a} + \frac{(g(x) - g(a))}{x - a} \right]$$
$$= \lim_{x \to a} \frac{f(x) - f(a)}{x - a} + \lim_{x \to a} \frac{g(x) - g(a)}{x - a} = f'(a) + g'(a).$$

The fifth equality follows from Theorem 8.13. $\square$

Example 10.13:

1. Define $g: [0, \infty) \to \mathbb{R}$ by $g(x) = \sqrt{x}$ and define $h: \mathbb{R} \to \mathbb{R}$ by $h(x) = \sin x$. In Example 10.10, we saw that for any $a \in (0, \infty)$, $g'(a) = \frac{1}{2\sqrt{a}}$ and $h'(a) = \cos a$. It follows from Theorem 10.12 that if $a \in (0, \infty)$, then the derivative of $(g + h)(x) = g(x) + h(x) = \sqrt{x} + \sin x$ at $x = a$ is
$$(g + h)'(a) = g'(a) + h'(a) = \frac{1}{2\sqrt{a}} + \cos a.$$

2. In Problem 1 below, you will be asked to prove that the derivative of a constant is 0. For example, if $f: \mathbb{R} \to \mathbb{R}$ is defined by $f(x) = 2$, then for any $a \in \mathbb{R}$, $f'(a) = 0$. Let $g: \mathbb{R} \to \mathbb{R}$ be defined by $g(x) = 2 + \sin x$. Combining the result just mentioned with Example 10.10 and Theorem 10.12, we see that for any $a \in \mathbb{R}$, we have $g'(a) = 0 + \cos a = \cos a$.

3. If $f: \mathbb{R} \to \mathbb{R}$ is a linear function, say $f(x) = mx + b$, then for any $a \in \mathbb{R}$, $f'(a) = m$. This can be verified right from the definition of the derivative:
$$f'(a) = \lim_{x \to a} \frac{f(x) - f(a)}{x - a} = \lim_{x \to a} \frac{(mx + b) - (ma + b)}{x - a} = \lim_{x \to a} \frac{mx - ma}{x - a} = \lim_{x \to a} \frac{m(x - a)}{x - a}$$
$$= m \cdot \lim_{x \to a} \frac{x - a}{x - a} = m \cdot 1 = m.$$

If we let $g: [0, \infty) \to \mathbb{R}$ be defined by $g(x) = \sqrt{x} + 5x - 6$, we see that for any $a \in (0, \infty)$,
$$g'(a) = \frac{1}{2\sqrt{a}} + 5.$$

Note that $\sqrt{x} + 5x - 6 = \sqrt{x} + 5x + (-6)$.

The rule for taking derivatives of products is not quite as straightforward as the rule for taking derivatives of sums. Nonetheless, it is not difficult.

Theorem 10.14: Let $A, B \subseteq \mathbb{R}$ and suppose that $f: A \to \mathbb{R}$, $g: B \to \mathbb{R}$ are differentiable at $a \in A \cap B$. Then the function $fg: A \cap B \to \mathbb{R}$ defined by $(fg)(x) = f(x)g(x)$ is differentiable at a and
$$(fg)'(a) = f(a)g'(a) + f'(a)g(a).$$

Proof:
$$\begin{aligned}
(fg)'(a) &= \lim_{x \to a} \frac{(fg)(x) - (fg)(a)}{x - a} = \lim_{x \to a} \frac{\big(f(x)g(x)\big) - \big(f(a)g(a)\big)}{x - a} \\
&= \lim_{x \to a} \frac{f(x)g(x) - f(x)g(a) + f(x)g(a) - f(a)g(a)}{x - a} \\
&= \lim_{x \to a} \left[\frac{f(x)g(x) - f(x)g(a)}{x - a} + \frac{f(x)g(a) - f(a)g(a)}{x - a} \right] \\
&= \lim_{x \to a} \frac{f(x)g(x) - f(x)g(a)}{x - a} + \lim_{x \to a} \frac{f(x)g(a) - f(a)g(a)}{x - a} \\
&= \lim_{x \to a} \frac{f(x)\big(g(x) - g(a)\big)}{x - a} + \lim_{x \to a} \frac{\big(f(x) - f(a)\big)g(a)}{x - a} \\
&= \lim_{x \to a} \left[f(x) \cdot \frac{g(x) - g(a)}{x - a} \right] + \lim_{x \to a} \left[\frac{f(x) - f(a)}{x - a} \cdot g(a) \right] \\
&= \lim_{x \to a} f(x) \cdot \lim_{x \to a} \left[\frac{g(x) - g(a)}{x - a} \right] + \lim_{x \to a} \left[\frac{f(x) - f(a)}{x - a} \right] \cdot \lim_{x \to a} g(a) \\
&= f(a)g'(a) + f'(a)g(a). \quad \square
\end{aligned}$$

Notes: (1) Observe how we used the Standard Advanced Calculus Trick (SACT) here (see Note 7 following the proof of Theorem 4.11) to rewrite the expression $f(x)g(x) - f(a)g(a)$ in the more useful form $f(x)g(x) - f(x)g(a) + f(x)g(a) - f(a)g(a)$. We subtracted and added $f(x)g(a)$ in order to generate $g(x) - g(a)$ and $f(x) - f(a)$. The expression $f(x)g(x) - f(x)g(a)$ factors nicely as $f(x)\big(g(x) - g(a)\big)$ allowing the definition of $g'(a)$ to appear. Similarly, the expression $f(x)g(a) - f(a)g(a)$ factors as $\big(f(x) - f(a)\big)g(a)$ allowing the definition of $f'(a)$ to appear.

(2) The dedicated reader should write down the appropriate definition or theorem that justifies each equality provided in the proof of Theorem 10.14.

(3) The formula given in Theorem 10.14 is known as the **product rule**.

Example 10.15: As in part 1 of Example 10.13, define $g: [0, \infty) \to \mathbb{R}$ and $h: \mathbb{R} \to \mathbb{R}$ by $g(x) = \sqrt{x}$ and $h(x) = \sin x$. In Example 10.10, we saw that for any $a \in (0, \infty)$, $g'(a) = \frac{1}{2\sqrt{a}}$ and $h'(a) = \cos a$. It follows from Theorem 10.14 that the derivative of $(gh)(x) = g(x)h(x) = (\sqrt{x})(\sin x)$ at $x = a$ (for any $a \in (0, \infty)$), is
$$(gh)'(a) = g(a)h'(a) + g'(a)h(a) = (\sqrt{a})(\cos a) + \left(\frac{1}{2\sqrt{a}}\right)(\sin a).$$

Theorem 10.16: Let $n \in \mathbb{Z}^+$ and define $f: \mathbb{R} \to \mathbb{R}$ by $f(x) = x^n$. Then the function f is differentiable at each $a \in \mathbb{R}$ and $f'(a) = na^{n-1}$.

Proof: We will prove the theorem by induction on $n \in \mathbb{Z}^+$.

Base Case ($k = 1$): $f'(a^1) = f'(a) = \lim_{x \to a} \frac{x-a}{x-a} = \lim_{x \to a} 1 = 1 = 1a^0 = 1a^{1-1}$.

Inductive Step: Let k be a natural number and assume that $f'(a^k) = ka^{k-1}$. Then by Theorem 10.14, we have $f'(a^{k+1}) = f'(a^k \cdot a) = a^k \cdot 1 + ka^{k-1} \cdot a = a^k + ka^k = (1+k)a^k = (k+1)a^{(k+1)-1}$.

By the Principle of Mathematical Induction, $f'(a) = na^{n-1}$ for all natural numbers $n \in \mathbb{Z}^+$. □

Note: The formula given in Theorem 10.16 is known as the **power rule**.

Example 10.17:

1. By Theorem 10.16, we have $\frac{d}{dx}[x] = \frac{d}{dx}[x^1] = 1x^{1-1} = x^0 = 1$. This also follows from part 3 of Example 10.13.

2. By Theorem 10.16, $\frac{d}{dx}[x^2] = 2x^{2-1} = 2x^1 = 2x$ and $\frac{d}{dx}[x^3] = 3x^2$. Using part 1 above, Theorem 10.12, and part 2 of Example 10.13, we have $\frac{d}{dx}[x^3 + x^2 + x + 1] = 3x^2 + 2x + 1$.

3. In Problem 4 below, you will be asked to prove that the derivative of a constant times a function is the constant times the derivative of the function. For example, if $f : \mathbb{R} \to \mathbb{R}$ is defined by $f(x) = 3x^5$, then $f'(x) = 3 \cdot \frac{d}{dx}[x^5] = 3 \cdot 5x^4 = 15x^4$. Let $g : \mathbb{R} \to \mathbb{R}$ be defined by $g(x) = 7x^{100} - 3x^{50} + 2x^4 - 7$. Then for any $a \in \mathbb{R}$, we have
$$g'(a) = 700a^{99} - 150a^{49} + 8a^3$$

Theorem 10.18: Let $A, B \subseteq \mathbb{R}$, suppose that $f : A \to \mathbb{R}$, $g : B \to \mathbb{R}$ are differentiable at $a \in A \cap B$, and assume that $g(a) \neq 0$. Let $C = (A \cap B) \setminus \{x \in B \mid g(x) = 0\}$. Then the function $f/g : C \to \mathbb{R}$ defined by $(f/g)(x) = \frac{f(x)}{g(x)}$ is differentiable at a and
$$(f/g)'(a) = \frac{g(a)f'(a) - f(a)g'(a)}{(g(a))^2}.$$

Notes: (1) Theorem 10.18 can be proved in a way very similar to the proof of the product rule (Theorem 10.14). Alternatively, the quotient rule can be proved by directly applying both the product rule (Theorem 10.14) and the chain rule (see Theorem 10.20). You will be asked to prove Theorem 10.18 in Problem 7 below.

(2) The formula given in Theorem 10.18 is known as the **quotient rule**.

(3) **Negative exponents** give an alternative way of expressing quotients. In general, $x^{-n} = \frac{1}{x^n}$, and more generally, $(f(x))^{-n} = \frac{1}{(f(x))^n}$. So, for example, $x^{-1} = \frac{1}{x^1} = \frac{1}{x}$ and $x^{-5} = \frac{1}{x^5}$. Using the notation of negative exponents, we see that the quotient $\frac{f(x)}{g(x)}$ could be written as $f(x)(g(x))^{-1}$ and the quotient rule could be written as $(f/g)'(a) = [g(a)f'(a) - f(a)g'(a)] \cdot (g(a))^{-2}$.

(4) Notice that C, the domain of f/g, consists of all real numbers $x \in A \cap B$ such that $g(x) \neq 0$.

Example 10.19:

1. $\frac{d}{dx}[\csc x] = \frac{d}{dx}\left[\frac{1}{\sin x}\right] = \frac{(\sin x)\cdot 0 - 1\cdot \cos x}{\sin^2 x} = \frac{-\cos x}{\sin^2 x} = -\frac{\cos x}{\sin x}\cdot\frac{1}{\sin x} = -\cot x \cdot \csc x.$

2. We can use Theorem 10.18 together with Theorem 10.16 to show that the power rule $\frac{d}{dx}[x^n] = nx^{n-1}$ holds for all $n \in \mathbb{Z}$. Theorem 10.16 already provides the result for all $n \in \mathbb{Z}^+$. Now, for $n \in \mathbb{Z}^+$ and $x \neq 0$, we have $\frac{d}{dx}[x^{-n}] = \frac{d}{dx}\left[\frac{1}{x^n}\right] = \frac{x^n \cdot 0 - 1(nx^{n-1})}{(x^n)^2} = \frac{-nx^{n-1}}{x^{2n}} = -nx^{-n-1}$ (see Note 3 above). This shows that the power rule holds whenever the exponent is a negative integer. Finally, if $n = 0$ and $x \neq 0$, then $\frac{d}{dx}[x^0] = \frac{d}{dx}[1] = 0$. Since $0x^{-1} = 0$ for all $x \neq 0$, we see that the power rule holds when the exponent is 0 as well.

Theorem 10.20: Let $A \subseteq \mathbb{R}$ and suppose that $f: A \to \mathbb{R}$ is differentiable at $a \in A$. Also, let $B \subseteq \mathbb{R}$ with $f[A] \subseteq B$ and suppose that $g: B \to \mathbb{R}$ is differentiable at $f(a)$. Then $g \circ f: A \to \mathbb{R}$ is differentiable at a and

$$(g \circ f)'(a) = g'(f(a))f'(a).$$

Note: The formula given in Theorem 10.20 is known as the **chain rule**.

Analysis: As a first attempt to prove the chain rule, observe that for $x \neq a$ and $f(x) \neq f(a)$, we have

$$\frac{g(f(x)) - g(f(a))}{x - a} = \frac{g(f(x)) - g(f(a))}{f(x) - f(a)} \cdot \frac{f(x) - f(a)}{x - a}$$

So, if there is a neighborhood N of a such that $f(x) \neq f(a)$ for all $x \in N$, then we have

$$\lim_{x \to a} \frac{g(f(x)) - g(f(a))}{x - a} = \lim_{x \to a} \frac{g(f(x)) - g(f(a))}{f(x) - f(a)} \cdot \lim_{x \to a} \frac{f(x) - f(a)}{x - a}$$

Unfortunately, it is not quite clear at this point that such a neighborhood N exists. We have not ruled out the possibility that $f(x) = f(a)$ for x-values that get arbitrarily close to a. Therefore, we cannot be certain that $\lim_{x \to a} \frac{g(f(x)) - g(f(a))}{f(x) - f(a)}$ is a finite real number.

To work around this, we will begin with a preliminary lemma.

Lemma 10.21: Let $A \subseteq \mathbb{R}$ and suppose that $f: A \to \mathbb{R}$ is differentiable at $a \in A$. Then there is a function $g: A \to \mathbb{R}$ such that $g(a) = 0$, g is continuous at a, and

$$f(x) = f(a) + [f'(a) + g(x)](x - a).$$

Proof: Define $g: A \to \mathbb{R}$ by

$$g(x) = \begin{cases} \dfrac{f(x) - f(a)}{x - a} - f'(a) & \text{if } x \neq a. \\ 0 & \text{if } x = a. \end{cases}$$

If $x \neq a$, then $g(x) = \frac{f(x)-f(a)}{x-a} - f'(a)$, and so, $f'(a) + g(x) = \frac{f(x)-f(a)}{x-a}$. Multiplying by $x - a$ yields $[f'(a) + g(x)](x - a) = f(x) - f(a)$. So, $f(x) = f(a) + [f'(a) + g(x)](x - a)$. If $x = a$, then we have $f(a) + [f'(a) + g(a)](a - a) = f(a) + [f'(a) + 0] \cdot 0 = f(a) + 0 = f(a)$. Also, we have $g(a) = 0$ and $\lim_{x \to a} g(x) = \lim_{x \to a} \left[\frac{f(x)-f(a)}{x-a} - f'(a) \right] = \lim_{x \to a} \frac{f(x)-f(a)}{x-a} - \lim_{x \to a} f'(a) = f'(a) - f'(a) = 0$. By Problem 5 from Problem Set 8, g is continuous at a. □

We are now ready to prove the chain rule.

Proof of Theorem 10.20: Since g is differentiable at $f(a)$, by Lemma 10.21, there is a function $h: B \to \mathbb{R}$ such that $h(f(a)) = 0$, h is continuous at $f(a)$, and
$$g(z) = g(f(a)) + [g'(f(a)) + h(z)](z - f(a)).$$

If we let $z = f(x)$, we get
$$g(f(x)) = g(f(a)) + [g'(f(a)) + h(f(x))](f(x) - f(a)).$$

So, we have
$$(g \circ f)(x) - (g \circ f)(a) = [g'(f(a)) + h(f(x))](f(x) - f(a)).$$

Dividing each side of this equation by $x - a$ gives us the following equation for $x \neq a$.
$$\frac{(g \circ f)(x) - (g \circ f)(a)}{x - a} = [g'(f(a)) + h(f(x))] \cdot \frac{f(x) - f(a)}{x - a}.$$

Since f is differentiable at a, by Theorem 10.11, f is continuous at a. Since we also have that h is continuous at $f(a)$, by Problem 8 from Problem Set 8, $h \circ f$ is continuous at a. So, by Problem 5 from Problem Set 8, $\lim_{x \to a} [h(f(x))] = h(f(a))$. It follows that
$$\lim_{x \to a} \frac{(g \circ f)(x) - (g \circ f)(a)}{x - a} = \lim_{x \to a} [g'(f(a)) + h(f(x))] \cdot \lim_{x \to a} \frac{f(x) - f(a)}{x - a}$$
$$= \left(\lim_{x \to a} [g'(f(a))] + \lim_{x \to a} [h(f(x))] \right) \cdot f'(a) = (g'(f(a)) + 0) \cdot f'(a) = g'(f(a))f'(a). \quad \square$$

Example 10.22:

1. Let $f(x) = \sin \sqrt{x}$. By Example 10.10 and the chain rule (Theorem 10.20), we have
$$f'(x) = (\cos \sqrt{x}) \cdot \frac{1}{2\sqrt{x}} = \frac{\cos \sqrt{x}}{2\sqrt{x}}.$$

2. Recall from Note 3 before Example 3.19 that if $n \in \mathbb{Z}^+$, $x^{\frac{1}{n}} = \sqrt[n]{x}$. More generally, if $\frac{a}{b} \in \mathbb{Q}$ with $b > 0$, then we define $x^{\frac{a}{b}} = (x^a)^{\frac{1}{b}} = \sqrt[b]{x^a}$. For example, we have $x^{\frac{3}{2}} = (x^3)^{\frac{1}{2}} = \sqrt{x^3}$ and $x^{\frac{-5}{17}} = (x^{-5})^{\frac{1}{17}} = \sqrt[17]{x^{-5}} = \sqrt[17]{\frac{1}{x^5}} = \frac{1}{\sqrt[17]{x^5}}$. In what follows, we will be using the fact that if $\frac{a}{b} \in \mathbb{Q}$, then $\left(x^{\frac{a}{b}}\right)^b = x^a$. The dedicated reader should verify this.

We can use the chain rule (Theorem 10.20) together with part 2 of Example 10.19 to show that the power rule $\frac{d}{dx}[x^n] = nx^{n-1}$ holds for all $n \in \mathbb{Q}$ and $x \neq 0$. If $n \in \mathbb{Q}$, then there are integers a and b with $b \neq 0$ such that $n = \frac{a}{b}$. Since $x^a = \left(x^{\frac{a}{b}}\right)^b = (x^n)^b$, we have $\frac{d}{dx}[x^a] = \frac{d}{dx}[(x^n)^b]$. Differentiating each side of this equation gives us $ax^{a-1} = b(x^n)^{b-1} \cdot \frac{d}{dx}[x^n]$. It follows that

$$\frac{d}{dx}[x^n] = \frac{ax^{a-1}}{b(x^n)^{b-1}} = \frac{a}{b} \cdot \frac{x^{a-1}}{x^{n(b-1)}} = \frac{a}{b} \cdot \frac{x^{a-1}}{x^{nb-n}} = \frac{a}{b} \cdot x^{(a-1)-(nb-n)} = \frac{a}{b} \cdot x^{a-1-nb+n}$$

$$= \frac{a}{b} x^{a-1-\frac{a}{b} \cdot b + \frac{a}{b}} = \frac{a}{b} x^{a-1-a+\frac{a}{b}} = \frac{a}{b} x^{\frac{a}{b}-1} = nx^{n-1}.$$

The Mean Value Theorem

Let $A \subseteq \mathbb{R}$, let $f: A \to \mathbb{R}$, and let $c \in A$. We say that $f(c)$ is a **relative maximum** (or **local maximum**) for f if there is a neighborhood N of c with $N \subseteq A$ such that $\forall x \in N(f(x) \leq f(c))$.

Similarly, we say that $f(c)$ is a **relative minimum** (or **local minimum**) for f if there is a neighborhood N of c with $N \subseteq A$ such that $\forall x \in N(f(x) \geq f(c))$.

If $f(c)$ is a relative maximum or a relative minimum, but we do not wish to specify which one, we may say that $f(c)$ is a **relative extremum** (or **local extremum**).

Informally, $f(c)$ is a relative extremum of f if $f(c)$ is either the largest or smallest value that the function attains "near c."

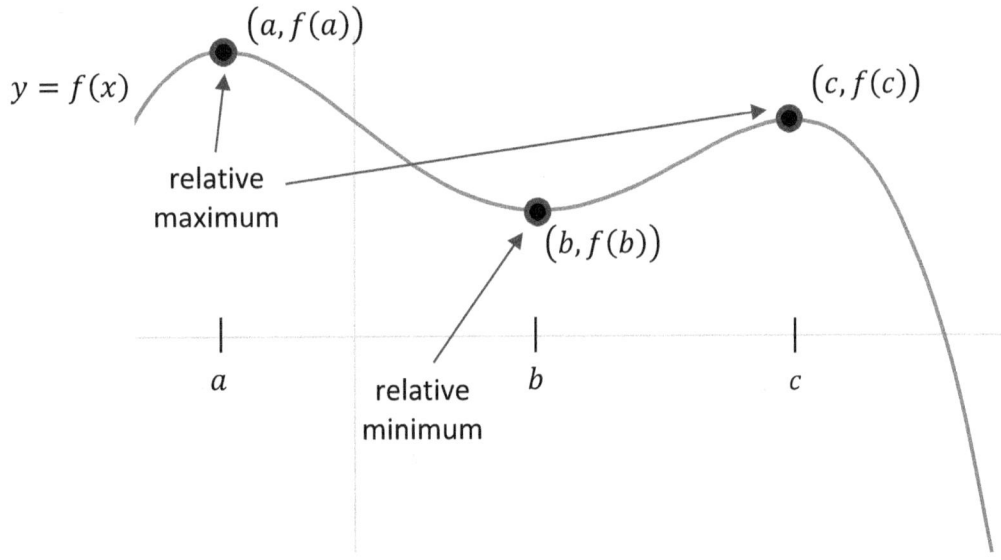

Note: You may want to compare the definition of relative extremum with the definition of absolute extremum given in Note 1 following the Maximum-Minimum Theorem (Corollary 9.7).

Theorem 10.23: Suppose that $f(c)$ is a relative maximum for a function $f: A \to \mathbb{R}$. Then either $f'(c) = 0$ or f is not differentiable at c.

Proof: Suppose that $f(c)$ is a relative maximum for $f: A \to \mathbb{R}$ and f is differentiable at c. First, let's suppose toward contradiction that $f'(c) > 0$. In other words, we have

$$\lim_{x \to c} \frac{f(x) - f(c)}{x - c} > 0.$$

By Problem 12 from Problem Set 8, there is a deleted neighborhood N of c such that $\frac{f(x)-f(c)}{x-c} > 0$ for all $x \in N$. Now, if $x \in N$ and $x > c$, then $x - c > 0$, and therefore,

$$f(x) - f(c) = \frac{f(x) - f(c)}{x - c}(x - c) > 0.$$

So, for $x \in N$ with $x > c$, we have $f(x) > f(c)$, contradicting our assumption that $f(c)$ is a relative maximum for f. So, it is impossible to have $f'(c) > 0$.

Next, let's suppose toward contradiction that $f'(c) < 0$. Then we have

$$\lim_{x \to c} \frac{f(x) - f(c)}{x - c} < 0.$$

Therefore, it follows that

$$\lim_{x \to c} \frac{f(c) - f(x)}{x - c} = -\left[\lim_{x \to c} \frac{f(x) - f(c)}{x - c}\right] > 0.$$

Again, by Problem 12 from Problem Set 8, there is a deleted neighborhood N of c such that $\frac{f(c)-f(x)}{x-c} > 0$ for all $x \in N$. Now, if $x \in N$ and $x < c$, then $x - c < 0$, and therefore,

$$f(c) - f(x) = \frac{f(c) - f(x)}{x - c}(x - c) < 0.$$

So, for $x \in N$ with $x < c$, we have $f(c) < f(x)$, once again contradicting our assumption that $f(c)$ is a relative maximum for f. So, it is also impossible to have $f'(c) < 0$.

Since f is differentiable at c, we must have $f'(c) = 0$. □

Note: If we replace the word "maximum" in Theorem 10.23 by the word "minimum," we get an analogous result. You will be asked to prove this result in Problem 8 below.

Theorem 10.24 (Rolle's Theorem): Suppose that $[a, b] \subseteq A \subseteq \mathbb{R}$, $f: A \to \mathbb{R}$ is continuous on $[a, b]$ and differentiable on (a, b), and $f(a) = f(b)$. Then there is $c \in (a, b)$ such that $f'(c) = 0$.

Proof: Let $k = f(a)$ (so that $k = f(b)$ as well). If $f(x) = 0$ for all $x \in (a, b)$, then by Problem 1 below, for all $c \in (a, b)$, $f'(c) = 0$.

Otherwise, there is $x \in (a, b)$ such that $f(x) > k$ or $f(x) < k$. First assume that there is $x \in (a, b)$ such that $f(x) > k$. By the Maximum-Minimum Theorem (Corollary 9.7), we have $f(c) = \sup f[[a, b]] \in f[[a, b]]$. Since $f(a) = f(b) = k$ and $f(c) > k$, it follows that $c \in (a, b)$. Therefore, f is differentiable at c. By Theorem 10.23, $f'(c) = 0$.

In the case where there is $x \in (a,b)$ such that $f(x) < k$, define $g: A \to \mathbb{R}$ by $g(x) = -x$. Let $d \in (a,b)$ satisfy $f(d) < k$. Then $g(d) = -f(d) > k$. By the paragraph above, there is $c \in (a,b)$ such that $g'(c) = 0$. Then $f'(c) = -g'(c) = -0 = 0$. □

Theorem 10.25 (Mean Value Theorem): Suppose that $[a,b] \subseteq A \subseteq \mathbb{R}$, $f: A \to \mathbb{R}$ is continuous on $[a,b]$, and f is differentiable on (a,b). Then there is $c \in (a,b)$ such that
$$f'(c) = \frac{f(b) - f(a)}{b - a}.$$

A geometric interpretation of the Mean Value Theorem says that if a function is "sufficiently nice" on the interval $[a,b]$ (continuous and differentiable as specified in the theorem) then there is a c between a and b, such that the slope of the tangent line at $(c, f(c))$ is equal to the slope of the secant line passing through $(a, f(a))$ and $(b, f(b))$. The idea can be visualized as follows:

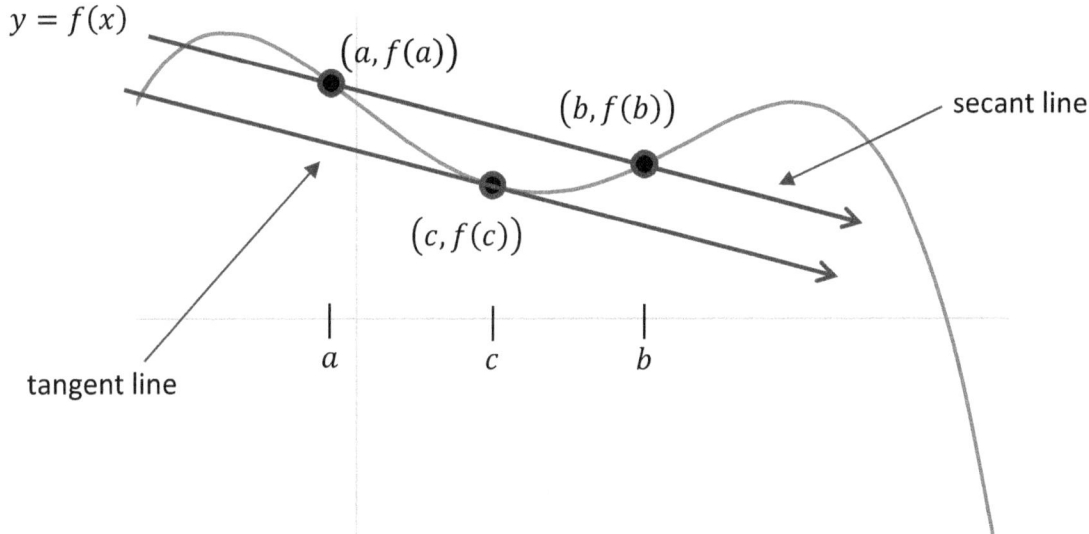

Proof: Define $g: A \to \mathbb{R}$ by $g(x) = f(x) - f(a) - \frac{f(b)-f(a)}{b-a}(x-a)$. Then g is continuous on $[a,b]$ and differentiable on (a,b) (Check this!). Also, $g(a) = f(a) - f(a) - \frac{f(b)-f(a)}{b-a}(a-a) = 0$ and $g(b) = f(b) - f(a) - \frac{f(b)-f(a)}{b-a}(b-a) = f(b) - f(a) - (f(b) - f(a)) = 0$. By Rolles's Theorem (Theorem 10.24), there is $c \in (a,b)$ such that $g'(c) = 0$. But $g'(c) = f'(c) - \frac{f(b)-f(a)}{b-a}$. Therefore, we have $f'(c) - \frac{f(b)-f(a)}{b-a} = 0$, or equivalently, $f'(c) = \frac{f(b)-f(a)}{b-a}$. □

Let $A \subseteq \mathbb{R}$. If $f: A \to \mathbb{R}$ satisfies $\forall x, y \in A(x < y \to f(x) < f(y))$, then we say that f is a **strictly increasing function** on A. Similarly, if f satisfies $\forall x, y \in A(x < y \to f(x) > f(y))$, then we say that f is a **strictly decreasing function** on A. We will sometimes use the expression **strictly monotone** (or **strictly monotonic**) on A to indicate that a function is either strictly increasing or strictly decreasing.

Note: We have the following analogous definitions of functions that are **monotone** (or **monotonic**) on A. f is **increasing** (or **nondecreasing**) on A if $\forall x, y \in A(x < y \to f(x) \leq f(y))$ and **decreasing** (or **nonincreasing**) on A if $\forall x, y \in A(x < y \to f(x) \geq f(y))$.

Corollary 10.26: Suppose that $(a, b) \subseteq A \subseteq \mathbb{R}$ and $f: A \to \mathbb{R}$ is differentiable on (a, b).

1. If $f'(x) > 0$ for all $x \in (a, b)$, then f is strictly increasing on (a, b).
2. If $f'(x) < 0$ for all $x \in (a, b)$, then f is strictly decreasing on (a, b).
3. If $f'(x) = 0$ for all $x \in (a, b)$, then f is constant on (a, b).

Proof: Let $x, y \in (a, b)$ with $x < y$. Then f is differentiable on $[x, y]$. By Theorem 10.11, f is continuous on $[x, y]$. By the Mean Value Theorem (Theorem 10.25), there is $c \in (x, y)$ such that $f'(c) = \frac{f(y) - f(x)}{y - x}$. So, $f(y) - f(x) = f'(c)(y - x)$. Since $x < y$, it follows that $y - x > 0$. So, under condition 1 above, we have $f(y) - f(x) > 0$, or equivalently, $f(x) < f(y)$. Under condition 2 above, we have $f(y) - f(x) < 0$, or equivalently, $f(x) > f(y)$. Under condition 3 above, we have $f(y) - f(x) = 0$, or equivalently $f(x) = f(y)$. Since $x, y \in (a, b)$ with $x < y$ were arbitrary, all three results follow. □

We will need the following more general form of the Mean Value Theorem to prove L'Hôpital's rule.

Theorem 10.27 (Generalized Mean Value Theorem): Suppose that $[a, b] \subseteq A \subseteq \mathbb{R}$ and $f, g: A \to \mathbb{R}$ are functions that are continuous on $[a, b]$ and differentiable on (a, b). Then there is $c \in (a, b)$ such that

$$[f(b) - f(a)]g'(c) = [g(b) - g(a)]f'(c).$$

Note: The proof of Theorem 10.27 is similar to the proof of Theorem 10.25. You will be asked to provide the details in Problem 12 below.

L'Hôpital's Rule

L'Hôpital's rule can be used to evaluate limits of the "form" $\frac{0}{0}$ and $\frac{\infty}{\infty}$ (these are called **indeterminate forms**). For example, suppose that we want to evaluate $\lim_{x \to 0} \frac{x^3 + 2x^2 - x}{\sin x}$. Since $\lim_{x \to 0}(x^3 + 2x^2 - x) = 0$ and $\lim_{x \to 0} \sin x = 0$, we will say that $\lim_{x \to 0} \frac{x^3 + 2x^2 - x}{\sin x}$ **has the form** $\frac{0}{0}$. The main idea of L'Hôpital's rule is that we take the derivative of each of the numerator and denominator separately and then the limit of the original quotient is equal to the limit of this new quotient. So, for example, we have

$$\lim_{x \to 0} \frac{x^3 + 2x^2 - x}{\sin x} = \lim_{x \to 0} \frac{3x^2 + 4x - 1}{\cos x} = \frac{-1}{1} = -1.$$

Since the proof of the most general form of L'Hôpital's rule is a bit involved, we will break this rule up into several "versions."—the baby version, three standard versions, and three infinite versions.

Theorem 10.28 (L'Hôpital's Rule – Baby Version): Suppose that $(a, b) \subseteq A \subseteq \mathbb{R}$, $f, g: A \to \mathbb{R}$ are differentiable on (a, b), and f' and g' are continuous on (a, b). Let $c \in (a, b)$ with $f(c) = g(c) = 0$ and $g'(c) \neq 0$. Then

$$\lim_{x \to c} \frac{f(x)}{g(x)} = \lim_{x \to c} \frac{f'(x)}{g'(x)} = \frac{f'(c)}{g'(c)}.$$

Proof:

$$\lim_{x \to c} \frac{f(x)}{g(x)} = \lim_{x \to c} \frac{f(x) - 0}{g(x) - 0} = \lim_{x \to c} \frac{f(x) - f(c)}{g(x) - g(c)} = \lim_{x \to c} \frac{\frac{f(x) - f(c)}{x - c}}{\frac{g(x) - g(c)}{x - c}} = \frac{\lim_{x \to c} \frac{f(x) - f(c)}{x - c}}{\lim_{x \to c} \frac{g(x) - g(c)}{x - c}} = \frac{f'(c)}{g'(c)}.$$

Since f' and g' are continuous on (a,b), we have $\lim_{x \to c} f'(x) = f'(c)$ and $\lim_{x \to c} g'(x) = g'(c)$. Therefore, we also have

$$\lim_{x \to c} \frac{f'(x)}{g'(x)} = \frac{\lim_{x \to c} f'(x)}{\lim_{x \to c} g'(x)} = \frac{f'(c)}{g'(c)}.$$

□

Notes: (1) This baby version of L'Hôpital's rule is enough to justify the computation that we did right before the theorem. Specifically, if we let $f, g: \mathbb{R} \to \mathbb{R}$ be defined by $f(x) = x^3 + 2x^2 - x$ and $g(x) = \sin x$, then f and g are differentiable everywhere, and $f'(x) = 3x^2 + 4x - 1$ and $g'(x) = \cos x$ are continuous everywhere. Letting $c = 0$, we have $f(0) = g(0) = 0$. Therefore, by the baby version of L'Hôpital's rule (Theorem 10.28), we have

$$\lim_{x \to 0} \frac{x^3 + 2x^2 - x}{\sin x} = \lim_{x \to 0} \frac{3x^2 + 4x - 1}{\cos x} = \frac{-1}{1} = -1.$$

(2) Consider the following computations (in Problem 3 below, you will show that $\frac{d}{dx}[\cos x] = -\sin x$).

$$\lim_{x \to 0} \frac{x^3 + 2x^2}{1 - \cos x} = \lim_{x \to 0} \frac{3x^2 + 4x}{\sin x} = \lim_{x \to 0} \frac{6x + 4}{\cos x} = \frac{4}{1} = 4$$

Are these computations justified by Theorem 10.28? The answer is **no**! If we let $f, g: \mathbb{R} \to \mathbb{R}$ be defined by $f(x) = x^3 + 2x^2$ and $g(x) = 1 - \cos x$, then $g'(x) = \sin x$ and $g'(0) = 0$. The fact that $g'(0) = 0$ means that we **cannot** use Theorem 10.28 to justify the first equality given above. However, it does turn out that the equality is perfectly legal. Although the baby version of L'Hôpital's rule does not give us this result, we will see shortly that the standard version does.

(3) By replacing limits with one-sided limits in Theorem 10.28 and its proof, under the conditions given in the theorem, we get the following two results as well:

$$\lim_{x \to c^-} \frac{f(x)}{g(x)} = \lim_{x \to c^-} \frac{f'(x)}{g'(x)} = \frac{f'(c)}{g'(c)}$$

$$\lim_{x \to c^+} \frac{f(x)}{g(x)} = \lim_{x \to c^+} \frac{f'(x)}{g'(x)} = \frac{f'(c)}{g'(c)}$$

Theorem 10.29 (L'Hôpital's Rule – Standard Version A): Suppose that $(a, b) \subseteq A \subseteq \mathbb{R}$, $f, g: A \to \mathbb{R}$ are continuous on A and differentiable on (a, b), $\lim_{x \to a^+} f(x) = \lim_{x \to a^+} g(x) = 0$, and $g'(x) \neq 0$ for all $x \in (a, b)$. If $\lim_{x \to a^+} \frac{f'(x)}{g'(x)}$ exists, then

$$\lim_{x \to a^+} \frac{f(x)}{g(x)} = \lim_{x \to a^+} \frac{f'(x)}{g'(x)}.$$

Proof: If $a \in A$, then since f and g are continuous on A and $\lim_{x \to a^+} f(x) = \lim_{x \to a^+} g(x) = 0$, we must have $f(a) = 0$ and $g(a) = 0$. In general, let $B = A \cup \{a\}$ and define $F, G: B \to \mathbb{R}$ as follows:

$$F(x) = \begin{cases} f(x) & \text{if } x \in A \\ 0 & \text{if } x = a \end{cases} \qquad G(x) = \begin{cases} g(x) & \text{if } x \in A \\ 0 & \text{if } x = a \end{cases}$$

Note that if $a \in A$, then $B = A$, $F = f$, and $G = g$. So, there is no harm in simply assuming that $a \in A$ and $f(a) = g(a) = 0$. It follows that f and g are continuous on $[a,b)$.

For any $x \in (a,b)$, we see that f and g are continuous on $[a,x]$ and differentiable on (a,x). By the Mean Value Theorem (Theorem 10.25), there is $c \in (a,x)$ such that $g'(c) = \frac{g(x)-g(a)}{x-a} = \frac{g(x)}{x-a}$. Since $g'(c) \neq 0$, we must also have $g(x) \neq 0$. By the Generalized Mean Value Theorem (Theorem 10.27), there is $c \in (a,x)$ such that $f(x)g'(c) = g(x)f'(c)$, or equivalently, $\frac{f(x)}{g(x)} = \frac{f'(c)}{g'(c)}$. Since $a < c < x$, as x approaches a from the right, so does c. So, we have

$$\lim_{x \to a^+} \frac{f(x)}{g(x)} = \lim_{x \to a^+} \frac{f'(c)}{g'(c)} = \lim_{c \to a^+} \frac{f'(c)}{g'(c)} = \lim_{x \to a^+} \frac{f'(x)}{g'(x)}.$$

Note that the last equality above is true because we simply changed the name of the variable. □

There is nothing special about taking limits from the right. We can modify the theorem slightly to get a version for limits from the left.

Theorem 10.30 (L'Hôpital's Rule – Standard Version B): Suppose that $(a,b) \subseteq A \subseteq \mathbb{R}$, $f,g: A \to \mathbb{R}$ are continuous on A and differentiable on (a,b), $\lim_{x \to b^-} f(x) = \lim_{x \to b^-} g(x) = 0$, and $g'(x) \neq 0$ for all $x \in (a,b)$. If $\lim_{x \to b^-} \frac{f'(x)}{g'(x)}$ exists, then

$$\lim_{x \to b^-} \frac{f(x)}{g(x)} = \lim_{x \to b^-} \frac{f'(x)}{g'(x)}.$$

The proof of Theorem 10.30 is very similar to the proof of Theorem 10.29. You will be asked to provide the details in Problem 6 below.

Theorems 10.29 and 10.30 can be combined to get an analogous theorem about two-sided limits.

Theorem 10.31 (L'Hôpital's Rule – Standard Version C): Suppose that $(a,c) \cup (c,b) \subseteq A \subseteq \mathbb{R}$, $f,g: A \to \mathbb{R}$ are continuous on A and differentiable on $(a,c) \cup (c,b)$, $\lim_{x \to c} f(x) = \lim_{x \to c} g(x) = 0$, and $g'(x) \neq 0$ for all $x \in (a,c) \cup (c,b)$. If $\lim_{x \to c} \frac{f'(x)}{g'(x)}$ exists, then

$$\lim_{x \to c} \frac{f(x)}{g(x)} = \lim_{x \to c} \frac{f'(x)}{g'(x)}.$$

Proof: By Theorem 10.29, $\lim_{x \to c^+} \frac{f(x)}{g(x)} = \lim_{x \to c^+} \frac{f'(x)}{g'(x)}$. By Theorem 10.30, $\lim_{x \to c^-} \frac{f(x)}{g(x)} = \lim_{x \to c^-} \frac{f'(x)}{g'(x)}$. Therefore, we have $\lim_{x \to c} \frac{f(x)}{g(x)} = \lim_{x \to c} \frac{f'(x)}{g'(x)}$. □

Note: Theorem 10.31 can be used to justify the computations made in Note 2 following Theorem 10.28. First note that we can use the baby version of L'Hôpital's rule (Theorem 10.28) or standard version C (Theorem 10.31) to get that

$$\lim_{x \to 0} \frac{3x^2 + 4x}{\sin x} = \lim_{x \to 0} \frac{6x + 4}{\cos x} = \frac{4}{1} = 4.$$

Next, if we let $f, g: \mathbb{R} \to \mathbb{R}$ be defined by $f(x) = x^3 + 2x^2$ and $g(x) = 1 - \cos x$, then f and g are differentiable everywhere, $\lim_{x \to 0} f(x) = \lim_{x \to 0} g(x) = 0$, and $g'(x) \neq 0$ for all $x \in (-1, 0) \cup (0, 1)$. Now, $f'(x) = 3x^2 + 4x$ and $g'(x) = \sin x$, and we just saw above that $\lim_{x \to 0} \frac{f'(x)}{g'(x)} = \lim_{x \to 0} \frac{3x^2 + 4x}{\sin x} = 4$. Therefore, by Theorem 10.31, we have

$$\lim_{x \to 0} \frac{x^3 + 2x^2}{1 - \cos x} = \lim_{x \to 0} \frac{3x^2 + 4x}{\sin x} = 4.$$

(2) The computations given in Note 1 above required us to take a derivative of a derivative, otherwise known as a **second derivative**. For example, if we let $f(x) = x^3 + 2x^2$, then the derivative of f is $f'(x) = 3x^2 + 4x$ and the second derivative of f is $f''(x) = 6x + 4$. Taking this further, we see that the **third derivative** of f is $f'''(x) = 6$ and the **fourth derivative** of f is $f^{(4)}(x) = 0$. In general, for $n \in \mathbb{N}$, we use the notation $f^{(n)}(x)$ for the **nth derivative** of the function f, where $f^{(0)}(x) = f(x)$ and $f^{(1)}(x), f^{(2)}(x)$, and $f^{(3)}(x)$ are usually abbreviated as $f'(x), f''(x)$, and $f'''(x)$, respectively. We will sometimes refer to the derivative $f'(x)$ as the **first derivative**.

(3) Recall from the Note following theorem 10.11 that we can also use the notation $\frac{d}{dx}$ to indicate that we are taking the derivative (or first derivative) of a function. More generally, the notation $\frac{d^n}{dx^n}$ can be used to indicate that we are taking the nth derivative of a function. For example, we saw in Note 2 above that $\frac{d}{dx}[x^3 + 2x^2] = 3x^2 + 4x$, $\frac{d^2}{dx^2}[x^3 + 2x^2] = 6x + 4$, $\frac{d^3}{dx^3}[x^3 + 2x^2] = 6$, and $\frac{d^4}{dx^4}[x^3 + 2x^2] = 0$. In general, for $n \in \mathbb{N}$ with $n \geq 4$, we have $\frac{d^n}{dx^n}[x^3 + 2x^2] = 0$.

You will be asked to prove the following infinite versions of L'Hôpital's Rule in Problem 18 below.

Theorem 10.32 (L'Hôpital's Rule – Infinite Version A): Suppose that $(a, c) \cup (c, b) \subseteq A \subseteq \mathbb{R}$, $f, g: A \to \mathbb{R}$ are continuous on A and differentiable on $(a, c) \cup (c, b)$, $\lim_{x \to c} f(x) = \lim_{x \to c} g(x) = \infty$ (either or both can also be $-\infty$) and $g'(x) \neq 0$ for all $x \in (a, c) \cup (c, b)$. If $\lim_{x \to c} \frac{f'(x)}{g'(x)}$ exists, then

$$\lim_{x \to c} \frac{f(x)}{g(x)} = \lim_{x \to c} \frac{f'(x)}{g'(x)}.$$

Theorem 10.33 (L'Hôpital's Rule – Infinite Version B): Suppose that $(a, \infty) \subseteq A \subseteq \mathbb{R}$, $f, g: A \to \mathbb{R}$ are continuous on A and differentiable on (a, ∞), $\lim_{x \to \infty} f(x) = \lim_{x \to \infty} g(x) = \infty$ (either or both limits can also be $-\infty$), and $g'(x) \neq 0$ for all $x \in (a, \infty)$. If $\lim_{x \to \infty} \frac{f'(x)}{g'(x)}$ exists, then

$$\lim_{x \to \infty} \frac{f(x)}{g(x)} = \lim_{x \to \infty} \frac{f'(x)}{g'(x)}.$$

Theorem 10.34 (L'Hôpital's Rule – Infinite Version C): Suppose that $(a, \infty) \subseteq A \subseteq \mathbb{R}$, $f, g: A \to \mathbb{R}$ are continuous on A and differentiable on (a, ∞), $\lim_{x \to \infty} f(x) = \lim_{x \to \infty} g(x) = 0$, and $g'(x) \neq 0$ for all $x \in (a, \infty)$. If $\lim_{x \to \infty} \frac{f'(x)}{g'(x)}$ exists, then

$$\lim_{x \to \infty} \frac{f(x)}{g(x)} = \lim_{x \to \infty} \frac{f'(x)}{g'(x)}.$$

Note: There are also analogues of Theorems 10.33 and 10.34 with limits that approach $-\infty$.

L'Hôpital's rule can also be used to evaluate limits of the forms $0 \cdot \infty$, $\infty - \infty$, 0^0, 1^∞, and ∞^0 (these are **indeterminate forms**). We will analyze the first two of these now and the other three in Lesson 12.

If $\lim_{x \to c} f(x) = 0$ and $\lim_{x \to c} g(x) = \infty$ (or $-\infty$), then we say that $\lim_{x \to c}[f(x)g(x)]$ has the form $0 \cdot \infty$ and we can rewrite this limit as $\lim_{x \to c} \left[\frac{f(x)}{\frac{1}{g(x)}}\right]$ (which has the form $\frac{0}{0}$) or $\lim_{x \to c} \left[\frac{g(x)}{\frac{1}{f(x)}}\right]$ (which has the form $\frac{\infty}{\infty}$). This observation gives us two methods for evaluating limits of the form $0 \cdot \infty$. One-sided limits can be handled the same way. We consider the forms $0 \cdot \infty$, $\infty \cdot 0$, $0 \cdot (-\infty)$, and $(-\infty) \cdot 0$ to be the same.

Example 10.35: Let's compute $\lim_{x \to \infty} x \sin \frac{1}{x}$. Since $\lim_{x \to \infty} x = \infty$ and $\lim_{x \to \infty} \sin \frac{1}{x} = \sin\left(\lim_{x \to \infty} \frac{1}{x}\right) = \sin 0 = 0$, we see that this limit has the form $\infty \cdot 0 = 0 \cdot \infty$. So, we can compute this limit as follows:

$$\lim_{x \to \infty} x \sin \frac{1}{x} = \lim_{x \to \infty} \frac{\sin \frac{1}{x}}{\frac{1}{x}} = \lim_{x \to \infty} \frac{-\frac{1}{x^2} \cos \frac{1}{x}}{-\frac{1}{x^2}} = \lim_{x \to \infty} \cos \frac{1}{x} = \cos\left(\lim_{x \to \infty} \frac{1}{x}\right) = \cos 0 = 1.$$

For the first equality, we changed the given limit of the form $\infty \cdot 0$ to one of the form $\frac{0}{0}$. For the second equality, we applied infinite version C of L'Hôpital's rule (Theorem 10.34).

If $\lim_{x \to c} f(x) = \infty$ and $\lim_{x \to c} g(x) = \infty$ (or both limits are $-\infty$), then we say that $\lim_{x \to c}[f(x) - g(x)]$ has the form $\infty - \infty$. We can usually use basic algebra to rewrite this limit as one of the form $\frac{0}{0}$ or $\frac{\infty}{\infty}$. Note that limits of the form $\infty + \infty$ and $-\infty - \infty$ are straightforward and do not require L'Hôpital's rule.

Example 10.36:

1. Let's compute $\lim_{x \to 0}\left[\frac{1}{x} - \csc x\right]$. Since $\lim_{x \to 0} \frac{1}{x} = \infty$ and $\lim_{x \to 0} \csc x = \infty$, we see that the given limit has the form $\infty - \infty$. So, we can compute this limit as follows:

$$\lim_{x \to 0}\left[\frac{1}{x} - \csc x\right] = \lim_{x \to 0}\left[\frac{1}{x} - \frac{1}{\sin x}\right] = \lim_{x \to 0}\left[\frac{\sin x - x}{x \sin x}\right] = \lim_{x \to 0}\left[\frac{\cos x - 1}{x \cos x + \sin x}\right]$$
$$= \lim_{x \to 0}\left[\frac{-\sin x}{-x \sin x + \cos x + \cos x}\right] = \lim_{x \to 0}\left[\frac{-\sin x}{-x \sin x + 2 \cos x}\right] = \frac{-0}{-0 + 2 \cdot 1} = \frac{0}{2} = 0.$$

For the first equality, we simply rewrote $\csc x$ in its equivalent form $\frac{1}{\sin x}$. For the second equality, we changed the given limit of the form $\infty - \infty$ to one of the form $\frac{0}{0}$. For the third equality, we applied standard version C of L'Hôpital's rule (Theorem 10.31). For the fourth equality, we could have used either the baby version or standard version C of L'Hôpital's rule (Theorem 10.28 or Theorem 10.31). The rest is straightforward.

2. $\lim_{x \to 0}\left[\frac{1}{x} + \csc x\right]$ is much easier to compute than the previous example (part 1 above). Since $\lim_{x \to 0} \frac{1}{x} = \infty$ and $\lim_{x \to 0} \csc x = \infty$, we see that the given limit has the form $\infty + \infty$. So, the answer is simply ∞. L'Hôpital's rule is not required here, and more importantly, **it is not allowed!**

Problem Set 10

Full solutions to these problems are available for free download here:
www.SATPrepGet800.com/RABQXZ

LEVEL 1

1. Let $r \in \mathbb{R}$ and define $f: \mathbb{R} \to \mathbb{R}$ by $f(x) = r$. Prove that for any $a \in \mathbb{R}$, $f'(a) = 0$.

2. Define a continuous function $f: \mathbb{R} \to \mathbb{R}$ that is not differentiable on $\mathbb{R}$.

3. Compute the derivatives of $\cos x$, $\tan x$, $\sec x$, and $\cot x$.

LEVEL 2

4. Let $A \subseteq \mathbb{R}$, let $k \in \mathbb{R}$, and let $f: A \to \mathbb{R}$ be differentiable at $a \in A$. Prove that the function $kf: A \to \mathbb{R}$ defined by $(kf)(x) = k \cdot f(x)$ is differentiable at a and $(kf)'(a) = k \cdot f'(a)$.

5. Suppose that $(a, b) \subseteq A \subseteq \mathbb{R}$, $f, g: A \to \mathbb{R}$ are differentiable on (a, b), and $f'(x) = g'(x)$ for all $x \in (a, b)$. Prove that there is a constant k such that for all $x \in (a, b)$, $g(x) = f(x) + k$.

6. Suppose that $(a, b) \subseteq A \subseteq \mathbb{R}$, $f, g: A \to \mathbb{R}$ are continuous on A and differentiable on (a, b), $\lim_{x \to b^-} f(x) = \lim_{x \to b^-} g(x) = 0$, $g'(x) \neq 0$ for all $x \in (a, b)$, and $\lim_{x \to b^-} \frac{f'(x)}{g'(x)}$ exists. Prove that

$$\lim_{x \to b^-} \frac{f(x)}{g(x)} = \lim_{x \to b^-} \frac{f'(x)}{g'(x)}.$$

This is standard version B of L'Hôpital's Rule (Theorem 10.30).

LEVEL 3

7. Let $A, B \subseteq \mathbb{R}$, suppose that $f: A \to \mathbb{R}$, $g: B \to \mathbb{R}$ are differentiable at $a \in A \cap B$, and assume that $g(a) \neq 0$. Let $C = (A \cap B) \setminus \{x \in B \mid g(x) = 0\}$. Prove that the function $f/g: C \to \mathbb{R}$ defined by $(f/g)(x) = \frac{f(x)}{g(x)}$ is differentiable at a and $(f/g)'(a) = \frac{g(a)f'(a) - f(a)g'(a)}{(g(a))^2}$. This is the **quotient rule** (Theorem 10.18).

8. Suppose that $f(c)$ is a relative minimum for a function $f: A \to \mathbb{R}$. Prove that either $f'(c) = 0$ or f is not differentiable at c.

9. Use the Mean Value Theorem to prove that $-x \leq \sin x \leq x$ for all $x \in [0, \infty)$.

10. Suppose that $(a, b) \subseteq A \subseteq \mathbb{R}$ and $f: A \to \mathbb{R}$ is differentiable on (a, b). Prove each of the following:

    (i) If $f'(x) \geq 0$ for all $x \in (a, b)$, then f is increasing on (a, b).

    (ii) If $f'(x) \leq 0$ for all $x \in (a, b)$, then f is decreasing on (a, b).

11. Prove the following special case of infinite version A of **L'Hôpital's Rule:** Suppose that $(a,c) \cup (c,b) \subseteq A \subseteq \mathbb{R}$, $f, g: A \to \mathbb{R}$ are continuous on A and differentiable on $(a,c) \cup (c,b)$, $\lim_{x \to c} \frac{f(x)}{g(x)} = L$, where L is a finite nonzero real number, $\lim_{x \to c} f(x) = \lim_{x \to c} g(x) = \infty$ (either or both can also be $-\infty$), and $g'(x) \neq 0$ for all $x \in (a,c) \cup (c,b)$. If $\lim_{x \to c} \frac{f'(x)}{g'(x)}$ exists, then
$$\lim_{x \to c} \frac{f(x)}{g(x)} = \lim_{x \to c} \frac{f'(x)}{g'(x)}.$$

LEVEL 4

12. Suppose that $[a,b] \subseteq A \subseteq \mathbb{R}$ and $f, g: A \to \mathbb{R}$ are functions that are continuous on $[a,b]$ and differentiable on (a,b). Prove that there is $c \in (a,b)$ such that
$$[f(b) - f(a)]g'(c) = [g(b) - g(a)]f'(c).$$

This is the **Generalized Mean Value Theorem** (Theorem 10.27).

13. Suppose that $f: (a,b) \to \mathbb{R}$ is differentiable on (a,b) with $f'(x) > 0$ for all $x \in (a,b)$. Prove that f is invertible. Let $g: B \to \mathbb{R}$ be the inverse of f. Prove that g is differentiable on B and $g'(f(x)) = \frac{1}{f'(x)}$ for all $x \in (a,b)$.

14. A function $f: A \to \mathbb{R}$ that is differentiable on A is said to be **uniformly differentiable** on A if
$$\forall \epsilon > 0 \; \exists \delta > 0 \; \forall x, y \in A \left(0 < |y - x| < \delta \to \left| \frac{f(y) - f(x)}{y - x} - f'(x) \right| < \epsilon \right).$$

Let $g: [a,b] \to \mathbb{R}$ be differentiable on $[a,b]$ with g' continuous on $[a,b]$. Prove that g is uniformly differentiable on $[a,b]$.

LEVEL 5

15. Provide a counterexample to show that the following statement is false: If $f: \mathbb{R} \to \mathbb{R}$ satisfies $f(0) = 0$ and $f'(0) = 1$, then there is a neighborhood N of 0 such that f is strictly monotonic on N.

16. Suppose that $f: (0, \infty) \to \mathbb{R}$ is differentiable on $(0, \infty)$, $f'(x)$ is uniformly continuous on $(0, \infty)$, and $\lim_{x \to \infty} f(x)$ is a finite real number. Prove that $\lim_{x \to \infty} f'(x) = 0$.

CHALLENGE PROBLEMS

17. Suppose that $[a,b] \subseteq A \subseteq \mathbb{R}$, $f: A \to \mathbb{R}$ is differentiable on $[a,b]$, and d is between $f'(a)$ and $f'(b)$. Prove that there is $c \in (a,b)$ such that $f'(c) = d$. This result is known as **Darboux's Theorem**.

18. Prove the three infinite versions of L'Hôpital's Rule (Theorems 10.32, 10.33, and 10.34).

LESSON 11
RIEMANN INTEGRATION

Area Under a Curve

Let $A \subseteq \mathbb{R}$, let $f: A \to \mathbb{R}$, and let $[a, b] \subseteq A$. For any $c \in [a, b]$, the rectangle with a base of length $\Delta x = b - a$ and a height of length $f(c)$ is called a **Riemann rectangle** for f.

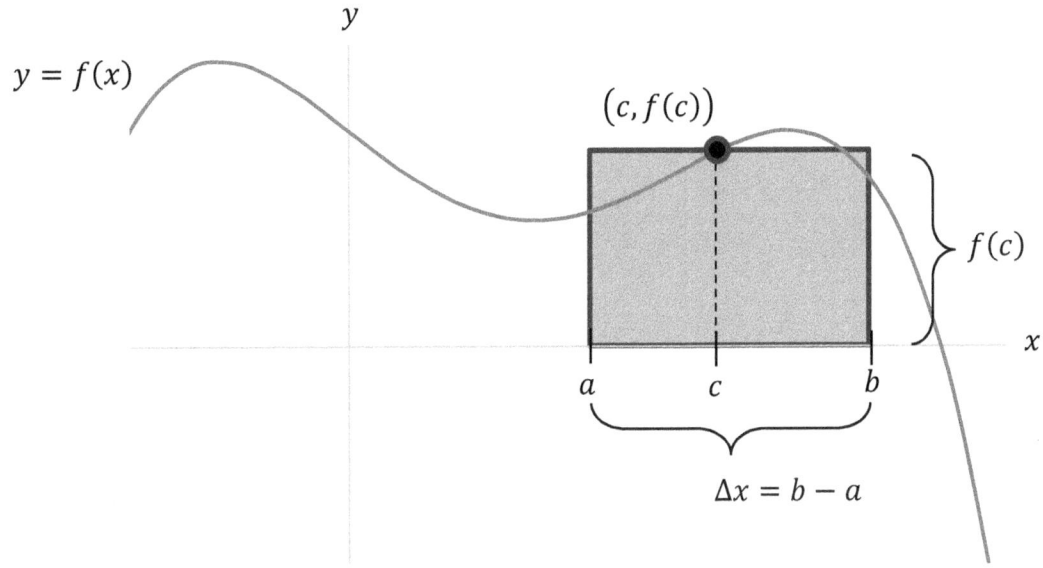

Notes: (1) The symbol Δ being used in the expression Δx is the capital Greek letter "Delta." It is often used to represent "change." Here we are using Δx to represent the change in moving from a to b or the distance between a and b.

(2) From now on, we will sometimes refer to the graph of a function as a **curve**.

(3) Let's use the notation $R_f([a, b], c)$ for the Riemann rectangle with base $\Delta x = b - a$ and height $f(c)$. The area of a rectangle is $A = $ (base) $\cdot$ (height). It follows that the area of the Riemann rectangle $R_f([a, b], c)$ is $A = (b - a) \cdot f(c) = \Delta x \cdot f(c) = f(c) \cdot \Delta x$.

(4) The area of the Riemann rectangle $R_f([a, b], c)$ gives us a rough approximation to the area under the graph of the function f between $x = a$ and $x = b$ (at least for functions with "sufficiently nice" graphs). There are infinitely many (in fact uncountably many) choices we could make for the real number c that is used to form the height of the Riemann rectangle.

(5) If $c = a$, then we will call $R_f([a, b], a)$ a **left Riemann rectangle** for f and if $c = b$, we will call $R_f([a, b], b)$ a **right Riemann rectangle** for f.

If $f(c)$ is the absolute minimum for f on $[a, b]$ (assuming it exists), then we will call $R_f([a, b], c)$ a **lower Riemann rectangle** for f and if $f(c)$ is the absolute maximum for f on $[a, b]$ (assuming it exists), then we will call $R_f([a, b], c)$ an **upper Riemann rectangle** for f.

For the function drawn above, the left Riemann rectangle is the same as the lower Riemann rectangle (this will **not** always be the case). We draw this rectangle below.

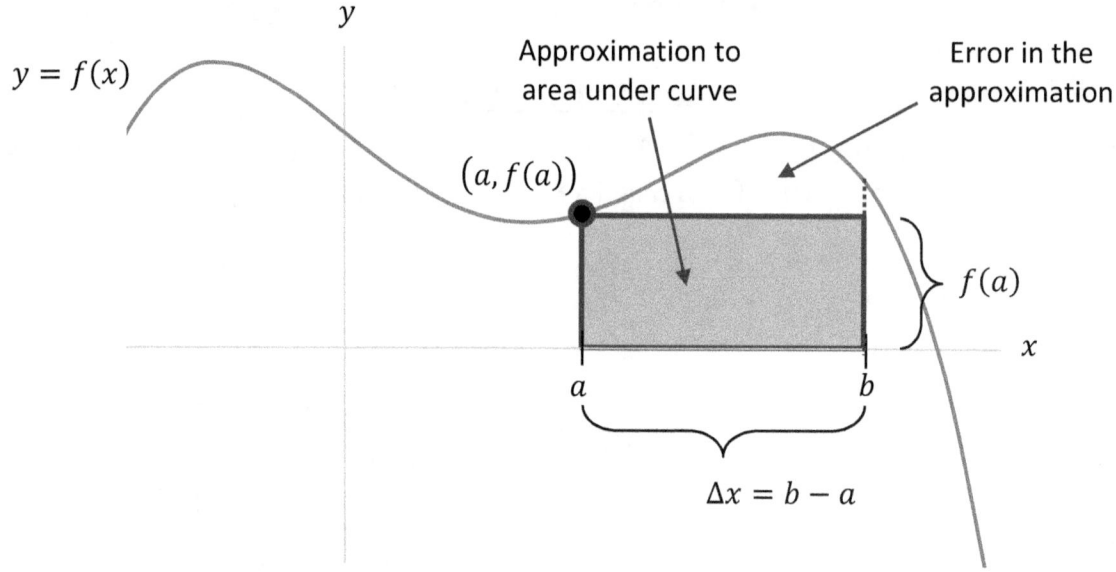

The dedicated reader should draw two additional pictures displaying the right Riemann rectangle for f and the upper Riemann rectangle for f. Note that these two rectangles will **not** be the same for the function drawn above.

Example 11.1: Define $f: \mathbb{R} \to \mathbb{R}$ by $f(x) = x^2$. Let's compute the area of various Riemann rectangles for f.

1. The lower Riemann rectangle $R_f([1,2], 1)$ has area $f(1) \cdot \Delta x = 1^2(2-1) = 1$. Note that in this case, this is also a left Riemann rectangle.

2. The upper Riemann rectangle $R_f([1,2], 2)$ has area $f(2) \cdot \Delta x = 2^2(2-1) = 4$. Note that in this case, this is also a right Riemann rectangle.

    To the right we see a picture of the function f together with the two Riemann rectangles $R_f([1,2], 1)$ and $R_f([1,2], 2)$.

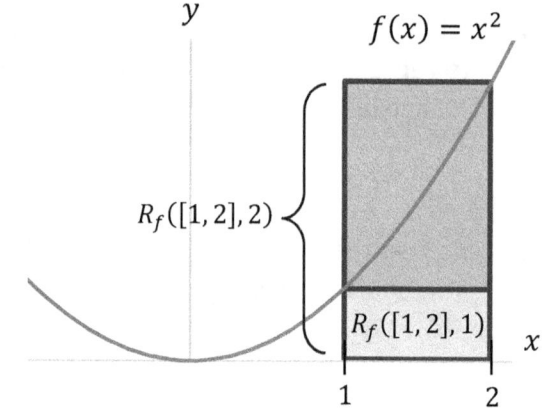

3. The lower Riemann rectangle $R_f([1, 1.5], 1)$ has area $f(1) \cdot \Delta x = 1^2(1.5 - 1) = 0.5$. Note that this is also a left Riemann rectangle.

4. The lower Riemann rectangle $R_f([1.5, 2], 1.5)$ has area $f(1.5) \cdot \Delta x = 1.5^2(2 - 1.5) = 1.125$. Note that this is also a left Riemann rectangle.

    To the right we see a picture of the function f together with the two left Riemann rectangles $R_f([1, 1.5], 1)$ and $R_f([1.5, 2], 1.5)$.

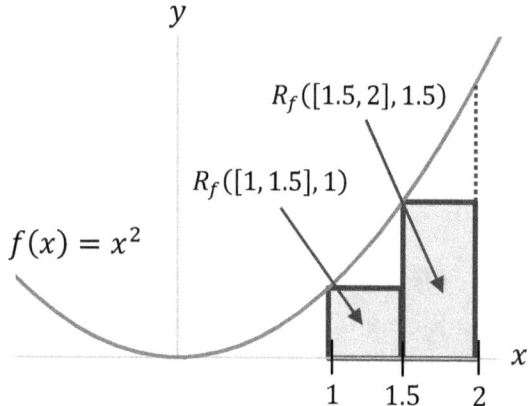

Notes: (1) In the upper figure above on the right, we see that the area L of the lower Riemann rectangle $R_f([1,2],1)$ provides us with an approximation to the area under the graph of $f(x) = x^2$ from $x = 1$ to $x = 2$. This area is certainly an underestimate because the top of the rectangle lies below the curve. The area U of the upper Riemann rectangle $R_f([1,2],2)$ provides a different approximation that is an overestimate because the top of the rectangle lies above the curve. In particular, we have $L \leq U$. Indeed, we have $L = 1$, $U = 4$, and $1 \leq 4$. The actual area under the curve lies somewhere between these two numbers.

(2) In the lower figure above on the right, we have two lower Riemann rectangles. If we let L_1 be the area of $R_f([1,1.5],1)$ and L_2 be the area of $R_f([1.5,2],1.5)$, then $L_1 + L_2 = 0.5 + 1.125 = 1.625$ (this computation is an example of a **Riemann sum**, which will be defined shortly). Notice that this sum provides a better approximation to the area under the curve than $L = 1$ from Note 1 above. By using two rectangles instead of just one, we have reduced the amount of error in our approximation. This approximation still provides an underestimate of the actual area under the curve because the top of each rectangle lies under the curve. So, we see that the actual area under the curve is at least 1.625.

(3) The reader is encouraged to draw a figure showing the two upper Riemann rectangles $R_f([1,1.5],1.5)$ and $R_f([1.5,2],2)$. The area of $R_f([1,1.5],1.5)$ is $U_1 = 1.5^2(1.5-1) = 1.125$ and the area of $R_f([1.5,2],2)$ is $U_2 = 2^2(2-1.5) = 2$. So, $U_1 + U_2 = 1.125 + 2 = 3.125$ (this computation is another example of a Riemann sum). This sum provides a better approximation to the area under the curve than $U = 4$ from Note 1 above. This approximation still provides an overestimate of the actual area under the curve because the top of each rectangle lies above the curve. Combining this with Note 2, we see that the actual area under the curve lies between 1.625 and 3.125.

(4) By using more and more Riemann rectangles, we can get better and better approximations to the area under a curve. A sum of the areas of lower Riemann rectangles will always give an underestimate of the actual area under the curve. A sum of the areas of upper Riemann rectangles will always give an overestimate of the actual area under the curve. These sums are examples of Riemann sums.

A **partition** of the closed interval $[a,b]$ is a set of intervals $\{[x_0,x_1],[x_1,x_2],[x_2,x_3],\ldots,[x_{n-1},x_n]\}$, where $a = x_0 < x_1 < x_2 < x_3 < \cdots < x_{n-1} < x_n = b$.

Notes: (1) The definition of partition given here is different from the definition of partition given in Lesson 2, although the ideas are similar. In this new definition, each element of the partition is a closed interval. Also, the partition is **not** pairwise disjoint. For example, $[x_0,x_1] \cap [x_1,x_2] = \{x_1\} \neq \emptyset$. It is unfortunate that these two definitions are given the same name. Since both usages of the word partition are standard, it seems we are stuck with them.

(2) Let $\{[x_0,x_1],[x_1,x_2],[x_2,x_3],\ldots,[x_{n-1},x_n]\}$ be a partition of $[a,b]$. The individual elements of the partition $[x_0,x_1],[x_1,x_2],[x_2,x_3],\ldots,[x_{n-1},x_n]$ are called **subintervals** of $[a,b]$.

(3) Notice that the partition $\{[x_0,x_1],[x_1,x_2],[x_2,x_3],\ldots,[x_{n-1},x_n]\}$ of $[a,b]$ consists of n subintervals. The first subinterval is $[x_0,x_1]$ and the **length** of the first subinterval is $\Delta x_1 = x_1 - x_0$. The second subinterval is $[x_1,x_2]$ and it has length $\Delta x_2 = x_2 - x_1$. In general, if i is a natural number between 1 and n, inclusive, then the ith subinterval $[x_{i-1},x_i]$ has length $\Delta x_i = x_i - x_{i-1}$.

(4) Let $\{[x_0, x_1], [x_1, x_2], [x_2, x_3], \ldots, [x_{n-1}, x_n]\}$ be a partition of $[a, b]$. If we add up the lengths of all the subintervals in the partition, we get the length of $[a, b]$. In other words, we have

$$\Delta x_1 + \Delta x_2 + \cdots + \Delta x_n = (x_1 - x_0) + (x_2 - x_1) + \cdots + (x_n - x_{n-1}) = x_n - x_0 = b - a.$$

We can abbreviate the above equalities using **Sigma notation** as follows:

$$\sum_{i=1}^{n} \Delta x_i = \sum_{i=1}^{n} (x_i - x_{i-1}) = x_n - x_0 = b - a.$$

(5) The symbol $\sum$ used in Note 4 above is the Greek letter Sigma. In mathematics, this symbol is often used to denote a sum. $\sum$ is generally used to abbreviate a very large sum or a sum of unknown length by specifying what a typical term of the sum looks like. Let's look at a simpler example first before we analyze the more complicated one in Note 4 above:

$$\sum_{k=1}^{5} k^2 = 1^2 + 2^2 + 3^2 + 4^2 + 5^2 = 1 + 4 + 9 + 16 + 25 = 55.$$

The expression "$k = 1$" written underneath the symbol indicates that we get the first term of the sum by replacing k by 1 in the given expression. When we replace k by 1 in the expression k^2, we get 1^2.

For the second term, we simply increase k by 1 to get $k = 2$. So, we replace k by 2 to get $k^2 = 2^2$.

We continue in this fashion, increasing k by 1 each time until we reach the number written above the symbol. In this case, that is $k = 5$.

(6) Let's now get back to the expressions that we're interested in. First, we have the following:

$$\sum_{i=1}^{n} \Delta x_i = \Delta x_1 + \Delta x_2 + \cdots + \Delta x_n$$

To see this, note once again that the expression "$i = 1$" written underneath the symbol $\sum$ indicates that we get the first term of the sum by replacing i by 1 in the given expression. When we replace i by 1 in the expression Δx_i, we get Δx_1.

For the second term, we simply increase i by 1 to get $i = 2$. So, we replace i by 2 to get Δx_2.

We continue in this fashion, increasing i by 1 each time until we reach the number written above the symbol. In this case, that is $i = n$. So, the last term is Δx_n.

See if you can now use the same reasoning to explain the following equality.

$$\sum_{i=1}^{n} (x_i - x_{i-1}) = (x_1 - x_0) + (x_2 - x_1) + \cdots + (x_n - x_{n-1})$$

(7) We will abbreviate the partition $\{[x_0, x_1], [x_1, x_2], [x_2, x_3], \ldots, [x_{n-1}, x_n]\}$ as $\{[x_{i-1}, x_i]\}_{i=1}^{n}$.

Example 11.2: Let's look at some partitions of the closed interval $[0, 1]$.

1. $\{[0, 1]\}$ is the only partition of $[0, 1]$ with one element.

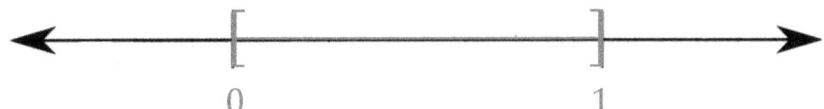

2. For any $c \in (0, 1)$, the set $\{[0, c], [c, 1]\}$ is a partition of $[0, 1]$. For example, $\left\{\left[0, \frac{1}{2}\right], \left[\frac{1}{2}, 1\right]\right\}$ is the only partition of $[0, 1]$ into two subintervals of equal length (they each have length $\frac{1}{2}$).

As another example, $\left\{\left[0, \frac{1}{4}\right], \left[\frac{1}{4}, 1\right]\right\}$ is a partition of $[0, 1]$ into two subintervals with lengths $\frac{1}{4}$ and $1 - \frac{1}{4} = \frac{3}{4}$, respectively.

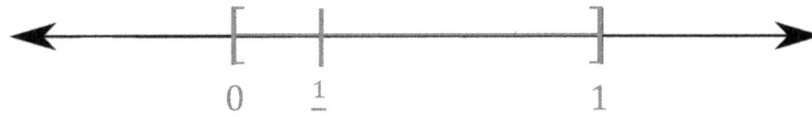

3. $\left\{\left[0, \frac{1}{3}\right], \left[\frac{1}{3}, \frac{2}{3}\right], \left[\frac{2}{3}, 1\right]\right\}$ is the only partition of $[0, 1]$ into three subintervals of equal length (they each have length $\frac{1}{3}$).

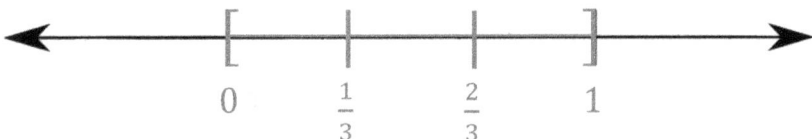

4. For each $n \in \mathbb{Z}^+$, $\left\{\left[0, \frac{1}{n}\right], \left[\frac{1}{n}, \frac{2}{n}\right], \ldots, \left[\frac{n-1}{n}, 1\right]\right\} = \left\{\left[\frac{i-1}{n}, \frac{i}{n}\right]\right\}_{i=1}^{n}$ is the only partition of $[0, 1]$ into n subintervals of equal length (they each have length $\frac{1}{n}$).

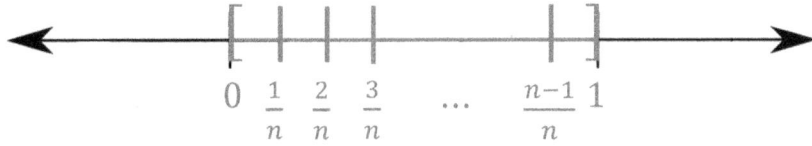

Let $A \subseteq \mathbb{R}$, let $f: A \to \mathbb{R}$, let $[a, b] \subseteq A$, and let $P = \{[x_{i-1}, x_i]\}_{i=1}^{n} = \{[x_0, x_1], [x_1, x_2], \ldots, [x_{n-1}, x_n]\}$ be a partition of $[a, b]$. For each $i = 1, 2, \ldots, n$, let $c_i \in [x_{i-1}, x_i]$ and let $R_f([x_{i-1}, x_i], c_i)$ be the Riemann rectangle with a base of length $\Delta x_i = x_i - x_{i-1}$ and a height of length $f(c_i)$. The sum S of the areas of these Riemann rectangles is called a **Riemann sum** for f over P. Specifically, we have

$$S = \sum_{i=1}^{n} f(c_i) \Delta x_i = f(c_1) \Delta x_1 + f(c_2) \Delta x_2 + \cdots + f(c_n) \Delta x_n \quad \text{or}$$

$$S = \sum_{i=1}^{n} f(c_i)(x_i - x_{i-1}) = f(c_1)(x_1 - x_0) + f(c_2)(x_2 - x_1) + \cdots + f(c_n)(x_n - x_{n-1}).$$

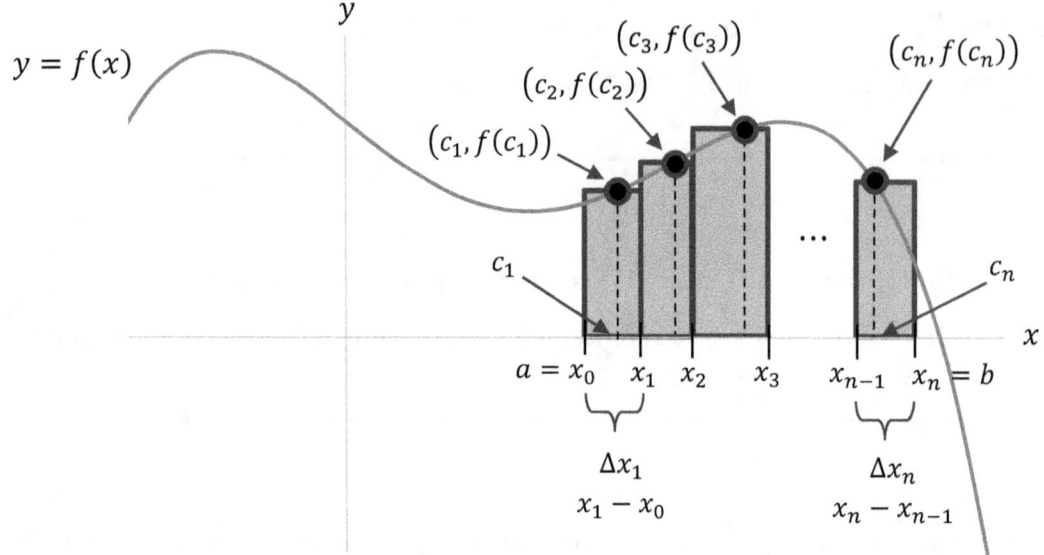

Notes: (1) The above picture displays four Riemann rectangles. From left to right, these rectangles are $R_f([x_0, x_1], c_1), R_f([x_1, x_2], c_2), R_f([x_2, x_3], c_3)$, and $R_f([x_{n-1}, x_n], c_n)$. The ellipses between the third and fourth rectangle indicate that there may be more Riemann rectangles that we are not displaying. Observe that the rectangles do not need to all be the same width.

(2) The leftmost Riemann rectangle, $R_f([x_0, x_1], c_1)$ has a base of $\Delta x_1 = x_0 - x_1$ and a height of $f(c_1)$ (note that c_1 can be visualized as the real number between x_0 and x_1, precisely where the dashed line hits the x-axis). The area of this Riemann rectangle is $f(c_1) \cdot (x_1 - x_0) = f(c_1) \cdot \Delta x_1$.

(3) As in Note 2, the area of the second Riemann rectangle, $R_f([x_1, x_2], c_2)$, is $f(c_2) \cdot \Delta x_2$, the area of the third Riemann rectangle, $R_f([x_2, x_3], c_3)$, is $f(c_3) \cdot \Delta x_3$, and so on.

(4) The real numbers $c_1, c_2, \ldots, c_n$ are sometimes called **tags**. To form a Riemann sum for a function f defined on $[a, b]$, we do the following:

  (i)   Choose a partition $P = \{[x_{i-1}, x_i]\}_{i=1}^{n}$ of $[a, b]$.
  (ii) For each subinterval $[x_{i-1}, x_i] \in P$, choose a tag $c_i \in [x_{i-1}, x_i]$.
  (iii) For each subinterval $[x_{i-1}, x_i] \in P$ and tag $c_i \in [x_{i-1}, x_i]$, form the Riemann rectangle $R_f([x_{i-1}, x_i], c_i)$ with base $\Delta x_i = x_i - x_{i-1}$ and height $f(c_i)$.
  (iv) Add up the areas of all these Riemann rectangles.

(5) The Riemann sum $\sum f(c_i)\Delta x_i = f(c_1)\Delta x_1 + f(c_2)\Delta x_2 + \cdots + f(c_n)\Delta x_n$ gives us an approximation to the area under the graph of the function f between $x = a$ and $x = b$ (at least for functions with "sufficiently nice" graphs). We can get better approximations to this area by using more Riemann rectangles and choosing rectangles with smaller widths.

(6) If $c_i = x_{i-1}$ for each $i = 1, 2, \ldots, n$, then we will call the corresponding Riemann sum a **left Riemann sum** for f and if $c_i = x_i$ for each $i = 1, 2, \ldots, n$, we will call the corresponding Riemann sum a **right Riemann sum** for f.

If $f(c_i)$ is the absolute minimum for f on $[x_{i-1}, x_i]$ for each $i = 1, 2, \ldots, n$, then we will call the corresponding Riemann sum a **lower Riemann sum** for f and if $f(c_i)$ is the absolute maximum for f on $[x_{i-1}, x_i]$ for each $i = 1, 2, \ldots, n$, then we will call the corresponding Riemann sum an **upper Riemann sum** for f. We will use the notation $L_f(P)$ to represent a lower Riemann sum for f and $U_f(P)$ to represent an upper Riemann sum for f, where P is the partition being used.

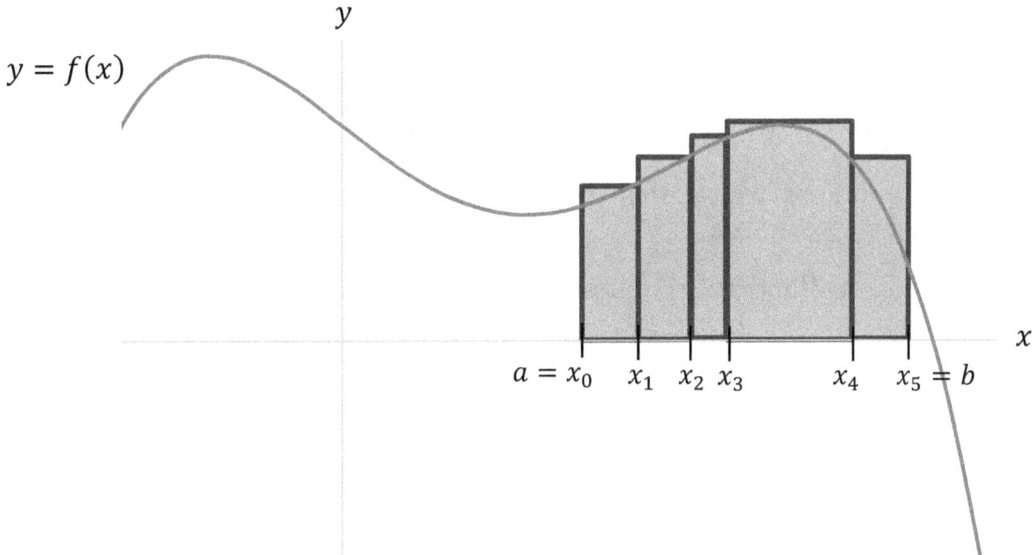

In the above figure, we have partitioned $[a, b]$ into 5 subintervals and drawn an upper Riemann rectangle for each interval. Observe that the three leftmost rectangles are also right Riemann rectangles, the fourth rectangle from the left is neither a left nor a right Riemann rectangle, and the rightmost rectangle is a left Riemann rectangle. The corresponding Riemann sum is

$$U_f(P) = f(x_1)\Delta x_1 + f(x_2)\Delta x_2 + f(x_3)\Delta x_3 + f(c_4)\Delta x_4 + f(x_4)\Delta x_5.$$

Look carefully at the tags and compare them to the heights of the rectangles in the figure. In particular, note that c_4 is neither a left nor a right endpoint of the interval $[x_3, x_4]$.

The dedicated reader should draw three additional pictures, one displaying all lower Riemann rectangles, one displaying all left Riemann rectangles, and one displaying all right Riemann rectangles.

(7) We will provide more general definitions of $L_f(P)$ and $U_f(P)$ in the next section.

(8) If all subintervals in a partition $P = \{[x_{i-1}, x_i]\}_{i=1}^n$ of $[a, b]$ have equal length, we will usually write Δx for the length of each subinterval instead of $\Delta x_1, \Delta x_2, \ldots, \Delta x_n$. In this case, $\Delta x = \frac{b-a}{n}$.

Example 11.3: Define $f: \mathbb{R} \to \mathbb{R}$ by $f(x) = x^2$ and consider the interval $[1, 2]$.

1. We already computed several Riemann sums for this function in Example 11.1 above. For example, for the trivial partition $P = \{[1, 2]\}$, we saw that the lower Riemann sum for f is $L_f(P) = f(1) \cdot \Delta x = 1^2 \cdot 1 = 1$ and the upper Riemann sum for f is $U_f(P) = f(2) \cdot \Delta x = 4$.

    Similarly, for the partition $Q = \{[1, 1.5], [1.5, 2]\}$, we saw that the lower Riemann sum for f is
    $$L_f(Q) = f(1) \cdot \Delta x_1 + f(1.5) \cdot \Delta x_2 = 1^2(1.5 - 1) + 1.5^2(2 - 1.5) = 0.5 + 1.125 = 1.625$$

and the upper Riemann sum for f is

$$U_f(Q) = f(1.5) \cdot \Delta x_1 + f(2) \cdot \Delta x_2 = 1.5^2(1.5 - 1) + 2^2(2 - 1.5) = 1.125 + 2 = 3.125.$$

See the figures to the right of Example 11.1 for visualizations of some of these Riemann sums.

Also, by Note 8 above, we have $\Delta x_1 = \Delta x_2 = \Delta x = \frac{2-1}{2} = \frac{1}{2} = 0.5$. So, we could have computed these sums more briefly as follows:

$$L_f(Q) = f(1) \cdot \Delta x + f(1.5) \cdot \Delta x = 1^2 \cdot 0.5 + 1.5^2 \cdot 0.5 = 0.5 + 1.125 = 1.625$$

$$U_f(Q) = f(1.5) \cdot \Delta x + f(2) \cdot \Delta x = 1.5^2 \cdot 0.5 + 2^2 \cdot 0.5 = 1.125 + 2 = 3.125$$

From these computations, we see that the area under the graph of $f(x) = x^2$ from $x = 1$ to $x = 2$ is between 1.625 and 3.125.

2. Let's now partition $[1, 2]$ into four equal subintervals. Let R be the following partition of $[1, 2]$:

$$R = \{[1, 1.25], [1.25, 1.5], [1.5, 1.75], [1.75, 2]\}$$

For this partition, we have $\Delta x = \frac{2-1}{4} = \frac{1}{4} = 0.25$ and the lower and upper Riemann sums are

$$\boldsymbol{L_f(R)} = f(1) \cdot \Delta x + f(1.25) \cdot \Delta x + f(1.5)\Delta x + f(1.75)\Delta x$$
$$= 1^2 \cdot 0.25 + 1.25^2 \cdot 0.25 + 1.5^2 \cdot 0.25 + 1.75^2 \cdot 0.25 = \boldsymbol{1.96875}$$

$$\boldsymbol{U_f(R)} = f(1.25) \cdot \Delta x + f(1.5) \cdot \Delta x + f(1.75)\Delta x + f(2)\Delta x$$
$$= 1.25^2 \cdot 0.25 + 1.5^2 \cdot 0.25 + 1.75^2 \cdot 0.25 + 2^2 \cdot 0.25 = \boldsymbol{2.71875}$$

From these computations, we see that the area under the graph of $f(x) = x^2$ from $x = 1$ to $x = 2$ is between 1.96875 and 2.71875.

Let's draw a picture showing the lower Riemann rectangles and the upper Riemann rectangles used in the above computations.

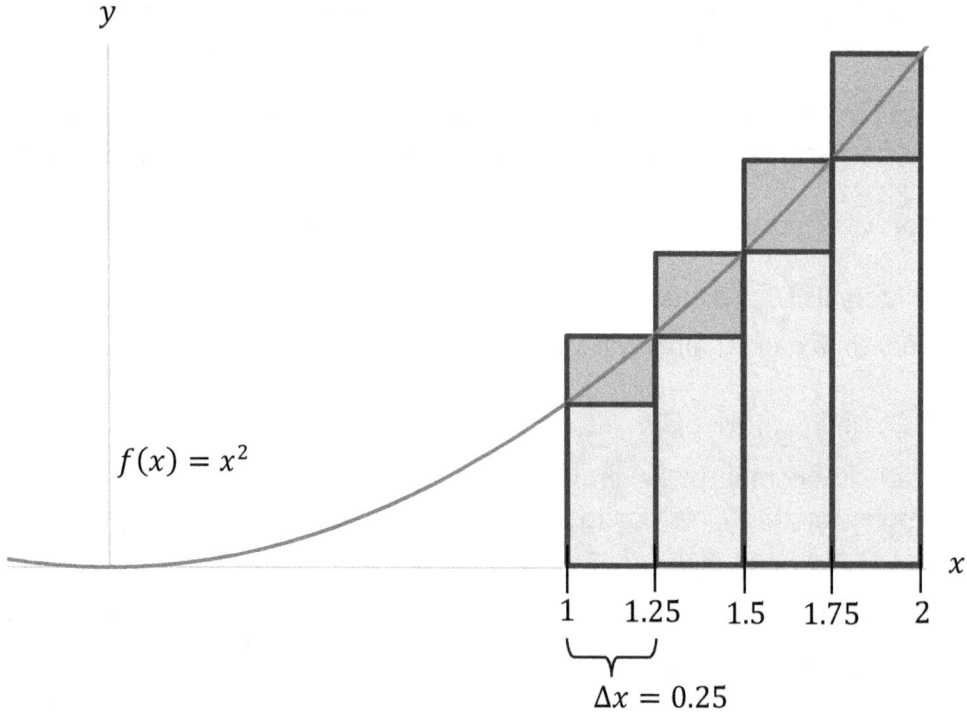

3. Let's partition $[1, 2]$ into n equal subintervals, for an arbitrary positive integer n.

$$R = \left\{\left[1, 1+\frac{1}{n}\right], \left[1+\frac{1}{n}, 1+\frac{2}{n}\right], \ldots, \left[1+\frac{n-1}{n}, 2\right]\right\} = \left\{\left[1+\frac{i-1}{n}, 1+\frac{i}{n}\right]\right\}_{i=1}^{n}$$

For this partition, we have $\Delta x = \frac{2-1}{n} = \frac{1}{n}$ and the upper Riemann sum is

$$U_f(R) = \sum_{i=1}^{n} f(x_i) \Delta x \quad \text{(notice that } c_i = x_i, \text{ the right endpoint of the interval)}$$

$$= \sum_{i=1}^{n} x_i^2 \cdot \frac{1}{n} \quad \left(\text{because } f(x) = x^2 \text{ and } \Delta x = \frac{2-1}{n} = \frac{1}{n}\right)$$

$$= \sum_{i=1}^{n} \left(1 + \frac{i}{n}\right)^2 \cdot \frac{1}{n} \quad \left(\text{because the right endpoint of the } i\text{th subinterval is } 1 + \frac{i}{n}\right)$$

$$= \frac{1}{n} \sum_{i=1}^{n} \left(1 + \frac{i}{n}\right)\left(1 + \frac{i}{n}\right) = \frac{1}{n} \sum_{i=1}^{n} \left[1 + \frac{2i}{n} + \left(\frac{i}{n}\right)^2\right] \quad \left(\text{because } \sum_{i=1}^{n} r a_i = r \sum_{i=1}^{n} a_i\right)$$

$$= \frac{1}{n} \left(\sum_{i=1}^{n} 1 + \sum_{i=1}^{n} \frac{2i}{n} + \sum_{i=1}^{n} \left(\frac{i}{n}\right)^2\right) \quad \left(\text{because } \sum_{i=1}^{n} (a_i + b_i) = \sum_{i=1}^{n} a_i + \sum_{i=1}^{n} b_i\right)$$

$$= \frac{1}{n} \left(\sum_{i=1}^{n} 1 + \frac{2}{n} \sum_{i=1}^{n} i + \frac{1}{n^2} \sum_{i=1}^{n} i^2\right) \quad \left(\text{because } \sum_{i=1}^{n} r a_i = r \sum_{i=1}^{n} a_i\right)$$

$$= \frac{1}{n} \left(n + \frac{2}{n} \cdot \frac{n(n+1)}{2} + \frac{1}{n^2} \cdot \frac{n(n+1)(2n+1)}{6}\right) \quad \text{(by Problem 1 in Problem Set 4)}$$

$$= 1 + \frac{n+1}{n} + \frac{2n^2 + 3n + 1}{6n^2} = \frac{6n^2}{6n^2} + \frac{6n^2 + 6n}{6n^2} + \frac{2n^2 + 3n + 1}{6n^2} = \frac{14n^2 + 9n + 1}{6n^2}.$$

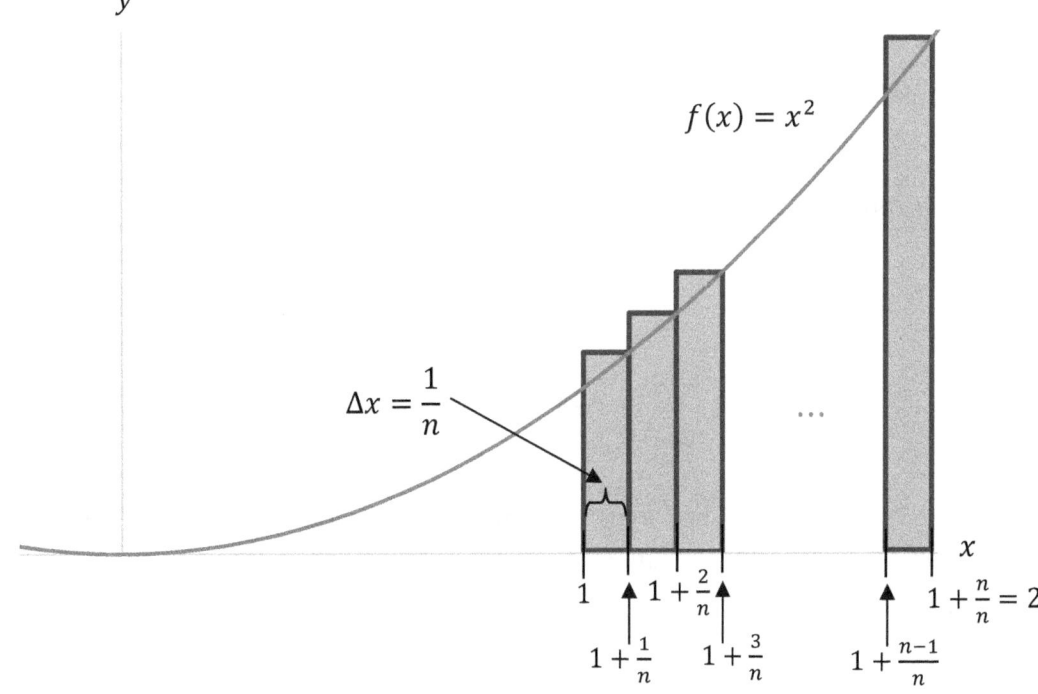

Notes: (1) Since the function $f: \mathbb{R} \to \mathbb{R}$ defined by $f(x) = x^2$ is increasing on $[1, 2]$, it follows that the lower Riemann sums for f on this interval are the same as the left Riemann sums on this interval. Similarly, the upper Riemann sums on this interval are the same as the right Riemann sums on this interval. Note that this is **not** true in general. It is true for this example only because the function is increasing on the given interval.

Similarly, if a function g is decreasing on a given interval $[a, b]$, then the lower Riemann sums of f on $[a, b]$ are equal to the right Riemann sums and the upper Riemann sums of f on $[a, b]$ are equal to the left Riemann sums.

(2) Given a partition $P = \{[x_{i-1}, x_i]\}_{i=1}^n$ of an interval $[a, b]$, we always have $L_f(P) \leq S \leq U_f(P)$ for any Riemann sum S. In other words, given a "sufficiently nice" function f defined on an interval $[a, b]$ and a partition P of $[a, b]$, the lower Riemann sum always gives the smallest possible value of all Riemann sums and the upper Riemann sum always gives the largest possible value of all Riemann sums for f using the partition P. It follows that if A is the exact area under the graph of f from $x = a$ to $x = b$, then $L_f(P) \leq A \leq U_f(P)$.

For example, for the function $f: \mathbb{R} \to \mathbb{R}$ defined by $f(x) = x^2$, we saw in part 2 of Example 11.3 above that $1.96875 \leq A \leq 2.71875$, where A is the area under the graph of f from $x = 1$ to $x = 2$.

(3) In part 3 of Example 11.3 above, we used the following facts about sums:

$$\sum_{i=1}^n r a_i = r \sum_{i=1}^n a_i$$

$$\sum_{i=1}^n (a_i + b_i) = \sum_{i=1}^n a_i + \sum_{i=1}^n b_i$$

$$\sum_{i=1}^n 1 = n$$

These formulas can be proved by using the Principle of Mathematical Induction. You will be asked to write out the proofs in Problem 1 below.

(4) We also used the following formulas in part 3 of Example 11.3:

$$\sum_{i=1}^n i = 1 + 2 + \cdots + n = \frac{n(n+1)}{2}$$

$$\sum_{i=1}^n i^2 = 1^2 + 2^2 + \cdots + n^2 = \frac{n(n+1)(2n+1)}{6}$$

These formulas were established in parts (ii) and (iii) of Problem 1 from Problem Set 4. In that problem, we added a 0 to each formula. However, adding 0 doesn't change the result.

(5) For the function $f: \mathbb{R} \to \mathbb{R}$ defined by $f(x) = x^2$ and any positive integer n, if we use the partition $R = \left\{\left[1 + \frac{i-1}{n}, 1 + \frac{i}{n}\right]\right\}_{i=1}^{n}$, we showed in part 3 of Example 11.3 that

$$U_f(R) = \frac{14n^2 + 9n + 1}{6n^2}.$$

Similarly, we have the following (this computation is more tedious):

$$L_f(R) = \sum_{i=1}^{n} f(x_{i-1})\Delta x = \sum_{i=1}^{n} x_{i-1}^2 \cdot \frac{1}{n} = \sum_{i=1}^{n} \left(1 + \frac{i-1}{n}\right)^2 \cdot \frac{1}{n}$$

$$= \frac{1}{n}\sum_{i=1}^{n}\left[\left(1 + \frac{i-1}{n}\right)\left(1 + \frac{i-1}{n}\right)\right] = \frac{1}{n}\sum_{i=1}^{n}\left[1 + 2\left(\frac{i-1}{n}\right) + \left(\frac{i-1}{n}\right)^2\right]$$

$$= \frac{1}{n}\left(\sum_{i=1}^{n} 1 + \sum_{i=1}^{n} 2\left(\frac{i-1}{n}\right) + \sum_{i=1}^{n}\left(\frac{i-1}{n}\right)^2\right) = \frac{1}{n}\left(\sum_{i=1}^{n} 1 + \frac{2}{n}\sum_{i=1}^{n}(i-1) + \frac{1}{n^2}\sum_{i=1}^{n}(i-1)^2\right)$$

$$= \frac{1}{n}\left(\sum_{i=1}^{n} 1 + \frac{2}{n}\sum_{i=1}^{n} i - \frac{2}{n}\sum_{i=1}^{n} 1 + \frac{1}{n^2}\sum_{i=1}^{n}(i^2 - 2i + 1)\right)$$

$$= \frac{1}{n}\left(\sum_{i=1}^{n} 1 + \frac{2}{n}\sum_{i=1}^{n} i - \frac{2}{n}\sum_{i=1}^{n} 1 + \frac{1}{n^2}\sum_{i=1}^{n} i^2 - \frac{2}{n^2}\sum_{i=1}^{n} i + \frac{1}{n^2}\sum_{i=1}^{n} 1\right)$$

$$= \frac{1}{n}\left(n + \frac{2}{n} \cdot \frac{n(n+1)}{2} - \frac{2}{n} \cdot n + \frac{1}{n^2} \cdot \frac{n(n+1)(2n+1)}{6} - \frac{2}{n^2} \cdot \frac{n(n+1)}{2} + \frac{1}{n^2} \cdot n\right)$$

$$= \frac{1}{n}\left(n + (n+1) - 2 + \frac{(n+1)(2n+1)}{6n} - \frac{n+1}{n} + \frac{1}{n}\right)$$

$$= \frac{1}{n}\left(2n - 1 + \frac{(n+1)(2n+1)}{6n} - \frac{n+1}{n} + \frac{1}{n}\right) = 2 - \frac{1}{n} + \frac{(n+1)(2n+1)}{6n^2} - \frac{n+1}{n^2} + \frac{1}{n^2}$$

$$= \frac{12n^2}{6n^2} - \frac{6n}{6n^2} + \frac{(n+1)(2n+1)}{6n^2} - \frac{6n+6}{6n^2} + \frac{6}{6n^2}$$

$$= \frac{12n^2 - 6n + 2n^2 + 3n + 1 - 6n - 6 + 6}{6n^2}$$

$$= \frac{14n^2 - 9n + 1}{6n^2}.$$

It follows from Note 2 above that for the function $f: \mathbb{R} \to \mathbb{R}$ defined by $f(x) = x^2$, for every positive integer n, we have

$$\frac{14n^2 - 9n + 1}{6n^2} \leq A \leq \frac{14n^2 + 9n + 1}{6n^2},$$

where A is the area under the graph of f from $x = 1$ to $x = 2$.

(6) Using L'Hôpital's rule, we have the following:

$$\lim_{n\to\infty} \frac{14n^2 + 9n + 1}{6n^2} = \lim_{n\to\infty} \frac{28n + 9}{12n} = \frac{28}{12} = \frac{7}{3}$$

$$\lim_{n\to\infty} \frac{14n^2 - 9n + 1}{6n^2} = \lim_{n\to\infty} \frac{28n - 9}{12n} = \frac{28}{12} = \frac{7}{3}$$

It follows that the exact area under the graph of f from $x = 1$ to $x = 2$ is $\frac{7}{3}$.

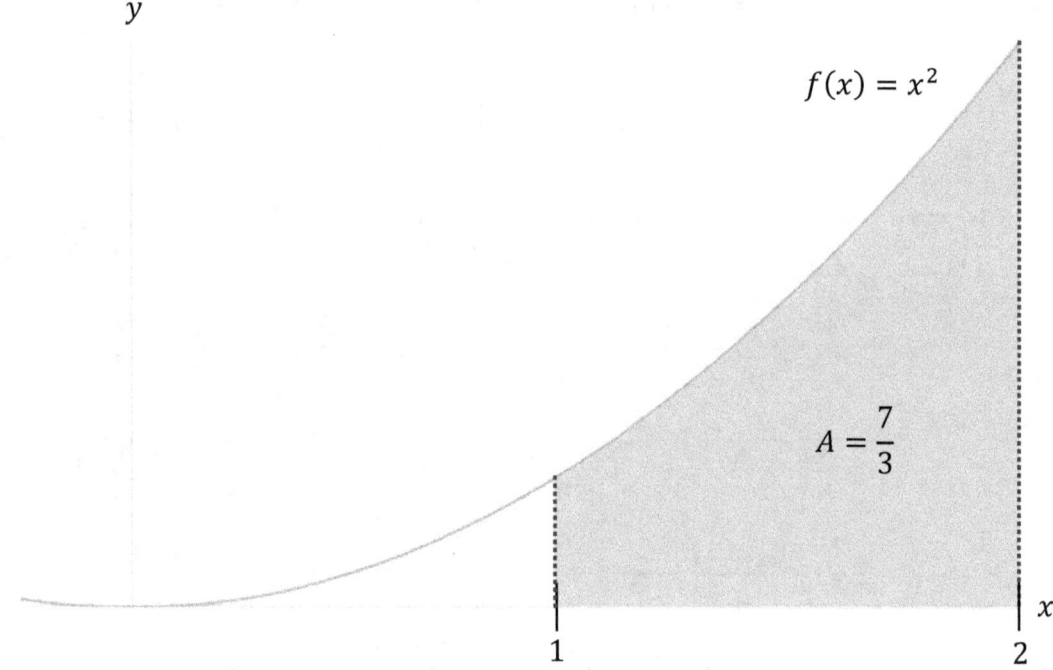

The Riemann Integral

We will now be restricting our attention to functions that are defined and bounded on a closed interval.

Recall that a function $f: [a, b] \to \mathbb{R}$ is **bounded** on $[a, b]$ if $\{f(x) \mid x \in [a, b]\}$ is bounded.

If f is bounded, then $m = \inf\{f(x) \mid x \in [a, b]\}$ and $M = \sup\{f(x) \mid x \in [a, b]\}$ both exist by the Completeness Property of the real numbers.

Note: From now on, when we use the notation $f: [a, b] \to \mathbb{R}$, we will allow for the possibility that f is defined on a set A with $[a, b] \subseteq A$. For example, when we say $f: [a, b] \to \mathbb{R}$ is bounded on $[a, b]$, the function f may itself be defined for real numbers outside the interval $[a, b]$, and in fact, as a whole, the function does **not** even need to be bounded. It needs to be bounded only when we restrict it to $[a, b]$.

In Note 6 following Example 11.2, we provided definitions for lower and upper Riemann sums $L_f(P)$ and $U_f(P)$, where we choose tags $c_i \in [x_{i-1}, x_i]$ such that $f(c_i)$ is the absolute minimum (for $L_f(P)$) or absolute maximum (for $U_f(P)$). If f is a continuous function, then by the Maximum-Minimum Theorem (Corollary 9.7), the tags can be chosen this way. However, for a bounded function with discontinuities, this may not always be possible.

Example 11.4: Consider the function $f: [0, 1] \to \mathbb{R}$ defined as follows:

$$f(x) = \begin{cases} x & \text{if } 0 \leq x < \frac{1}{2}. \\ 0 & \text{if } \frac{1}{2} \leq x \leq 1. \end{cases}$$

This function is bounded on $[0, 1]$, but it has no absolute maximum on $[0, 1]$. Therefore, for the trivial partition $P = \{[0, 1]\}$, the upper Riemann sum $U_f(P)$ (as presently defined) does not exist.

However, by the Completeness Property of $\mathbb{R}$, $\sup\{f(x) \mid x \in [0, 1]\}$ does exist for this function f. It is equal to $\frac{1}{2}$. Therefore, it seems reasonable that we should take $U_f(P) = \frac{1}{2}(1 - 0) = \frac{1}{2}$.

With Example 11.4 as motivation, we generalize our definitions of lower and upper Riemann sums as follows:

Let $f: [a, b] \to \mathbb{R}$ be a bounded function and let $P = \{[x_{i-1}, x_i]\}_{i=1}^{n}$ be a partition of $[a, b]$. For each $i = 1, 2, \ldots, n$, we let $m_i = \inf\{f(x) \mid x \in [x_{i-1}, x_i]\}$ and $M_i = \sup\{f(x) \mid x \in [x_{i-1}, x_i]\}$. Also, let $\Delta x_i = x_i - x_{i-1}$. Then we define the **lower Darboux sum** $L_f(P)$ and the **upper Darboux sum** $U_f(P)$ as follows:

$$L_f(P) = \sum_{i=1}^{n} m_i \Delta x_i = m_1 \Delta x_1 + m_2 \Delta x_2 + \cdots + m_n \Delta x_n$$

$$U_f(P) = \sum_{i=1}^{n} M_i \Delta x_i = M_1 \Delta x_1 + M_2 \Delta x_2 + \cdots + M_n \Delta x_n$$

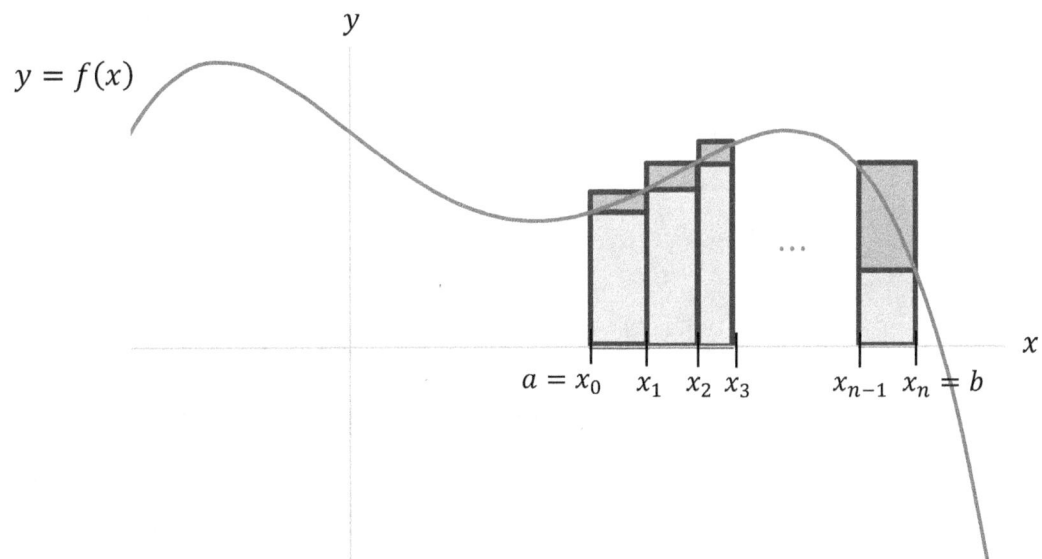

Notes: (1) In the figure above, each m_i and M_i are actually in the range of the function f. For an arbitrary bounded function, this does not need to be true.

(2) If $m_i \in \operatorname{ran} f$ for all $i = 1, 2, \ldots, n$, then the lower Darboux sum $L_f(P) = \sum m_i \Delta x_i$ is also a lower Riemann sum. Similarly, if $M_i \in \operatorname{ran} f$ for all $i = 1, 2, \ldots, n$, then the upper Darboux sum $U_f(P) = \sum M_i \Delta x_i$ is also an upper Riemann sum. In this lesson, we may abuse language a bit and use the names "Riemann" and "Darboux" interchangeably. As it turns out, the theories of Riemann integration and Darboux integration are equivalent, and so, not much harm can come from this.

Theorem 11.5: Let $f: [a, b] \to \mathbb{R}$ be a bounded function, let $m = \inf\{f(x) \mid x \in [a, b]\}$, let $M = \sup\{f(x) \mid x \in [a, b]\}$, and let P be a partition of $[a, b]$. Then

$$m(b - a) \leq L_f(P) \leq U_f(P) \leq M(b - a).$$

Analysis: Each of the four expressions in the sequence of inequalities above can be thought of as an area. For example, $m(b - a)$ is the area of the rectangle with base the length of $[a, b]$ and height m, where m is the "bottom" of the graph of f between a and b. Similarly, $M(b - a)$ is the area of the rectangle with base the length of $[a, b]$ and height M, where M is the "top" of the graph of f between a and b. Each of $L_f(P)$ and $U_f(P)$ is a sum of areas of rectangles, as described above.

A geometric interpretation of Theorem 11.5 can be visualized as shown in the picture below. Observe how each lower Riemann rectangle is at least as high as m, the height of each upper Riemann rectangle is at least as high as the height of each lower Riemann rectangle, and M is at least as high as the height of each upper Riemann rectangle.

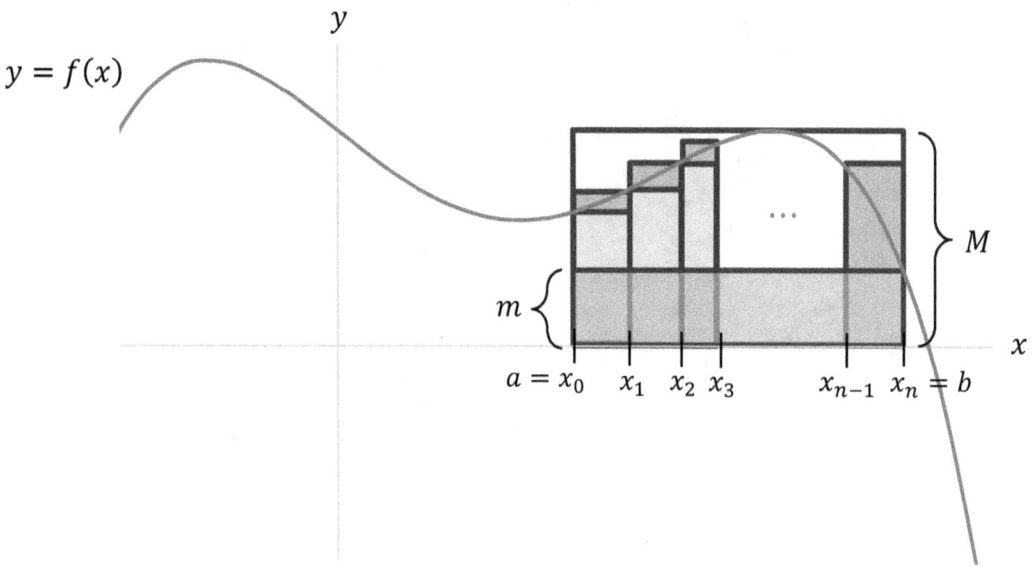

Proof: Let f, m, and M be as given in the theorem and let $P = \{[x_{i-1}, x_i]\}_{i=1}^{n}$ be the given partition. For each $i = 1, 2, \ldots, n$, let $m_i = \inf\{f(x) \mid x \in [x_{i-1}, x_i]\}$ and let $M_i = \sup\{f(x) \mid x \in [x_{i-1}, x_i]\}$.

For each $i = 1, 2, \ldots, n$, we have $\{f(x) \mid x \in [x_{i-1}, x_i]\} \subseteq \{f(x) \mid x \in [a, b]\}$. By Problem 3 below, for each $i = 1, 2, \ldots, n$, we have $m \leq m_i \leq M_i \leq M$. Also, we have

$$\sum_{i=1}^{n} \Delta x_i = \sum_{i=1}^{n}(x_i - x_{i-1}) = (x_1 - x_0) + (x_2 - x_1) + \cdots + (x_n - x_{n-1}) = x_n - x_0 = b - a.$$

It follows that

$$m(b-a) = m\sum_{i=1}^{n} \Delta x_i = \sum_{i=1}^{n} m\Delta x_i \le \sum_{i=1}^{n} m_i \Delta x_i \le \sum_{i=1}^{n} M_i \Delta x_i \le \sum_{i=1}^{n} M\Delta x_i = M\sum_{i=1}^{n} \Delta x_i = M(b-a).$$

So, $m(b-a) \le L_f(P) \le U_f(P) \le M(b-a)$, as desired. $\square$

Let $f: [a,b] \to \mathbb{R}$ be a bounded function. By Theorem 11.5, the sets $\{L_f(P) \mid P \text{ is a partition of } [a,b]\}$ and $\{U_f(P) \mid P \text{ is a partition of } [a,b]\}$ are bounded. Therefore, by the Completeness Property of $\mathbb{R}$, we can make the following definitions:

$$\overline{\int_a^b} f(x)\,dx = \inf\{U_f(P) \mid P \text{ is a partition of } [a,b]\}$$

$$\underline{\int_a^b} f(x)\,dx = \sup\{L_f(P) \mid P \text{ is a partition of } [a,b]\}$$

$\overline{\int_a^b} f(x)\,dx$ is called the **upper Riemann integral of f** (or the **upper Darboux integral** of f) over $[a,b]$ and $\underline{\int_a^b} f(x)\,dx$ is called the **lower Riemann integral of f** (or the **lower Darboux integral** of f) over $[a,b]$.

If $\overline{\int_a^b} f(x)\,dx = \underline{\int_a^b} f(x)\,dx$, then we say that f is **Riemann integrable** (or **Darboux integrable**) on $[a,b]$ and we define $\int_a^b f(x)\,dx$ to be this common value. In this case, we will call $\int_a^b f(x)\,dx$ the **Riemann integral** (or **Darboux integral**) of f over $[a,b]$.

Notes: (1) In Note 6 following Example 11.3, we computed the area under the graph of a function by considering the limiting value of upper and lower Riemann sums as the partitions became "more refined." To be more precise, if $P = \{\{[x_{i-1}, x_i]\}_{i=1}^{n}\}$ and $Q = \{\{[y_{i-1}, y_i]\}_{i=1}^{m}\}$ are partitions of $[a,b]$, we say that Q is a **refinement** of P if $\{x_0, x_1, \ldots x_n\} \subseteq \{y_0, y_1, \ldots, y_m\}$. Notice that this definition implies that $m \ge n$. For example, if $P = \{[0,1]\}$ and $Q = \{[0,\frac{1}{2}], [\frac{1}{2},1]\}$, then Q is a refinement of P because $\{0,1\} \subseteq \{0, \frac{1}{2}, 1\}$.

(2) If $P = \{\{[x_{i-1}, x_i]\}_{i=1}^{n}\}$ and $Q = \{\{[y_{i-1}, y_i]\}_{i=1}^{m}\}$ are partitions of $[a,b]$, then we define $P * Q$ to be the partition $P * Q = \{\{[z_{i-1}, z_i]\}_{i=1}^{k}\}$, where $\{z_0, z_1, \ldots z_k\} = \{x_0, x_1, \ldots x_n\} \cup \{y_0, y_1, \ldots, y_m\}$ and $z_0 < z_1 < \cdots < z_k$. For example, if $P = \{[0,\frac{1}{2}], [\frac{1}{2},1]\}$ and $Q = \{[0,\frac{1}{3}], [\frac{1}{3},\frac{2}{3}], [\frac{2}{3},1]\}$, then we have $P * Q = \{[0,\frac{1}{3}], [\frac{1}{3},\frac{1}{2}], [\frac{1}{2},\frac{2}{3}], [\frac{2}{3},1]\}$. Observe that $P * Q$ is the smallest partition that is a refinement of both P and Q.

(3) If $f: [a,b] \to \mathbb{R}$ is a bounded function and P and Q are partitions of $[a,b]$ with Q a refinement of P, then $L_f(P) \le L_f(Q)$ and $U_f(Q) \le U_f(P)$. You will be asked to prove this in Problem 10 below.

(4) If $f:[a,b] \to \mathbb{R}$ is a bounded function, then $\underline{\int_a^b} f(x)\,dx \leq \overline{\int_a^b} f(x)\,dx$. You will be asked to prove this in Problem 11 below. You may want to use Notes 2 and 3 above in the proof.

(5) Let $f:[a,b] \to \mathbb{R}$ be a bounded function and for each $n \in \mathbb{Z}^+$, let P_n be the partition consisting of n equal subintervals of $[a,b]$. In other words, if we let $x_i = a + \frac{i(b-a)}{n}$, then $P_n = \{[x_{i-1}, x_i]\}_{i=1}^n$. In this case, we have $\underline{\int_a^b} f(x)\,dx = \lim_{n \to \infty} L_f(P_n)$ and $\overline{\int_a^b} f(x)\,dx = \lim_{n \to \infty} U_f(P_n)$. You will be asked to prove this in Problem 18 below.

(6) If $f:[a,b] \to \mathbb{R}$ is a bounded function, then for any $c \in [a,b]$, we define $\int_c^c f(x)\,dx = 0$ and we define $\int_b^a f(x)\,dx$ to be the negative of $\int_a^b f(x)\,dx$. That is, we define

$$\int_b^a f(x)\,dx = -\int_a^b f(x)\,dx.$$

Example 11.6:

1. Define $f:[1,2] \to \mathbb{R}$ by $f(x) = x^2$. Using Note 6 following Example 11.3 together with Note 5 above, we have

$$\underline{\int_1^2} f(x)\,dx = \lim_{n \to \infty} \frac{14n^2 - 9n + 1}{6n^2} = \frac{7}{3} \quad \text{and} \quad \overline{\int_1^2} f(x)\,dx = \lim_{n \to \infty} \frac{14n^2 + 9n + 1}{6n^2} = \frac{7}{3}.$$

    It follows that f is Riemann integrable on $[1,2]$ and

$$\int_1^2 f(x)\,dx = \frac{7}{3}.$$

    Also, by Note 6 above, we have

$$\int_2^1 f(x)\,dx = -\frac{7}{3}.$$

2. Define $g:[0,1] \to \mathbb{R}$ by $g(x) = \begin{cases} 1 & \text{if } x \text{ is rational.} \\ 0 & \text{if } x \text{ is irrational.} \end{cases}$ Let $P = \{[x_{i-1}, x_i]\}_{i=1}^n$ be any partition of $[0,1]$. Then for each $i = 1, 2, \ldots, n$, we have $m_i = 0$ and $M_i = 1$ (this follows from the Density Theorem (Theorem 6.17) and Problem 19 from Problem Set 6). Therefore, we have

$$L_f(P) = \sum_{i=1}^n m_i \Delta x_i = \sum_{i=1}^n 0 \cdot \Delta x_i = 0 \quad \text{and} \quad U_f(P) = \sum_{i=1}^n M_i \Delta x_i = \sum_{i=1}^n 1 \cdot \Delta x_i = 1 - 0 = 1.$$

    It follows that

$$\underline{\int_a^b} g(x)\,dx = 0 \quad \text{and} \quad \overline{\int_a^b} g(x)\,dx = 1.$$

    Since $\underline{\int_a^b} g(x)\,dx \neq \overline{\int_a^b} g(x)\,dx$, f is **not** Riemann integrable on $[0,1]$.

Theorem 11.7: Let $f: [a,b] \to \mathbb{R}$ be a continuous function. Then f is Riemann integrable on $[a,b]$.

Proof: By the Heine-Borel Theorem (Theorem 7.28), $[a,b]$ is compact. It follows from Problem 10 in Problem Set 9 that f is uniformly continuous on $[a,b]$.

Let $\epsilon > 0$. Since f is uniformly continuous on $[a,b]$, there is $\delta > 0$ such that for all $x, y \in [a,b]$,
$$|x - y| < \delta \to |f(x) - f(y)| < \frac{\epsilon}{b-a}.$$

By the Archimedean Property of $\mathbb{R}$ (Theorem 6.16), there is $n \in \mathbb{N}$ such that $n > \frac{b-a}{\delta}$, or equivalently, $\frac{b-a}{n} < \delta$. For each $i = 0, 1, 2, \ldots, n$, let $x_i = a + \frac{i(b-a)}{n}$ and let $P = \{[x_{i-1}, x_i]\}_{i=1}^n$. Then $\Delta x_i = x_i - x_{i-1} = \frac{b-a}{n} < \delta$. So, for each $i = 1, 2, \ldots, n$, if $x, y \in [x_{i-1}, x_i]$, then $|x - y| \le \Delta x_i < \delta$ and it follows that $f(x) - f(y) \le |f(x) - f(y)| < \frac{\epsilon}{b-a}$.

By the Maximum-Minimum Theorem (Corollary 9.7), if we let $m_i = \inf\{f(x) \mid x \in [x_{i-1}, x_i]\}$ and $M_i = \sup\{f(x) \mid x \in [x_{i-1}, x_i]\}$, there are $c_i, d_i \in [x_{i-1}, x_i]$ with $f(c_i) = m_i$ and $f(d_i) = M_i$. Then, we have $M_i - m_i = f(d_i) - f(c_i) < \frac{\epsilon}{b-a}$. It follows that

$$\overline{\int_a^b} f(x)\,dx - \underline{\int_a^b} f(x)\,dx \le U_f(P) - L_f(P) = \sum_{i=1}^n M_i \Delta x_i - \sum_{i=1}^n m_i \Delta x_i$$

$$= \sum_{i=1}^n (M_i - m_i) \Delta x_i < \sum_{i=1}^n \frac{\epsilon}{b-a} \Delta x_i = \frac{\epsilon}{b-a} \sum_{i=1}^n \Delta x_i = \frac{\epsilon}{b-a}(b-a) = \epsilon.$$

Also, by Note 4 before Example 11.6 above (or Problem 11 below), $\overline{\int_a^b} f(x)\,dx - \underline{\int_a^b} f(x)\,dx \ge 0$. Since $\epsilon > 0$ was arbitrary, by Problem 15 from Problem Set 6, we have $\overline{\int_a^b} f(x)\,dx - \underline{\int_a^b} f(x)\,dx = 0$, or equivalently, $\overline{\int_a^b} f(x)\,dx = \underline{\int_a^b} f(x)\,dx$. So, f is Riemann integrable on $[a,b]$. □

Does the converse of Theorem 11.7 hold? In other words, is every Riemann integrable function on $[a,b]$ continuous? The answer is no! In Problem 16 below, you will be asked to prove that every bounded function $f: [a,b] \to \mathbb{R}$ that has finitely many discontinuities is Riemann integrable on $[a,b]$. In Problem 13, you will be asked to prove that every monotonic function $f: [a,b] \to \mathbb{R}$ is Riemann integrable on $[a,b]$. In Problem 17, you will be asked to find a Riemann integrable function $f: [0,1] \to \mathbb{R}$ with infinitely many discontinuities that is not monotonic.

Integration Rules

Theorem 11.8: Let $f: [a,b] \to \mathbb{R}$ be defined by $f(x) = r$. Then f is Riemann integrable on $[a,b]$ and
$$\int_a^b f(x)\,dx = r(b-a).$$

This theorem can be interpreted geometrically as follows: the integral of a constant function with constant value r over the interval $[a, b]$ is just the area of the rectangle with base length $b - a$ and height r.

For example,
$$\int_3^{11} 4\, dx = 4(11 - 3) = 4 \cdot 8 = 32.$$

Proof: Let $m = \inf\{f(x) \mid x \in [a,b]\}$ and $M = \sup\{f(x) \mid x \in [a,b]\}$. Then $m = M = r$. By Problem 11 below, we have
$$r(b - a) \leq \underline{\int_a^b} f(x)\, dx \leq \overline{\int_a^b} f(x)\, dx \leq r(b - a).$$

Therefore, $\overline{\int_a^b} f(x)\, dx = \underline{\int_a^b} f(x)\, dx = r(b - a)$. So, f is Riemann integrable on $[a, b]$ and
$$\int_a^b f(x)\, dx = r(b - a).$$
$\square$

Note: The conclusion of Theorem 11.8 still holds if $b < a$. In this case, we can use Theorem 11.8 together with Note 6 before Example 11.6 to get
$$\int_a^b f(x)\, dx = -\int_b^a f(x)\, dx = -r(a - b) = r(b - a).$$

For example, we have
$$\int_{11}^3 4\, dx = -\int_3^{11} 4\, dx = -4(11 - 3) = -4 \cdot 8 = -32.$$

Theorem 11.9: Let $f, g : [a, b] \to \mathbb{R}$ be Riemann integrable on $[a, b]$. Then $f + g$ is Riemann integrable on $[a, b]$ and
$$\int_a^b (f + g)(x)\, dx = \int_a^b f(x)\, dx + \int_a^b g(x)\, dx.$$

Theorem 11.9 is a bit harder to prove than one might initially expect. To make the proof more manageable, we will prove two preliminary lemmas.

Lemma 11.10: Let $f, g : [a, b] \to \mathbb{R}$ be a bounded function. Then f is Riemann integrable on $[a, b]$ if and only if for every $\epsilon > 0$, there is a partition P of $[a, b]$ such that $U_f(P) - L_f(P) < \epsilon$.

Proof: First assume that f is Riemann integrable on $[a, b]$.

Let $\epsilon > 0$. Since $\underline{\int_a^b} f(x)\,dx = \sup\{L_f(P) \mid P \text{ is a partition of } [a,b]\}$, it follows that $\underline{\int_a^b} f(x)\,dx - \frac{\epsilon}{2}$ is **not** an upper bound of $\{L_f(P) \mid P \text{ is a partition of } [a,b]\}$. Therefore, there is a partition P of $[a,b]$ such that $\underline{\int_a^b} f(x)\,dx - \frac{\epsilon}{2} < L_f(P)$. Since $\overline{\int_a^b} f(x)\,dx = \inf\{U_f(P) \mid P \text{ is a partition of } [a,b]\}$, it follows that $\overline{\int_a^b} f(x)\,dx + \frac{\epsilon}{2}$ is **not** a lower bound of $\{U_f(P) \mid P \text{ is a partition of } [a,b]\}$. Therefore, there is a partition Q of $[a,b]$ such that $\overline{\int_a^b} f(x)\,dx + \frac{\epsilon}{2} > U_f(Q)$. Let $R = P \star Q$ (see Note 2 before Example 11.6 above). By Problem 10 below, we have $L_f(P) \leq L_f(R)$ and $U_f(R) \leq U_f(Q)$. Therefore,

$$\underline{\int_a^b} f(x)\,dx - \frac{\epsilon}{2} < L_f(P) \leq L_f(R) \leq U_f(R) \leq U_f(Q) < \overline{\int_a^b} f(x)\,dx + \frac{\epsilon}{2}.$$

Since f is Riemann integrable, we have $\overline{\int_a^b} f(x)\,dx = \underline{\int_a^b} f(x)\,dx$, and so,

$$U_f(R) - L_f(R) < \overline{\int_a^b} f(x)\,dx + \frac{\epsilon}{2} - L_f(R) = \underline{\int_a^b} f(x)\,dx + \frac{\epsilon}{2} - L_f(R)$$

$$= \left(\underline{\int_a^b} f(x)\,dx - \frac{\epsilon}{2}\right) + \frac{\epsilon}{2} + \frac{\epsilon}{2} - L_f(R) < L_f(R) + \frac{\epsilon}{2} + \frac{\epsilon}{2} - L_f(R) = \epsilon.$$

To prove the converse, let $\epsilon > 0$ and choose a partition P of $[a,b]$ such that $U_f(P) - L_f(P) < \epsilon$. Then we have

$$\overline{\int_a^b} f(x)\,dx - \underline{\int_a^b} f(x)\,dx \leq U_f(P) - L_f(P) < \epsilon.$$

Also, by Note 4 before Example 11.6 above, $\overline{\int_a^b} f(x)\,dx - \underline{\int_a^b} f(x)\,dx \geq 0$. Since $\epsilon > 0$ was arbitrary, by Problem 15 from Problem Set 6, we have $\overline{\int_a^b} f(x)\,dx - \underline{\int_a^b} f(x)\,dx = 0$, or equivalently, $\overline{\int_a^b} f(x)\,dx = \underline{\int_a^b} f(x)\,dx$. So, f is Riemann integrable on $[a,b]$. $\square$

Lemma 11.11: Let $f, g \colon [a,b] \to \mathbb{R}$ be a bounded function and let P be a partition of $[a,b]$. Then $L_{f+g}(P) \geq L_f(P) + L_g(P)$ and $U_{f+g}(P) \leq U_f(P) + U_g(P)$.

Proof: Suppose that $P = \{[x_{i-1}, x_i]\}_{i=1}^n$ and that we have the following:

$$L_f(P) = \sum_{i=1}^n m_i \Delta x_i \qquad U_f(P) = \sum_{i=1}^n M_i \Delta x_i \qquad L_g(P) = \sum_{i=1}^n k_i \Delta x_i \qquad U_g(P) = \sum_{i=1}^n K_i \Delta x_i$$

$$L_{f+g}(P) = \sum_{i=1}^n h_i \Delta x_i \qquad U_{f+g}(P) = \sum_{i=1}^n H_i \Delta x_i$$

Then we have the following:

$$m_i = \inf\{f(x) \mid x \in [x_{i-1}, x_i]\} \quad M_i = \sup\{f(x) \mid x \in [x_{i-1}, x_i]\}$$
$$k_i = \inf\{g(x) \mid x \in [x_{i-1}, x_i]\} \quad K_i = \sup\{g(x) \mid x \in [x_{i-1}, x_i]\}$$
$$h_i = \inf\{f(x) + g(x) \mid x \in [x_{i-1}, x_i]\} \quad H_i = \sup\{f(x) + g(x) \mid x \in [x_{i-1}, x_i]\}$$

By Problem 5 below, we have $h_i \geq m_i + k_i$ and $H_i \leq M_i + K_i$. Therefore, we have the following:

$$L_{f+g}(P) = \sum_{i=1}^n h_i \Delta x_i \geq \sum_{i=1}^n (m_i + k_i)\Delta x_i = \sum_{i=1}^n m_i \Delta x_i + \sum_{i=1}^n k_i \Delta x_i = L_f(P) + L_g(P)$$

$$U_{f+g}(P) = \sum_{i=1}^n H_i \Delta x_i \leq \sum_{i=1}^n (M_i + K_i)\Delta x_i = \sum_{i=1}^n M_i \Delta x_i + \sum_{i=1}^n K_i \Delta x_i = U_f(P) + U_g(P)$$

□

Proof of Theorem 11.9: Let $\epsilon > 0$. By Lemma 11.10, there are partitions P and Q such that $U_f(P) - L_f(P) < \frac{\epsilon}{2}$ and $U_g(Q) - L_g(Q) < \frac{\epsilon}{2}$. Let $R = P * Q$ (see Note 2 before Example 11.6 above). By Problem 10 below, we have $L_f(P) \leq L_f(R)$, $L_g(Q) \leq L_g(R)$, $U_f(R) \leq U_f(P)$, and $U_g(R) \leq U_g(Q)$. It follows that $U_f(R) - L_f(R) \leq U_f(P) - L_f(P) < \frac{\epsilon}{2}$ and $U_g(R) - L_g(R) \leq U_g(Q) - L_g(Q) < \frac{\epsilon}{2}$. So, by Lemma 11.11, we have

$$U_{f+g}(R) - L_{f+g}(R) \leq [U_f(R) + U_g(R)] - [L_f(R) + L_g(R)]$$
$$= [U_f(R) - L_f(R)] + [U_g(R) - L_g(R)] < \frac{\epsilon}{2} + \frac{\epsilon}{2} = \epsilon.$$

By Lemma 11.10 once again, $f + g$ is Riemann integrable on $[a, b]$.

Now, if P is any partition of $[a, b]$, then by Lemma 11.11, $L_{f+g}(P) \geq L_f(P) + L_g(P)$. Therefore, by Problem 12 below, we have

$$\int_a^b (f + g)(x)\, dx = \underline{\int_a^b} (f + g)(x)\, dx = \sup\{L_{f+g}(P) \mid P \text{ is a partition of } [a, b]\}$$
$$\geq \sup\{L_f(P) + L_g(P) \mid P \text{ is a partition of } [a, b]\}$$
$$= \sup\{L_f(P) \mid P \text{ is a partition of } [a, b]\} + \sup\{L_g(P) \mid P \text{ is a partition of } [a, b]\}$$
$$= \underline{\int_a^b} f(x)\, dx + \underline{\int_a^b} g(x)\, dx = \int_a^b f(x)\, dx + \int_a^b g(x)\, dx.$$

So,

$$\int_a^b (f + g)(x)\, dx \geq \int_a^b f(x)\, dx + \int_a^b g(x)\, dx.$$

Similarly, if P is any partition of $[a, b]$, then by Lemma 11.11, $U_{f+g}(P) \leq U_f(P) + U_g(P)$. Therefore, again by Problem 12 below, we have

$$\overline{\int_a^b} (f+g)(x)\,dx = \overline{\int_a^b} (f+g)(x)\,dx = \inf\{U_{f+g}(P) \mid P \text{ is a partition of } [a,b]\}$$

$$\leq \inf\{U_f(P) + U_g(P) \mid P \text{ is a partition of } [a,b]\}$$

$$= \inf\{U_f(P) \mid P \text{ is a partition of } [a,b]\} + \inf\{U_g(P) \mid P \text{ is a partition of } [a,b]\}$$

$$= \overline{\int_a^b} f(x)\,dx + \overline{\int_a^b} g(x)\,dx = \int_a^b f(x)\,dx + \int_a^b g(x)\,dx$$

So,

$$\overline{\int_a^b} (f+g)(x)\,dx \leq \int_a^b f(x)\,dx + \int_a^b g(x)\,dx.$$

It follows that

$$\int_a^b (f+g)(x)\,dx = \int_a^b f(x)\,dx + \int_a^b g(x)\,dx.$$

$\square$

Notes: (1) The proof of Theorem 11.9 also shows each of the following:

$$\underline{\int_a^b} (f+g)(x)\,dx \geq \underline{\int_a^b} f(x)\,dx + \underline{\int_a^b} g(x)\,dx$$

$$\overline{\int_a^b} (f+g)(x)\,dx \leq \overline{\int_a^b} f(x)\,dx + \overline{\int_a^b} g(x)\,dx$$

(2) The inequalities in Note 1 above **cannot** be replaced by equations. As an example, define $g, h: [0,1] \to \mathbb{R}$ by $g(x) = \begin{cases} 1 & \text{if } x \text{ is rational} \\ 0 & \text{if } x \text{ is irrational} \end{cases}$ and $h(x) = \begin{cases} 0 & \text{if } x \text{ is rational.} \\ 1 & \text{if } x \text{ is irrational.} \end{cases}$ We saw in part 2 of Example 11.6 that

$$\underline{\int_0^1} g(x)\,dx = 0 \quad \text{and} \quad \overline{\int_0^1} g(x)\,dx = 1.$$

Similar computations show that

$$\underline{\int_0^1} h(x)\,dx = 0 \quad \text{and} \quad \overline{\int_0^1} h(x)\,dx = 1.$$

Now, $g + h$ is just the constant function defined by $(g+h)(x) = 1$. Since $g + h$ is Riemann integrable on $[0, 1]$, we have the following:

$$\underline{\int_0^1} (g+h)(x)\,dx = \int_0^1 (g+h)(x)\,dx = \int_0^1 dx = 1 > 0 + 0 = \underline{\int_0^1} g(x)\,dx + \underline{\int_0^1} h(x)\,dx$$

$$\overline{\int_0^1} (g+h)(x)\,dx = \int_0^1 (g+h)(x)\,dx = \int_0^1 dx = 1 < 1 + 1 = \overline{\int_0^1} g(x)\,dx + \overline{\int_0^1} h(x)\,dx$$

Fundamental Theorem of Calculus

The Fundamental Theorem of Calculus describes a surprising relationship between differentiation and integration. It comes in two forms.

Theorem 11.12 (Fundamental Theorem of Calculus – First Form): Let $f: [a,b] \to \mathbb{R}$ be continuous on $[a,b]$ and define $F: [a,b] \to \mathbb{R}$ by $F(x) = \int_a^x f(t)\, dt$. Then F is uniformly continuous on $[a,b]$ and $F'(c) = f(c)$ for all $c \in (a,b)$.

Proof: Since $[a,b]$ is closed and bounded, by the Heine-Borel Theorem (Theorem 7.28), $[a,b]$ is compact. So, by Theorem 9.4, $f[A]$ is compact. Once again, by the Heine-Borel theorem, f is bounded on $[a,b]$. Therefore, there is $M > 0$ such that $|f(x)| \leq M$ for all $x \in [a,b]$.

Let $\epsilon > 0$ and let $\delta = \frac{\epsilon}{M}$. If $x, y \in [a,b]$ with $|x - y| < \delta$, then using Problems 6, 9, and 15 below, we have

$$|F(x) - F(y)| = \left| \int_a^x f(t)\, dt - \int_a^y f(t)\, dt \right| = \left| \int_a^x f(t)\, dt + \int_y^a f(t)\, dt \right|$$

$$= \left| \int_y^a f(t)\, dt + \int_a^x f(t)\, dt \right| = \left| \int_y^x f(t)\, dt \right|$$

$$\leq \left| \int_y^x |f(t)|\, dt \right| \leq \left| \int_y^x M\, dt \right|$$

$$= \left| M \int_y^x dt \right| = |M| \left| \int_y^x dt \right| \text{ (Check this!)} = M|x - y| < M\delta = \frac{M\epsilon}{M} = \epsilon.$$

Thus, F is uniformly continuous on $[a,b]$.

Now, let $c \in (a,b)$. If we also have $x \in (a,b)$, then

$$\frac{F(x) - F(c)}{x - c} = \frac{\int_a^x f(t)\, dt - \int_a^c f(t)\, dt}{x - c} = \frac{\int_a^x f(t)\, dt + \int_c^a f(t)\, dt}{x - c}$$

$$= \frac{\int_c^a f(t)\, dt + \int_a^x f(t)\, dt}{x - c} = \frac{1}{x - c} \int_c^x f(t)\, dt.$$

Let $\epsilon > 0$. Since f is continuous at c, we can choose $\delta > 0$ so that for $t \in [a,b]$, we have

$$|t - c| < \delta \to |f(t) - f(c)| < \frac{\epsilon}{2}.$$

Now,

$$|f(t) - f(c)| < \frac{\epsilon}{2} \leftrightarrow -\frac{\epsilon}{2} < f(t) - f(c) < \frac{\epsilon}{2} \leftrightarrow f(c) - \frac{\epsilon}{2} < f(t) < f(c) + \frac{\epsilon}{2}.$$

If $x \in [a,b]$ with $x > c$, by Problem 6 below, we have

216

$$\int_c^x \left(f(c) - \frac{\epsilon}{2}\right) dt \le \int_c^x f(t)\, dt \le \int_c^x \left(f(c) + \frac{\epsilon}{2}\right) dt$$

Since $f(c) - \frac{\epsilon}{2}$ and $f(c) + \frac{\epsilon}{2}$ are constant, by Theorem 11.8, we have

$$(x - c)\left(f(c) - \frac{\epsilon}{2}\right) \le \int_c^x f(t)\, dt \le (x - c)\left(f(c) + \frac{\epsilon}{2}\right).$$

Since $x > c$, we have $x - c > 0$, and so,

$$f(c) - \frac{\epsilon}{2} \le \frac{1}{x - c} \int_c^x f(t)\, dt \le f(c) + \frac{\epsilon}{2}.$$

If $x \in [a, b]$ with $x < c$, then again by Problem 6 below, we have

$$\int_x^c \left(f(c) - \frac{\epsilon}{2}\right) dt \le \int_x^c f(t)\, dt \le \int_x^c \left(f(c) + \frac{\epsilon}{2}\right) dt$$

Again, since $f(c) - \frac{\epsilon}{2}$ and $f(c) + \frac{\epsilon}{2}$ are constant, by Theorem 11.8, we have

$$(c - x)\left(f(c) - \frac{\epsilon}{2}\right) \le \int_x^c f(t)\, dt \le (c - x)\left(f(c) + \frac{\epsilon}{2}\right).$$

Since $x < c$, we have $c - x > 0$, and so,

$$f(c) - \frac{\epsilon}{2} \le \frac{1}{c - x} \int_x^c f(t)\, dt \le f(c) + \frac{\epsilon}{2}.$$

Since $x - c = -(c - x)$ and $\int_c^x f(t)\, dt = -\int_x^c f(t)\, dt$, we have

$$f(c) - \frac{\epsilon}{2} \le \frac{1}{x - c} \int_c^x f(t)\, dt \le f(c) + \frac{\epsilon}{2}.$$

We have now shown that for all $x \in [a, b]$,

$$f(c) - \frac{\epsilon}{2} \le \frac{1}{x - c} \int_c^x f(t)\, dt \le f(c) + \frac{\epsilon}{2}.$$

Therefore,

$$\left| \frac{F(x) - F(c)}{x - c} - f(c) \right| = \left| \left(\frac{1}{x - c} \int_c^x f(t)\, dt \right) - f(c) \right| \le \frac{\epsilon}{2} < \epsilon.$$

It follows that

$$F'(c) = \lim_{x \to c} \frac{F(x) - F(c)}{x - c} = f(c).$$

□

Notes: (1) To show that F is uniformly continuous on $[a, b]$, we used only the fact that $f: [a, b] \to \mathbb{R}$ is bounded and Riemann integrable on $[a, b]$ (together with Problems 6, 9, and 15 below). So, if $f: [a, b] \to \mathbb{R}$ is any function that is bounded and Riemann integrable on $[a, b]$, then the function $F: [a, b] \to \mathbb{R}$ defined by $F(x) = \int_a^x f(t)\, dt$ is uniformly continuous on $[a, b]$.

In the special case that $f: [a, b] \to \mathbb{R}$ is continuous on $[a, b]$, we saw in the first part of the proof of Theorem 11.12 that f will be bounded on $[a, b]$. It then follows from Theorem 11.7 and Problem 15 below that f is Riemann integrable on any interval of the form $[x, y]$ with $x, y \in [a, b]$.

(2) If f is bounded and Riemann integrable on $[a, b]$, but not necessarily continuous on the whole interval $[a, b]$, then for each $c \in (a, b)$ for which f is continuous at c, we still have that F is differentiable at c and $F'(c) = f(c)$. This follows directly from the proof of Theorem 11.12 (we used only the continuity at c and not continuity on the whole interval in the proof).

(3) Theorem 11.12 makes no mention of what happens at the endpoints of $[a, b]$. Let's make the following definitions:

$$F'_+(c) = \lim_{x \to c^+} \frac{F(x) - F(c)}{x - c} \qquad F'_-(c) = \lim_{x \to c^-} \frac{F(x) - F(c)}{x - c}$$

We will call $F'_+(c)$ the **derivative of F from the right at c** and we call $F'_-(c)$ the **derivative of F from the left at c**.

If we replace c by a in the second part of the proof above, then for $x \in (a, b)$, we get

$$\frac{F(x) - F(a)}{x - a} = \frac{\int_a^x f(t)\, dt - \int_a^a f(t)\, dt}{x - a} = \frac{\int_a^x f(t)\, dt - 0}{x - a} = \frac{1}{x - a} \int_a^x f(t)\, dt.$$

We then see that if $\epsilon > 0$ is given, we can choose $\delta > 0$ so that for $t \in [a, b]$, we have

$$t - a < \delta \to |f(t) - f(a)| < \frac{\epsilon}{2}.$$

Replacing c by a in the rest of the argument above and considering only the case $x > a$, we get

$$\left| \frac{F(x) - F(a)}{x - a} - f(a) \right| = \left| \left(\frac{1}{x - a} \int_a^x f(t)\, dt \right) - f(a) \right| \leq \frac{\epsilon}{2} < \epsilon.$$

This shows that $F'_+(a) = f(a)$.

Similar reasoning can be used to show that $F'_-(b) = f(b)$.

Example 11.13:

1. Define $f: [0, 1] \to \mathbb{R}$ by $f(x) = x$ and $F: [0, 1] \to \mathbb{R}$ by $F(x) = \int_0^x t\, dt$. By the first form of the Fundamental Theorem of Calculus (Theorem 11.12), F is uniformly continuous on $[0, 1]$ and $F'(x) = x$ for all $x \in (0, 1)$.

Let's also check this directly (without using Theorem 11.12). Let $n \in \mathbb{Z}^+$, for each $i = 1, 2, \ldots n$, let $x_i = \frac{ix}{n}$, and let $P = \{[x_{i-1}, x_i]\}_{i=1}^n$. Then we have

$$U_f(P) = \sum_{i=1}^n x_i \Delta x_i = \sum_{i=1}^n \frac{ix}{n} \cdot \frac{x}{n} = \frac{x^2}{n^2} \sum_{i=1}^n i = \frac{x^2}{n^2} \cdot \frac{n(n+1)}{2} = \frac{x^2(n+1)}{2n}.$$

Since f is Riemann integrable on $[0, 1]$ (because it is continuous on $[0, 1]$), by Note 5 before Example 11.6 (or Problem 18 below),

$$F(x) = \int_0^x t \, dt = \overline{\int_0^x} t \, dt = \lim_{n \to \infty} U_f(P_n) = \lim_{n \to \infty} \frac{x^2(n+1)}{2n} = \frac{x^2}{2} = \frac{1}{2}x^2.$$

So, $F'(x) = \frac{1}{2}(2x) = x$. Observe that the result also holds for $x = 0$ and $x = 1$.

2. Define $f: [0, 3] \to \mathbb{R}$ by $f(x) = \begin{cases} x & \text{if } 0 \leq x \leq 1 \\ x + 1 & \text{if } 1 < x \leq 3 \end{cases}$ and $F: [0, 3] \to \mathbb{R}$ by $F(x) = \int_0^x f(t) \, dt$. Since f is bounded and piecewise continuous on $[0, 3]$, by Problem 16 below, it is Riemann integrable on $[0, 3]$. Therefore, by Note 1 following Theorem 11.12, F is uniformly continuous on $[0, 1]$ and $F'(x) = f(x) = \begin{cases} x & \text{if } 0 \leq x < 1 \\ x + 1 & \text{if } 1 < x \leq 3 \end{cases}$ for all $x \in (0, 1) \cup (1, 3)$.

Furthermore, by Note 3 following Theorem 11.12, $F'_+(0) = f(0) = 0$ and $F'_-(3) = f(3) = 4$.

I leave it to the reader to check that $F(x) = \int_0^x f(t) \, dt = \begin{cases} \frac{x^2}{2} & \text{if } 0 \leq x \leq 1 \\ \frac{x^2}{2} + x - 1 & \text{if } 1 < x \leq 3 \end{cases}$,

$F'(x) = f(x)$ for all $x \in (0, 1) \cup (1, 3)$, $F'_+(0) = 0 = f(0)$, and $F'_-(3) = 4 = f(3)$.

Observe that although F is uniformly continuous on $[0, 3]$, F is **not** differentiable at $x = 1$. Indeed, $F'_-(1) = 1$ and $F'_+(1) = 2$.

Theorem 11.14 (Fundamental Theorem of Calculus – Second Form): Let $F: [a, b] \to \mathbb{R}$ be a continuous function on $[a, b]$ that is differentiable on (a, b). Let $f: [a, b] \to \mathbb{R}$ be a Riemann integrable function on $[a, b]$ such that $F'(x) = f(x)$ for all $x \in (a, b)$. Then

$$\int_a^b f(x) \, dx = F(b) - F(a).$$

Proof: Let $P = \{[x_{i-1}, x_i]\}_{i=1}^n$ be a partition of $[a, b]$. For each $i = 1, 2, \ldots, n$, by the Mean Value Theorem (Theorem 10.25), there is $c_i \in (x_{i-1}, x_i)$ such that

$$F'(c_i) = \frac{F(x_i) - F(x_{i-1})}{x_i - x_{i-1}} = \frac{F(x_i) - F(x_{i-1})}{\Delta x_i},$$

or equivalently, $f(c_i) \Delta x_i = F(x_i) - F(x_{i-1})$. Now, we have

$$\sum_{i=1}^{n} f(c_i)\Delta x_i$$

$$= \sum_{i=1}^{n} \left(F(x_i) - F(x_{i-1})\right)$$

$$= \left(F(x_1) - F(x_0)\right) + \left(F(x_2) - F(x_1)\right) + \cdots + \left(F(x_n) - F(x_{n-1})\right)$$

$$= F(x_n) - F(x_0) = F(b) - F(a).$$

If we let $m_i = \inf\{f(x) \mid x \in [x_{i-1}, x_i]\}$ and $M_i = \sup\{f(x) \mid x \in [x_{i-1}, x_i]\}$, then $m_i \leq f(c_i) \leq M_i$, and so, we have

$$L_f(P) = \sum_{i=1}^{n} m_i \Delta x_i \leq \sum_{i=1}^{n} f(c_i)\Delta x_i \leq \sum_{i=1}^{n} M_i \Delta x_i = U_f(P)$$

or equivalently, $L_f(P) \leq F(b) - F(a) \leq U_f(P)$.

Since P was an arbitrary partition of $[a, b]$, we have the following:

$$\underline{\int_a^b} f(x)\, dx = \sup\{L_f(P) \mid P \text{ is a partition of } [a, b]\} \leq F(b) - F(a)$$

$$\overline{\int_a^b} f(x)\, dx = \inf\{U_f(P) \mid P \text{ is a partition of } [a, b]\} \geq F(b) - F(a)$$

Since f is Riemann integrable on $[a, b]$, we have

$$\int_a^b f(x)\, dx = \underline{\int_a^b} f(x)\, dx \leq F(b) - F(a) \quad \text{and} \quad \int_a^b f(x)\, dx = \overline{\int_a^b} f(x)\, dx \geq F(b) - F(a).$$

So,

$$\int_a^b f(x)\, dx = F(b) - F(a).$$

□

Note: If $f: [a, b] \to \mathbb{R}$ is Riemann integrable on $[a, b]$ and $F: [a, b] \to \mathbb{R}$ satisfies $F'(x) = f(x)$ for all $x \in (a, b)$, then we say that F is an **antiderivative** of f on $[a, b]$.

Example 11.15:

1. Define $f: [0, 1] \to \mathbb{R}$ by $f(x) = x$. Since $\frac{d}{dx}\left[\frac{1}{2}x^2\right] = \frac{1}{2}(2x) = x$, we see that the function $F: [0, 1] \to \mathbb{R}$ defined by $F(x) = \frac{1}{2}x^2$ is an antiderivative of f on $[0, 1]$. By the second form of the Fundamental Theorem of Calculus (Theorem 11.14), we have

$$\int_0^1 x\, dx = F(1) - F(0) = \frac{1}{2} \cdot 1^2 - \frac{1}{2} \cdot 0^2 = \frac{1}{2}.$$

2. The second form of the Fundamental Theorem of Calculus (Theorem 11.14) **cannot** be used to compute the following integral:

$$\int_0^1 \frac{1}{x^2}\, dx$$

The problem is that the function $f\colon [0,1] \to \mathbb{R}$ defined by $f(x) = \frac{1}{x^2}$ is **not** continuous on $[0,1]$ (there is a discontinuity at 0). This is an example of an **improper integral**. This type of integral will be explored in Lesson 13.

Theorem 11.16 (Change of Variables): Let $f\colon [a,b] \to \mathbb{R}$ be continuous on $[a,b]$, let $g\colon [c,d] \to \mathbb{R}$ have a continuous derivative on $[c,d]$, and suppose that $g([c,d]) \subseteq [a,b]$. Then

$$\int_c^d f(g(x))g'(x)\, dx = \int_{g(c)}^{g(d)} f(u)\, du.$$

Proof: Since f, g, and g' are continuous, by Theorem 8.14 and Problems 5 and 8 from Problem Set 8, the function $h\colon [c,d] \to \mathbb{R}$ defined by $h(x) = f(g(x))g'(x)$ is continuous on $[c,d]$. So, by Theorem 11.7, h is Riemann integrable on $[c,d]$. Since f is continuous on any subinterval of $[a,b]$, again by Theorem 11.7, it is Riemann integrable on any subinterval of $[a,b]$. Define $F\colon [g(c), g(d)] \to \mathbb{R}$ by

$$F(t) = \int_{g(c)}^t f(u)\, du.$$

By the first form of the Fundamental Theorem of Calculus (Theorem 11.12), F is continuous on $[a,b]$, F is differentiable on $(g(c), g(d))$, and $F'(t) = f(t)$ for all $t \in (g(c), g(d))$.

By Problem 8 from Problem Set 8, $F \circ g$ is continuous on $[c,d]$. By the chain rule (Theorem 10.20), $F \circ g$ is differentiable on (c,d) and $(F \circ g)'(x) = F'(g(x))g'(x) = f(g(x))g'(x)$.

Therefore, by the second form of the Fundamental Theorem of Calculus (Theorems 11.12 and 11.14),

$$\int_{g(c)}^{g(d)} f(u)\, du = F(g(d)) - F(g(c)) = \int_c^d (F \circ g)'(x)\, dx = \int_c^d f(g(x))g'(x)\, dx.$$

$\square$

Example 11.17: Define $f\colon [3,4] \to \mathbb{R}$ by $f(x) = x^5$ and $g\colon [0,1] \to \mathbb{R}$ by $g(x) = x^2 + 3$. Then $g'(x) = 2x$, and we see that f, g, and g' are continuous and $g([0,1]) = [3,4]$. By Theorem 11.16,

$$\int_0^1 (x^2+3)^5 (2x)\, dx = \int_3^4 u^5\, du = \left[\frac{u^6}{6}\right]_3^4 = \frac{1}{6}(4^6 - 3^6) = \frac{1}{6}(4096 - 729) = \frac{3367}{6}.$$

Note: Since $F(u) = \frac{u^6}{6}$ is an antiderivative of $f(u) = u^5$, by the second form of the Fundamental Theorem of Calculus, $\int_3^4 u^5\, du = \int_3^4 f(u)\, du = F(4) - F(3) = \frac{1}{6}(4^6 - 3^6)$. The expression $\left[\frac{u^6}{6}\right]_3^4$ above is being used to indicate that we will substitute 4 and 3 into the expression $\frac{u^6}{6}$ and then subtract.

Problem Set 11

Full solutions to these problems are available for free download here:
www.SATPrepGet800.com/RABQXZ

LEVEL 1

1. Prove each of the following:

   (i) For each $n \in \mathbb{Z}^+$,
   $$\sum_{i=1}^{n} 1 = n.$$

   (ii) Let $n \in \mathbb{Z}^+$ and let $r, a_1, a_2, \ldots, a_n \in \mathbb{R}$. Then
   $$\sum_{i=1}^{n} r a_i = r \sum_{i=1}^{n} a_i.$$

   (iii) Let $n \in \mathbb{Z}^+$. For each $i = 1, 2, \ldots, n$, let $a_i, b_i \in \mathbb{R}$. Then
   $$\sum_{i=1}^{n} (a_i + b_i) = \sum_{i=1}^{n} a_i + \sum_{i=1}^{n} b_i.$$

2. Let $f: [0, 3] \to \mathbb{R}$ be defined by $f(x) = x$.

   (i) Let $P = \{[0,1], [1,2], [2,3]\}$. Compute $L_f(P)$ and $U_f(P)$.

   (ii) Let $n \in \mathbb{Z}^+$ and let P_n be the partition of $[0, 3]$ into n equal subintervals. Compute $L_f(P_n)$ and $U_f(P_n)$.

   (iii) Compute $\underline{\int_a^b} f(x)\, dx$ and $\overline{\int_a^b} f(x)\, dx$. (You may use Problem 18 below.)

   (iv) Is f Riemann integrable on $[0, 3]$? If so, compute $\int_0^3 f(x)\, dx$. If not, explain why.

LEVEL 2

3. Let A and B be nonempty bounded sets of real numbers with $A \subseteq B$. Prove that
   $$\inf B \leq \inf A \leq \sup A \leq \sup B.$$

4. Let A be a nonempty bounded set of real numbers, let $c \geq 0$, and let $B = \{cx \mid x \in A\}$. Prove that $\sup B = c \sup A$ and $\inf B = c \inf A$.

5. Let $f, g: [a, b] \to \mathbb{R}$ be bounded functions, let $A = \{f(x) \mid x \in [a, b]\}$, $B = \{g(x) \mid x \in [a, b]\}$, and $C = \{f(x) + g(x) \mid x \in [a, b]\}$. Prove that
   $$\sup C \leq \sup A + \sup B \quad \text{and} \quad \inf C \geq \inf A + \inf B.$$

6. Let $f, g: [a, b] \to \mathbb{R}$ be bounded functions that are Riemann integrable on $[a, b]$ and let $f(x) \leq g(x)$ for all $x \in [a, b]$. Prove that
$$\int_a^b f(x)\, dx \leq \int_a^b g(x)\, dx.$$

LEVEL 3

7. Let A and B be nonempty bounded sets of real numbers and let $A + B = \{a + b \mid a \in A, b \in B\}$. Prove that $\sup(A + B) = \sup A + \sup B$ and $\inf(A + B) = \inf A + \inf B$.

8. Let $f: [a, b] \to \mathbb{R}$ be continuous on $[a, b]$. Prove that there is $c \in [a, b]$ such that
$$\int_a^b f(x)\, dx = f(c)(b - a).$$

This result is known as the **Mean Value Theorem for Integrals**.

9. Let $f: [a, b] \to \mathbb{R}$ be a bounded function that is Riemann integrable on $[a, b]$ and let $c \in \mathbb{R}$. Prove that cf is Riemann integrable on $[a, b]$ and
$$\int_a^b cf(x)\, dx = c \int_a^b f(x)\, dx.$$

LEVEL 4

10. Let $f: [a, b] \to \mathbb{R}$ be a bounded function and let P and Q be partitions of $[a, b]$ with Q a refinement of P. Prove that $L_f(P) \leq L_f(Q)$ and $U_f(Q) \leq U_f(P)$.

11. Let $f: [a, b] \to \mathbb{R}$ be a bounded function, let $m = \inf\{f(x) \mid x \in [a, b]\}$, and let $M = \sup\{f(x) \mid x \in [a, b]\}$. Prove that
$$m(b - a) \leq \underline{\int_a^b} f(x)\, dx \leq \overline{\int_a^b} f(x)\, dx \leq M(b - a).$$

12. Let $f, g: [a, b] \to \mathbb{R}$ be bounded functions, let $A = \{L_f(P) \mid P \text{ is a partition of } [a, b]\}$, $B = \{L_g(P) \mid P \text{ is a partition of } [a, b]\}$, and $C = \{L_f(P) + L_g(P) \mid P \text{ is a partition of } [a, b]\}$. Prove that $\sup C = \sup A + \sup B$. Then prove an analogous result involving upper sums.

LEVEL 5

13. Let $f: [a, b] \to \mathbb{R}$ be a monotonic function. Prove that f is Riemann integrable on $[a, b]$.

14. If P is a partition of $[a, b]$, we define the **norm** of P, written $\|P\|$ to be the maximum length of a subinterval in P. Prove that a bounded function $f: [a, b] \to \mathbb{R}$ is Riemann integrable if and only if there is $L \in \mathbb{R}$ such that the following holds: For every $\epsilon > 0$ there is $\delta > 0$ such that whenever P is a partition of $[a, b]$ with $\|P\| < \delta$ and S is any Riemann sum over P, then $|S - L| < \epsilon$. You may use Problem 19 below.

15. Let $f:[a,b] \to \mathbb{R}$ be a bounded function that is Riemann integrable on $[a,b]$. Let $c, d, e \in [a,b]$. Prove that

$$\int_c^e f(x)\,dx = \int_c^d f(x)\,dx + \int_d^e f(x)\,dx.$$

16. Let $f:[a,b] \to \mathbb{R}$ be a bounded function with finitely many discontinuities. Prove that f is Riemann integrable on $[a,b]$.

CHALLENGE PROBLEMS

17. Define a Riemann integrable function $f:[0,1] \to \mathbb{R}$ with infinitely many discontinuities that is not monotonic.

18. Let $f:[a,b] \to \mathbb{R}$ be a bounded function and for each $n \in \mathbb{Z}^+$, let $P_n = \{[x_{i-1}, x_i]\}_{i=1}^n$, where $x_i = a + \frac{i(b-a)}{n}$. Prove that $\underline{\int_a^b} f(x)\,dx = \lim_{n \to \infty} L_f(P_n)$ and $\overline{\int_a^b} f(x)\,dx = \lim_{n \to \infty} U_f(P_n)$.

19. Let $f:[a,b] \to \mathbb{R}$ be Riemann integrable on $[a,b]$. Prove that for every $\epsilon > 0$, there is $\delta > 0$ such that whenever P is a partition of $[a,b]$ with $\|P\| < \delta$, we have

$$\int_a^b f(x)\,dx - \epsilon < L_f(P) \le U_f(P) < \int_a^b f(x)\,dx + \epsilon$$

20. Let $g:[a,b] \to \mathbb{R}$ be an increasing function. Let $P = \{[x_{i-1}, x_i]\}_{i=1}^n$ be a partition of $[a,b]$ and let $\Delta g_i = g(x_i) - g(x_{i-1})$. We make the following additional definitions:

$$U_f(P;g) = \sum_{i=1}^n M_i \Delta g_i = M_1 \Delta g_1 + M_2 \Delta g_2 + \cdots + M_n \Delta g_n$$

$$L_f(P;g) = \sum_{i=1}^n m_i \Delta g_i = m_1 \Delta g_1 + m_2 \Delta g_2 + \cdots + m_n \Delta g_n$$

$$\overline{\int_a^b} f(x)\,dg(x) = \inf\{U_f(P;g) \mid P \text{ is a partition of } [a,b]\}$$

$$\underline{\int_a^b} f(x)\,dg(x) = \sup\{L_f(P;g) \mid P \text{ is a partition of } [a,b]\}$$

If $\overline{\int_a^b} f(x)\,dg(x) = \underline{\int_a^b} f(x)\,dg(x)$, then we say that f is **Riemann-Stieltjes integrable** with respect to g on $[a,b]$ and we define $\int_a^b f(x)\,dg(x)$ to be this common value. In this case, we will call $\int_a^b f(x)\,dg(x)$ the **Riemann-Stieltjes integral** of f with respect to g over $[a,b]$.

Suppose that $f:[a,b] \to \mathbb{R}$ is bounded on $[a,b]$, f has only finitely many discontinuities on $[a,b]$, $g:[a,b] \to \mathbb{R}$ is increasing, and g is continuous at each discontinuity of f. Prove that f is Riemann-Stieltjes integrable with respect to g on $[a,b]$.

Lesson 12
Logarithmic and Exponential Functions

The Natural Logarithmic Function

Let $f: (0, \infty) \to \mathbb{R}$ be defined by

$$f(x) = \frac{1}{x}.$$

If $b \in (0, \infty)$, then by Problems 2, 5, 6, and 15 in Problem Set 8,

$$\lim_{x \to b} f(x) = \lim_{x \to b} \frac{1}{x} = \frac{\lim_{x \to b} 1}{\lim_{x \to b} x} = \frac{1}{b}.$$

So, by Problem 5 in Problem Set 8, f is continuous at b. Since $b \in (0, \infty)$ was arbitrary, f is continuous on $(0, \infty)$.

It follows that if $x \geq 1$, then f is continuous on $[1, x]$, and therefore, by Theorem 11.7, f is Riemann integrable on $[1, x]$. Similarly, if $0 < x < 1$, then f is Riemann integrable on $[x, 1]$. Therefore, we can define $F: (0, 1) \to \mathbb{R}$ as follows:

$$F(x) = \int_1^x \frac{1}{t} \, dt$$

Note that if $0 < x < 1$, then by definition, we have

$$F(x) = \int_1^x \frac{1}{t} \, dt = -\int_x^1 \frac{1}{t} \, dt.$$

We call this function F the **natural logarithmic function** and we write $F(x) = \ln x$.

Example 12.1:

1. Let's use the definition of $\ln x$ to compute $\ln 1$:

$$\ln 1 = \int_1^1 \frac{1}{t} \, dt = 0.$$

2. Let's analyze $\ln 2$ next. We have

$$\ln 2 = \int_1^2 \frac{1}{t} \, dt.$$

If $t \in [1, 2]$, then $\frac{1}{2} \leq \frac{1}{t} \leq 1$. So, by Problem 11 from Problem Set 11 together with the fact that $f(x) = \frac{1}{x}$ is integrable on $[1, 2]$, we have $\frac{1}{2} \leq \ln 2 \leq 1$.

We can improve these bounds by partitioning $[1, 2]$ into subintervals and computing lower and upper sums. For example, if we let $P = \{[1, 1.5], [1.5, 2]\}$, then we have

$$L_f(P) = \frac{1}{1.5} \cdot \frac{1}{2} + \frac{1}{2} \cdot \frac{1}{2} = \frac{7}{12} \quad \text{and} \quad U_f(P) = \frac{1}{1} \cdot \frac{1}{2} + \frac{1}{1.5} \cdot \frac{1}{2} = \frac{5}{6}.$$

So, we see that $\frac{7}{12} \leq \ln 2 \leq \frac{5}{6}$. This provides more information than the application of Problem 11 from Problem Set 11 mentioned above.

By computing lower and upper sums using partitions with more and more subintervals, we can approximate the value of $\ln 2$ to any degree of accuracy we choose. Of course, there is nothing special about $\ln 2$. We can do the same thing for $\ln x$, where x is any positive real number.

By the first form of the Fundamental Theorem of Calculus (Theorem 11.12), for all $x \in (0, \infty)$ we have

$$\frac{d}{dx}[\ln x] = \frac{d}{dx}\left[\int_1^x \frac{1}{t}\, dt\right] = \frac{1}{x}.$$

Example 12.2:

1. Let $f(x) = \ln(x^5 + 1)$. By the chain rule (Theorem 10.20),

$$f'(x) = \frac{1}{x^5 + 1} \cdot 5x^4 = \frac{5x^4}{x^5 + 1}.$$

More generally, if $h: A \to \mathbb{R}$ is differentiable on A, $h[A] \subseteq (0, \infty)$, and $f(x) = \ln(h(x))$, then by the chain rule (Theorem 10.20), we have

$$f'(x) = \frac{h'(x)}{h(x)}.$$

2. Define $f: \mathbb{R}^* \to \mathbb{R}$ by $f(x) = \ln|x|$. We have

$$f(x) = \ln|x| = \begin{cases} \ln x & \text{if } x > 0. \\ \ln(-x) & \text{if } x < 0. \end{cases}$$

If $x > 0$, then $f'(x) = \frac{d}{dx}[\ln x] = \frac{1}{x}$. If $x < 0$, then $f'(x) = \frac{d}{dx}[\ln(-x)] = \frac{-1}{-x} = \frac{1}{x}$ (by the chain rule). It follow that for all $x \neq 0$,

$$f'(x) = \frac{d}{dx}[\ln|x|] = \frac{1}{x}.$$

Therefore, for any $a, b \in \mathbb{R}$ with $a < b$, as long as $0 \notin [a, b]$, by the second form of the Fundamental Theorem of Calculus (Theorem 11.14), we have

$$\int_a^b \frac{1}{x}\, dx = \ln|b| - \ln|a|.$$

3. Let's evaluate the following integral through a change of variables (Theorem 11.16):

$$\int_1^2 \frac{\ln x}{x}\, dx$$

If we let $f(x) = x$ and $g(x) = \ln x$, then $f(g(x)) = f(\ln x) = \ln x$ and $g'(x) = \frac{1}{x}$. So, we have

$$\int_1^2 \frac{\ln x}{x}\, dx = \int_1^2 (\ln x) \frac{1}{x}\, dx = \int_1^2 f(g(x)) g'(x)\, dx = \int_0^{\ln 2} f(u)\, du$$

$$= \int_0^{\ln 2} u\, du = \left[\frac{u^2}{2}\right]_0^{\ln 2} = \frac{(\ln 2)^2}{2} - \frac{0^2}{2} = \frac{(\ln 2)^2}{2}.$$

The following more streamlined procedure is often used when applying a change of variables (Theorem 11.16): We let $u = \ln x$. Then $du = \frac{1}{x} dx$ (we say that the **differential of u** is equal to the product of $\frac{1}{x}$ and the **differential of x**). So,

$$\int_1^2 \frac{\ln x}{x}\, dx = \int_1^2 (\ln x) \frac{1}{x}\, dx = \int_0^{\ln 2} u\, du = \left[\frac{u^2}{2}\right]_0^{\ln 2} = \frac{(\ln 2)^2}{2} - \frac{0^2}{2} = \frac{(\ln 2)^2}{2}.$$

From now on, we will usually use this more streamlined process when applying Theorem 11.16.

Theorem 12.3: For all $a, b \in (0, \infty)$, $\ln(ab) = \ln a + \ln b$.

Proof: If we let $u = bt$, then $du = b\, dt$, and by Theorem 11.16 and Problem 15 in Problem Set 11, we have

$$\ln a = \int_1^a \frac{1}{t}\, dt = \int_1^a \frac{1}{bt} \cdot b\, dt = \int_b^{ab} \frac{1}{u}\, du = \int_b^1 \frac{1}{u}\, du + \int_1^{ab} \frac{1}{u}\, du$$

$$= \int_1^{ab} \frac{1}{u}\, du - \int_1^b \frac{1}{u}\, du = \ln ab - \ln b.$$

Therefore, $\ln ab = \ln a + \ln b$. $\square$

Notes: (1) To apply Theorem 11.16 more formally, we let $f(t) = \frac{1}{t}$ and $g(t) = bt$. Then $f(g(t)) = \frac{1}{bt}$ and $g'(t) = b$. So, we have

$$\ln a = \int_1^a \frac{1}{t}\, dt = \int_1^a \frac{1}{bt} \cdot b\, dt = \int_1^a f(g(t)) g'(t)\, dt = \int_{g(1)}^{g(a)} f(u)\, du = \int_b^{ba} \frac{1}{u}\, du.$$

(2) Notice our use of SACT (see Note 7 following the proof of Theorem 4.11) in the second equality above. We multiplied and divided by b inside the integral to make the change of variables work.

(3) We used Problem 15 from Problem Set 11 to write

$$\int_b^{ab} \frac{1}{u}\, du = \int_b^1 \frac{1}{u}\, du + \int_1^{ab} \frac{1}{u}\, du.$$

Theorem 12.4: For all $q \in \mathbb{Q}$ and $a \in (0, \infty)$, $\ln(a^q) = q \ln a$.

Proof: Let $a \in (0, \infty)$. We first prove by induction that for all $n \in \mathbb{N}$, $\ln(a^n) = n \ln a$.

Base case ($k = 0$): $\ln(a^0) = \ln 1 = 0 = 0 \ln a$.

Inductive Step: Let $k \in \mathbb{N}$ and assume that $\ln(a^k) = k \ln a$. By Theorem 12.3, we have
$$\ln(a^{k+1}) = \ln(a^k \cdot a) = \ln(a^k) + \ln a = k \ln a + \ln a = (k+1) \ln a.$$

By the Principle of Mathematical Induction, for all $n \in \mathbb{N}$, $\ln(a^n) = n \ln a$.

By Theorem 12.3, we have $\ln a + \ln \frac{1}{a} = \ln\left(a \cdot \frac{1}{a}\right) = \ln 1 = 0$. So, $\ln a^{-1} = \ln \frac{1}{a} = -\ln a$. It follows that for $n \in \mathbb{N}$, $\ln a^{-n} = \ln(a^n)^{-1} = -\ln a^n = -n \ln a$. So, the result holds for all integers.

Now, if $n \in \mathbb{Z}^*$, we have $\ln a = \ln\left(a^{\frac{1}{n}}\right)^n = n \ln\left(a^{\frac{1}{n}}\right)$. So, for $n \in \mathbb{Z}^*$, we have $\ln\left(a^{\frac{1}{n}}\right) = \frac{1}{n} \ln a$.

Finally, let $q \in \mathbb{Q}$. Then there are $n \in \mathbb{Z}$ and $m \in \mathbb{Z}^*$ with $q = \frac{n}{m}$. So,
$$\ln a^q = \ln\left(a^{\frac{n}{m}}\right) = \ln\left(a^{\frac{1}{m}}\right)^n = n \ln\left(a^{\frac{1}{m}}\right) = \frac{n}{m} \ln a = q \ln a.$$

So, the result holds for all $q \in \mathbb{Q}$. $\square$

Note: It is a simple consequence of Theorems 12.3 and 12.4 that for all $a, b \in (0, \infty)$, we have
$$\ln\left(\frac{a}{b}\right) = \ln a - \ln b.$$

Indeed, we have $\ln\left(\frac{a}{b}\right) = \ln(ab^{-1}) = \ln a + \ln(b^{-1}) = \ln a + (-1)\ln b = \ln a - \ln b$.

Example 12.5: By Theorems 12.3 and 12.4 and the Note following Theorem 12.4, we have
$$\ln \frac{(x^3 + 7)^2 \cos x}{x^5} = \ln(x^3 + 7)^2 + \ln \cos x - \ln x^5 = 2\ln(x^3 + 7) + \ln \cos x - 5 \ln x.$$

Theorem 12.6: The natural logarithmic function is a strictly increasing bijection from $(0, \infty)$ to $\mathbb{R}$.

Proof: Since $\frac{d}{dx}[\ln x] = \frac{1}{x} > 0$ for all $x \in (0, \infty)$, it follows from Corollary 10.26 (part 1) that $\ln x$ is strictly increasing on $(0, \infty)$.

To see that $\ln x$ is injective, let $x, y \in (0, \infty)$ with $x \neq y$. Without loss of generality we can assume that $x < y$. Since $\ln x$ is strictly increasing, $\ln x < \ln y$. In particular, $\ln x \neq \ln y$.

To see that $\ln x$ is surjective, first let $y \in (0, \infty)$. By part 2 of Example 12.1, $\frac{1}{2} \leq \ln 2 \leq 1$. In particular, $\ln 2 > 0$. By the Archimedean Property of $\mathbb{R}$ (Theorem 6.16), there is $n \in \mathbb{N}$ such that $n > \frac{y}{\ln 2}$.

Since $\ln 2 > 0$, by Theorem 12.4, we have $\ln 2^n = n \ln 2 > y$.

By the first form of the Fundamental Theorem of Calculus (Theorem 11.12), $\ln x$ is continuous on $(1, 2^n)$, and so, by the Intermediate Value Theorem (Corollary 9.9), $\ln[(1, 2^n)]$ is an interval. Since $\ln 1 = 0 < y < \ln 2^n$, there is $x \in (1, 2^n)$ such that $\ln x = y$. This shows that every positive real number is in the range of $\ln x$.

Now, if $y > 0$ and $\ln x = y$ for some $x > 0$, then $\frac{1}{x} > 0$ and by Theorem 12.4, we have $\ln \frac{1}{x} = \ln(x^{-1}) = -\ln x = -y$. This shows that every negative real number is in the range of $\ln x$.

Finally, since $\ln 1 = 0$, it follows that 0 is in the range of $\ln x$. So, $\ln x$ is surjective. $\square$

Theorem 12.7: $\lim_{x \to \infty} \ln x = \infty$ and $\lim_{x \to 0^+} \ln x = -\infty$.

Proof: Let $M > 0$. By the same argument given in the proof of Theorem 12.6, we can find $n \in \mathbb{N}$ such that $\ln 2^n > M$. Since $\ln x$ is strictly increasing, for all $x > 2^n$, we have $\ln x > \ln 2^n > M$. Therefore,

$$\forall M > 0 \; \exists K > 0 \; (x > K \to \ln x > M).$$

Note that the value of K we are using here is $K = 2^n$.

By the solution to part (iv) of Problem 16 from Problem Set 8, $\lim_{x \to \infty} \ln x = \infty$.

Now, $\lim_{x \to 0^+} \ln x = \lim_{x \to 0^+} \ln \left(\frac{1}{x}\right)^{-1} = \lim_{x \to 0^+} \left[-\ln \frac{1}{x}\right] = -\lim_{x \to 0^+} \left[\ln \frac{1}{x}\right] = -\lim_{x \to \infty} \ln x$ (Check this!) $= -\infty$. $\square$

Note: We now have enough information to draw a sketch of the function $F(x) = \ln x$. It looks as follows:

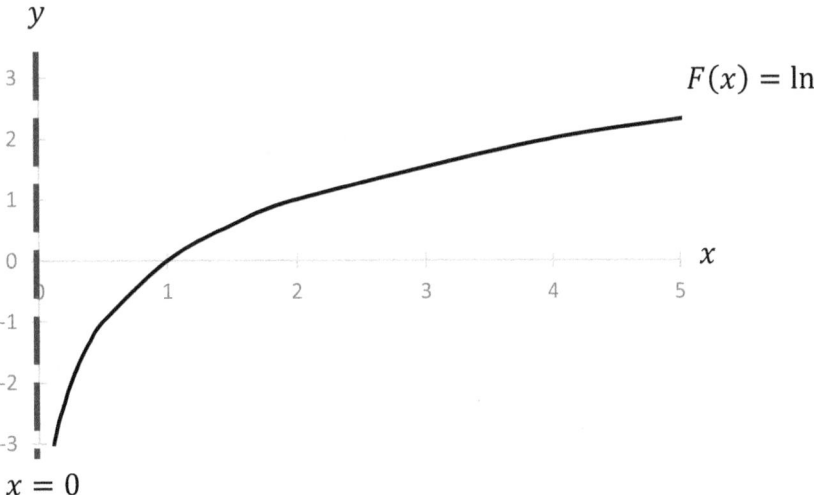

Notice how the graph has a vertical asymptote of $x = 0$ because $\lim_{x \to 0^+} \ln x = -\infty$. Also observe that $\ln 1 = 0$ and the function is a continuous strictly increasing bijection from $(0, \infty)$ onto $\mathbb{R}$.

Theorem 12.8: Let $F: (0, \infty) \to \mathbb{R}$ be a function such that $F(1) = 0$ and $\frac{d}{dx}[F(x)] = \frac{1}{x}$ for all $x \in (0, \infty)$. Then $F(x) = \ln x$ for all $x \in (0, \infty)$.

Proof: First note that by definition, for all $x \in (0, \infty)$, we have

$$\ln x = \int_1^x \frac{1}{t} \, dt.$$

Let $x \in (0, \infty)$. First assume that $x \geq 1$. Since $F: [1, x] \to \mathbb{R}$ is differentiable on $[1, x]$, by Theorem 10.11, F is continuous on $[1, x]$. So, by the second form of the Fundamental Theorem of Calculus (Theorem 11.14), we have

$$\int_1^x \frac{1}{t} \, dt = F(x) - F(1) = F(x).$$

Therefore,

$$F(x) = \int_1^x \frac{1}{t} \, dt = \ln x.$$

Now, if $0 < x < 1$, then since $F: [x, 1] \to \mathbb{R}$ is differentiable on $[x, 1]$, by Theorem 10.11, F is continuous on $[x, 1]$. So, again by the second form of the Fundamental Theorem of Calculus (Theorem 11.14), we have

$$\int_x^1 \frac{1}{t} \, dt = F(1) - F(x) = -F(x).$$

Therefore,

$$F(x) = -\int_x^1 \frac{1}{t} \, dt = \int_1^x \frac{1}{t} \, dt = \ln x.$$

So, we have verified that for all $x \in (0, \infty)$, $F(x) = \ln x$. □

The Natural Exponential Function

The function $F: (0, \infty) \to \mathbb{R}$ defined by $F(x) = \ln x$ is a bijection (by Theorem 12.6). Therefore, it is invertible. Let $E: \mathbb{R} \to (0, \infty)$ be the inverse of F.

Furthermore, since $F(x) = \ln x$ is differentiable on $(0, \infty)$ and $F'(x) = \frac{1}{x} > 0$ for all $x \in (0, \infty)$, by Problem 13 in Problem Set 10, E is differentiable on $\mathbb{R}$. So, by Theorem 10.11, E is continuous on $\mathbb{R}$.

We call this function E the **natural exponential function** and we write $E(x) = e^x$.

Example 12.9:
1. Since $\ln 1 = 0$, it follows that $e^0 = 1$.
2. Since $\ln x$ and e^x are inverses of each other, by Theorem 3.8, we have $\ln e^x = x$ for all $x \in \mathbb{R}$ and $e^{\ln x} = x$ for all $x \in (0, \infty)$. For example, $\ln e = \ln e^1 = 1$ and $e^{\ln 2} = 2$.

Let $y \in \mathbb{R}$. Since $\ln x$ is a bijection from $(0, \infty)$ to $\mathbb{R}$, there is $x \in (0, \infty)$ such that $y = \ln x$ (or equivalently, $x = e^y$). By Problem 13 in Problem Set 10, we have

$$\frac{d}{dx}[e^y] = \frac{1}{\frac{d}{dx}[\ln x]} = \frac{1}{\left(\frac{1}{x}\right)} = x = e^y.$$

So, e^x is equal to its own derivative! This observation together with Corollary 10.26 (part 1) and the fact that the range of e^x is $(0, \infty)$ shows that e^x is a strictly increasing bijection from $\mathbb{R}$ to $(0, \infty)$.

It also follows from the second form of the Fundamental theorem of Calculus (Theorem 11.14) that for any $a, b \in \mathbb{R}$, $\int_a^b e^x \, dx = e^b - e^a$.

Example 12.10:

1. Let $f(x) = xe^{7x}$. By the product rule (Theorem 10.14) and the chain rule (Theorem 10.20),
$$f'(x) = xe^{7x} \cdot 7 + 1 \cdot e^{7x} = e^{7x}(7x + 1).$$

2. Let's evaluate the following integral through a change of variables (Theorem 11.16):
$$\int_0^2 x^3 e^{x^4} \, dx$$

   If we let $u = x^4$, then $du = 4x^3 dx$. Using Theorem 11.16, we get
$$\int_0^2 x^3 e^{x^4} \, dx = \frac{1}{4} \int_0^2 e^{x^4} \cdot 4x^3 dx = \frac{1}{4} \int_0^{16} e^u \, du = \left[\frac{1}{4} e^u\right]_0^{16} = \frac{1}{4} e^{16} - \frac{1}{4} e^0 = \frac{1}{4}(e^{16} - 1).$$

Theorem 12.11: For all $a, b \in \mathbb{R}$, $e^{a+b} = e^a \cdot e^b$.

Proof: Let $a, b \in \mathbb{R}$. Since $F(x) = \ln x$ is a bijection from $(0, \infty)$ to $\mathbb{R}$ (by Theorem 12.6), there are $c, d \in (0, \infty)$ with $a = \ln c$ and $b = \ln d$. Then $c = e^a$, $d = e^b$, and by Theorem 12.3, we have
$$e^{a+b} = e^{\ln c + \ln d} = e^{\ln cd} = cd = e^a \cdot e^b.$$ □

Theorem 12.12: For all $q \in \mathbb{Q}$ and $a \in \mathbb{R}$, $e^{qa} = (e^a)^q$.

Proof: Let $q \in \mathbb{Q}$ and $a \in \mathbb{R}$. Since $F(x) = \ln x$ is a bijection from $(0, \infty)$ to $\mathbb{R}$ (by Theorem 12.6), there is $b \in (0, \infty)$ with $a = \ln b$. Then $b = e^a$ and by Theorem 12.4, we have
$$e^{qa} = e^{q \ln b} = e^{\ln b^q} = b^q = (e^a)^q.$$ □

Note: It is a simple consequence of Theorems 12.11 and 12.12 that for all $a, b \in \mathbb{R}$, we have
$$e^{a-b} = \frac{e^a}{e^b}.$$

Indeed, we have $e^{a-b} = e^{a+(-b)} = e^a e^{-b} = e^a (e^b)^{-1} = \frac{e^a}{e^b}$.

Example 12.13: By Theorems 12.11 and 12.12 and the Note following Theorem 12.12, we have

$$\frac{e^5(e^3)^2}{e^7} = \frac{e^5 e^6}{e^7} = e^{5+6-7} = e^4.$$

Theorem 12.14: $\lim\limits_{x \to \infty} e^x = \infty$ and $\lim\limits_{x \to -\infty} e^x = 0$.

Proof: Let $M > 0$. Since e^x is a bijection from $\mathbb{R}$ to $(0, \infty)$, there is $K \in \mathbb{R}$ such that $e^K = M$. Since e^x is strictly increasing, for all $x > K$, we have $e^x > e^K = M$. Therefore,

$$\forall M > 0 \, \exists K > 0 \, (x > K \to e^x > M).$$

By the solution to part (iv) of Problem 16 from Problem Set 8, $\lim\limits_{x \to \infty} e^x = \infty$.

Now, let $\epsilon > 0$. Since e^x is a bijection from $\mathbb{R}$ to $(0, \infty)$, there is $a \in \mathbb{R}$ such that $e^a = \epsilon$. Since e^x is strictly increasing, for all $x < a$, we have $e^x < e^a = \epsilon$. If $a \neq 0$, let $K = |a|$. If $a = 0$, let $K = 1$. Since $e^x = |e^x - 0|$, we have

$$\forall \epsilon > 0 \, \exists K > 0 \, (x < -K \to |e^x - 0| < \epsilon).$$

By the solution to part (iii) of Problem 16 from Problem Set 8, $\lim\limits_{x \to -\infty} e^x = 0$. □

Note: We now have enough information to draw a sketch of the function $E(x) = e^x$. It looks as follows:

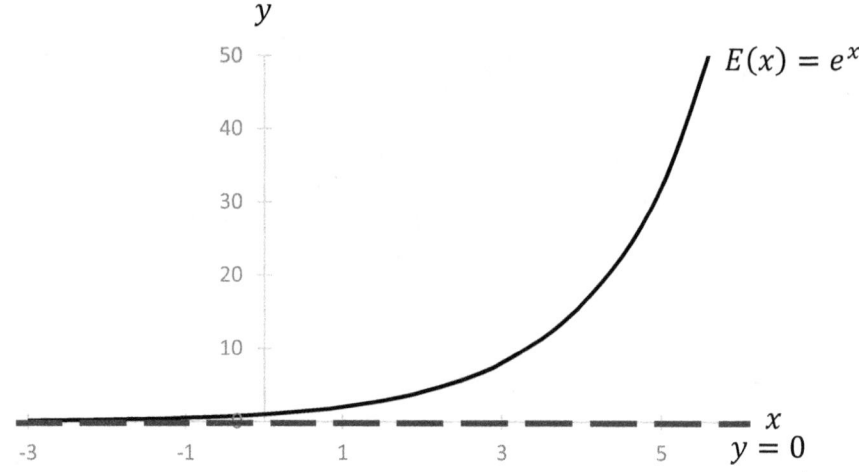

Notice how the graph has a horizontal asymptote of $y = 0$ because $\lim\limits_{x \to -\infty} e^x = 0$. Also observe that $e^0 = 1$ and the function is a continuous strictly increasing bijection from $\mathbb{R}$ onto $(0, \infty)$.

Theorem 12.15: Let $G: \mathbb{R} \to (0, \infty)$ be a function such that $G(0) = 1$ and $G'(x) = G(x)$ for all $x \in \mathbb{R}$. Then $G(x) = e^x$ for all $x \in \mathbb{R}$.

Proof: By Theorem 10.14 (the product rule), we have

$$\frac{d}{dx}[e^{-x} G(x)] = e^{-x} G'(x) + (e^{-x})(-1)G(x) = e^{-x} G(x) - e^{-x} G(x) = 0.$$

Since $\frac{d}{dx}[0] = 0$, by problem 5 in Problem Set 10, there is a constant k such that for all $x \in \mathbb{R}$, $e^{-x}G(x) = 0 + k$. Since $e^0 = 1$ and $G(0) = 1$, it follows that $k = e^0 G(0) = 1 \cdot 1 = 1$. Therefore, we have $e^{-x}G(x) = 1$. So, $G(x) = 1 \cdot G(x) = e^0 \cdot G(x) = e^{x-x} \cdot G(x) = e^x e^{-x} G(x) = e^x \cdot 1 = e^x$ for all $x \in \mathbb{R}$. □

General Exponential and Logarithmic Functions

Up to this point, given $a \in (0, \infty)$ and $q \in \mathbb{Q}$, we have defined what we mean by a^q (see part 2 of Example 10.22). If $a > 0$, since $e^{\ln a} = a$, by Theorem 12.4, we can write $a^q = e^{\ln a^q} = e^{q \ln a}$.

With this information, we are finally ready to extend the definition of exponentiation to include exponents that are irrational. If $a > 0$, then notice that the expression $e^{q \ln a}$ makes perfect sense, even if q is irrational. Therefore, for $a > 0$ and $b \in \mathbb{R}$, we define a^b as follows:

$$a^b = e^{b \ln a}.$$

Notes: (1) Allowing a to be negative takes us outside the subject of real analysis. For example, the expression $(-1)^{\frac{1}{2}}$ is equal to $\sqrt{-1}$, which is **not** a real number. Therefore, in this book we do not attempt to define a^b for $a < 0$ and b irrational.

(2) What about $a = 0$? Well, we can define 0^b to be 0, as long as $b \neq 0$. The expression 0^0 is an indeterminate form and will be discussed below.

Since $F(x) = \ln x$ is continuous on $(0, \infty)$ and e^x is continuous on $\mathbb{R}$, for each $a > 0$, the function $H_a: \mathbb{R} \to (0, 1)$ defined by $H_a(x) = a^x$ is continuous on $(0, \infty)$.

If $a > 0$ and $a \neq 1$, we call $H_a(x) = a^x$ the **general exponential function with base** a.

Example 12.16:

1. If $a = 1$, then for all $x \in \mathbb{R}$, we have $1^x = e^{x \ln 1} = e^0 = 1$. So, $H_1(x) = 1^x$ is the constant function that takes on the value 1 for all $x \in \mathbb{R}$. Therefore, we do **not** consider $1^x = 1$ to be an exponential function.

2. We can now perform computations such as $2^{\sqrt{5}} = e^{\sqrt{5} \ln 2}$. We also have $2^{\sqrt{4}} = 2^2 = 4$ and $2^{\sqrt{9}} = 2^3 = 8$. Since $H_2(x) = 2^x$ is a strictly increasing function (because $E(x) = e^x$ is strictly increasing and $\ln 2 > 0$), we have $4 < 2^{\sqrt{5}} < 9$.

Using the chain rule, we see that if $a > 0$, then

$$\frac{d}{dx}[a^x] = \frac{d}{dx}[e^{x \ln a}] = e^{x \ln a}(\ln a) = a^x \ln a.$$

It follows from the second form of the Fundamental Theorem of Calculus (Theorem 11.14) that for any $a > 0$ and $c, d \in \mathbb{R}$ with $c \neq d$, we have

$$\int_c^d a^x \, dx = \frac{1}{\ln a} \int_c^d a^x (\ln a) dx = \frac{1}{\ln a}(a^d - a^c) = \frac{a^d}{\ln a} - \frac{a^c}{\ln a}.$$

We can now extend Theorems 12.4 and 12.12 to include irrational exponents.

Theorem 12.17: For all $r \in \mathbb{R}$ and $a \in (0, \infty)$, $\ln(a^r) = r \ln a$.

Proof: Let $r \in \mathbb{R}$, let $a \in (0, \infty)$, and let (q_n) be a sequence of rational numbers such that $q_n \to r$. By Theorem 12.4, for each $n \in \mathbb{N}$, we have $\ln(a^{q_n}) = q_n \ln a$. Since $\ln(a^x)$ and $x \ln a$ are continuous at r, we have

$$\ln(a^r) = \ln\left(a^{\lim_{n \to \infty} q_n}\right) = \lim_{n \to \infty} \ln(a^{q_n}) = \lim_{n \to \infty} (q_n \ln a) = \left(\lim_{n \to \infty} q_n\right) \ln a = r \ln a.$$

Here we have used Theorem 9.1 (1→3). □

Theorem 12.18: For all $r, a \in \mathbb{R}$, $e^{ra} = (e^a)^r$.

Proof: Let $r, a \in \mathbb{R}$ and let (q_n) be a sequence of rational numbers such that $q_n \to r$. By Theorem 12.12, for each $n \in \mathbb{N}$, we have $e^{q_n a} = (e^a)^{q_n}$. Since e^{xa} and $(e^a)^x$ are continuous at r, we have

$$e^{ra} = e^{\left(\lim_{n \to \infty} q_n\right)a} = \lim_{n \to \infty} e^{q_n a} = \lim_{n \to \infty} (e^a)^{q_n} = (e^a)^{\lim_{n \to \infty} q_n} = (e^a)^r.$$

Here we have used Theorem 9.1 (1→3) just as we did in the proof of Theorem 12.17. □

Notes: (1) It is now quite straightforward to verify that the usual laws of exponents hold for all real values of the exponents. Let $a > 0$ and let $x, y \in \mathbb{R}$. Then by Theorem 12.11, we have

$$a^{x+y} = e^{(x+y)\ln a} = e^{x \ln a + y \ln a} = e^{x \ln a} e^{y \ln a} = a^x a^y.$$

By Theorem 12.18, we have

$$(a^x)^y = \left(e^{x \ln a}\right)^y = e^{y(x \ln a)} = e^{(xy) \ln a} = a^{xy}.$$

Combining the previous two results gives us

$$a^{x-y} = a^{x+(-y)} = a^x a^{-y} = a^x (a^y)^{-1} = \frac{a^x}{a^y}.$$

If $a > 0$ and $a \neq 1$, then the general exponential function $f: \mathbb{R} \to (0, \infty)$ defined by $f(x) = a^x$ is injective because its derivative $f'(x) = a^x \ln a$ is either always positive (if $a > 1$), in which case it is strictly increasing, or it's derivative is always negative (if $a < 1$), in which case it is strictly decreasing. Therefore, $f(x) = a^x$ is invertible.

If we let $g: (0, \infty) \to \mathbb{R}$ be the inverse of f, then we call g the **general logarithmic function with base a** and we write $g(x) = \log_a x$.

Logarithmic Differentiation

In the expression $y = f(x)$, we say that x is the **independent variable** and y (or $f(x)$) is the **dependent variable**. Furthermore, in this case, we say that $f(x)$ is an **explicit function**.

For example, in the expression $y = x^2 + 1$, the dependent variable, y, is an explicit function of the independent variable x. If we rewrite the equation in the form $x^2 - y + 1 = 0$, then we will say that the dependent variable y is an **implicit function** of x.

When y is an implicit function of x, we can still find $\frac{dy}{dx}$, often using the chain rule, even if it is not possible to rewrite y as an explicit function of x. We refer to this process as **implicit differentiation**.

Example 12.19:

1. If $x^2 - y + 1 = 0$, we can find $\frac{dy}{dx}$ by differentiating each side of the given equation. We get $2x - \frac{dy}{dx} + 0 = 0$ (the derivative of 1 is 0 and the derivative of 0 is 0). So, $\frac{dy}{dx} = 2x$. Observe that we get the same answer if we first rewrite the given equation explicitly as $y = x^2 + 1$ and then differentiate.

2. If $e^{xy} = x$, then differentiating each side of the equation yields $e^{xy}\left(x \cdot \frac{dy}{dx} + 1 \cdot y\right) = 1$. So, we have $xe^{xy} \cdot \frac{dy}{dx} + ye^{xy} = 1$. Thus, $xe^{xy} \cdot \frac{dy}{dx} = 1 - ye^{xy}$, and therefore,

$$\frac{dy}{dx} = \frac{1 - ye^{xy}}{xe^{xy}}.$$

3. If $\ln y = \frac{x}{y}$, then differentiating each side of the equation yields

$$\frac{1}{y} \cdot \frac{dy}{dx} = \frac{y \cdot 1 - x \cdot \frac{dy}{dx}}{y^2}.$$

Multiplying each side of this equation by y^2 gives us $y \cdot \frac{dy}{dx} = y - x \cdot \frac{dy}{dx}$. Therefore, we have $y \cdot \frac{dy}{dx} + x \cdot \frac{dy}{dx} = y$, or equivalently, $\frac{dy}{dx}(y + x) = y$. Dividing each side of this equation by $y + x$ yields

$$\frac{dy}{dx} = \frac{y}{y + x}.$$

Logarithmic differentiation is an important application of implicit differentiation that allows us to differentiate functions involving products, quotients, and powers that might be very difficult otherwise. Given an equation defining y as a function of x (either explicitly or implicitly), the procedure for logarithmic differentiation is to first take the natural logarithm of each side of the given equation, then to apply Theorem 12.3 and/or Theorem 12.4 (sometimes Theorem 12.17 will be required), and then to differentiate each side of the equation implicitly. We will usually solve the resulting equation for $\frac{dy}{dx}$ and rewrite y in terms of x if possible.

Example 12.20:

1. Let's differentiate the following function using logarithmic differentiation:

$$y = \frac{x^{\frac{5}{3}}\sqrt{x^5 + 2}}{(5x + 1)^7}$$

We first take the natural logarithm of each side of the equation and apply Theorems 12.3 and 12.4 (as well as the Note following Theorem 12.4).

$$\ln y = \ln \frac{x^{\frac{5}{3}}\sqrt{x^5+2}}{(5x+1)^7} = \ln x^{\frac{5}{3}} + \ln \sqrt{x^5+2} - \ln(5x+1)^7$$

$$= \frac{5}{3}\ln x + \frac{1}{2}\ln(x^5+2) - 7\ln(5x+1)$$

We now differentiate each side of the equation using implicit differentiation on the left-hand side:

$$\frac{1}{y} \cdot \frac{dy}{dx} = \frac{5}{3} \cdot \frac{1}{x} + \frac{1}{2} \cdot \frac{5x^4}{x^5+2} - 7 \cdot \frac{5}{5x+1}$$

Solving for $\frac{dy}{dx}$ and replacing y by $\frac{x^{\frac{5}{3}}\sqrt{x^5+2}}{(5x+1)^7}$ results in the following:

$$\frac{dy}{dx} = y\left(\frac{5}{3x} + \frac{5x^4}{2(x^5+2)} - \frac{35}{5x+1}\right) = \frac{x^{\frac{5}{3}}\sqrt{x^5+2}}{(5x+1)^7}\left(\frac{5}{3x} + \frac{5x^4}{2(x^5+2)} - \frac{35}{5x+1}\right)$$

2. Let's differentiate the function $y = x^x$ using logarithmic differentiation. We first take the natural logarithm of each side of the equation and apply Theorem 12.17:

$$\ln y = \ln x^x = x \ln x$$

We now differentiate each side of the equation using implicit differentiation on the left-hand side and the product rule on the right-hand side:

$$\frac{1}{y} \cdot \frac{dy}{dx} = x \cdot \frac{1}{x} + 1 \cdot \ln x = 1 + \ln x$$

Solving for $\frac{dy}{dx}$ and replacing y by x^x results in the following:

$$\frac{dy}{dx} = y(1 + \ln x) = x^x(1 + \ln x)$$

3. In part 2 of Example 10.22, we showed that the power rule $\frac{d}{dx}[x^n] = nx^{n-1}$ holds for all $n \in \mathbb{Q}$. We can now use logarithmic differentiation to show that the power rule holds for all $n \in \mathbb{R}$. We begin by letting $y = x^n$. We now take the natural logarithm of each side of the equation and apply Theorem 12.17:

$$\ln y = \ln x^n = n \ln x \quad (x > 0)$$

Let's now differentiate each side of the equation using implicit differentiation on the left-hand side:

$$\frac{1}{y} \cdot \frac{dy}{dx} = n \cdot \frac{1}{x}$$

Solving for $\frac{dy}{dx}$ and replacing y by x^n gives us the desired result:

$$\frac{dy}{dx} = y \cdot \frac{n}{x} = x^n \cdot \frac{n}{x} = n \cdot \frac{x^n}{x} = nx^{n-1}$$

At the end of Lesson 10 it was mentioned that L'Hôpital's rule could be used to evaluate limits of the forms 0^0, 1^∞, and ∞^0. With logarithmic differentiation at our disposal, we are finally equipped to handle these forms.

Example 12.21:

1. Let's compute $\lim_{x \to 0^+} x^x$. Since $\lim_{x \to 0^+} x = 0$, we see that this limit has the form 0^0. We begin by writing $y = x^x$ and then taking the natural logarithm of each side of this equation to get

$$\ln y = \ln x^x = x \ln x = \frac{\ln x}{\frac{1}{x}}.$$

Note that for the second equality, we applied Theorem 12.17. For the third equality, we applied SACT (see Note 7 following the proof of Theorem 4.11) in a new way. We took the reciprocal of x to get $\frac{1}{x}$ and then took the reciprocal again to get $\frac{1}{\frac{1}{x}}$. Since the reciprocal of a reciprocal is just the original expression, we see that $\frac{1}{\frac{1}{x}} = x$.

So, why did we rewrite the expression this way? Well, $\lim_{x \to 0^+} \ln x = -\infty$ and $\lim_{x \to 0^+} \frac{1}{x} = \infty$. So, this new expression has the form $\frac{\infty}{\infty}$ and we can apply L'Hôpital's rule as usual to get

$$\lim_{x \to 0^+} \ln y = \lim_{x \to 0^+} \frac{\ln x}{\frac{1}{x}} = \lim_{x \to 0^+} \frac{\frac{1}{x}}{-\frac{1}{x^2}} = \lim_{x \to 0^+} \frac{1}{x}(-x^2) = \lim_{x \to 0^+} (-x) = 0.$$

L'Hôpital's rule was applied in the second equality above.

Since $y = e^{\ln y}$, we have

$$\lim_{x \to 0^+} x^x = \lim_{x \to 0^+} y = \lim_{x \to 0^+} e^{\ln y} = e^{\lim_{x \to 0^+} \ln y} = e^0 = 1.$$

The third equality above is true because e^x is a continuous function and $\lim_{x \to 0^+} \ln y$ exists. See the Note below for further explanation.

2. Let's compute $\lim_{x \to 1} x^{\frac{1}{x-1}}$. Since $\lim_{x \to 1} x = 1$ and $\lim_{x \to 1} \frac{1}{x-1} = \infty$ (technically speaking, the limit is $-\infty$ from the left and $+\infty$ from the right), we see that this limit has the form 1^∞. We begin once again by writing $y = x^{\frac{1}{x-1}}$ and then taking the natural logarithm of each side of this equation to get

$$\ln y = \ln x^{\frac{1}{x-1}} = \left(\frac{1}{x-1}\right) \ln x = \frac{\ln x}{x-1}.$$

Note that for the second equality, we applied Theorem 12.17.

Since, $\lim_{x \to 1} \ln x = 0$ and $\lim_{x \to 1} x - 1 = 0$, this new expression has the form $\frac{0}{0}$ and we can apply L'Hôpital's rule as usual to get

$$\lim_{x \to 1} \ln y = \lim_{x \to 1} \frac{\ln x}{x-1} = \lim_{x \to 1} \frac{\frac{1}{x}}{1} = \lim_{x \to 1} \frac{1}{x} = 1.$$

L'Hôpital's rule was applied in the second equality above.

Once again, since $y = e^{\ln y}$, we have

$$\lim_{x \to 1} x^{\frac{1}{x-1}} = \lim_{x \to 1} y = \lim_{x \to 1} e^{\ln y} = e^{\lim_{x \to 1} \ln y} = e^1 = e.$$

The third equality above is true because e^x is a continuous function and $\lim_{x \to 1} \ln y$ exists. See the Note below for details.

3. Let's compute $\lim_{x \to \infty} x^{\frac{1}{x}}$. Since $\lim_{x \to \infty} x = \infty$ and $\lim_{x \to \infty} \frac{1}{x} = 0$, we see that this limit has the form ∞^0.

We begin once again by writing $y = x^{\frac{1}{x}}$ and then taking the natural logarithm of each side of this equation to get

$$\ln y = \ln x^{\frac{1}{x}} = \frac{1}{x} \ln x = \frac{\ln x}{x}.$$

Note that for the second equality, we applied Theorem 12.17.

Since, $\lim_{x \to \infty} \ln x = \infty$ and $\lim_{x \to \infty} x = \infty$, this new expression has the form $\frac{\infty}{\infty}$ and we can apply L'Hôpital's rule as usual to get

$$\lim_{x \to \infty} \ln y = \lim_{x \to \infty} \frac{\ln x}{x} = \lim_{x \to \infty} \frac{\frac{1}{x}}{1} = \lim_{x \to \infty} \frac{1}{x} = 0.$$

L'Hôpital's rule was applied in the second equality above.

Once again, since $y = e^{\ln y}$,

$$\lim_{x \to \infty} x^{\frac{1}{x}} = \lim_{x \to \infty} y = \lim_{x \to \infty} e^{\ln y} = e^{\lim_{x \to \infty} \ln y} = e^0 = 1.$$

The third equality above is true because e^x is a continuous function and $\lim_{x \to \infty} \ln y$ exists. See the Note below for further explanation.

Note: Let $f: A \to \mathbb{R}$, $g: \mathbb{R} \to \mathbb{R}$ with g continuous on $\mathbb{R}$, let $a \in A$ and suppose that $\lim_{x \to a} f(x)$ exists. Then $\lim_{x \to a} g(f(x)) = g\left(\lim_{x \to a} f(x)\right)$. To see this, suppose that $\lim_{x \to a} f(x) = L$ and let $\epsilon > 0$. Since g is continuous at L, there is $\delta_1 > 0$ such that $|y - L| < \delta_1$ implies $|g(y) - g(L)| < \epsilon$. Since $\lim_{x \to a} f(x) = L$, there is $\delta_2 > 0$ such that $0 < |x - a| < \delta_2$ implies $|f(x) - L| < \delta_1$. Now, suppose that $0 < |x - a| < \delta_2$. Then $|f(x) - L| < \delta_1$. It follows that $|g(f(x)) - g(L)| < \epsilon$. Since $\epsilon > 0$ was arbitrary, $\lim_{x \to a} g(f(x)) = g(L) = g\left(\lim_{x \to a} f(x)\right)$.

In all three parts of Example 12.21, we have $f: A \to \mathbb{R}$ and $g: \mathbb{R} \to \mathbb{R}$ defined by $f(x) = \ln y$ (where y is a function of x) and $g(x) = e^x$. In part 2, we have $L = 1$. It follows that

$$\lim_{x \to 1} e^{\ln y} = \lim_{x \to 1} g(f(x)) = g\left(\lim_{x \to 1} f(x)\right) = e^{\lim_{x \to 1} \ln y}.$$

By modifying the above argument just a bit, we can attain similar results for one-sided limits and infinite limits that will justify the analogous computations made in parts 1 and 3, respectively. The dedicated reader should write out the details.

Problem Set 12

Full solutions to these problems are available for free download here:
www.SATPrepGet800.com/RABQXZ

LEVEL 1

1. Let $a \in (0,1) \cup (1, \infty)$. Prove that for all $x \in (0, \infty)$, $\log_a x = \frac{\ln x}{\ln a}$.

2. Let $a, b \in (0,1) \cup (1, \infty)$. Prove that for all $x \in (0, \infty)$, $\log_a x = \frac{\log_b x}{\log_b a}$.

3. Let $a \in (0,1) \cup (1, \infty)$ and let $b, c \in (0, \infty)$. Prove that $\log_a bc = \log_a b + \log_a c$.

4. Let $a \in (0,1) \cup (1, \infty)$, let $b \in (0, \infty)$, and let $r \in \mathbb{R}$. Prove that $\log_a(b^r) = r \log_a b$.

LEVEL 2

5. Use Problem 11 from Problem Set 11 to prove that $\frac{1}{3} \leq \ln 1.5 \leq \frac{1}{2}$.

6. Use Riemann sums to prove that $\frac{25}{66} < \ln 1.5 < \frac{47}{110}$.

7. Use a geometric argument to prove that $\ln 1.5 < \frac{5}{12}$

LEVEL 3

8. Let $x \in \mathbb{R}$. Prove that $e^x = \lim_{h \to 0}(1+hx)^{\frac{1}{h}}$.

9. Let $x \in \mathbb{R}$. Prove that $e^x = \lim_{h \to \infty}\left(1 + \frac{x}{h}\right)^h$.

10. Prove that for each $n \in \mathbb{Z}^+$ with $n \geq 2$,
$$\frac{1}{2} + \frac{1}{3} + \cdots + \frac{1}{n} < \ln n < 1 + \frac{1}{2} + \frac{1}{3} + \cdots + \frac{1}{n-1}.$$

11. Prove that for all $x \in \mathbb{R}$, $1 + x \leq e^x$.

12. Prove that the following inequality holds:
$$\frac{4}{3} \leq \int_0^1 e^{x^2}\, dx \leq e.$$

Level 4

13. Prove that for all $n \in \mathbb{N}$ and all $x \in [0, \infty)$,

$$e^x \geq \sum_{k=0}^{n} \frac{x^k}{k!}$$

(where $0! = 1$ and $k! = 1 \cdot 2 \cdots k$ for all natural numbers $k \geq 1$).

Level 5

14. For each $k = 1, 2, \ldots, n$, let $c_k \in (0, \infty)$, prove the following inequality:

$$(c_1 c_2 \cdots c_n)^{\frac{1}{n}} \leq \frac{c_1 + c_2 + \cdots c_n}{n}$$

15. Prove that for all $n \in \mathbb{Z}^+$ and all $x \in [0, \infty)$,

$$\ln(x+1) = \sum_{k=1}^{n} \frac{(-1)^{k+1} x^k}{k} + \int_0^x \frac{(-1)^n t^n}{t+1} \, dt.$$

Challenge Problems

16. Prove that e is an irrational number.

17. Prove that for each $n \in \mathbb{Z}^+ \setminus \{1\}$, $\ln n$ is an irrational number.

LESSON 13
IMPROPER INTEGRATION

Improper Integrals

In Lesson 11, we learned a process called Riemann integration. In that lesson we restricted our attention to bounded functions defined on closed and bounded intervals. In this lesson, we will extend the notion of integration to arbitrary functions (possibly unbounded) defined on arbitrary intervals (half-open, open, and possibly infinite).

Let $I \subseteq A \subseteq \mathbb{R}$ with I an interval and let $f: A \to \mathbb{R}$. We will say that f is **locally integrable** on I if for any closed interval $[a, b] \subseteq I$, f is Riemann integrable on $[a, b]$.

Note: Recall that there are nine different types of intervals (see Lesson 2). In the definition above, I can be any one of these nine types of intervals.

Example 13.1:

1. The function $f: \mathbb{R} \to \mathbb{R}$ defined by $f(x) = x^2$ is bounded and Riemann integrable on $[0, 1]$. By Problem 15 from Problem Set 11, f is Riemann integrable on every closed interval that is a subset of $[0, 1]$. Therefore, f is also locally integrable on $[0, 1]$.

    More generally, if $I \subseteq A \subseteq \mathbb{R}$ with I a closed interval and $f: A \to \mathbb{R}$ is bounded on I, then f is locally integrable on I if and only if f is Riemann integrable on I.

2. Since the function $f(x) = x^2$ is bounded and Riemann integrable on all closed intervals $[a, b]$, f is locally integrable on every interval. In particular, f is locally integrable on $(-\infty, \infty)$.

3. The function $g: (-\infty, 0) \cup (0, \infty)$ defined by $g(x) = \frac{1}{x}$ is locally integrable on $(-\infty, 0)$ and $(0, \infty)$.

4. The function $h: [1, \infty)$ defined by $h(x) = \sqrt{x-1}$ is locally integrable on $[1, \infty)$.

Let $a \in \mathbb{R}$, let $b \in \mathbb{R} \cup \{\infty\}$, let $f: [a, b) \to \mathbb{R}$ and suppose that f is locally integrable on $[a, b)$. We define the **improper integral** of f over $[a, b)$ to be

$$\int_a^b f(x)\,dx = \lim_{c \to b^-} \int_a^c f(x)\,dx$$

assuming that the given left-hand limit exists. If $b = \infty$, then we define b^- to also be ∞. If the given left-hand limit is a finite number, we say that the improper integral **converges** (and we write $\int_a^b f(x)\,dx \downarrow$). Otherwise, we say that the improper integral **diverges** (and we write $\int_a^b f(x)\,dx \uparrow$).

Similarly, let $b \in \mathbb{R}$, let $a \in \mathbb{R} \cup \{-\infty\}$, let $f: (a, b] \to \mathbb{R}$ and suppose that f is locally integrable on $(a, b]$. We define the **improper integral** of f over $(a, b]$ to be

$$\int_a^b f(x)\,dx = \lim_{c \to a^+} \int_c^b f(x)\,dx$$

assuming that the given right-hand limit exists. If $a = -\infty$, then we define a^+ to also be $-\infty$. If the given right-hand limit is a finite number, we say that the improper integral **converges** (and we write $\int_a^b f(x)\, dx \downarrow$). Otherwise, we say that the improper integral **diverges** (and we write $\int_a^b f(x)\, dx \uparrow$).

Finally, let $a \in \mathbb{R} \cup \{-\infty\}$, let $b \in \mathbb{R} \cup \{\infty\}$, let $f: (a, b) \to \mathbb{R}$ and suppose that f is locally integrable on (a, b). We define the **improper integral** of f over (a, b) to be

$$\int_a^b f(x)\, dx = \int_a^c f(x)\, dx + \int_c^b f(x)\, dx$$

where $a < c < b$, assuming that both improper integrals on the right converge (in which case, we say that the improper integral on the left **converges** and we write $\int_a^b f(x)\, dx \downarrow$). If either one (or both) of the integrals on the right diverge, then we say that the improper integral on the left **diverges** (and we write $\int_a^b f(x)\, dx \uparrow$).

Note: In Problem 2 below, you will be asked to prove that the convergence or divergence of $\int_a^b f(x)\, dx$ as well as the value of the integral if it converges is unaffected if we change the value of c that we choose between a and b.

Example 13.2:

1. Define $f: [1, 2) \to \mathbb{R}$ by $f(x) = x^2$. If $1 \leq c < 2$, then by the second form of the Fundamental Theorem of Calculus (Theorem 11.14), we have

$$\int_1^c x^2\, dx = \left[\frac{x^3}{3}\right]_1^c = \frac{1}{3}(c^3 - 1).$$

It follows that the improper integral of $f(x) = x^2$ over $[1, 2)$ is

$$\int_1^2 x^2\, dx = \lim_{c \to 2^-} \int_1^c x^2\, dx = \lim_{c \to 2^-} \left[\frac{1}{3}(c^3 - 1)\right] = \frac{1}{3}(2^3 - 1) = \frac{1}{3}(8 - 1) = \frac{7}{3}.$$

So, this improper integral converges to $\frac{7}{3}$. Notice that this value agrees with the Riemann integral of $f(x) = x^2$ over $[1, 2]$ (see part 1 of Example 11.6).

In fact, whenever f is a bounded function that is Riemann integrable on $[a, b]$, the improper integral of f on $[a, b), (a, b]$, or (a, b) agrees with the Riemann integral on $[a, b]$.

2. The function $g: [1, \infty) \to \mathbb{R}$ defined by $g(x) = \frac{1}{x^2}$ is locally integrable on $[1, \infty)$. If $c > 1$, then by the second form of the Fundamental Theorem of Calculus (Theorem 11.14), we have

$$\int_1^c \frac{1}{x^2}\, dx = \int_1^c x^{-2}\, dx = \left[\frac{x^{-1}}{-1}\right]_1^c = \left[\frac{-1}{x}\right]_1^c = -\frac{1}{c} - \left(\frac{-1}{1}\right) = 1 - \frac{1}{c}.$$

It follows that the improper integral of $g(x) = \frac{1}{x^2}$ over $[1, \infty)$ converges and the value of the integral is

$$\int_1^\infty \frac{1}{x^2} dx = \lim_{c\to\infty} \int_1^c \frac{1}{x^2} dx = \lim_{c\to\infty} \left[1 - \frac{1}{c}\right] = 1.$$

3. Let $p \in \mathbb{R}$. The function $g_p: [1, \infty) \to \mathbb{R}$ defined by $g_p(x) = \frac{1}{x^p}$ is locally integrable on $[1, \infty)$. First assume that $p \neq 1$. If $c > 1$, then by the second form of the Fundamental Theorem of Calculus (Theorem 11.14), we have

$$\int_1^c \frac{1}{x^p} dx = \int_1^c x^{-p} dx = \left[\frac{x^{1-p}}{1-p}\right]_1^c = \frac{c^{1-p}}{1-p} - \frac{1}{1-p} = \frac{c^{1-p} - 1}{1-p}.$$

It follows that the improper integral of $g_p(x) = \frac{1}{x^p}$ over $[1, \infty)$ is

$$\int_1^\infty \frac{1}{x^p} dx = \lim_{c\to\infty} \int_1^c \frac{1}{x^p} dx = \lim_{c\to\infty} \left[\frac{c^{1-p} - 1}{1-p}\right] = \frac{1}{1-p} \lim_{c\to\infty} [c^{1-p} - 1].$$

If $p > 1$, then $\lim_{c\to\infty} [c^{1-p} - 1] = 0 - 1 = -1$, and so, the improper integral converges. We have

$$\int_1^\infty \frac{1}{x^p} dx = \frac{1}{1-p} \lim_{c\to\infty} [c^{1-p} - 1] = \frac{-1}{1-p} = \frac{1}{p-1}.$$

For example, we have each of the following:

$$\int_1^\infty \frac{1}{x^2} dx = \frac{1}{2-1} = 1 \text{ (as we saw in part 2 above)}$$

$$\int_1^\infty \frac{1}{x^3} dx = \frac{1}{3-1} = \frac{1}{2}$$

$$\int_1^\infty \frac{1}{x^{\frac{3}{2}}} dx = \frac{1}{\frac{3}{2}-1} = \frac{1}{\frac{1}{2}} = 2$$

If $p < 1$, then $\lim_{c\to\infty} [c^{1-p} - 1] = \infty$, and so,

$$\int_1^\infty \frac{1}{x^p} dx = \frac{1}{1-p} \lim_{c\to\infty} [c^{1-p} - 1] = \infty \text{ (and so, the improper integral diverges)}.$$

If $p = 1$, then for $c > 1$, we have

$$\int_1^c \frac{1}{x^p} dx = \int_1^c \frac{1}{x} dx = \ln c - \ln 1 = \ln c.$$

It follows that the improper integral of $g_1(x) = \frac{1}{x}$ over $[1, \infty)$ is

$$\int_1^\infty \frac{1}{x} dx = \lim_{c\to\infty} \int_1^c \frac{1}{x} dx = \lim_{c\to\infty} [\ln c] = \infty \text{ (and so, the improper integral diverges)}.$$

4. Consider the function $h: \mathbb{R} \to \mathbb{R}$ defined by $h(x) = x$. Then h is locally integrable on $(-\infty, \infty)$ and for any $r \in \mathbb{R}^+$, by the second form of the Fundamental Theorem of Calculus, we have

$$\int_{-r}^r x \, dx = \frac{r^2}{2} - \frac{(-r)^2}{2} = \frac{r^2}{2} - \frac{r^2}{2} = 0.$$

Therefore,
$$\lim_{r \to \infty} \int_{-r}^{r} x \, dx = 0.$$

Nonetheless, the improper integral of h over the interval $(-\infty, \infty)$ diverges. To see this, remember that by definition, we have

$$\int_{-\infty}^{\infty} x \, dx = \int_{-\infty}^{0} x \, dx + \int_{0}^{\infty} x \, dx$$

and if either one (or both) of the integrals on the right diverge, then the improper integral on the left diverges. In this case, both integrals on the right diverge. For example, we have

$$\int_{0}^{\infty} x \, dx = \lim_{c \to \infty} \int_{0}^{c} x \, dx = \lim_{c \to \infty} \left[\frac{x^2}{2}\right]_{0}^{c} = \lim_{c \to \infty} \left[\frac{c^2}{2}\right] = \infty.$$

Integration by Parts

We will now introduce an integration method that can sometimes be helpful for integrating products of different "types" of functions. The five most common types are logarithmic, inverse trigonometric, algebraic, trigonometric, and exponential.

Theorem 13.3 (Integration by Parts): Let $u, v: [a, b] \to \mathbb{R}$ be differentiable functions on $[a, b]$ with u' and v' continuous on $[a, b]$. Then

$$\int_{a}^{b} u(x)v'(x) \, dx = u(b)v(b) - u(a)v(a) - \int_{a}^{b} v(x)u'(x) \, dx.$$

Proof: Since u and v are differentiable on $[a, b]$, by Theorem 10.11, they are continuous on $[a, b]$. By Theorem 8.14 and Problem 5 from Problem Set 8, uv' and $u'v$ are continuous on $[a, b]$. By Theorem 8.13, $uv' + u'v$ is continuous on $[a, b]$. Therefore, by Theorem 11.9, $uv' + u'v$ is Riemann integrable on $[a, b]$.

By the product rule (Theorem 10.14), we have $\frac{d}{dx}[u(x)v(x)] = u(x)v'(x) + u'(x)v(x)$. It follows once again from Theorem 11.9 that

$$\int_{a}^{b} \frac{d}{dx}[u(x)v(x)] \, dx = \int_{a}^{b} [u(x)v'(x) + u'(x)v(x)] \, dx = \int_{a}^{b} u(x)v'(x) \, dx + \int_{a}^{b} u'(x)v(x) \, dx.$$

By the second form of the Fundamental Theorem of Calculus (Theorem 11.14), we have

$$\int_{a}^{b} \frac{d}{dx}[u(x)v(x)] \, dx = u(b)v(b) - u(a)v(a).$$

It follows that

$$\int_{a}^{b} u(x)v'(x) \, dx = \int_{a}^{b} \frac{d}{dx}[u(x)v(x)] \, dx - \int_{a}^{b} u'(x)v(x) \, dx$$
$$= u(b)v(b) - u(a)v(a) - \int_{a}^{b} u'(x)v(x) \, dx. \quad \square$$

Note: By letting $du = u'(x)dx$ and $dv = v'(x)dx$, we can write the integration by parts formula in the following abbreviated form:

$$\int_a^b u(x)\,dv = [u(x)v(x)]_a^b - \int_a^b v(x)\,du$$

Example 13.4: Let's use integration by parts (Theorem 13.3) to evaluate the following integral:

$$\int_0^\pi x \cos 3x \, dx$$

If we let $u(x) = x$ and $dv = \cos 3x \, dx$, then we have $du = dx$ and $v(x) = \frac{1}{3}\sin 3x$. Therefore,

$$\int_0^\pi x \cos 3x \, dx = \left[(x)\left(\frac{1}{3}\sin 3x\right)\right]_0^\pi - \int_0^\pi \frac{1}{3}\sin 3x \, dx = 0 + \left[\frac{1}{9}\cos 3x\right]_0^\pi$$

$$= \frac{1}{9}(\cos 3\pi - \cos 0) = \frac{1}{9}(-1-1) = -\frac{2}{9}.$$

Notes: (1) We can find an antiderivative of $x \cos 3x$ using the following shortcut method called **tabular integration by parts** (see Note 2 below for details on how to use this table). Here we will describe this method informally. For a formal treatment, see Problem 12 below.

+	x	$\cos 3x$
−	1	$\frac{1}{3}\sin 3x$
+	0	$-\frac{1}{9}\cos 3x$

Following the arrows gives us an antiderivative of $\frac{x}{3}\sin 3x + \frac{1}{9}\cos 3x$. We then apply the second form of the Fundamental Theorem of Calculus (Theorem 11.14) to get

$$\int_0^\pi x \cos 3x \, dx = \left[\frac{x}{3}\sin 3x + \frac{1}{9}\cos 3x\right]_0^\pi = \left(\frac{\pi}{3}\sin 3\pi + \frac{1}{9}\cos 3\pi\right) - \left(\frac{0}{3}\sin 0 + \frac{1}{9}\cos 0\right)$$

$$= \left(0 - \frac{1}{9}\right) - \left(0 + \frac{1}{9}\right) = -\frac{1}{9} - \frac{1}{9} = -\frac{2}{9}.$$

(2) Let's carefully go over how the table in Note 1 above was generated.

In the first column we simply alternate signs starting with a plus sign.

In the middle column we put our choice for $u(x)$ and we differentiate as we go down the column.

In the third column we put our choice for dv, and we integrate as we go down the column.

In this particular example we stop at the third row because we get a 0 in the middle column.

Finally we follow the arrow pattern as seen in Note 1 above to write down the desired antiderivative. As we follow each arrow, we multiply, and then we add up the results from the individual arrows.

(3) How do we figure out how to choose u and dv? There are no absolute rules, but as a general guideline, it is helpful to memorize the mnemonic LIATE (although the T and E in "LIATE" are generally interchangeable: LIAET usually works just as well).

L stands for Logarithmic, I stands for Inverse Trigonometric, A stands for Algebraic, T stands for Trigonometric, and E stands for exponential.

As a first attempt, it's not a bad idea to choose the "leftmost" letter for $u(x)$ and the "rightmost" letter for dv.

In Example 13.4 above, we chose the algebraic function x for $u(x)$, and the trigonometric function $\cos 3x$ for dv.

(4) Recall from Lesson 3 that the class of "algebraic functions" includes polynomials and rational functions, but is more general in the sense that roots can appear in the functions as well. For example, $\sqrt{x}$ and $\frac{\left(\sqrt[5]{x}+3x^2\right)}{x^{\frac{2}{7}}-5}$ are algebraic, but not polynomial or rational.

(5) For an example of an inverse trigonometric function, see Problem 9 below.

Example 13.5: Let's use integration by parts (Theorem 13.3) to evaluate the following integral:

$$\int_0^{\frac{\pi}{2}} e^{3x} \cos 2x \, dx$$

If we let $u(x) = e^{3x}$ and $dv = \cos 2x \, dx$, then we have $du = 3e^{3x} dx$ and $v(x) = \frac{1}{2}\sin 2x$. Therefore,

$$\int_0^{\frac{\pi}{2}} e^{3x} \cos 2x \, dx = \left[(e^{3x})\left(\frac{1}{2}\sin 2x\right)\right]_0^{\frac{\pi}{2}} - \int_0^{\frac{\pi}{2}} \left(\frac{1}{2}\sin 2x\right)(3e^{3x}) \, dx$$

$$= 0 - \frac{3}{2}\int_0^{\frac{\pi}{2}} e^{3x} \sin 2x \, dx = -\frac{3}{2}\int_0^{\frac{\pi}{2}} e^{3x} \sin 2x \, dx.$$

We now repeat this procedure with $\int_0^{\frac{\pi}{2}} e^{3x} \sin 2x \, dx$. We let $u(x) = e^{3x}$, and $dv = \sin 2x \, dx$. It then follows that $du = 3e^{3x} dx$ and $v(x) = -\frac{1}{2}\cos 2x$. So, we have

$$\int_0^{\frac{\pi}{2}} e^{3x} \sin 2x \, dx = \left[(e^{3x})\left(-\frac{1}{2}\cos 2x\right)\right]_0^{\frac{\pi}{2}} - \int_0^{\frac{\pi}{2}} \left(-\frac{1}{2}\cos 2x\right)(3e^{3x}) \, dx$$

$$= -\frac{1}{2}e^{\frac{3\pi}{2}}\cos \pi + \frac{1}{2}\cos 0 + \frac{3}{2}\int_0^{\frac{\pi}{2}} e^{3x} \cos 2x \, dx = \frac{1}{2}e^{\frac{3\pi}{2}} + \frac{1}{2} + \frac{3}{2}\int_0^{\frac{\pi}{2}} e^{3x} \cos 2x \, dx.$$

Putting these two results together, we have

$$\int_0^{\frac{\pi}{2}} e^{3x} \cos 2x \, dx = -\frac{3}{2} \int_0^{\frac{\pi}{2}} e^{3x} \sin 2x \, dx = -\frac{3}{2}\left(\frac{1}{2} e^{\frac{3\pi}{2}} + \frac{1}{2} + \frac{3}{2} \int_0^{\frac{\pi}{2}} e^{3x} \cos 2x \, dx\right)$$

$$= -\frac{3}{4} e^{\frac{3\pi}{2}} - \frac{3}{4} - \frac{9}{4} \int_0^{\frac{\pi}{2}} e^{3x} \cos 2x \, dx$$

Adding $\frac{9}{4} \int_0^{\frac{\pi}{2}} e^{3x} \cos 2x \, dx$ to each side of this last equation gives

$$\frac{13}{4} \int_0^{\frac{\pi}{2}} e^{3x} \cos 2x \, dx = -\frac{3}{4} e^{\frac{3\pi}{2}} - \frac{3}{4}.$$

Multiplying each side of this last equation by $\frac{4}{13}$ yields

$$\int_0^{\frac{\pi}{2}} e^{3x} \cos 2x \, dx = -\frac{3}{13} e^{\frac{3\pi}{2}} - \frac{3}{13}.$$

Notes: (1) As in Note 1 following Example 13.4, we can use tabular integration by parts to evaluate the integral more easily (see Note 2 below for details on how to use this table).

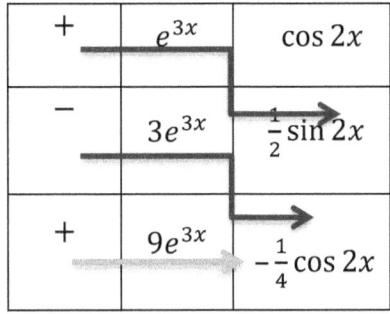

Following the arrows gives us an antiderivative of

$$\frac{1}{2} e^{3x} \sin 2x + \frac{3}{4} e^{3x} \cos 2x + \int -\frac{9}{4} e^{3x} \cos 2x \, dx.$$

We then apply the second form of the Fundamental Theorem of Calculus (Theorem 11.14) to get

$$\int_0^{\frac{\pi}{2}} e^{3x} \cos 2x \, dx = \left[\frac{1}{2} e^{3x} \sin 2x + \frac{3}{4} e^{3x} \cos 2x\right]_0^{\frac{\pi}{2}} - \frac{9}{4} \int_0^{\frac{\pi}{2}} e^{3x} \cos 2x \, dx$$

$$= \left(\frac{1}{2} e^{\frac{3\pi}{2}} \sin \pi + \frac{3}{4} e^{\frac{3\pi}{2}} \cos \pi\right) - \left(\frac{1}{2} e^0 \sin 0 + \frac{3}{4} e^0 \cos 0\right) - \frac{9}{4} \int_0^{\frac{\pi}{2}} e^{3x} \cos 2x \, dx$$

$$= \left(0 - \frac{3}{4} e^{\frac{3\pi}{2}}\right) - \left(0 + \frac{3}{4}\right) - \frac{9}{4} \int_0^{\frac{\pi}{2}} e^{3x} \cos 2x \, dx.$$

We add $\frac{9}{4}\int_0^{\frac{\pi}{2}} e^{3x} \cos 2x \, dx$ to each side of this last equation to get

$$\frac{13}{4}\int_0^{\frac{\pi}{2}} e^{3x} \cos 2x \, dx = -\frac{3}{4}e^{\frac{3\pi}{2}} - \frac{3}{4}.$$

Multiplying each side of this last equation by $\frac{4}{13}$ yields

$$\int_0^{\frac{\pi}{2}} e^{3x} \cos 2x \, dx = -\frac{3}{13}e^{\frac{3\pi}{2}} - \frac{3}{13}.$$

(2) Let's carefully go over how the table in Note 1 above was generated.

In the first column we simply alternate signs starting with a plus sign.

In the middle column we put our choice for $u(x)$, and we differentiate as we go down the column.

In the third column we put our choice for dv, and we integrate as we go down the column.

In this particular example we stop at the third row since the expression $\cos 2x$ appears for the second time in the third column (more specifically, we can stop when we see a function in the third column that is a constant multiple of a function that appears above it).

Finally we follow the arrow pattern as seen in Note 1 above to write down the desired antiderivative. Notice that the bottom arrow that points from left to right leads to an integral expression.

Example 13.6: Let's use integration by parts (Theorem 13.3) to evaluate the following improper integral:

$$\int_{-\infty}^{0} xe^x \, dx$$

If we let $u(x) = x$ and $dv = e^x$, then we have $du = dx$ and $v(x) = e^x$. Therefore,

$$\int_{-\infty}^{0} xe^x \, dx = \lim_{c \to -\infty} \int_c^0 xe^x \, dx = \lim_{c \to -\infty} \left([xe^x]_c^0 - \int_c^0 e^x \, dx\right) = \lim_{c \to -\infty}(0 - ce^c - (e^0 - e^c))$$

$$= \lim_{c \to -\infty}(-ce^c - 1 + e^c) = \lim_{c \to -\infty}\frac{-c}{e^{-c}} - \lim_{c \to -\infty} 1 + \lim_{c \to -\infty} e^c$$

$$= \lim_{c \to -\infty}\frac{-1}{-e^{-c}} - \lim_{c \to -\infty} 1 + \lim_{c \to -\infty} e^c = 0 - 1 + 0 = -1.$$

Notes: (1) By Theorem 12.14, we have $\lim_{c \to -\infty} e^c = 0$.

(2) The dedicated reader should check carefully that $\lim_{c \to -\infty}\frac{-c}{e^{-c}}$ has the form $\frac{\infty}{\infty}$. Therefore, by L'Hôpital's rule, we have $\lim_{c \to -\infty}\frac{-c}{e^{-c}} = \lim_{c \to -\infty}\frac{-1}{-e^{-c}} = \lim_{c \to -\infty} e^c$ (Check this!). This last limit is 0 by Theorem 12.14 again.

Comparison Tests

Sometimes it may be difficult or even impossible to find the exact value of an improper integral. Nonetheless, it is usually still very useful to know if the improper integral converges or diverges. The remaining theorems in this lesson will provide us with a few "convergence tests" that can be used on a case by case basis to provide us with this information. It is worth stressing that these tests will never help us find the exact value that an improper integral converges to—their only purpose is to tell us if the improper integral converges or diverges.

Theorem 13.7 (Comparison Test): Let $a \in \mathbb{R}$, $b \in \mathbb{R} \cup \{\infty\}$, and let $f, g: [a, b) \to \mathbb{R}$ be locally integrable on $[a, b)$. Assume that for all $x \in [a, b)$, we have $0 \leq f(x) \leq g(x)$. Then we have the following:

$$\text{if } \int_a^b g(x)\, dx \text{ converges, then } \int_a^b f(x)\, dx \text{ converges;}$$

$$\text{if } \int_a^b f(x)\, dx \text{ diverges, then } \int_a^b g(x)\, dx \text{ diverges.}$$

Proof: Let $x \in [a, b)$. Since $f(t) \leq g(t)$ for all $t \in [a, x]$, by Problem 6 from Problem Set 11, we have

$$\int_a^x f(t)\, dt \leq \int_a^x g(t)\, dt.$$

Therefore, we have $\sup\{\int_a^x f(t)\, dt \mid t \in [a, b)\} \leq \sup\{\int_a^x g(t)\, dt \mid t \in [a, b)\}$. Since $f(x) \geq 0$ and $g(x) \geq 0$ for all $x \in [a, b)$, by Problem 10 below, we have

$$\int_a^b f(x)\, dx = \sup\{\int_a^x f(t)\, dt \mid x \in [a, b)\};$$

$$\int_a^b g(x)\, dx = \sup\{\int_a^x g(t)\, dt \mid x \in [a, b)\}.$$

If $\int_a^b g(x)\, dx$ converges, say $\int_a^b g(x)\, dx = L$ for some finite real number L, then we have $\int_a^b f(x)\, dx = \sup\{\int_a^x f(t)\, dt \mid t \in [a, b)\} \leq \sup\{\int_a^x g(t)\, dt \mid x \in [a, b)\} = \int_a^b g(x)\, dx = L$, and so, $\int_a^b f(x)\, dx$ is also a finite real number.

The second statement is the contrapositive of the first, and therefore, is true as well. □

Note: See the analysis after Theorem 1.28 for more explanation on the contrapositive.

Example 13.8:

1. Let's use the Comparison Test (Theorem 13.7) to show that $\int_1^\infty e^{-x^2}\, dx$ converges. First note that for $x \geq 1$, we have $x^2 \geq x$, and therefore, $-x^2 \leq -x$. Since e^x is a strictly increasing function, for $x \geq 1$, we have $e^{-x^2} \leq e^{-x}$. By Theorem 12.14, we have

$$\int_1^\infty e^{-x}\, dx = \lim_{c \to \infty} \int_1^c e^{-x}\, dx = \lim_{c \to \infty} [-e^{-x}]_1^c = \lim_{c \to \infty} [e^{-1} - e^{-c}] = e^{-1} - \lim_{c \to \infty} e^{-c} = e^{-1} - 0 = e^{-1}.$$

Therefore, $\int_1^\infty e^{-x}\,dx$ converges. By Theorem 13.7, $\int_1^\infty e^{-x^2}\,dx$ converges as well.

In fact, if a is any real number, then $\int_a^\infty e^{-x^2}\,dx$ converges. If $a \geq 1$, then the previous argument works. If $a < 1$, we can write

$$\int_a^\infty e^{-x^2}\,dx = \int_a^1 e^{-x^2}\,dx + \int_1^\infty e^{-x^2}\,dx.$$

Since the first integral on the right-hand side of the equation is a regular Riemann integral and e^{-x^2} is continuous on $\mathbb{R}$, by Theorem 11.7, it converges to some real number L. We showed above that the second integral on the right-hand side converges to some real number, say M. It follows that $\int_a^\infty e^{-x^2}\,dx$ converges to $L + M$.

2. Let's use the Comparison Test (Theorem 13.7) to show that $\int_1^\infty \frac{x+3}{\sqrt{x^4-2x}}\,dx$ diverges. We have $x + 3 \geq x$. We also have $\sqrt{x^4 - 2x} \leq \sqrt{x^4} = x^2$, and therefore, $\frac{1}{\sqrt{x^4-2x}} \geq \frac{1}{x^2}$. It follows that $\frac{x+3}{\sqrt{x^4-2x}} \geq \frac{x}{x^2} = \frac{1}{x}$. By part 3 of Example 13.2, $\int_1^\infty \frac{1}{x}\,dx$ diverges. So, by Theorem 13.7, $\int_1^\infty \frac{x+3}{\sqrt{x^4-2x}}\,dx$ diverges as well.

Theorem 13.9 (Limit Comparison Test): Let $a \in \mathbb{R}$, $b \in \mathbb{R} \cup \{\infty\}$, and let $f, g: [a, b) \to \mathbb{R}$ be locally integrable on $[a, b)$. Assume that there is a subinterval $[k, b) \subseteq [a, b)$ such that for all $x \in [k, b)$, we have $f(x) \geq 0$ and $g(x) > 0$. Also assume that there is $L \in \mathbb{R} \cup \{\infty\}$ such that

$$\lim_{x \to b^-} \frac{f(x)}{g(x)} = L. \quad \text{(Recall that } \infty^- = \infty.\text{)}$$

Then we have the following:

if $L \in \mathbb{R}^+$, then $\int_a^b f(x)\,dx \downarrow$ if and only if $\int_a^b g(x)\,dx \downarrow$.

if $L = 0$ and $\int_a^b g(x)\,dx \downarrow$, then $\int_a^b f(x)\,dx \downarrow$.

if $L = \infty$ and $\int_a^b g(x)\,dx = \infty$, then $\int_a^b f(x)\,dx = \infty$.

Notes: (1) You will be asked to prove Theorem 13.9 in Problem 14 below.

(2) By Problem 10 below, if $f(x) \geq 0$ for all $x \in [k, b)$, then the statement $\int_a^b f(x)\,dx = \infty$ is equivalent to the statement $\int_a^b f(x)\,dx \uparrow$. In other words, for functions that are *eventually* nonnegative (nonnegative for all $x \in [k, b)$, where $[k, b)$ is some subset of $[a, b)$), the integral approaching ∞ is equivalent to the integral diverging.

(3) The claim made in Note 2 above is not true for general f. If the function f takes on both positive and negative values throughout $[a, b)$, then it is possible for the integral to diverge for other reasons. See Problem 1 below for an example.

Example 13.10:

1. Let's use the Limit Comparison Test (Theorem 13.9) to show that $\int_1^\infty \frac{1}{x+1} dx$ diverges. By part 3 of Example 13.2, $\int_1^\infty \frac{1}{x} dx$ diverges. We also have

$$\lim_{x \to \infty} \frac{\frac{1}{x+1}}{\frac{1}{x}} = \lim_{x \to \infty} \left[\frac{1}{x+1} \cdot \frac{x}{1} \right] = 1.$$

   By Theorem 13.9, $\int_1^\infty \frac{1}{x+1} dx$ diverges.

2. More generally, we can use the Limit Comparison Test (Theorem 13.9) to show that $\int_1^\infty \frac{1}{(x+1)^p} dx$ converges if $p > 1$ and diverges if $p < 1$. This follows immediately from part 3 of Example 13.2 together with the fact that

$$\lim_{x \to \infty} \frac{\frac{1}{(x+1)^p}}{\frac{1}{x^p}} = \lim_{x \to \infty} \left[\frac{1}{(x+1)^p} \cdot \frac{x^p}{1} \right] = 1.$$

3. Let's use the Limit Comparison Test (Theorem 13.9) to show that $\int_1^\infty \ln x \, dx$ diverges. We will once again use the fact that $\int_1^\infty \frac{1}{x} dx$ diverges. We have

$$\lim_{x \to \infty} \frac{\ln x}{\frac{1}{x}} = \lim_{x \to \infty} [x \ln x] = \infty.$$

   It follows from the third part of the Limit Comparison Test (Theorem 13.9) that $\int_1^\infty \ln x \, dx = \infty$, and so, it diverges.

   In Problem 5 below, you will be asked to use integration by parts to show that $\int_1^\infty \ln x \, dx$ diverges.

Absolute and Conditional Convergence

Let $a \in \mathbb{R}$, let $b \in \mathbb{R} \cup \{\infty\}$, and let $f: [a, b) \to \mathbb{R}$ be locally integrable on $[a, b)$. If the improper integral $\int_a^b |f(x)| dx$ converges, then we say that $\int_a^b f(x) dx$ **converges absolutely** (or is **absolutely convergent**) on $[a, b)$. We may also say that f is **absolutely integrable** on $[a, b)$.

Does every absolutely convergent improper integral actually converge? Although it is not completely obvious, the answer to this question is yes. You will be asked to prove this in Problem 8 below.

Does every convergent improper integral converge absolutely? The answer to this question is no. We will see examples of improper integrals that are convergent, but not absolutely convergent below

If $\int_a^b f(x) dx$ converges, but does not converge absolutely, we say that $\int_a^b f(x) dx$ **converges conditionally** (or is **conditionally convergent**) on $[a, b)$. We may also say that f is **conditionally integrable** on $[a, b)$.

Example 13.11:

1. The improper integral $\int_1^\infty \frac{\sin x}{x^2} dx$ converges absolutely. To see this, note that for $x \geq 1$, $\left|\frac{\sin x}{x^2}\right| = \frac{|\sin x|}{x^2} \leq \frac{1}{x^2}$ and $\int_1^\infty \frac{1}{x^2} dx$ converges by part 3 of Example 13.2. The result then follows from the Comparison Test (Theorem 13.7).

2. In Problem 6 below, you will be asked to generalize the result in part 1 above by proving that for any real number $p > 1$, $\int_1^\infty \frac{\sin x}{x^p} dx$ converges absolutely.

3. In Problem 11 below, you will be asked to prove that for $0 < p \leq 1$, $\int_1^\infty \frac{\sin x}{x^p} dx$ converges conditionally.

Theorem 13.12 (Dirichlet's Test): Let $a \in \mathbb{R}$, let $b \in \mathbb{R} \cup \{\infty\}$, and let $f, g: [a, b) \to \mathbb{R}$ be functions such that f is continuous on $[a, b)$, g is differentiable on $[a, b)$, g' is absolutely integrable on $[a, b)$, and $\lim_{x \to b^-} g(x) = 0$. Suppose also that the function $F: [a, b) \to \mathbb{R}$ defined by $F(x) = \int_a^x f(t) dt$ is bounded on $[a, b)$. Then $\int_a^b f(x) g(x) dx$ converges.

You will be asked to prove Theorem 13.12 in Problem 13 below.

Example 13.13:

1. Let's use Dirichlet's Test (Theorem 13.12) to show that $\int_1^\infty x \sin x \, dx$ diverges. To see this, assume towards contradiction that $\int_1^\infty x \sin x \, dx$ converges. Define $f, g, F: [1, \infty) \to \mathbb{R}$ by $f(x) = x \sin x$, $g(x) = \frac{1}{x}$, and $F(x) = \int_a^x t \sin t \, dt$. Then f is continuous on $[1, \infty)$ and g is differentiable on $[1, \infty)$ with $g'(x) = -\frac{1}{x^2}$. By part 3 of Example 13.2, g' is absolutely integrable on $[1, \infty)$. We also have $\lim_{x \to \infty} g(x) = 0$. By Problem 10 below, F is bounded on $[1, \infty)$. So, by Dirichlet's Test, $\int_1^\infty (x \sin x) \left(\frac{1}{x}\right) dx = \int_1^\infty \sin x \, dx$ converges. However, we have

$$\int_1^\infty \sin x \, dx = \lim_{c \to \infty} \int_1^c \sin x \, dx = \lim_{c \to \infty} [-\cos x]_1^c = \lim_{c \to \infty} (\cos 1 - \cos c) = \cos 1 - \lim_{c \to \infty} \cos c.$$

For every $n \in \mathbb{Z}$, we have $\cos 2n\pi = 1$ and $\cos \frac{(2n+1)\pi}{2} = 0$. It follows that $\lim_{c \to \infty} \cos c$ does not exist, and therefore, $\int_1^\infty \sin x \, dx$ diverges, a contradiction. So, $\int_1^\infty x \sin x \, dx$ diverges.

2. A similar argument can be used to show that for any $p > 0$, $\int_1^\infty x^p \sin x \, dx$ diverges. In this case, assuming that $\int_1^\infty x^p \sin x \, dx$ converges, we let $f(x) = x^p \sin x$, $g(x) = \frac{1}{x^p}$, and $F(x) = \int_a^x t^p \sin t \, dt$. Then as in part 1 above, $f, g,$ and F satisfy the hypotheses of Dirichlet's Test (Check this!), and so, $\int_1^\infty (x^p \sin x) \left(\frac{1}{x^p}\right) dx = \int_1^\infty \sin x \, dx$ converges, leaving us with the same contradiction as in part 1 above.

Problem Set 13

Full solutions to these problems are available for free download here:
www.SATPrepGet800.com/RABQXZ

LEVEL 1

1. Prove that $\int_0^\infty \cos x \, dx$ diverges, but not to ∞ or $-\infty$.

2. Let $a \in \mathbb{R} \cup \{-\infty\}$, let $b \in \mathbb{R} \cup \{\infty\}$, let $c \in (a, b)$, let $f: (a, b) \to \mathbb{R}$ be locally integrable on (a, b), and suppose that the integrals $\int_a^c f(x) \, dx$ and $\int_c^b f(x) \, dx$ are both finite real numbers. Prove that for all $d \in (a, b)$,

$$\int_a^c f(x) \, dx + \int_c^b f(x) \, dx = \int_a^d f(x) \, dx + \int_d^b f(x) \, dx.$$

LEVEL 2

3. Let $c, d \in \mathbb{R}$, let I be an interval with endpoints a and b ($a < b$, a can be $-\infty$, and b can be ∞), and let $f, g: I \to \mathbb{R}$ be locally integrable on I such that the improper integrals of f and g over I converge. Prove that the improper integral of $cf + dg$ over I converges and

$$\int_a^b (cf + dg)(x) \, dx = c \int_a^b f(x) \, dx + d \int_a^b g(x) \, dx.$$

4. For each $p \in \mathbb{R}$, evaluate the following improper integral:

$$\int_0^1 \frac{1}{x^p} \, dx$$

5. Use integration by parts to show that the following improper integral diverges:

$$\int_0^\infty \ln x \, dx$$

6. Prove that for $p > 1$, $\int_1^\infty \frac{\sin x}{x^p} \, dx$ converges absolutely.

LEVEL 3

7. The function $\Gamma: (0, \infty) \to \mathbb{R}$ defined as follows is known as the **Gamma function**:

$$\Gamma(x) = \int_0^\infty t^{x-1} e^{-t} \, dt$$

Prove that for all $x \in (0, \infty)$, $\Gamma(x)$ converges.

8. Let $a \in \mathbb{R}$, let $b \in \mathbb{R} \cup \{\infty\}$, and let $f: [a, b) \to \mathbb{R}$ be locally integrable on $[a, b)$. If $\int_a^b f(x)\, dx$ converges absolutely, then $\int_a^b f(x)\, dx$ converges.

9. Let $f: \left(-\frac{\pi}{2}, \frac{\pi}{2}\right) \to \mathbb{R}$ be defined by $f(x) = \tan x$.

   (i) Prove that f is a continuous bijection from $\left(-\frac{\pi}{2}, \frac{\pi}{2}\right)$ onto $\mathbb{R}$.

   (ii) Let $g: \mathbb{R} \to \left(-\frac{\pi}{2}, \frac{\pi}{2}\right)$ be the inverse of f. We call the function g the arctangent function and we write $g(x) = \arctan x$. Prove that $g'(x) = \frac{1}{x^2+1}$ for all $x \in \mathbb{R}$.

   (iii) Determine if the following improper integral converges or diverges. If it converges, evaluate it.
   $$\int_{-\infty}^{\infty} \frac{1}{x^2 + 1}\, dx$$

LEVEL 4

10. Let $a \in \mathbb{R}$, let $b \in \mathbb{R} \cup \{\infty\}$, let $f: [a, b) \to \mathbb{R}$ be a bounded locally integrable function on $[a, b)$ such that for all $x \in [a, b)$, $f(x) \geq 0$. Prove that
$$\int_a^b f(x)\, dx = \sup\left\{\int_a^x f(t)\, dt \mid x \in [a, b)\right\}.$$

11. Prove that for $0 < p \leq 1$, $\int_1^{\infty} \frac{\sin x}{x^p}\, dx$ converges conditionally.

LEVEL 5

12. Let $n \in \mathbb{Z}^+$ and let $f, g: [a, b] \to \mathbb{R}$ be continuous functions such that $f^{(k)}$ exists for each $k = 1, 2, \ldots, n$, where $f^{(k)}$ is the kth derivative of f (and $f^{(0)} = f$). For each $k = 1, 2, \ldots, n$, let $g^{(-k)}$ be a kth antiderivative of g (in other words, $g^{(-(k+1))}$ is an antiderivative of $g^{(-k)}$). Prove that the following formula holds, where $f^{(0)} = f$:
$$\int_a^b f(x)g(x)\, dx = \left[\sum_{k=0}^{n-1} (-1)^k f^{(k)}(x) g^{(-(k+1))}(x)\right]_a^b + (-1)^n \int_a^b f^{(n)}(x) g^{(-n)}(x)\, dx$$

This formula is called **tabular integration by parts**.

13. Let $a \in \mathbb{R}$, let $b \in \mathbb{R} \cup \{\infty\}$, and let $f, g: [a, b) \to \mathbb{R}$ be functions such that f is continuous on $[a, b)$, g is differentiable on $[a, b)$, g' is absolutely integrable on $[a, b)$, and $\lim_{x \to b^-} g(x) = 0$. Suppose also that the function $F: [a, b) \to \mathbb{R}$ defined by $F(x) = \int_a^x f(t)\, dt$ is bounded on $[a, b)$. Prove that $\int_a^b f(x) g(x)\, dx$ converges. This result is known as **Dirichlet's Test** (Theorem 13.12).

14. Let $a \in \mathbb{R}$, let $b \in \mathbb{R} \cup \{\infty\}$, and let $f, g: [a, b) \to \mathbb{R}$ be locally integrable on $[a, b)$. Assume that there is a subinterval $[c, b) \subseteq [a, b)$ such that for all $x \in [c, b)$, we have $f(x) \geq 0$ and $g(x) > 0$. Also assume that there is $L \in \mathbb{R} \cup (\infty\}$ such that
$$\lim_{x \to b^-} \frac{f(x)}{g(x)} = L.$$

Prove each of the following:

$$\text{if } L \in \mathbb{R}^+, \text{then } \int_a^b f(x)\, dx \downarrow \text{ if and only if } \int_a^b g(x)\, dx \downarrow.$$

$$\text{if } L = 0 \text{ and } \int_a^b g(x)\, dx \downarrow, \text{then } \int_a^b f(x)\, dx \downarrow.$$

$$\text{if } L = \infty \text{ and } \int_a^b g(x)\, dx = \infty, \text{then } \int_a^b f(x)\, dx = \infty.$$

This result is known as the **Limit Comparison Test** (Theorem 13.9).

CHALLENGE PROBLEMS

15. Determine if the following statement is true or false. If it is true, prove it. If it is false, provide a counterexample.

    If $f: [0, \infty) \to \mathbb{R}$ is a continuous function such that $f(x) \geq 0$ for all $x \in [0, \infty)$, $\lim_{x \to \infty} f(x)$ does not exist, and f is unbounded on $[0, \infty)$, then $\int_0^\infty f(x)\, dx$ diverges.

16. Let $L \in \mathbb{R} \cup \{-\infty, \infty\}$ and let $f: [0, \infty)$ be a function such that $\lim_{x \to \infty} f(x) = L$, $\lim_{x \to \infty} f'(x) = 0$, and f'' is absolutely integrable on $[0, \infty)$. Determine all values of L for which $\int_0^\infty f(x)\, dx$ converges.

LESSON 14
SEQUENCES

Sequences of Real Numbers

Recall from Lesson 3 that a **real-valued sequence** is a function from $\mathbb{N}$ to $\mathbb{R}$. From now on, we will always assume that a sequence is real-valued, unless stated otherwise. If $f: \mathbb{N} \to \mathbb{R}$ is a sequence, we will represent the sequence as (s_n), where $s_n = f(n)$ is called the **nth term** of the sequence.

In Lesson 8 we learned about convergent sequences and Cauchy sequences. Let's review those definitions now:

A sequence (s_n) **converges** to $s \in \mathbb{R}$, written $\lim_{n \to \infty} s_n = s$ (or $s_n \to s$) if for every $\epsilon > 0$, there is a natural number K such that $n > K$ implies $|s_n - s| < \epsilon$. Symbolically, we can abbreviate this as
$$\forall \epsilon > 0 \; \exists K \in \mathbb{N}(n > K \to |s_n - s| < \epsilon).$$

If (s_n) does not converge to any real number, then we say that (s_n) **diverges** (or (s_n) is **divergent**).

Notes: (1) The idea is that if (s_n) converges to s, then the real numbers s_n get "closer and closer" to the real number s, as n gets larger and larger. If (s_n) converges to a real number, then we say that the sequence (s_n) is **convergent**.

(2) We may write $(s_n) \downarrow$ or $(s_n) \uparrow$ to indicate that (s_n) converges or diverges, respectively.

A sequence (s_n) is called a **Cauchy sequence** if for every $\epsilon > 0$, there is a natural number K such that $m \geq n > K$ implies $|s_m - s_n| < \epsilon$. Symbolically, we can abbreviate this as
$$\forall \epsilon > 0 \; \exists K \in \mathbb{N}(m \geq n > K \to |s_m - s_n| < \epsilon).$$

Note: The idea is that if (s_n) is a Cauchy sequence, then the real numbers s_n and s_m get "closer and closer" to *each other*, as n and m get larger and larger.

In Lesson 8 (Corollary 8.29), we saw that a sequence (s_n) converges to some $s \in \mathbb{R}$ if and only if it is a Cauchy sequence.

Example 14.1:

1. The sequence $(s_n) = (\sqrt{n+1} - \sqrt{n}) = (1, \sqrt{2} - 1, \sqrt{3} - \sqrt{2}, 2 - \sqrt{3}, \dots)$ converges to 0. To see this, let $\epsilon > 0$ and let K be a positive integer such that $K > \frac{1}{\epsilon^2}$ (we can find such a K by the Archimedean Property of $\mathbb{R}$). Then $\epsilon^2 > \frac{1}{K}$, and so, $\epsilon > \frac{1}{\sqrt{K}}$. Let $n > K$. We have

$$|s_n - 0| = |\sqrt{n+1} - \sqrt{n}| = \left|\frac{\sqrt{n+1} - \sqrt{n}}{1} \cdot \frac{\sqrt{n+1} + \sqrt{n}}{\sqrt{n+1} + \sqrt{n}}\right| = \left|\frac{n+1-n}{\sqrt{n+1} + \sqrt{n}}\right|$$

$$= \frac{1}{\sqrt{n+1} + \sqrt{n}} < \frac{1}{\sqrt{n} + \sqrt{n}} = \frac{1}{2\sqrt{n}} < \frac{1}{\sqrt{n}} < \frac{1}{\sqrt{K}} < \epsilon.$$

2. Let $r \in (0, 1)$. The sequence $(t_n) = (r^n) = (1, r, r^2, r^3, \ldots)$ converges to 0. To see this, let $\epsilon > 0$, let K be a positive integer such that $K > \frac{\ln \epsilon}{\ln r}$ (we can find such a K by the Archimedean Property of $\mathbb{R}$), and let $n > K$. So, $n > \frac{\ln \epsilon}{\ln r}$. Since $0 < r < 1$, $\ln r < 0$. Therefore, $n \ln r < \ln \epsilon$. By Theorem 12.4, $\ln r^n < \ln \epsilon$. Since $f(x) = e^x$ is strictly increasing and $e^{\ln x} = x$ for all $x \in (0, \infty)$ (see part 2 of Example 12.9), $r^n < \epsilon$. So, we have $|t_n - 0| = r^n < \epsilon$, as desired.

We can extend this result to all $r \in (-1, 1]$. If $r = 0$, then $(t_n) = (0^n) = (0) = (0, 0, 0, \ldots)$ is the constant zero sequence. This converges to 0 by part 2 of Example 8.23. Similarly, if $r = 1$, then $(t_n) = (1^n) = (1) = (1, 1, 1, \ldots)$ is the constant sequence consisting only of 1, and so, the sequence converges to 1.

If $r \in (0, 1)$, then $(t_n) = ((-r)^n) = (1, -r, r^2, -r^3, \ldots)$ also converges to 0 because $|t_n - 0| = |(-r)^n| = |(-1)^n r^n| = |(-1)^n||r^n| = r^n$. So, we can simply replace r by $|r|$ in the argument given above (in particular, we will choose $K > \frac{\ln \epsilon}{\ln |r|}$).

What if $r \in (-\infty, -1] \cup (1, \infty)$. In this case, the sequence $(t_n) = (r^n)$ diverges. First suppose that $r \in (-\infty, -1) \cup (1, \infty)$. So, $r < -1$ or $r > 1$. It follows that $|r| > 1$ and $|r - 1| > 0$. We will show that (r^n) is not a Cauchy sequence. To this end, let $\epsilon = |r - 1|$ and let $K \in \mathbb{N}$ be arbitrary. Then $|r^{K+2} - r^{K+1}| = |r^{K+1}(r - 1)| = |r|^{K+1}|r - 1| > 1^{K+1}|r - 1| = \epsilon$.

Therefore, (r^n) is not a Cauchy sequence, and so, by Theorem 8.25, (r^n) is divergent (we used the contrapositive of Theorem 8.25 here).

Finally, if $r = -1$, then $(t_n) = ((-1)^n) = (1, -1, 1, -1, \ldots)$. Let $\epsilon = 1$ and let $K \in \mathbb{N}$ be arbitrary. Then $|(-1)^{K+2} - (-1)^{K+1}| = 2 > 1 = \epsilon$. Therefore, $((-1)^n)$ is not a Cauchy sequence, and so, once again by Theorem 8.25, (r^n) is divergent.

3. The sequence $(x_n) = \left(\frac{5n+2}{n+3}\right) = \left(\frac{2}{3}, \frac{7}{4}, \frac{12}{5}, \frac{17}{6}, \ldots\right)$ converges to 5. To see this, let $\epsilon > 0$, let K be a positive integer such that $K > \frac{13}{\epsilon}$. (we can find such a K by the Archimedean Property of $\mathbb{R}$), and let $n > K$. Since $K > \frac{13}{\epsilon}$, we have $\frac{1}{K} < \frac{\epsilon}{13}$. Therefore,

$$|x_n - 5| = \left|\frac{5n+2}{n+3} - 5\right| = \left|\frac{5n+2}{n+3} - \frac{5n+15}{n+3}\right| = \frac{13}{n+3} < \frac{13}{n} < 13 \cdot \frac{1}{K} < 13 \cdot \frac{\epsilon}{13} = \epsilon.$$

The following theorem provides a method for verifying that a sequence converges to a specific value without having to go through the formality of an $\epsilon - K$ argument.

Theorem 14.2: Let $(s_n), (t_n)$ be sequences, let $s, M \in \mathbb{R}$ with $M > 0$, let $A \in \mathbb{N}$, suppose that $\lim_{n \to \infty} t_n = 0$, and suppose that $n > A$ implies $|s_n - s| \leq M|t_n|$. Then $\lim_{n \to \infty} s_n = s$.

Proof: Let $\epsilon > 0$. Since $\lim_{n \to \infty} t_n = 0$, there is $B \in \mathbb{N}$ such that $n > B$ implies $|t_n| = |t_n - 0| < \frac{\epsilon}{M}$. Let $K = \max\{A, B\}$ and let $n > K$. Since $n > A$, $|s_n - s| \leq M|t_n|$. Since $n > B$, $|t_n| < \frac{\epsilon}{M}$. Therefore, we have $|s_n - s| < M \cdot \frac{\epsilon}{M} = \epsilon$. So, $\lim_{n \to \infty} s_n = s$. □

Example 14.3:

1. Consider the sequence $(s_n) = \left(\frac{2n-5}{3n^2+7n}\right)$. Since $2n - 5 \leq 2n$ and $3n^2 + 7n \geq 3n^2 + 3n$, we have

$$|s_n - 0| = \left|\frac{2n-5}{3n^2+7n}\right| \leq \left|\frac{2n}{3n^2+3n}\right| = \left|\frac{2n}{3n(n+1)}\right| = \left|\frac{2}{3(n+1)}\right| = \frac{2}{3}\left|\frac{1}{n+1}\right|$$

   Let $(t_n) = \left(\frac{1}{n+1}\right)$, $s = 0$, $M = \frac{2}{3}$, and $A = 0$. By part 1 of Example 8.23, $\lim_{n \to \infty} t_n = 0$. So, by Theorem 14.2, we have $\lim_{n \to \infty} \frac{2n-5}{3n^2+7n} = 0$.

2. Recall from part 2 of Example 3.11 that a **geometric sequence** has the form (ar^n), where a and r are real numbers. We saw in part 2 of Example 14.1 above that if $r \in (-1, 1]$, then the sequence $(t_n) = (r^n)$ converges to 0. Since $|ar^n - 0| = |a||r^n|$, it follows from Theorem 14.2 that $\lim_{n \to \infty} ar^n = 0$.

   If $r \in (-\infty, -1] \cup (1, \infty)$ and $a \neq 0$, then (ar^n) diverges. To see this, assume toward contradiction that $ar^n \to L$. Then $ar^n - L \to 0$. Since $\left|r^n - \frac{L}{a}\right| = \left|\frac{1}{a}ar^n - \frac{1}{a}L\right| = \left|\frac{1}{a}\right||ar^n - L|$, it follows from Theorem 14.2 that $r^n \to \frac{L}{a}$. However, by part 2 of Example 14.1, (r^n) diverges. This contradiction shows that (ra^n) diverges as well.

Monotone Sequences

A sequence is **increasing** if for all $n \in \mathbb{N}$, $x_n \leq x_{n+1}$. A sequence is **decreasing** if for all $n \in \mathbb{N}$, $x_n \geq x_{n+1}$. We will say that a sequence is **monotone** (or **monotonic**) if it is increasing or decreasing.

Notes: (1) We can visualize an increasing sequence as follows:

$$x_0 \leq x_1 \leq x_2 \leq \cdots \leq x_n \leq x_{n+1} \leq \cdots$$

(2) We can visualize a decreasing sequence as follows:

$$x_0 \geq x_1 \geq x_2 \geq \cdots \geq x_n \geq x_{n+1} \geq \cdots$$

Example 14.4:

1. The sequence $(s_n) = (n) = (0, 1, 2, 3, \ldots)$ is increasing.

2. The sequence $(t_n) = \left(\frac{1}{n+1}\right) = \left(1, \frac{1}{2}, \frac{1}{3}, \frac{1}{4}, \ldots\right)$ is decreasing.

3. Let $a > 0$. The sequence $(ar^n) = (a, ar, ar^2, \ldots)$ is increasing if $r \geq 1$ and decreasing if $0 < r \leq 1$. Note that if $r = 1$, then we get the constant sequence $(a) = (a, a, a, \ldots)$, which is both increasing and decreasing. If $r < 0$, then (ar^n) is not monotonic. For example, if $a = 1$ and $r = -1$, then $(ar^n) = ((-1)^n) = (1, -1, 1, -1, \ldots)$ is neither increasing nor decreasing.

Recall from Lesson 8 that a sequence (s_n) is **bounded** if the set $\{s_n\}$ is bounded. In other words, (s_n) is bounded if there is $M \in \mathbb{R}^+$ such that for all $n \in \mathbb{N}$, $|s_n| \leq M$ (see the discussion before the statement of Theorem 7.28).

Theorem 14.5 (Monotone Convergence Theorem): A monotone sequence converges if and only if it is bounded. Furthermore, an increasing bounded sequence converges to its supremum and a decreasing bounded sequence converges to its infimum.

Proof: Let (s_n) be a monotone sequence. If (s_n) converges, then by Problem 2 below, it is bounded.

Next, assume that (s_n) is increasing and bounded. Then (s_n) is bounded above. So, by the Completeness Property of $\mathbb{R}$, $s = \sup\{s_n \mid n \in \mathbb{N}\}$ exists. Let $\epsilon > 0$. Then $s - \epsilon$ is not an upper bound of $\{s_n \mid n \in \mathbb{N}\}$. So, there is $K \in \mathbb{N}$ with $s_K > s - \epsilon$. Since (s_n) is increasing, if $n > K$, then $s_n \geq s_K$, and so, $s_n > s - \epsilon$. Therefore, for $n > K$, we have

$$s - \epsilon < s_n \leq s < s + \epsilon.$$

Thus, $n > K \to |s_n - s| < \epsilon$. So, $\lim_{n \to \infty} s_n = s = \sup\{s_n \mid n \in \mathbb{N}\}$.

Finally, assume that (s_n) is decreasing and bounded. Then (s_n) is bounded below. So, by Problem 18 from Problem Set 6, $s = \inf\{s_n \mid n \in \mathbb{N}\}$ exists. Let $\epsilon > 0$. Then $s + \epsilon$ is not a lower bound of $\{s_n \mid n \in \mathbb{N}\}$. So, there is $K \in \mathbb{N}$ with $s_K < s + \epsilon$. Since (s_n) is decreasing, if $n > K$, then $s_n \leq s_K$, and so, $s_n < s + \epsilon$. Therefore, for $n > K$, we have

$$s - \epsilon < s \leq s_n < s + \epsilon.$$

Thus, $n > K \to |s_n - s| < \epsilon$. So, $\lim_{n \to \infty} s_n = s = \inf\{s_n \mid n \in \mathbb{N}\}$. □

Note: When proving that an increasing bounded sequence converges to its supremum, we actually used only the fact that the sequence is bounded above. Similarly, when proving that a decreasing bounded sequence converges to its infimum, we used only the fact that the sequence is bounded below.

Example 14.6:

1. Consider the sequence $(s_n) = \left(\frac{3}{5n+2}\right)$. First observe that for all $n \in \mathbb{N}$, $\frac{3}{5n+2} > 0$. It follows that (s_n) is bounded below. Furthermore, for $n \in \mathbb{N}$, we have $5(n+1) + 2 = 5n + 7 \geq 5n + 2$, and therefore, $s_{n+1} = \frac{3}{5(n+1)+2} \leq \frac{3}{5n+2} = s_n$. So, (s_n) is decreasing. By the Monotone Convergence Theorem (Theorem 14.5), $\lim_{n \to \infty} \frac{3}{5n+2} = \inf\left\{\frac{3}{5n+2} \mid n \in \mathbb{N}\right\}$. Let's show that $\inf\left\{\frac{3}{5n+2} \mid n \in \mathbb{N}\right\} = 0$. We have already observed that 0 is a lower bound of $\left\{\frac{3}{5n+2} \mid n \in \mathbb{N}\right\}$. Let $M > 0$. Then we have

$$\frac{3}{5n+2} < M \Leftrightarrow 3 < M(5n+2) = 5Mn + 2M \Leftrightarrow 3 - 2M < 5Mn \Leftrightarrow n > \frac{3 - 2M}{5M}.$$

By the Archimedean Property of $\mathbb{R}$ (Theorem 6.16), we can find $K \in \mathbb{N}$ with $K > \frac{3-2M}{5M}$, and so, for this value of K, $s_K = \frac{3}{5K+2} < M$. Therefore, M is **not** a lower bound of $\left\{\frac{3}{5n+2} \mid n \in \mathbb{N}\right\}$. Since $M > 0$ was arbitrary, it follows that $\inf\left\{\frac{3}{5n+2} \mid n \in \mathbb{N}\right\} = 0$.

2. Define the sequence (t_n) recursively as follows: Let $t_0 = 1$ and for $n \in \mathbb{N}$, let $t_{n+1} = \sqrt{5t_n}$. If we assume that (t_n) converges to $L \in \mathbb{R}$, then $L = \lim_{n\to\infty} t_{n+1} = \lim_{n\to\infty} \sqrt{5t_n} = \sqrt{5 \lim_{n\to\infty} t_n} = \sqrt{5L}$ (use Problem 16 below). Therefore, $L^2 = 5L$, and so, $L(L-5) = L^2 - 5L = 0$. So, $L = 0$ or $L = 5$. Does (t_n) actually converge to one of these values? If we show that the sequence is increasing and bounded above, then the Monotone Convergence Theorem (Theorem 14.5) guarantees that it does. Furthermore, since $t_0 = 1 > 0$, we know that $L \neq 0$, and therefore, $L = 5$.

   We first show by the Principle of Mathematical Induction that (t_n) is bounded above by 5. The base case is $t_0 = 1 \leq 5$. Assuming that $k \in \mathbb{N}$ and $t_k \leq 5$, we have $t_{k+1} = \sqrt{5t_n} \leq \sqrt{5 \cdot 5} = 5$.

   Next, we use the Principle of Mathematical Induction again to show that (t_n) is increasing. The base case is $t_0 = 1 \leq \sqrt{5} = \sqrt{5 \cdot 1} = \sqrt{5t_0} = t_1$. Assuming that $k \in \mathbb{N}$ and $t_k \leq t_{k+1}$, we have $t_{k+1} = \sqrt{5t_k} \leq \sqrt{5t_{k+1}} = t_{k+2}$.

   Since (t_n) is bounded above and increasing, by the Monotone Convergence Theorem, (t_n) converges to some real number L. As we have seen above, the only possibility is that $L = 5$.

3. Define the sequence (x_n) recursively as follows: Let $x_0 = 3$ and for $n \in \mathbb{N}$, let $x_{n+1} = 2x_n + 1$. If we assume that (x_n) converges to $L \in \mathbb{R}$, then $L = \lim_{n\to\infty} x_{n+1} = \lim_{n\to\infty} (2x_n + 1) = 2L + 1$. So, $L = -1$. In other words, if (x_n) converges, then it must converge to -1. However, this sequence does not converge. To see this, we will use the Monotone Convergence Theorem.

   We first show by the Principle of Mathematical Induction that (x_n) is **not** bounded above. Specifically, we will show that for all $n \in \mathbb{N}^+$, $x_n > n$. The base case is $x_1 = 2 \cdot 3 + 1 = 7 \geq 1$. Assuming that $k \in \mathbb{N}^+$ and $x_k > k$, we have $x_{k+1} = 2x_k + 1 > 2k + 1 > k + 1$.

   Next, we use the Principle of Mathematical Induction again to show that (x_n) is increasing. The base case is $x_0 = 3 \leq 7 = 2 \cdot 3 + 1 = x_1$. Assuming that $k \in \mathbb{N}$ and $x_k \leq x_{k+1}$, we have $x_{k+1} = 2x_k + 1 \leq 2x_{k+1} + 1 = x_{k+2}$.

   Since (x_n) is a monotone sequence that is unbounded, by the Monotone Convergence Theorem, (x_n) diverges. The conditional statement "if (x_n) converges, then it converges to -1" is still true (vacuously), but since the sequence does **not** converge, the value -1 that we found is not useful.

Note: In part 2 of Example 14.6 above, we used the following result: If $a, b \geq 0$, then $a \leq b$ implies $\sqrt{a} \leq \sqrt{b}$. The follows from part (iii) of Problem 11 from Problem Set 6.

Limit Superior and Limit Inferior

Let (s_n) be a bounded sequence. For each $n \in \mathbb{N}$, let $x_n = \sup\{s_k \mid k \geq n\}$ and $y_n = \inf\{s_k \mid k \geq n\}$. We define the **limit superior** of (s_n), written $\limsup s_n$ or $\overline{\lim} \, s_n$, to be $\lim_{n\to\infty} x_n$, and we define the **limit inferior** of (s_n), written $\liminf s_n$ or $\underline{\lim} \, s_n$, to be $\lim_{n\to\infty} y_n$.

If (s_n) is not bounded above, then $\limsup s_n = \infty$. If (s_n) is not bounded below, then $\liminf s_n = -\infty$.

Example 14.7:

1. Consider the sequence $(s_n) = ((-1)^n) = (1, -1, 1, -1, \ldots)$. We have
$$\limsup s_n = \lim_{n \to \infty}(\sup\{s_k \mid k \geq n\}) = \lim_{n \to \infty} 1 = 1$$
$$\liminf s_n = \lim_{n \to \infty}(\inf\{s_k \mid k \geq n\}) = \lim_{n \to \infty} (-1) = -1$$
In this example, $\limsup s_n = \sup\{s_n \mid n \in \mathbb{N}\}$ and $\liminf s_n = \inf\{s_n \mid n \in \mathbb{N}\}$.

2. Consider the sequence (t_n), where $t_n = \begin{cases} 1 - \frac{1}{n+1} & \text{if } n \text{ is even.} \\ 2 + \frac{1}{n+1} & \text{if } n \text{ is odd.} \end{cases}$ We can visualize this sequence as follows:
$$(t_n), = \left(1 - \frac{1}{1}, 2 + \frac{1}{2}, 1 - \frac{1}{3}, 2 + \frac{1}{4}, \ldots\right).$$
For every $m, n \in \mathbb{N}$, we have $1 - \frac{1}{m+1} < 1 < 2 < 2 + \frac{1}{n+1}$. Therefore,
$$\limsup s_n = \lim_{n \to \infty}(\sup\{s_k \mid k \geq n\}) = \lim_{n \to \infty}(\sup\{s_{2k+1} \mid k \geq n\}) = \lim_{n \to \infty}\left(2 + \frac{1}{2n+2}\right) = 2$$
$$\liminf s_n = \lim_{n \to \infty}(\inf\{s_k \mid k \geq n\}) = \lim_{n \to \infty}(\inf\{s_{2k} \mid k \geq n\}) = \lim_{n \to \infty}\left(1 - \frac{1}{2n+1}\right) = 1$$
In this example, $\limsup s_n \neq \sup\{s_n \mid n \in \mathbb{N}\}$ and $\liminf s_n \neq \inf\{s_n \mid n \in \mathbb{N}\}$. Indeed, we have $\sup\{s_n \mid n \in \mathbb{N}\} = 2 + \frac{1}{2} = 2.5$ and $\inf\{s_n \mid n \in \mathbb{N}\} = 1 - \frac{1}{1} = 0$.

Theorem 14.8: Let (s_n) be a bounded sequence. For each $n \in \mathbb{N}$, let $x_n = \sup\{s_k \mid k \geq n\}$ and $y_n = \inf\{s_k \mid k \geq n\}$. Then (x_n) is a bounded decreasing sequence and (y_n) is a bounded increasing sequence.

Proof: Let $n \in \mathbb{N}$. Since $\{s_k \mid k \geq n+1\} \subseteq \{s_k \mid k \geq n\}$, by Problem 3 from Problem Set 11, we have $x_{n+1} = \sup\{s_k \mid k \geq n+1\} \leq \sup\{s_k \mid k \geq n\} \leq x_n$. Since $n \in \mathbb{N}$ was arbitrary, (x_n) is decreasing. Similarly, by the same problem, we have $y_n = \inf\{s_k \mid k \geq n\} \leq \inf\{s_k \mid k \geq n+1\} = y_{n+1}$. Since $n \in \mathbb{N}$ was arbitrary, (y_n) is increasing.

Now, since (s_n) is bounded, by the Completeness Property of $\mathbb{R}$ and Problem 18 from Problem Set 6, $a = \inf\{s_k \mid k \in \mathbb{N}\}$ and $b = \sup\{s_k \mid k \in \mathbb{N}\}$ exist. For each $n \in \mathbb{N}$, $\{s_k \mid k \geq n\} \subseteq \{s_k \mid k \in \mathbb{N}\}$. It follows that for each $n \in \mathbb{N}$, a is a lower bound of $\{s_k \mid k \geq n\}$ and b is an upper bound of $\{s_k \mid k \geq n\}$. Therefore, for each $n \in \mathbb{N}$, $a \leq y_n \leq x_n \leq b$. So, (x_n) and (y_n) are bounded sequences. $\square$

Theorem 14.9: Let (s_n) be a bounded sequence. Then $\limsup s_n$ and $\liminf s_n$ are finite real numbers. Furthermore, if we let $x_n = \sup\{s_k \mid k \geq n\}$ and $y_n = \inf\{s_k \mid k \geq n\}$, then we have $\limsup s_n = \inf\{x_n \mid n \in \mathbb{N}\}$ and $\liminf s_n = \sup\{y_n \mid n \in \mathbb{N}\}$.

Proof: By Theorem 14.8, $x_n = \sup\{s_k \mid k \geq n\}$ is a bounded decreasing sequence and $y_n = \inf\{s_k \mid k \geq n\}$ is a bounded increasing sequence. Therefore, by the Monotone Convergence Theorem, $\limsup s_n = \lim_{n \to \infty} x_n = \inf\{x_n \mid n \in \mathbb{N}\}$ and $\liminf s_n = \lim_{n \to \infty} y_n = \sup\{y_n \mid n \in \mathbb{N}\}$. $\square$

Problem Set 14

Full solutions to these problems are available for free download here:
www.SATPrepGet800.com/RABQXZ

LEVEL 1

1. Let (s_n) be a convergent sequence. Prove that $\lim_{n\to\infty} s_n$ is unique.

2. Prove that a convergent sequence is bounded.

LEVEL 2

3. Define the sequence (s_n) recursively as follows: Let $s_0 = 3$ and for $n \in \mathbb{N}$, let $s_{n+1} = \frac{1}{3}(2s_n + 6)$. Prove that (s_n) converges and find the limit.

4. Let (s_n) and (t_n) be convergent sequences. Prove that $(s_n + t_n)$ converges and
$$\lim_{n\to\infty}(s_n + t_n) = \lim_{n\to\infty} s_n + \lim_{n\to\infty} t_n.$$

5. Let (s_n) be a sequence. We say that (s_n) diverges to ∞, written $\lim_{n\to\infty} s_n = \infty$, if for any $M \in \mathbb{R}$, there is $K \in \mathbb{N}$ such that $n > K$ implies $s_n > M$. Similarly, we say that (s_n) diverges to $-\infty$, written $\lim_{n\to\infty} s_n = -\infty$, if for any $M \in \mathbb{R}$, there is $K \in \mathbb{N}$ such that $n > K$ implies $s_n < M$. Prove that if (s_n) is an unbounded increasing sequence, then $\lim_{n\to\infty} s_n = \infty$ and if (s_n) is an unbounded decreasing sequence, then $\lim_{n\to\infty} s_n = -\infty$.

LEVEL 3

6. Let (s_n) be a sequence. For $k \in \mathbb{N}$, the sequence $(s_n)_{n=k}^{\infty} = (s_k, s_{k+1}, s_{k+2}, \ldots)$ is called the **k-tail** of the sequence (s_n). For example, the 0-tail of (s_n) is $(s_n)_{n=0}^{\infty} = (s_0, s_1, s_2, \ldots)$. This is just the original sequence. As another example, the 1-tail of (s_n) is $(s_n)_{n=1}^{\infty} = (s_1, s_2, s_3, \ldots)$. Let $k \in \mathbb{N}$. Prove that (s_n) converges if and only if the k-tail of (s_n) converges.

7. Let (s_n) and (t_n) be convergent sequences. Prove that $(s_n t_n)$ converges and
$$\lim_{n\to\infty} s_n t_n = \left(\lim_{n\to\infty} s_n\right)\left(\lim_{n\to\infty} t_n\right).$$

LEVEL 4

8. Prove that a sequence converges if and only if every one of its subsequences converges.

9. Let $(x_n), (y_n), (z_n)$ be sequences such that for all $n \in \mathbb{N}$, $x_n \leq y_n \leq z_n$. Prove that if (x_n) and (z_n) both converge to the same limit L, then (y_n) also converges to L. This result is known as the **Squeeze Theorem**.

10. Let $A \subseteq \mathbb{R}$ with $[k, \infty) \subseteq A$ for some $k \in \mathbb{N}$. Let $L \in \mathbb{R} \cup \{-\infty, \infty\}$ and let $f: A \to \mathbb{R}$ be a function such that $\lim_{x \to \infty} f(x) = L$. For each $n \in \mathbb{N}$ with $n \geq k$, let $s_n = f(n)$. Prove that $\lim_{n \to \infty} s_n = L$.

11. Prove that $\lim_{n \to \infty} n^{\frac{1}{n}} = 1$.

LEVEL 5

12. Let (s_n) and (t_n) be convergent sequences such that $\lim_{n \to \infty} t_n \neq 0$ and suppose that for all $n \in \mathbb{N}$, $t_n \neq 0$. Prove that $\left(\frac{s_n}{t_n}\right)$ converges and
$$\lim_{n \to \infty} \frac{s_n}{t_n} = \frac{\lim_{n \to \infty} s_n}{\lim_{n \to \infty} t_n}.$$

13. Let (s_n) be a bounded sequence, let $a = \liminf s_n$ and let $b = \limsup s_n$. Prove that there is a subsequence of (s_n) converging to a and another subsequence of (s_n) converging to b. Note that this provides a proof of the Bolzano-Weierstrauss Theorem (Theorem 8.27).

14. Let (s_n) be a bounded sequence. Prove that (s_n) converges if and only if $\limsup s_n = \liminf s_n$.

15. Determine with proof if the following statement is true or false: There is a sequence (x_n) such that for all $r \in \mathbb{R}$, (x_n) has a subsequence converging to r.

CHALLENGE PROBLEMS

16. Let (s_n) be a convergent sequence such that for all $n \in \mathbb{N}$, $s_n \geq 0$. Prove that for each $k \in \mathbb{N}^+$, $\left(\sqrt[k]{s_n}\right)$ converges and $\lim_{n \to \infty} \sqrt[k]{s_n} = \sqrt[k]{\lim_{n \to \infty} s_n}$.

17. Let $(s_n) = \left(\left(1 + \frac{1}{n}\right)^n\right)$. Use the Monotone Convergence Theorem to prove that (s_n) converges. What is the limit?

18. Let (s_n) be a sequence such that for all $n \in \mathbb{N}$, $s_n \neq 0$. Also, assume that $\lim_{n \to \infty} \left|\frac{s_{n+1}}{s_n}\right| = L$ for some $L \in \mathbb{R}$. Prove that if $L < 1$, then (s_n) converges to 0 and if $L > 1$, then (s_n) diverges. This result is known as the **Ratio Test for Sequences**.

19. Let (s_n) and (t_n) be bounded sequences. Suppose that there is $L \in \mathbb{R}^+$ such that $\lim_{n \to \infty} s_n = L$. Prove that $\limsup(s_n t_n) = L \limsup t_n$ and $\liminf(s_n t_n) = L \liminf t_n$.

LESSON 15
SERIES

Series of Real Numbers

Let (s_n) be a real-valued sequence. The **kth partial sum** of the sequence is

$$\sum_{n=0}^{k} s_n = s_0 + s_1 + \cdots + s_k.$$

So, the 0th partial sum is $x_0 = s_0$, the 1st partial sum is $x_1 = s_0 + s_1$, the 2nd partial sum is $x_2 = s_0 + s_1 + s_2$, and so on.

Observe that the partial sums themselves form a sequence (x_n). This sequence is called a **series**.

The real numbers $s_0, s_1, \ldots, s_n, \ldots$ are called the **terms** of the series (whereas the real numbers $x_0, x_1, \ldots, x_n, \ldots$ are the **partial sums** of the series). So, s_n is the **nth term** of the series (whereas $x_n = s_1 + s_2 + \cdots s_n$ is the **nth partial sum** of the series).

If $\lim_{n \to \infty} x_n$ is a finite real number L, then we say that the series **converges** to L or that the **sum** of the series is L (we may also say that the series is **convergent**). Otherwise, we say that the series **diverges** (or is **divergent**).

The following notation is often used to represent both the series (x_n) and the sum of the series $\lim_{n \to \infty} x_n$:

$$\sum_{n=0}^{\infty} s_n = s_0 + s_1 + \cdots$$

Notes: (1) In the expression s_n, the integer n is called an **index** (the plural of index is **indices**). For example s_0 has an index of 0 and s_1 has an index of 1. The series above starts with the index 0. In Lesson 11, when writing Riemann sums, we always started with an index of 1.

(2) A series can always be **reindexed** to begin with any index that we want. If we want to increase the starting index by j, then we simply replace n by $n - j$ inside the sum. This reindexed sum is identical to the original sum. Indeed, we have

$$\sum_{n=j}^{\infty} s_{n-j} = s_{j-j} + s_{(j+1)-j} + \cdots = s_0 + s_1 + \cdots = \sum_{n=0}^{\infty} s_n.$$

As a specific example, let's consider the following sum

$$\sum_{n=0}^{\infty} \frac{1}{(n+1)^2} = 1 + \frac{1}{2^2} + \frac{1}{3^2} + \cdots = 1 + \frac{1}{4} + \frac{1}{9} + \cdots$$

This series looks a bit nicer if we reindex it so that it begins with an index of 1. To do this, we simply replace n by $n-1$ to get

$$\sum_{n=1}^{\infty} \frac{1}{n^2} = 1 + \frac{1}{2^2} + \frac{1}{3^2} + \cdots = 1 + \frac{1}{4} + \frac{1}{9} + \cdots$$

Observe that although this series is more pleasant to look at, it is identical to the original series.

(3) We may sometimes use the notation $\sum s_n$ for a series if (i) it is already clear what the starting index is or (ii) the starting index is not important to whatever is being proved or discussed.

Example 15.1:

1. Consider the following series:

$$\sum_{n=1}^{\infty} n = 1 + 2 + 3 + \cdots$$

   The corresponding sequence of partial sums is the following:

   $x_1 = 1$

   $x_2 = 1 + 2 = 3$

   $x_3 = 1 + 2 + 3 = 6$

   $\vdots \qquad \vdots$

   $x_k = 1 + 2 + 3 \cdots + k = \dfrac{k(k+1)}{2}$ (by Problem 1(ii) from Problem Set 4)

   $\vdots \qquad \vdots \qquad \vdots$

   Since $\lim\limits_{k \to \infty} x_k = \lim\limits_{k \to \infty} \frac{k(k+1)}{2} = \infty$, it follows that the given series diverges.

   Notice that the sequence $(s_n) = (n)$ has the property that $\lim\limits_{n \to \infty} s_n \neq 0$. It turns out that this is enough to guarantee that the corresponding series with terms $s_1, s_2, \ldots$ diverges, as we will now see.

2. Consider the following general series:

$$\sum_{n=j}^{\infty} s_n = s_j + s_{j+1} + \cdots$$

   If $\lim\limits_{n \to \infty} s_n \neq 0$, then the series diverges. This result is known as the **Divergence Test** (or sometimes the **nth Term Test**). Let's verify this result with a proof by contrapositive (see the analysis after Theorem 1.28). Assume that the given series converges. If $x_n = s_0 + s_1 + \cdots + s_n$ is the nth partial sum of the series, then we have $s_n = x_n - x_{n-1}$. Since $\lim\limits_{n \to \infty} x_n$ and $\lim\limits_{n \to \infty} x_{n-1}$ are both equal to the sum of the series, we have $\lim\limits_{n \to \infty} x_n = \lim\limits_{n \to \infty} x_{n-1}$. Therefore,

$$\lim_{n \to \infty} s_n = \lim_{n \to \infty} (x_n - x_{n-1}) = \lim_{n \to \infty} x_n - \lim_{n \to \infty} x_{n-1} = 0.$$

In addition to the series given in part 1 above, here are a few more series that diverge by the Divergence Test:

$$\sum_{n=0}^{\infty} \frac{n^2}{3n^2+7} \qquad \sum_{n=1}^{\infty} \cos\frac{1}{n^2} \qquad \sum_{n=0}^{\infty} \sin\frac{n\pi}{2}$$

For the first series, we have $\lim_{n\to\infty} \frac{n^2}{3n^2+7} = \frac{1}{3}$. For the second, we have $\lim_{n\to\infty}\left(\cos\frac{1}{n^2}\right) = \cos 0 = 1$. For the third, we have that $\lim_{n\to\infty}\left(\sin\frac{n\pi}{2}\right)$ does not exist (the sequence $\left(\sin\frac{n\pi}{2}\right)$ oscillates among 0, 1, and -1). In all three cases, the limit of the underlying sequence is not 0, and so, by the Divergence Test, each of the given series diverges.

3. Recall from part 2 of Example 3.11 that a geometric sequence has the form (ar^n) or $(a, ar, ar^2, ar^3, ar^4, \ldots)$ for some real numbers a (the first term of the sequence) and r (the common ratio of the sequence). For example, if we let $a = 1$ and $r = \frac{1}{2}$, we get the geometric sequence $\left(\left(\frac{1}{2}\right)^n\right) = \left(\frac{1}{2^n}\right) = \left(1, \frac{1}{2}, \frac{1}{4}, \frac{1}{8}, \frac{1}{16}, \ldots\right)$. Similarly, if we let $a = 5$ and $r = \sqrt{2}$, we get the geometric sequence $\left(5(\sqrt{2})^n\right) = (5, 5\sqrt{2}, 10, 10\sqrt{2}, 20, \ldots)$. Corresponding to the geometric sequence $(ar^n) = (a, ar, ar^2, ar^3, ar^4, \ldots)$, we have the **geometric series**

$$\sum_{n=0}^{\infty} ar^n = ar^0 + ar^1 + ar^2 + \cdots = a + ar + ar^2 + \cdots$$

The geometric series converges if and only if $|r| < 1$. To see this, first observe that the nth partial sum of this series is $x_n = a + ar + ar^2 + \cdots + ar^n$. If we multiply each side of this equation by the common ratio r, we get $rx_n = ar + ar^2 + \cdots + ar^{n+1}$. Subtracting rx_n from x_n gives us $x_n - rx_n = a - ar^{n+1}$, or equivalently, $(1-r)x_n = a(1 - r^{n+1})$. Assuming that $r \neq 1$, we can divide by $1 - r$ to get $x_n = \frac{a(1-r^{n+1})}{1-r} = \frac{a}{1-r} - \frac{ar^{n+1}}{1-r}$. Therefore, $\left|x_n - \frac{a}{1-r}\right| = \left|\frac{a}{1-r}\right| \cdot |r^{n+1}|$. If $|r| < 1$ (or equivalently, $-1 < r < 1$), then by part 2 of Example 14.1, $\lim_{n\to\infty} r^{n+1} = 0$, and so, by Theorem 14.2, $x_n \to \frac{a}{1-r}$. If $r \leq -1$ or $r > 1$, then again by part 2 of Example 14.1, $(r^{n+1}) \uparrow$, and so, by the Divergence Test (part 2 of this example above), $(x_n) \uparrow$. If $r = 1$, then once again by part 2 of Example 14.1, $(r^{n+1}) = (1)$ converges to 1. So, by the Divergence Test again (part 2 of this example above), $(x_n) \uparrow$.

4. Consider the following series:

$$\sum_{n=1}^{\infty} \frac{1}{n(n+1)} = \frac{1}{1\cdot 2} + \frac{1}{2\cdot 3} + \frac{1}{3\cdot 4} + \cdots$$

Observe that $\frac{1}{n(n+1)} = \frac{1}{n} - \frac{1}{n+1}$ (we have $\frac{1}{n} - \frac{1}{n+1} = \frac{n+1}{n(n+1)} - \frac{n}{n(n+1)} = \frac{1}{n(n+1)}$). It follows that the nth partial sum of the series is given by the following **telescoping sum**:

$$x_n = \left(1 - \frac{1}{2}\right) + \left(\frac{1}{2} - \frac{1}{3}\right) + \left(\frac{1}{3} - \frac{1}{4}\right) + \cdots + \left(\frac{1}{n} - \frac{1}{n+1}\right) = 1 - \frac{1}{n+1}.$$

Therefore, the series converge to $\lim_{n \to \infty} \left(1 - \frac{1}{n+1}\right) = \mathbf{1}$.

5. For $p \in \mathbb{R}^+$, the following series is called a **p-series**:

$$\sum_{n=1}^{\infty} \frac{1}{n^p} = 1 + \frac{1}{2^p} + \frac{1}{3^p} + \cdots$$

Here are a few specific examples of p-series:

$$\sum_{n=1}^{\infty} \frac{1}{n^2} = 1 + \frac{1}{2^2} + \frac{1}{3^2} + \cdots \qquad \sum_{n=1}^{\infty} \frac{1}{\sqrt{n}} = 1 + \frac{1}{\sqrt{2}} + \frac{1}{\sqrt{3}} + \cdots \qquad \sum_{n=1}^{\infty} \frac{1}{n} = 1 + \frac{1}{2} + \frac{1}{3} + \cdots$$

The first series above is the 2-series, the second is the $\frac{1}{2}$-series (because $\sqrt{n} = n^{\frac{1}{2}}$), and the third is the 1-series. The 1 series is usually referred to as the **harmonic series**.

In Example 15.3 below, we will show that a p-series converges if and only if $p > 1$.

In general, it can be quite difficult to determine what a series converges to (if it does in fact converge). Therefore, our primary focus for the rest of this lesson will be to determine whether a given series converges or diverges without necessarily finding its sum explicitly (although we will sometimes indicate how to find good estimates for the sum). Unfortunately, there is no "one size fits all" rule that will tell us if a given series converges. Therefore, in what follows, we will learn about several different "convergence tests."

As an example, we already learned about the Divergence Test in part 2 of Example 15.1 above. Once again, this test says that given a series $\sum s_n$, if $\lim_{n \to \infty} s_n \neq 0$ (the limit of the underlying sequence is **not** 0), then the series diverges. This test alone shows us that a very large class of series diverges. It is very important to realize that the Divergence Test provides no information about a series $\sum s_n$ for which $\lim_{n \to \infty} s_n = 0$. If it did, we would already have all the answers and we wouldn't need any other tests.

The Integral Test

The Integral Test allows us to determine whether a series of the form $\sum f(n)$ converges or diverges by looking at the analogous improper integral. Specifically, we have the following:

Theorem 15.2 (Integral Test): Let $f: [k, \infty) \to \mathbb{R}$ be a decreasing locally integrable function such that $f(x) > 0$ for all $x \in [k, \infty)$. Then

$$\sum_{n=k}^{\infty} f(n) \text{ converges if and only if } \int_{k}^{\infty} f(x)\, dx \text{ converges.}$$

Analysis: Let's provide a geometrical motivation for why the Integral Test might be true. We can think of the series as adding up the areas of infinitely many Riemann rectangles for f, each rectangle having a base of length $\Delta x = 1$ (the endpoints of each subinterval forming the base will be two consecutive integers). For the purpose of illustration, let's start at $k = 1$. We will look at both upper and lower Riemann rectangles. Since the function is decreasing, upper Riemann rectangles are the same as left Riemann rectangles and lower Riemann rectangles are the same as right Riemann rectangles.

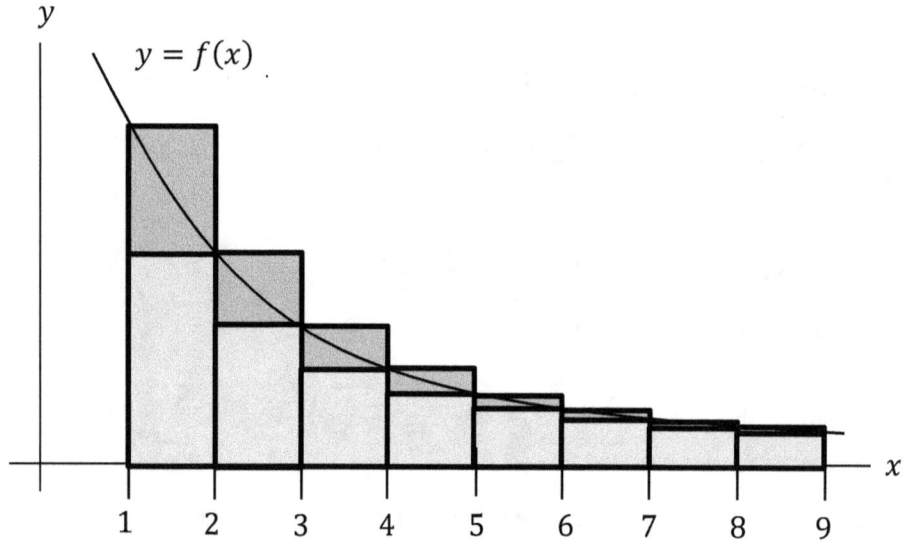

In the picture above, we started the first upper (left) Riemann rectangle over the subinterval $[1, 2]$. It has area $f(1) \cdot 1 = f(1)$. The sum of all the upper Riemann rectangles can be thought of as

$$\sum_{j=1}^{\infty} f(j) = f(1) + f(2) + f(3) + \cdots.$$

Similarly, the first lower (right) Riemann rectangle over the subinterval $[1, 2]$ has area $f(2) \cdot 1 = f(2)$ and the sum of all the lower Riemann rectangles is

$$\sum_{j=1}^{\infty} f(j+1) = f(2) + f(3) + f(4) + \cdots.$$

In each subinterval, the area under the graph of the function $y = f(x)$ is between the areas of the lower and upper Riemann rectangles. So, it should follow that

$$\sum_{j=1}^{\infty} f(j+1) \leq \int_{1}^{\infty} f(x)\, dx \leq \sum_{j=1}^{n} f(j).$$

From these inequalities, it seems reasonable that the series $\sum f(n)$ converges if and only if the corresponding improper integral does.

Note: Observe that

$$\sum_{j=1}^{n} f(j) = f(1) + f(2) + \cdots = f(1) + \sum_{j=1}^{n} f(j+1)$$

Since the two series differ by a finite real number, one of the series converges if and only if the other one does.

More generally, the convergence of a series depends only on the **tail** of the series (that is, the tail of the sequence of partial sums). A series converges if and only if all its tails converge (see Problem 6 from Problem Set 14).

Proof of Theorem 15.2: Let $j \in \mathbb{N}$ with $j \geq k$. Since f is decreasing on $[j, j+1]$, the upper Riemann rectangle for f on $[j, j+1]$ is the same as the left Riemann rectangle for f on $[j, j+1]$. Similarly, the lower and right Riemann rectangles for f on $[j, j+1]$ are the same. Since $\Delta x = (j+1) - j = 1$, it follows that the area of the lower (right) Riemann rectangle is $f(j+1)$ and the area of the upper (left) Riemann rectangle is $f(j)$. So,

$$f(j+1) \leq \int_j^{j+1} f(x)\, dx \leq f(j).$$

If $n \geq k$, we can add the areas of these Riemann rectangles to get

$$\sum_{j=k}^n f(j+1) \leq \sum_{j=k}^n \int_j^{j+1} f(x)\, dx \leq \sum_{j=k}^n f(j).$$

By Problem 15 in Problem Set 11 (and induction), we have

$$\sum_{j=k}^n f(j+1) \leq \int_k^{n+1} f(x)\, dx \leq \sum_{j=k}^n f(j).$$

First, suppose that the given improper integral converges. Since $f(x) > 0$ for all $x \in [k, \infty)$, we see that for each $n \in \mathbb{N}$ with $n \geq k$,

$$\sum_{j=k}^n f(j+1) \leq \int_k^{n+1} f(x)\, dx \leq \int_k^\infty f(x)\, dx.$$

So, $\left\{\sum_{j=k}^n f(j+1) \,\middle|\, n \in \mathbb{N}\right\}$ is an increasing sequence that is bounded above. By the Monotone Convergence Theorem (Theorem 14.5), we have that

$$\sum_{j=k}^\infty f(j+1) = \lim_{n\to\infty} \sum_{j=k}^n f(j+1) \text{ converges.}$$

Therefore, we have

$$\sum_{j=k}^\infty f(j) = f(k) + \left(\sum_{j=k}^\infty f(j+1)\right),$$

which is a finite number. Thus, the series on the left converges.

Now, suppose that the given improper integral diverges. Since $f(x) > 0$ for all $x \in [k, \infty)$, by the first form of the Fundamental Theorem of Calculus (Theorem 11.12) and Corollary 10.26, $F(x) = \int_k^x f(t)\, dt$ is a strictly increasing function. It follows that $\int_k^\infty f(x)\, dx = \lim_{b\to\infty} \int_k^b f(x)\, dx = \infty$.

Let $M > 0$. Since $\lim_{b\to\infty} \int_k^b f(x)\, dx = \infty$, there is $K \in \mathbb{N}$ such that $x > K \to \int_k^x f(t)\, dt > M$. In particular, if $n \in \mathbb{N}$ with $n > K$, then $\int_k^{n+1} f(x)\, dx > M$. So, if $n > K$, we have

$$\sum_{j=k}^{n} f(j) \geq \int_{k}^{n+1} f(x)\, dx > M.$$

Therefore,

$$\sum_{j=k}^{\infty} f(j) = \lim_{n \to \infty} \sum_{j=k}^{n} f(j) \text{ diverges.} \qquad \square$$

Note: It is important to realize that although Theorem 15.2 provides us with a test for convergence of a series, it does not tell us the value of the sum of the convergent series. In particular, the series most likely does **not** converge to the same value as the corresponding improper integral. However, the proof of the theorem does provide us with a method for approximating the sum of the series.

During the proof of Theorem 15.2, we arrived at the following inequalities:

$$\sum_{j=k}^{n} f(j+1) \leq \int_{k}^{n+1} f(x)\, dx \leq \sum_{j=k}^{n} f(j).$$

By reindexing the leftmost sum, we get

$$\boxed{\sum_{j=k+1}^{n} f(j) \leq \int_{k}^{n+1} f(x)\, dx} \leq \sum_{j=k}^{n} f(j).$$

By replacing k by $k+1$ in the rightmost inequality, we get

$$\boxed{\int_{k+1}^{n+1} f(x)\, dx \leq \sum_{j=k+1}^{n} f(j).}$$

Putting together the two expressions in rectangles above, we get

$$\int_{k+1}^{n+1} f(x)\, dx \leq \sum_{j=k+1}^{n} f(j) \leq \int_{k}^{n+1} f(x)\, dx.$$

Assuming that the series (or equivalently, the improper integral) converges, we can take the limit as n goes to ∞ to get the following estimate for the sum of the series:

$$\int_{k+1}^{\infty} f(x)\, dx \leq \sum_{j=k+1}^{\infty} f(j) \leq \int_{k}^{\infty} f(x)\, dx$$

Example 15.3:

1. The function $g : [1, \infty) \to \mathbb{R}$ defined by $g(x) = \frac{1}{x^2}$ is decreasing, locally integrable, and positive ($g(x) > 0$) on $[1, \infty)$. By part 2 of Example 13.2, we have

$$\int_{1}^{\infty} \frac{1}{x^2}\, dx = 1 \quad \text{(and in particular, the improper integral converges).}$$

By the Integral Test (Theorem 15.2), the following 2-series converges:

$$\sum_{n=1}^{\infty} \frac{1}{n^2} = 1 + \frac{1}{2^2} + \frac{1}{3^2} + \cdots$$

By the Note following the proof of Theorem 15.2, setting $k = 1$, we know that

$$\int_{2}^{\infty} \frac{1}{x^2} dx \leq \sum_{n=2}^{\infty} \frac{1}{n^2} \leq \int_{1}^{\infty} \frac{1}{x^2} dx.$$

Now, we have

$$\int_{2}^{\infty} \frac{1}{x^2} dx = \lim_{c \to \infty} \int_{2}^{c} \frac{1}{x^2} dx = \lim_{c \to \infty} \left[\frac{1}{2} - \frac{1}{c} \right] = \frac{1}{2}.$$

So, although, we do not know the exact value of the 2-series, we do know that

$$\frac{1}{2} \leq \sum_{n=2}^{\infty} \frac{1}{n^2} \leq 1.$$

Adding $\frac{1}{1^2} = 1$ (the "missing" first term of the series) to each expression, we get

$$\frac{3}{2} \leq \sum_{n=1}^{\infty} \frac{1}{n^2} \leq 2.$$

We can get even better upper and lower bounds for the series by increasing the value of k in the expression

$$\int_{k+1}^{\infty} f(x) dx \leq \sum_{j=k+1}^{\infty} f(j) \leq \int_{k}^{\infty} f(x) dx$$

For example, let's use $k = 2$. In this case, we get

$$\int_{3}^{\infty} \frac{1}{x^2} dx \leq \sum_{n=3}^{\infty} \frac{1}{n^2} \leq \int_{2}^{\infty} \frac{1}{x^2} dx.$$

From which it follows that

$$\frac{1}{3} \leq \sum_{n=2}^{\infty} \frac{1}{n^2} \leq \frac{1}{2}.$$

Adding $\frac{1}{1^2} + \frac{1}{2^2} = 1 + \frac{1}{4} = \frac{5}{4}$ (the "missing" sum from the beginning of the series) to each expression, we get

$$\frac{19}{12} \leq \sum_{n=1}^{\infty} \frac{1}{n^2} \leq \frac{7}{4}.$$

2. More generally, for $p > 0$, the function $g_p: [1, \infty) \to \mathbb{R}$ defined by $f(x) = \frac{1}{x^p}$ is decreasing, locally integrable and positive on $[1, \infty)$. By part 3 of Example 13.2, we have

$$\int_1^\infty \frac{1}{x^p} dx \begin{cases} \text{converges if } p > 1. \\ \text{diverges} \quad \text{if } 0 < p \leq 1. \end{cases}$$

By the Integral Test (Theorem 15.2), we have the following corresponding result for p-series.

$$\sum_{n=1}^\infty \frac{1}{n^p} \begin{cases} \text{converges if } p > 1. \\ \text{diverges} \quad \text{if } 0 < p \leq 1. \end{cases}$$

In particular, the harmonic series ($p = 1$) diverges.

For $p > 1$, we know from part 3 of Example 13.2 that

$$\int_1^\infty \frac{1}{x^p} dx = \frac{1}{p-1}.$$

By the Note following the proof of Theorem 15.2, we know that for $p > 1$, we have

$$\int_2^\infty \frac{1}{x^p} dx \leq \sum_{n=2}^\infty \frac{1}{n^p} \leq \int_1^\infty \frac{1}{x^p} dx.$$

Now, we have

$$\int_2^\infty \frac{1}{x^p} dx = \lim_{c \to \infty} \int_2^c \frac{1}{x^p} dx = \frac{1}{1-p} \lim_{c \to \infty} [c^{1-p} - 2^{1-p}] = \frac{-2^{1-p}}{1-p} = \frac{1}{2^{p-1}(p-1)}.$$

So, although, we do not know the exact value of the p-series, we do know that

$$\frac{1}{2^{p-1}(p-1)} \leq \sum_{n=2}^\infty \frac{1}{n^p} \leq \frac{1}{p-1}.$$

Adding $\frac{1}{1^p} = 1$ (the "missing" first term of the series) to each expression, we get

$$1 + \frac{1}{2^{p-1}(p-1)} \leq \sum_{n=1}^\infty \frac{1}{n^p} \leq 1 + \frac{1}{p-1}.$$

As a few examples, we have the following:

$$\frac{3}{2} \leq \sum_{n=1}^\infty \frac{1}{n^2} \leq 2 \qquad 1 + \frac{2}{\sqrt{2}} \leq \sum_{n=1}^\infty \frac{1}{\sqrt{n^3}} \leq 3 \qquad \frac{65}{64} \leq \sum_{n=1}^\infty \frac{1}{n^5} \leq \frac{5}{4}$$

The Alternating Series Test

A sequence (a_n) of nonzero real numbers is **alternating** if the series $((-1)^n a_n)$ has terms that are either all positive or all negative. The corresponding series with terms $a_0, a_1, a_2, \ldots$ is called an **alternating series**.

Example 15.4: The following series are examples of alternating series:

$$\sum_{n=1}^{\infty} (-1)^n n = -1 + 2 - 3 + \cdots \qquad \sum_{n=1}^{\infty} \frac{(-1)^{n+1}}{n} = 1 - \frac{1}{2} + \frac{1}{3} - \frac{1}{4} + \cdots$$

For the series on the left, if we let $a_n = (-1)^n n$, then we have $((-1)^n a_n) = ((-1)^n (-1)^n n) = (n)$, which has only positive terms. For the series on the right, if we let $b_n = \frac{(-1)^{n+1}}{n}$, then we have $((-1)^n b_n) = \left((-1)^n \frac{(-1)^{n+1}}{n}\right) = \left(\frac{(-1)^{2n+1}}{n}\right) = \left(-\frac{1}{n}\right)$, which has only negative terms.

Theorem 15.5 (Alternating Series Test): Let (s_n) be a decreasing sequence such that $s_n \to 0$ and for all $n \in \mathbb{N}$, $s_n > 0$. Then the following alternating series converges:

$$\sum_{n=0}^{\infty} (-1)^n s_n = s_0 - s_1 + s_2 - s_3 + \cdots$$

Proof: Let (x_n) be the sequence of partial sums of the series and consider the subsequence (x_{2n+1}). We have $x_1 = s_0 - s_1$, $x_3 = (s_0 - s_1) + (s_2 - s_3) = x_1 + (s_2 - s_3)$, and in general,

$$x_{2n+1} = (s_0 - s_1) + (s_2 - s_3) + \cdots + (s_{2n} - s_{2n+1}) = x_{2n-1} + (s_{2n} - s_{2n+1}).$$

Since (s_n) is decreasing, $s_{2n} \geq s_{2n+1}$. Therefore, $s_{2n} - s_{2n+1} \geq 0$, and so, for all $n \in \mathbb{N}$, we have

$$x_{2n+1} = x_{2n-1} + (s_{2n} - s_{2n+1}) \geq x_{2n-1}.$$

It follows that the subsequence (x_{2n+1}) is increasing.

We also have

$$x_{2n+1} = s_0 - (s_1 - s_2) - (s_3 - s_4) - \cdots - (s_{2n-1} - s_{2n}) - s_{2n+1} \leq s_0.$$

Therefore, the subsequence (x_{2n+1}) is bounded above.

By the Monotone Convergence Theorem (Theorem 14.5), there is $x \in \mathbb{R}$ such that $x_{2n+1} \to x$.

We will now show that $x_n \to x$. To see this, let $\epsilon > 0$. Since $x_{2n+1} \to x$, we can choose $K_1 \in \mathbb{N}$ such that $n > K_1$ implies $|x_{2n+1} - x| < \frac{\epsilon}{2}$. Since $s_n \to 0$, we can choose $K_2 \in \mathbb{N}$ such that $n > K_2$ implies $|s_n| < \frac{\epsilon}{2}$. Since $2n + 2 > n$ for all $n \in \mathbb{N}$ and (s_n) is decreasing, $n > K_2$ implies $|s_{2n+2}| \leq |s_n| < \frac{\epsilon}{2}$. Let $K = \max\{K_1, K_2\}$. Then $n > K$ implies

$$|x_{2n+2} - x| = |x_{2n+1} + s_{2n+2} - x| \leq |x_{2n+1} - x| + |s_{2n+2}| < \frac{\epsilon}{2} + \frac{\epsilon}{2} = \epsilon.$$

Let $K^* = 2K + 3$ and assume that $n > K^*$. There is $j \in \mathbb{N}$ such that $n = 2j + 1$ (if n is odd) or $n = 2j + 2$ (if n is even). In either case, $j > K$, and so, $|x_n - x| < \epsilon$. Since $\epsilon > 0$ was arbitrary, we have $x_n \to x$. Therefore, the given alternating series converges to x. $\square$

Notes: (1) If the alternating series given in the statement of Theorem 15.5 converges by the Alternating Series Test, as with any convergent series, we can estimate the sum of the series by computing the kth partial sum $x_k = s_0 - s_1 + s_2 - s_3 + \cdots + s_k$. As it turns out, s_{k+1} provides us with a bound on the error in the approximation. To see this, first note that by Problem 2 below, we have

$$\sum_{n=k+1}^{\infty} (-1)^n s_n = (-1)^{k+1} s_{k+1} + (-1)^{k+2} s_{k+2} + (-1)^{k+3} s_{k+3} + \cdots$$

$$= (-1)^{k+1}(s_{k+1} - s_{k+2} + s_{k+3} - s_{k+4} + \cdots).$$

Now, since (s_n) is a decreasing sequence, for all $k \in \mathbb{N}$, we have $s_k \geq s_{k+1}$, and so, $s_k - s_{k+1} \geq 0$. Therefore, by Problem 11 below, we have

$$\left| \sum_{n=0}^{\infty} (-1)^n s_n - \sum_{n=0}^{k} (-1)^n s_n \right| = \left| \sum_{n=k+1}^{\infty} (-1)^n s_n \right| = s_{k+1} - s_{k+2} + s_{k+3} - s_{k+4} + \cdots$$

$$= s_{k+1} - (s_{k+2} - s_{k+3}) - (s_{k+4} - s_{k+5}) - \cdots \leq s_{k+1}.$$

(2) Consider the following alternating series:

$$\sum_{n=0}^{\infty} (-1)^{n+1} s_n = -s_0 + s_1 - s_2 + s_3 - \cdots$$

This series is a constant multiple of the series given in the statement of Theorem 15.5. Indeed, we have

$$\sum_{n=0}^{\infty} (-1)^{n+1} s_n = -\sum_{n=0}^{\infty} (-1)^n s_n = -1 \cdot \sum_{n=0}^{\infty} (-1)^n s_n$$

It follows from Problem 2 below that the Alternating Series Test can be applied the same way if the first term of the given alternating series is negative.

(3) Since a series converges if and only if all of its tails converge (see the Note after the statement of Theorem 15.2), the Alternating Series Test can be applied as long as the sequence (s_n) is **eventually decreasing** (in other words, we can replace "decreasing" by "eventually decreasing"—the other conditions need to be present as well).

Formally, we can say that a sequence is **eventually decreasing** if some k-tail of the sequence is decreasing (see Problem 6 from Problem Set 14 for the definition of a k-tail of a sequence).

Similarly, we can say that a sequence is **eventually increasing** if some k-tail of the sequence is increasing.

Example 15.6:

1. The following series is known as the **alternating harmonic series**:

$$\sum_{n=1}^{\infty} \frac{(-1)^{n+1}}{n} = 1 - \frac{1}{2} + \frac{1}{3} - \frac{1}{4} + \cdots$$

Since $\left(\frac{1}{n}\right)$ is a decreasing sequence with $\frac{1}{n} > 0$ for all $n \in \mathbb{N}^+$ and $\frac{1}{n} \to 0$, the alternating harmonic series converges by the Alternating Series Test.

If we add up the first three terms of this series, we get $x_3 = s_1 - s_2 + s_3 = 1 - \frac{1}{2} + \frac{1}{3} = \frac{5}{6}$. So, the sum S of the alternating harmonic series is *approximately* equal to $\frac{5}{6}$. By Note 1 following Theorem 15.5, the maximum error in this approximation is given by $s_4 = \frac{1}{4}$. So we know for certain that $\left|S - \frac{5}{6}\right| \leq \frac{1}{4}$. So, $\frac{5}{6} - \frac{1}{4} \leq S \leq \frac{5}{6} + \frac{1}{4}$, or equivalently, $\frac{7}{12} \leq S \leq \frac{13}{12}$. We can get better approximations by taking larger partial sums. For example, if we want to guarantee that $|S - x_n|$ is no larger than $\frac{1}{100}$, we would compute $x_{99} = 1 - \frac{1}{2} + \frac{1}{3} - \frac{1}{4} + \cdots + \frac{1}{99}$.

2. More generally, for $p > 0$, the following series is called an **alternating p-series**:

$$\sum_{n=1}^{\infty} \frac{(-1)^{n+1}}{n^p} = 1 - \frac{1}{2^p} + \frac{1}{3^p} - \frac{1}{4^p} + \cdots$$

For $p > 0$, $\left(\frac{1}{n^p}\right)$ is a decreasing sequence with $\frac{1}{n^p} > 0$ for all $n \in \mathbb{N}^+$ and $\frac{1}{n^p} \to 0$. Therefore, for $p > 0$, the alternating p-series converges by the Alternating Series Test.

If $p \leq 0$, the corresponding p-series diverges by the Divergence Test (see part 2 of Example 15.1). As a specific example of this, if $p = -1$, the corresponding divergent p-series is

$$\sum_{n=1}^{\infty} \frac{(-1)^{n+1}}{n^{-1}} = \sum_{n=1}^{\infty} (-1)^{n+1} n = 1 - 2 + 3 - 4 + \cdots$$

3. Consider the following series:

$$\sum_{n=0}^{\infty} \frac{(-1)^n}{(2n-3)^2} = \frac{1}{9} - 1 + 1 - \frac{1}{9} + \frac{1}{25} - \frac{1}{49} + \cdots$$

The sequence $\left(\frac{1}{(2n-3)^2}\right)$ is **not** a decreasing sequence because $\frac{1}{9} < 1$. However, it is *eventually decreasing*. Indeed, it is decreasing for $n \geq 1$. Since $\frac{1}{(2n-3)^2} \to 0$, the given series converges by the Alternating Series Test together with Note 3 above.

Dirichlet's and Abel's Tests

The following convergence test is more general than the Alternating Series Test (Theorem 15.5), and therefore, it can be applied to a wider range of series.

Theorem 15.7 (Dirichlet's Test): Let (s_n) be a decreasing sequence such that $s_n \to 0$ and let (t_n) be another sequence whose sequence of partial sums is bounded. Then the following series converges:

$$\sum_{n=1}^{k} s_n t_n$$

To prove Theorem 15.7, we will first prove two preliminary lemmas. The first lemma is essentially just the Cauchy Criterion for Convergence (Theorem 8.28) reformulated in terms of series.

Lemma 15.8 (Cauchy Convergence Criterion for Series): The series $\sum s_n$ converges if and only if for every $\epsilon > 0$, there is $K \in \mathbb{N}$ such that $m \geq n > K$ implies

$$\left| \sum_{k=n+1}^{m} s_k \right| = |s_{n+1} + s_{n+2} + \cdots + s_m| < \epsilon.$$

Proof: Let (x_n) be the sequence of partial sums of the given series. $\sum s_n$ converges if and only if (x_n) converges if and only if (x_n) is a Cauchy sequence (by Corollary 8.29) if and only if for every $\epsilon > 0$, there is $K \in \mathbb{N}$ such that $m \geq n > K$ implies

$$\left| \sum_{k=n+1}^{m} s_k \right| = \left| \sum_{k=0}^{m} s_k - \sum_{k=0}^{n} s_k \right| = |x_m - x_n| < \epsilon. \qquad \square$$

The next lemma is sometimes called **summation by parts**. It is the series analogue of integration by parts (see Theorem 13.3).

Lemma 15.9 (Abel's Lemma): Let $(s_n)_{n=1}^{\infty}$ and $(t_n)_{n=1}^{\infty}$ be sequences and let $(y_n)_{n=1}^{\infty}$ be the sequence of partial sums of $(t_n)_{n=1}^{\infty}$. Also, define y_0 to be 0. Then

$$m > n \geq 0 \to \sum_{k=n+1}^{m} s_k t_k = (s_m y_m - s_{n+1} y_n) + \sum_{k=n+1}^{m-1} (s_k - s_{k+1}) y_k.$$

Notes: (1) If $a < b$, then we define the following series to be 0:

$$\sum_{k=b}^{a} x_k$$

This situation arises in the formula given in Lemma 15.9. For example, if $n = 2$ and $m = 3$, then we get

$$\sum_{k=3}^{3} s_k t_k = (s_3 y_3 - s_3 y_2) + \sum_{k=3}^{2} (s_k - s_{k+1}) y_k.$$

The left-hand side of the above equation is equal to $s_3 t_3$. With our convention of setting the rightmost sum equal to 0, we see that the right-hand side of the above equation is

$$s_3 y_3 - s_3 y_2 + 0 = s_3(y_3 - y_2) = s_3 t_3.$$

(2) Why do we make the extra definition of $y_0 = 0$ in Abel's Lemma? Well, if $n = 0$, the equation in the lemma reduces to

$$\sum_{k=1}^{m} s_k t_k = (s_m y_m - s_1 y_0) + \sum_{k=1}^{m-1} (s_k - s_{k+1}) y_k.$$

Since we decided to start our sequences at 1 in the lemma and y_0 appears in the equation, we need to explicitly say what we want it to be equal to. Of course, we will choose the value for y_0 that makes the equation true. In this case, setting y_0 equal to 0 gives us

$$\sum_{k=1}^{m} s_k t_k = s_m y_m + \sum_{k=1}^{m-1} (s_k - s_{k+1}) y_k.$$

Let's verify that the above equation is true for a specific value of m. To keep it simple, let's let $m = 2$. Since $y_1 = t_1$ and $y_2 = t_1 + t_2$, we have

$$s_2 y_2 + \sum_{k=1}^{1} (s_k - s_{k+1}) y_k = s_2 y_2 + (s_1 - s_2) y_1 = s_2(t_1 + t_2) + (s_1 - s_2) t_1$$

$$= s_2 t_1 + s_2 t_2 + s_1 t_1 - s_2 t_1 = s_1 t_1 + s_2 t_2 = \sum_{k=1}^{2} s_k t_k.$$

This shows that the given formula is true when $n = 0$ and $m = 2$.

Proof of Lemma 15.9: For all $k \in \mathbb{N}$, we have $t_k = y_k - y_{k-1}$. Therefore, if $m > n$, then

$$\sum_{k=n+1}^{m} s_k t_k = \sum_{k=n+1}^{m} s_k (y_k - y_{k-1})$$

$$= s_{n+1}(y_{n+1} - y_n) + s_{n+2}(y_{n+2} - y_{n+1}) + \cdots + s_m(y_m - y_{m-1})$$

$$= (s_{n+1} y_{n+1} - s_{n+1} y_n) + (s_{n+2} y_{n+2} - s_{n+2} y_{n+1}) + \cdots + (s_m y_m - s_m y_{m-1})$$

$$= (s_m y_m - s_{n+1} y_n) + s_{n+1} y_{n+1} + (s_{n+2} y_{n+2} - s_{n+2} y_{n+1}) + \cdots + (s_{m-1} y_{m-1} - s_{m-1} y_{m-2}) - s_m y_{m-1}$$

$$= (s_m y_m - s_{n+1} y_n) + (s_{n+1} - s_{n+2}) y_{n+1} + (s_{n+2} - s_{n+3}) y_{n+2} + \cdots + (s_{m-1} - s_m) y_{m-1}$$

$$= (s_m y_m - s_{n+1} y_n) + \sum_{k=n+1}^{m-1} (s_k - s_{k+1}) y_k. \qquad \square$$

We are now ready to prove Theorem 15.7.

Proof of Theorem 15.7: Let (y_n) be the sequence of partial sums of (t_n). Since (y_n) is bounded, there is $M \in \mathbb{N}^+$ such that $|y_n| \le M$ for all $n \in \mathbb{N}$. Since (s_n) is decreasing, for all $k \in \mathbb{N}$, we have $s_k \ge s_{k+1}$, or equivalently, $s_k - s_{k+1} \ge 0$. Since (s_n) is decreasing with limit 0, $s_n \ge 0$ for all $n \in \mathbb{N}$. So, by Abel's Lemma (Lemma 15.9), we have

$$\left|\sum_{k=n+1}^{m} s_k t_k\right| \le |s_m y_m - s_{n+1} y_n| + \sum_{k=n+1}^{m-1} |(s_k - s_{k+1}) y_k|$$

$$\le |s_m||y_m| + |s_{n+1}||y_n| + \sum_{k=n+1}^{m-1} |s_k - s_{k+1}||y_k|$$

$$\le s_m M + s_{n+1} M + \sum_{k=n+1}^{m-1} (s_k - s_{k+1}) M$$

$$= [(s_m + s_{n+1}) + (s_{n+1} - s_m)] M = 2 s_{n+1} M.$$

Let $\epsilon > 0$. Since $s_n \to 0$, we can choose $K \in \mathbb{N}$ such that $n > K$ implies $s_n = |s_n| < \frac{\epsilon}{2M}$. Then if $m \ge n > K$, we have

$$\left|\sum_{k=n+1}^{m} s_k t_k\right| \le 2 s_{n+1} M < 2 \cdot \frac{\epsilon}{2M} \cdot M = \epsilon.$$

By the Cauchy Convergence Criterion for Series (Lemma 15.8), the given series converges. □

Example 15.10:

1. Let's use Dirichlet's Test (Theorem 15.7) to show that the following series converges:

$$\sum_{n=1}^{\infty} \frac{\cos n}{n} = \cos 1 + \frac{\cos 2}{2} + \frac{\cos 3}{3} + \cdots$$

Let $(s_n) = \left(\frac{1}{n}\right)$ and $(t_n) = (\cos n)$. Observe that $\left(\frac{1}{n}\right)$ is a decreasing sequence and $\frac{1}{n} \to 0$. Now, the kth partial sum of (t_n) is $y_k = \cos 1 + \cos 2 + \cdots + \cos k$. By Problem 12 below, for each $k \in \mathbb{N}^+$, we have

$$y_k = \sum_{n=1}^{k} \cos n = \frac{\sin\left(k + \frac{1}{2}\right) - \sin\frac{1}{2}}{2 \sin\frac{1}{2}}$$

Since $|\sin x| \le 1$ for all $x \in \mathbb{R}$ and $\sin\frac{1}{2} \ne 0$,

$$|y_k| = \left|\frac{\sin\left(k + \frac{1}{2}\right) - \sin\frac{1}{2}}{2 \sin\frac{1}{2}}\right| \le \frac{1}{2\left|\sin\frac{1}{2}\right|}\left(\left|\sin\left(k + \frac{1}{2}\right)\right| + \left|\sin\frac{1}{2}\right|\right) \le \frac{(1+1)}{2\left|\sin\frac{1}{2}\right|} = \frac{1}{\left|\sin\frac{1}{2}\right|}.$$

Therefore, $|y_n|$ is bounded. The result now follows from Dirichlet's Test (Theorem 15.7).

2. More generally, if (s_n) is any decreasing sequence of positive real numbers such that $s_n \to 0$ and a is any real number such that a is not an integer multiple of 2π, then we can use Dirichlet's Test (Theorem 15.7) to show that the following series converges:

$$\sum_{n=1}^{\infty} s_n \cos na = s_0 \cos a + s_1 \cos 2a + s_2 \cos 3a + \cdots$$

The argument is very similar to what was done in part 1 above, and so, I leave the details to the reader. Note that Problem 12 below will need to be generalized as well.

3. The Alternating Series Test is an immediate consequence of Dirichlet's Test. To see this, let (s_n) be the sequence given in the hypothesis of the Alternating Series Test and let $(t_n) = (-1)^n$. Since every partial sum of (t_n) is either 0 or 1, the partial sums of (t_n) are bounded. Therefore, it follows from Dirichlet's Test that

$$\sum_{n=0}^{\infty} (-1)^n s_n \text{ converges.}$$

Theorem 15.11 (Abel's Test): Let (s_n) be a bounded monotone sequence and let (t_n) be another sequence whose sequence of partial sums converges. Then the following series converges:

$$\sum_{n=1}^{k} s_n t_n$$

Dirichlet's Test (Theorem 15.7) can be used to prove Abel's Test (Theorem 15.11). You will be asked to provide the details in Problem 13 below.

Example 15.12:

1. Let's use Abel's Test (Theorem 15.11) to show that the following series converges:

$$\sum_{n=1}^{\infty} \frac{(-1)^{n+1}}{n^{1-\frac{1}{n}}} = 1 - \frac{1}{2^{\frac{1}{2}}} + \frac{1}{3^{\frac{2}{3}}} - \frac{1}{4^{\frac{3}{4}}} + \cdots$$

We first rewrite the series in the following form:

$$\sum_{n=1}^{\infty} \frac{(-1)^{n+1}}{n^{1-\frac{1}{n}}} = \sum_{n=1}^{\infty} n^{\frac{1}{n}} \frac{(-1)^{n+1}}{n}$$

Let $(s_n) = n^{\frac{1}{n}}$ and $(t_n) = \frac{(-1)^{n+1}}{n}$. By part 1 of Example 15.6, the partial sums of (t_n) converge.

Let's show that (s_n) is *eventually* decreasing. Let $y = x^{\frac{1}{x}}$. We will use logarithmic differentiation. We first take the natural logarithm of each side of the equation and apply Theorem 12.17:

$$\ln y = \ln x^{\frac{1}{x}} = \frac{1}{x} \cdot \ln x = \frac{\ln x}{x}$$

We now differentiate each side of the equation using implicit differentiation on the left-hand side and the quotient rule on the right-hand side:

$$\frac{1}{y} \cdot \frac{dy}{dx} = \frac{x \cdot \frac{1}{x} - \ln x}{x^2} = \frac{1 - \ln x}{x^2}$$

Solving for $\frac{dy}{dx}$ and replacing y by x^x results in the following:

$$\frac{dy}{dx} = y\left(\frac{1 - \ln x}{x^2}\right) = x^x\left(\frac{1 - \ln x}{x^2}\right)$$

The right-hand side of the above equation is negative when $1 - \ln x < 0$, or equivalently, $\ln x > 1$. This last inequality is equivalent to $x = e^{\ln x} > e^1 = e$. By part 2 of Corollary 10.26, $x^{\frac{1}{x}}$ is strictly decreasing for $x > e$. In particular, for $n \in \mathbb{N}$ with $n > e$, $n^{\frac{1}{n}} > (n+1)^{\frac{1}{n+1}}$. So, the sequence (s_n) is eventually decreasing. Since $s_n > 0$ for all $n \in \mathbb{N}$, (s_n) is also bounded. Abel's Test (Theorem 15.11) now guarantees the convergence of the given series.

2. More generally, if (t_n) is a sequence whose partial sums converge, then Abel's Test guarantees the convergence of the following series:

$$\sum_{n=1}^{\infty} n^{\frac{1}{n}} \cdot t_n = t_1 + \sqrt{2}t_2 + \sqrt[3]{3}t_3 + \sqrt[4]{4}t_4 + \cdots$$

The argument is nearly identical to what we did in part 1 above. Therefore, I leave the details to the reader.

Comparison Tests

Theorem 15.13 (Comparison Test): Let (s_n) and (t_n) be sequences and suppose that there is $K \in \mathbb{N}$ such that $n > K \to 0 \leq s_n \leq t_n$. Then we have the following:

$$\text{if } \sum_{n=0}^{\infty} t_n \text{ converges, then } \sum_{n=0}^{\infty} s_n \text{ converges}$$

$$\text{if } \sum_{n=0}^{\infty} s_n \text{ diverges, then } \sum_{n=0}^{\infty} t_n \text{ diverges}$$

Proof: Suppose that $\sum t_n$ converges and let $\epsilon > 0$ be given. By the Cauchy Convergence Criterion for Series (Lemma 15.8), there is $K_1 \in \mathbb{N}$ such that $m \geq n > K_1$ implies

$$\left|\sum_{k=n+1}^{m} t_k\right| = |t_{n+1} + t_{n+2} + \cdots + t_m| = t_{n+1} + t_{n+2} + \cdots + t_m < \epsilon.$$

Let $K^* = \max\{K, K_1\}$. Then $m \geq n > K^*$ implies

$$\left|\sum_{k=n+1}^{m} s_k\right| \leq \left|\sum_{k=n+1}^{m} t_k\right| < \epsilon.$$

By the Cauchy Convergence Criterion for Series once again, $\sum s_n$ converges.

The second statement is the contrapositive of the first, and therefore, is true as well. □

Note: See the analysis after Theorem 1.28 for more explanation on the contrapositive.

Example 15.14:

1. Let's use the Comparison Test (Theorem 15.13) to show that the following series converges.

$$\sum_{n=0}^{\infty} \frac{1}{5^n + n}$$

Since $5^n + n \geq 5^n$ for all $n \in \mathbb{N}$, we have $\frac{1}{5^n+n} \leq \frac{1}{5^n} = \left(\frac{1}{5}\right)^n$ for all $n \in \mathbb{N}$.

The series $\sum t_n = \sum \left(\frac{1}{5}\right)^n$ is a convergent geometric series (because $0 < \frac{1}{5} < 1$). So, by the Comparison Test (Theorem 15.13), the given series converges.

2. Let's use the Comparison Test (Theorem 15.13) to show that the following series diverges.

$$\sum_{n=0}^{\infty} \frac{n^2 + 5}{\sqrt{n^5 - 3n}}$$

We have $n^2 + 5 \geq n^2$. We also have $\sqrt{n^5 - 3n} \leq \sqrt{n^5} = n^{\frac{5}{2}}$, and therefore, $\frac{n^2+5}{\sqrt{n^5-3n}} \geq \frac{n^2}{n^{\frac{5}{2}}} = \frac{1}{n^{\frac{1}{2}}}$.

By part 2 of Example 15.3, $\sum \frac{1}{n^{\frac{1}{2}}}$ diverges. So, by Theorem 15.13, the given series diverges too.

Theorem 15.15 (Limit Comparison Test): Let (s_n) and (t_n) be sequences and suppose that there is $K \in \mathbb{N}$ such that $n > K$ implies that $s_n \geq 0$ and $t_n > 0$. Also, suppose that there is $L \in \mathbb{R} \cup \{\infty\}$ such that

$$\lim_{n \to \infty} \frac{s_n}{t_n} = L.$$

Then we have the following:

if $L \in \mathbb{R}^+$, then $\sum_{n=0}^{\infty} s_n$ converges if and only if $\sum_{n=0}^{\infty} t_n$ converges.

if $L = 0$ and $\sum_{n=0}^{\infty} t_n$ converges, then $\sum_{n=0}^{\infty} s_n$ converges.

if $L = \infty$ and $\sum_{n=0}^{\infty} t_n$ diverges, then $\sum_{n=0}^{\infty} s_n$ diverges.

You will be asked to prove Theorem 15.15 in Problem 16 below.

Example 15.16:

1. Let's use the Limit Comparison Test (Theorem 15.15) to show that the following series converges.

$$\sum_{n=0}^{\infty} \frac{1}{5^n - n}$$

Let $(s_n) = \left(\frac{1}{5^n - n}\right)$ and let $(t_n) = \left(\left(\frac{1}{5}\right)^n\right)$. Then we have

$$\lim_{n \to \infty} \frac{s_n}{t_n} = \lim_{n \to \infty} \frac{\frac{1}{5^n - n}}{\left(\frac{1}{5}\right)^n} = \lim_{n \to \infty} \left(\frac{1}{5^n - n}\right)(5^n) = \lim_{n \to \infty} \left(\frac{5^n}{5^n - n}\right) = 1.$$

Since $1 \neq 0$ and $\sum t_n = \sum \left(\frac{1}{5}\right)^n$ is a convergent geometric series (because $0 < \frac{1}{5} < 1$), by the Limit Comparison Test (Theorem 15.15), the given series converges.

Observe that the Limit Comparison Test is much easier to apply that the Comparison Test in this example. The most natural choice for comparison, namely $\sum \left(\frac{1}{5}\right)^n$, cannot be used in the Comparison Test for the following reasoning. Since $5^n - n \leq 5^n$ for all $n \in \mathbb{N}$, it follows that $\frac{1}{5^n - n} \geq \frac{1}{5^n}$. So, the Comparison Test tells us that "if $\sum \left(\frac{1}{5}\right)^n$ diverges, then $\sum \frac{1}{5^n - n}$ diverges." Due to the fact that $\sum \left(\frac{1}{5}\right)^n$ converges, the Comparison Test does not give us any useful information.

2. Let $p \in \mathbb{R}$ and consider the following series.

$$\sum_{n=0}^{\infty} \frac{1}{(n+1)^p}$$

We can use the Limit Comparison Test (Theorem 15.15) to show that the series converges if $p > 1$ and diverges if $0 < p \leq 1$.

Let $(s_n) = \left(\frac{1}{(n+1)^p}\right)$ and let $(t_n) = \left(\frac{1}{n^p}\right)$. Then we have

$$\lim_{n \to \infty} \frac{s_n}{t_n} = \lim_{n \to \infty} \frac{\frac{1}{(n+1)^p}}{\frac{1}{n^p}} = \lim_{n \to \infty} \left[\frac{1}{(n+1)^p} \cdot \frac{n^p}{1}\right] = 1.$$

The result then follows from part 2 of Example 15.3 and the Limit Comparison Test.

3. Consider the following series.

$$\sum_{n=0}^{\infty} \ln n$$

We can use the Limit Comparison Test (Theorem 15.15) to show that this series diverges. We will use the fact that $\sum \frac{1}{n}$ diverges. Let $(s_n) = (\ln n)$ and let $(t_n) = \left(\frac{1}{n}\right)$. Then we have

$$\lim_{n \to \infty} \frac{\ln n}{\frac{1}{n}} = \lim_{n \to \infty} [n \ln n] = \infty.$$

It follows from the third part of the Limit Comparison Test (Theorem 15.15) that the given series diverges.

Absolute and Conditional Convergence

Let (s_n) be a sequence. If the series $\sum |s_n|$ converges, then we say that $\sum s_n$ **converges absolutely** (or is **absolutely convergent**).

In Problem 7 below, you will be asked to prove that an absolutely convergent series is convergent. On the other hand, it is **not** true that every convergent series is absolutely convergent. If a series converges, but does not converge absolutely, then we say that the series **converges conditionally** (or is **conditionally convergent**).

Example 15.17:

1. Consider the alternating 2-series:

$$\sum_{n=1}^{\infty} \frac{(-1)^{n+1}}{n^2} = 1 - \frac{1}{2^2} + \frac{1}{3^2} - \frac{1}{4^2} + \cdots$$

By part 1 of Example 15.3,

$$\sum_{n=1}^{\infty} \left|\frac{(-1)^{n+1}}{n^2}\right| = \sum_{n=1}^{\infty} \frac{1}{n^2} = 1 + \frac{1}{2^2} + \frac{1}{3^2} + \cdots \text{ converges.}$$

Therefore, the alternating 2-series is absolutely convergent.

2. Consider the alternating harmonic series (or alternating 1-series):

$$\sum_{n=1}^{\infty} \frac{(-1)^{n+1}}{n} = 1 - \frac{1}{2} + \frac{1}{3} - \frac{1}{4} + \cdots$$

By part 2 of Example 15.3,

$$\sum_{n=1}^{\infty} \left|\frac{(-1)^{n+1}}{n}\right| = \sum_{n=1}^{\infty} \frac{1}{n} = 1 + \frac{1}{2} + \frac{1}{3} + \cdots \text{ diverges.}$$

Therefore, the harmonic series is **not** absolutely convergent. However, by part 1 of Example 15.6, the alternating harmonic series converges. Therefore, the alternating harmonic series is conditionally convergent.

3. Let $p > 0$ and consider the alternating p-series:

$$\sum_{n=1}^{\infty} \frac{(-1)^{n+1}}{n^p} = 1 - \frac{1}{2^p} + \frac{1}{3^p} - \frac{1}{4^p} + \cdots$$

By part 2 of Example 15.3,

$$\sum_{n=1}^{\infty} \left|\frac{(-1)^{n+1}}{n^p}\right| = \sum_{n=1}^{\infty} \frac{1}{n^p} \begin{cases} \text{converges if } p > 1. \\ \text{diverges } \text{ if } 0 < p \leq 1. \end{cases}$$

So, for $p > 1$, the alternating p-series is absolutely convergent. By part 2 of Example 15.6, for $0 < p \leq 1$, the alternating p-series is conditionally convergent.

The series $\sum t_n$ is said to be a **rearrangement** of the series $\sum s_n$ if there is a bijection $f: \mathbb{N} \to \mathbb{N}$ such that $t_k = s_{f(k)}$ for all $k \in \mathbb{N}$.

Note: In the definition of rearrangement it is implicit that both series are starting with an index of 0. Since all series can be reindexed to begin with an index of 0, the definition can be applied to series beginning at any index. See Note 2 in the beginning of this lesson for information on reindexing a series.

Example 15.18: Consider the alternating harmonic series:

$$\sum_{n=1}^{\infty} \frac{(-1)^{n+1}}{n} = 1 - \frac{1}{2} + \frac{1}{3} - \frac{1}{4} + \cdots$$

In part 2 of Example 15.17, we showed that this series converges conditionally. Let L be the sum of the series (By Problem 15 below, $L = \ln 2$). So, we have

$$1 - \frac{1}{2} + \frac{1}{3} - \frac{1}{4} + \cdots = L$$

If we multiply each side of this equation by $\frac{1}{2}$, we get

$$\frac{1}{2} - \frac{1}{4} + \frac{1}{6} - \frac{1}{8} + \cdots = \frac{1}{2}L$$

Let's insert zeros between consecutive terms of this last series and add it to the original series:

$$0 + \frac{1}{2} + 0 - \frac{1}{4} + 0 + \frac{1}{6} + 0 - \frac{1}{8} + \cdots = \frac{1}{2}L$$

$$\underline{1 - \frac{1}{2} + \frac{1}{3} - \frac{1}{4} + \frac{1}{5} - \frac{1}{6} + \frac{1}{7} - \frac{1}{8} + \cdots = L}$$

$$1 + \frac{1}{3} - \frac{1}{2} + \frac{1}{5} + \frac{1}{7} - \frac{1}{4} + \cdots = \frac{3}{2}L$$

The resulting series is a rearrangement of the original series. Observe that this rearrangement has a different sum than the original series. This example shows that a rearrangement of a conditionally convergent series does not necessarily converge to the same sum as the original series.

In fact, in Problem 17 below, you will be asked to prove that if a series $\sum s_n$ is conditionally convergent and r is any real number, then there is a rearrangement of $\sum s_n$ that converges to r. In Problem 18 below, you will be asked to prove that if a series $\sum s_n$ is conditionally convergent, then there is a rearrangement of $\sum s_n$ that diverges.

Absolutely convergent series are much more well-behaved, as we will now see.

Theorem 15.19 (Rearrangement Theorem): Let $\sum s_n$ be an absolutely convergent series and let $\sum t_n$ be a **rearrangement** of $\sum s_n$. Then $\sum t_n$ converges to the same value as $\sum s_n$.

To prove Theorem 15.19, we will first prove a preliminary lemma. This lemma is essentially just the Monotone Convergence Theorem reformulated in terms of series.

Lemma 15.20: Let (s_n) be a sequence such that $s_n \geq 0$ for all $n \in \mathbb{N}$. Then $\sum s_n$ converges if and only if the sequence of partial sums of (s_n) is bounded above.

Proof: Since $s_n \geq 0$ for all $n \in \mathbb{N}$, the sequence of partial sums of (s_n) is increasing. So, by the Monotone Convergence Theorem (Theorem 14.5), $\sum s_n$ converges if and only if the sequence of partial sums of (s_n) converges if and only if the sequence of partial sums of (s_n) is bounded if and only if the sequence of partial sums of (s_n) is bounded above. $\square$

Proof of Theorem 15.19: For each $n \in \mathbb{N}$, let

$$S_n = \sum_{i=0}^{n} |s_i| = |s_0| + |s_1| + \cdots + |s_n| \quad \text{and} \quad T_n = \sum_{i=0}^{n} |t_i| = |t_0| + |t_1| + \cdots + |t_n|.$$

For each $n \in \mathbb{N}$, there is $k_n \in \mathbb{N}$ such that $\{t_0, t_1, \ldots, t_n\} \subseteq \{s_0, s_1, \ldots, s_{k_n}\}$, and so, $T_n \leq S_{n_k}$. Since $\sum s_n$ is absolutely convergent, by Lemma 15.20, $\{S_n \mid x \in \mathbb{N}\}$ is bounded above. Therefore, $\{T_n \mid x \in \mathbb{N}\}$ is bounded above. So, again by Lemma 15.20, $\sum t_n$ is absolutely convergent.

Now, let $X = \sum s_n$, let $Y = \sum t_n$, and for each $n \in \mathbb{N}$, let

$$X_n = \sum_{i=0}^{n} s_i = s_0 + s_1 + s_2 + \cdots + s_n \quad \text{and} \quad Y_n = \sum_{i=0}^{n} t_i = t_0 + t_1 + t_2 + \cdots + t_n.$$

Let $\epsilon > 0$. Since $\sum |s_n|$ converges, by the Cauchy Convergence Criterion for Series, there is $K \in \mathbb{N}$ such that $m \geq n > K$ implies

$$\left| \sum_{k=n+1}^{m} |s_k| \right| = \sum_{k=n+1}^{m} |s_k| = |s_{n+1}| + |s_{n+2}| + \cdots + |s_m| < \epsilon.$$

In particular, for each $j \in \mathbb{N}$ with $j \geq K + 2$, we have

$$\sum_{k=K+2}^{j} |s_k| = |s_{K+2}| + |s_{K+3}| + \cdots + |s_j| < \epsilon.$$

Choose $K^* \in \mathbb{N}$ so that $\{s_0, s_1, \ldots, s_{K+1}\} \subseteq \{t_0, t_1, \ldots, t_{K^*}\}$.

If $n > K^*$, then X_n and Y_n both include the terms $s_0, s_1, \ldots, s_{K+1}$. It follows that there is $j \in \mathbb{N}$ such that

$$|X_n - Y_n| \leq \sum_{k=K+2}^{j} |s_k| < \epsilon.$$

Since $\epsilon > 0$ was arbitrary, $\lim_{n \to \infty} (X_n - Y_n) = 0$. It follows that

$$\sum_{i=0}^{\infty} s_i = \lim_{n \to \infty} X_n = \lim_{n \to \infty} Y_n = \sum_{i=0}^{\infty} t_i.$$

Therefore, $\sum t_n$ converges to the same value as $\sum s_n$. □

Tests for Absolute Convergence

We will finish this lesson with two more important convergence tests: the Ratio Test and the Root Test. These two tests are used to determine if a series is absolutely convergent.

Theorem 15.21 (Ratio Test): Let (s_n) be a sequence with $s_n \neq 0$ for all $n \in \mathbb{N}$ and let

$$L = \lim_{n \to \infty} \left| \frac{s_{n+1}}{s_n} \right|.$$

If L is a finite real number, then we have the following:

$$\text{if } L < 1, \text{ then } \sum_{n=0}^{\infty} s_n \text{ converges absolutely.}$$

$$\text{if } L > 1, \text{ then } \sum_{n=0}^{\infty} s_n \text{ diverges.}$$

Note: If L turns out to be 0, then nothing can be concluded from the ratio test. When this happens, you should try a different test from this lesson.

To prove Theorem 15.21, we will use the following two Lemmas, which you will be asked to prove in Problems 8 and 9 below.

Lemma 15.22: Let (s_n) be a sequence with $s_n \neq 0$ for all $n \in \mathbb{N}$ and suppose that there is $r \in (0, 1)$ and $K \in \mathbb{N}$ such that $n > K \to \left| \frac{s_{n+1}}{s_n} \right| < r$. Then $\sum s_n$ converges absolutely.

Lemma 15.23: Let (s_n) be a sequence with $s_n \neq 0$ for all $n \in \mathbb{N}$ and suppose that there is $K \in \mathbb{N}$ such that $n > K \to \left| \frac{s_{n+1}}{s_n} \right| > 1$. Then $\sum s_n$ diverges.

Proof of Theorem 15.21: First suppose that $L < 1$. Let r be any real number such that $L < r < 1$. Then $0 < r - L < 1 - L < 1$ (because $L \geq 0$). Since $\lim_{n \to \infty} \left| \frac{s_{n+1}}{s_n} \right| = L$, there is $K \in \mathbb{N}$ such that

$$n > K \rightarrow \left| \left| \frac{s_{n+1}}{s_n} \right| - L \right| < r - L,$$

or equivalently, $L - r < \left| \frac{s_{n+1}}{s_n} \right| - L < r - L$. Adding L, we get $2L - r < \left| \frac{s_{n+1}}{s_n} \right| < r$. In particular,

$$n > K \rightarrow \left| \frac{s_{n+1}}{s_n} \right| < r.$$

By Lemma 15.22, $\sum s_n$ converges absolutely.

Now, suppose that $L > 1$. Then $L - 1 > 0$. Since $\lim_{n \to \infty} \left| \frac{s_{n+1}}{s_n} \right| = L$, there is $K \in \mathbb{N}$ such that

$$n > K \rightarrow \left| \left| \frac{s_{n+1}}{s_n} \right| - L \right| < L - 1,$$

or equivalently, $1 - L < \left| \frac{s_{n+1}}{s_n} \right| - L < L - 1$. Adding L, we get $1 < \left| \frac{s_{n+1}}{s_n} \right| < 2L - 1$. In particular,

$$n > K \rightarrow \left| \frac{s_{n+1}}{s_n} \right| > 1.$$

By Lemma 15.23, $\sum s_n$ diverges. □

Example 15.24:

1. Let's use the ratio test to determine if the following series converges:

$$\sum_{n=1}^{\infty} \frac{5^n}{n!} = 5 + \frac{5^2}{2!} + \frac{5^3}{3!} + \frac{5^4}{4!} + \cdots$$

If we let $s_n = \frac{5^n}{n!}$, then we have

$$\lim_{n \to \infty} \left| \frac{s_{n+1}}{s_n} \right| = \lim_{n \to \infty} \left| \frac{\frac{5^{n+1}}{(n+1)!}}{\frac{5^n}{n!}} \right| = \lim_{n \to \infty} \left| \frac{5^{n+1}}{(n+1)!} \cdot \frac{n!}{5^n} \right| = \lim_{n \to \infty} \left| \frac{5^{n+1}}{5^n} \cdot \frac{n!}{(n+1)n!} \right| = \lim_{n \to \infty} \left(\frac{5}{n+1} \right) = 0.$$

Since $L = \lim_{n \to \infty} \left| \frac{s_{n+1}}{s_n} \right| = 0 < 1$, by the Ratio Test, the given series is absolutely convergent.

2. Let's use the Ratio Test to determine if the following series converges:

$$\sum_{n=1}^{\infty} \frac{n^n}{n!} = 1 + \frac{2^2}{2!} + \frac{3^3}{3!} + \frac{4^4}{4!} + \cdots$$

If we let $s_n = \frac{n^n}{n!}$, then we have

$$\lim_{n\to\infty}\left|\frac{s_{n+1}}{s_n}\right| = \lim_{n\to\infty}\left|\frac{\frac{(n+1)^{n+1}}{(n+1)!}}{\frac{n^n}{n!}}\right| = \lim_{n\to\infty}\left|\frac{(n+1)^{n+1}}{(n+1)!}\cdot\frac{n!}{n^n}\right| = \lim_{n\to\infty}\left|\frac{(n+1)^n}{n^n}\cdot\frac{(n+1)n!}{(n+1)!}\right|$$

$$= \lim_{n\to\infty}\left|\left(\frac{n+1}{n}\right)^n\right| = \lim_{n\to\infty}\left|\left(1+\frac{1}{n}\right)^n\right| = \lim_{n\to\infty}\left(1+\frac{1}{n}\right)^n = e.$$

The reader should check carefully that $\lim_{n\to\infty}\left(1+\frac{1}{n}\right)^n = e$ (use Problem 9 from Problem Set 12).

Since $L = \lim_{n\to\infty}\left|\frac{s_{n+1}}{s_n}\right| = e > 1$, by the Ratio Test, the given series is divergent.

3. Let's attempt to use the Ratio Test to determine if the following series converges:

$$\sum_{n=1}^{\infty}\frac{(-1)^{n+1}\sqrt{n}}{n+1} = \frac{1}{2} - \frac{\sqrt{2}}{3} + \frac{\sqrt{3}}{4} - \frac{\sqrt{4}}{5} + \cdots$$

If we let $s_n = \frac{(-1)^{n+1}\sqrt{n}}{n+1}$, then we have

$$\lim_{n\to\infty}\left|\frac{s_{n+1}}{s_n}\right| = \lim_{n\to\infty}\left|\frac{\frac{(-1)^{n+2}\sqrt{n+1}}{n+2}}{\frac{(-1)^{n+1}\sqrt{n}}{n+1}}\right| = \lim_{n\to\infty}\left(\frac{\sqrt{n+1}}{n+2}\cdot\frac{n+1}{\sqrt{n}}\right) = \lim_{n\to\infty}\left(\frac{n\sqrt{n}}{n\sqrt{n}}\right) = 1.$$

Since $L = \lim_{n\to\infty}\left|\frac{s_{n+1}}{s_n}\right| = 1$, the Ratio Test is inconclusive.

A straightforward application of the Limit Comparison Test (using the $\frac{1}{2}$-series for comparison) shows that this series is **not** absolutely convergent. The Alternating Series Test can be used to show that this series is conditionally convergent. I leave the details to the reader.

Theorem 15.25 (Root Test): Let (s_n) be a sequence with $s_n \neq 0$ for all $n \in \mathbb{N}$ and let

$$L = \limsup |s_n|^{\frac{1}{n}}$$

Then we have the following:

$$\text{if } L < 1, \text{ then } \sum_{n=0}^{\infty} s_n \text{ converges absolutely.}$$

$$\text{if } L > 1, \text{ then } \sum_{n=0}^{\infty} s_n \text{ diverges.}$$

The proof of the Root Test (Theorem 15.25) is similar to the proof of the Ratio Test (Theorem 15.21), and so, I leave it as an exercise (see Problem 10 below).

Example 15.26:

1. Let's use the root test to determine if the following series converges:

$$\sum_{n=1}^{\infty} \left(\frac{n+2}{2n}\right)^n = \left(\frac{3}{2}\right)^1 + \left(\frac{4}{4}\right)^2 + \left(\frac{5}{6}\right)^3 + \left(\frac{3}{4}\right)^4 + \cdots$$

If we let $s_n = \left(\frac{n+2}{2n}\right)^n$, then we have

$$\limsup |s_n|^{\frac{1}{n}} = \limsup \left|\left(\frac{n+2}{2n}\right)^n\right|^{\frac{1}{n}} = \limsup \left(\frac{n+2}{2n}\right) = \frac{1}{2}.$$

Since $L = \lim_{n \to \infty} \left|\frac{s_{n+1}}{s_n}\right| = \frac{1}{2} < 1$, by the Root Test, the given series is absolutely convergent.

2. For each $n \in \mathbb{N}$, let

$$s_n = \begin{cases} \dfrac{1}{3^n} & \text{if } n \text{ is odd.} \\ \dfrac{1}{3^{n+2}} & \text{if } n \text{ is even.} \end{cases}$$

Let's use the Root Test to determine if the following series converges:

$$\sum_{n=0}^{\infty} s_n = \frac{1}{3^2} + \frac{1}{3} + \frac{1}{3^4} + \frac{1}{3^3} + \cdots$$

We have $\limsup |s_n|^{\frac{1}{n}} = \frac{1}{3} < 1$, and so, by the Root Test, the given series is absolutely convergent.

3. Let's use the Root Test to determine if the following series converges:

$$\sum_{n=1}^{\infty} \frac{n^n}{7^{5n-1}} = \frac{1}{7^4} + \frac{2^2}{7^9} + \frac{3^3}{7^{14}} + \frac{4^4}{7^{19}} + \cdots$$

If we let $s_n = \frac{n^n}{7^{5n-1}}$, then we have

$$\limsup |s_n|^{\frac{1}{n}} = \limsup \left|\frac{n^n}{7^{5n-1}}\right|^{\frac{1}{n}} = \limsup \frac{n}{7^{5-\frac{1}{n}}} = \limsup \frac{n}{7^5 7^{-\frac{1}{n}}} = \limsup \frac{n\sqrt[n]{7}}{7^5}$$

$$= \frac{1}{7^5}(\limsup \sqrt[n]{7})(\limsup n) = \frac{1}{7^5}(1)(\limsup n) \text{ (Check this!)} = \infty.$$

Since $L = \lim_{n \to \infty} \left|\frac{s_{n+1}}{s_n}\right| = \infty$, by the Root Test, the given series is divergent.

Problem Set 15

Full solutions to these problems are available for free download here:
www.SATPrepGet800.com/RABQXZ

LEVEL 1

1. Prove that the sum of a convergent series is unique.

2. Let $k \in \mathbb{N}$ and suppose that S and T are finite real numbers such that
$$S = \sum_{n=k}^{\infty} s_n \quad \text{and} \quad T = \sum_{n=k}^{\infty} t_n$$
Prove that if $c, d \in \mathbb{R}$, then
$$\sum_{n=k}^{\infty} (cs_n + dt_n) = cS + dT.$$

LEVEL 2

3. Use the Root Test to prove that the following series converges:
$$\sum_{n=0}^{\infty} s_n, \text{ where } s_n = \begin{cases} \dfrac{1}{2^n} & \text{if } n \text{ is odd.} \\ 0 & \text{if } n \text{ is even.} \end{cases}$$

4. Suppose that we apply the Root Test to the series $\sum s_n$ and we are able to conclude that either the series converges or diverges. Will the Ratio Test necessarily provide the same information? (Compare this to Problem 14 below.)

5. Prove that a conditionally convergent series contains infinitely many positive terms and infinitely many negative terms.

6. Let $\sum s_n$ be a conditionally convergent series. For each $n \in \mathbb{N}$, let $s_n^+ = \frac{s_n + |s_n|}{2}$ and $s_n^- = \frac{s_n - |s_n|}{2}$. Prove that the series $\sum s_n^+$ consisting of only the positive terms from $\sum s_n$ is divergent and the series $\sum s_n^-$ consisting of only the negative terms from $\sum s_n$ is divergent.

LEVEL 3

7. Prove that if a series converges absolutely, then it converges.

8. Let (s_n) be a sequence with $s_n \neq 0$ for all $n \in \mathbb{N}$ and suppose that there is $r \in (0, 1)$ and $K \in \mathbb{N}$ such that $n > K \to \left|\frac{s_{n+1}}{s_n}\right| \leq r$. Prove that $\sum s_n$ converges absolutely. This is Lemma 15.22.

9. Let (s_n) be a sequence with $s_n \neq 0$ for all $n \in \mathbb{N}$ and suppose that there is $K \in \mathbb{N}$ such that $n > K \to \left|\frac{s_{n+1}}{s_n}\right| > 1$. Prove that $\sum s_n$ diverges. This is Lemma 15.23.

10. Let (s_n) be a sequence with $s_n \neq 0$ for all $n \in \mathbb{N}$ and let
$$L = \limsup |s_n|^{\frac{1}{n}}$$
Prove each of the following:
$$\text{if } L < 1, \text{ then } \sum_{n=0}^{\infty} s_n \text{ converges absolutely.}$$
$$\text{if } L > 1, \text{ then } \sum_{n=0}^{\infty} s_n \text{ diverges.}$$
This is the **Root Test** (Theorem 15.25).

11. Consider the series $\sum_{n=k}^{\infty} s_n$. Let (n_j) be an increasing sequence of integers with $n_0 = k$. For each $j \in \mathbb{N}^+$, let
$$t_j = \sum_{n=n_j}^{n_{j+1}-1} s_n = s_{n_j} + s_{n_j+1} + \cdots + s_{n_{j+1}-1}.$$
The series $\sum_{j=0}^{\infty} t_n$ is called a **regrouping** of $\sum_{n=k}^{\infty} s_n$.

Prove that if $\sum_{n=k}^{\infty} s_n$ is a convergent series, then any regrouping of $\sum_{n=k}^{\infty} s_n$ converges to the same value. Is there an analogous result for divergent series?

LEVEL 4

12. Prove that for each $k \in \mathbb{N}^+$,
$$\sum_{n=1}^{k} \cos n = \frac{\sin\left(k+\frac{1}{2}\right) - \sin\frac{1}{2}}{2\sin\frac{1}{2}}.$$

13. Let (s_n) be a bounded monotone sequence and let (t_n) be another sequence whose sequence of partial sums converges. Then the following series converges:
$$\sum_{n=1}^{k} s_n t_n$$
This is **Abel's Test** (Theorem 15.11).

LEVEL 5

14. Suppose that we apply the Ratio Test to the series $\sum s_n$ and we are able to conclude that either the series converges or diverges. Prove that the Root Test will provide the same information.

15. Prove that the alternating harmonic series converges to ln 2. That is

$$\sum_{n=1}^{\infty} \frac{(-1)^{n+1}}{n} = 1 - \frac{1}{2} + \frac{1}{3} - \frac{1}{4} + \cdots = \ln 2.$$

16. Let (s_n) and (t_n) be sequences and suppose that there is $K \in \mathbb{N}$ such that $n > K$ implies that $s_n \geq 0$ and $t_n > 0$. Also, suppose that there is $L \in \mathbb{R} \cup \{\infty\}$ such that

$$\lim_{n \to \infty} \frac{s_n}{t_n} = L.$$

Prove each of the following:

if $L \neq 0$, then $\sum_{n=0}^{\infty} s_n$ converges if and only if $\sum_{n=0}^{\infty} t_n$ converges.

if $L = 0$ and $\sum_{n=0}^{\infty} t_n$ converges, then $\sum_{n=0}^{\infty} s_n$ converges.

if $L = \infty$ and $\sum_{n=0}^{\infty} t_n$ diverges, then $\sum_{n=0}^{\infty} s_n$ diverges.

This is the **Limit Comparison Test** (Theorem 15.15).

CHALLENGE PROBLEMS

17. Let $\sum s_n$ be a conditionally convergent series and let $c \in \mathbb{R}$. Prove that there is a rearrangement of $\sum s_n$ that converges to c.

18. Let $\sum s_n$ be a conditionally convergent series. Prove that there is a rearrangement of $\sum s_n$ that diverges.

19. Let (s_n) be a sequence with $s_n \neq 0$ for all $n \in \mathbb{N}$ and let

$$L = \lim_{n \to \infty} n\left(\frac{s_n}{s_{n+1}} - 1\right).$$

Assuming that L is a finite real number, prove that (i) $L > 1$ implies that $\sum s_n$ converges and (ii) $L < 1$ implies that $\sum s_n$ diverges. This result is known as **Raabe's Test**.

LESSON 16
SEQUENCES AND SERIES OF FUNCTIONS

Sequences of Functions

Let $A \subseteq \mathbb{R}$ and for each $n \in \mathbb{N}$, let $f_n: A \to \mathbb{R}$. Then (f_n) is called a **sequence of functions on A**. If $B \subseteq A$, we say that this sequence of functions **converges pointwise to f on B**, written $\lim_{n \to \infty} f_n = f$ on B (or $f_n \to f$ on B), if for every $x \in B$, $\lim_{n \to \infty} f_n(x) = f(x)$.

Example 16.1:

1. For each $n \in \mathbb{N}$, let $f_n: \mathbb{R} \to \mathbb{R}$ be defined by $f_n(x) = x^n$. Let $f(x) = \begin{cases} 0 & \text{if } -1 < x < 1. \\ 1 & \text{if } \quad x = 1. \end{cases}$ Then $\lim_{n \to \infty} f_n = f$ on $(-1, 1]$. This follows from part 2 of Example 14.1. Observe that if $x \leq -1$ or $x > 1$, then $(f_n(x))$ diverges (again by part 2 of Example 14.1).

    Notice that although each function f_n is continuous everywhere, the pointwise limit of (f_n) has a discontinuity at $x = 1$. This example shows that the pointwise limit of a sequence of continuous functions is **not** necessarily continuous. Similarly, this example also shows that the pointwise limit of a sequence of differentiable functions is **not** necessarily differentiable.

2. Let $\mathbb{Q} \cap [0, 1] = \{q_0, q_1, q_2, \dots\}$ be the set of rational numbers such that each rational number in $[0, 1]$ appears exactly once ($\mathbb{Q}$ can be written this way because it is countable—see part (ii) of Problem 7 in Problem Set 5). For each $n \in \mathbb{N}$, define $f_n: [0, 1] \to \mathbb{R}$ by $f_n(x) = \begin{cases} 1 & \text{if } x \in \{q_0, \dots, q_n\}. \\ 0 & \text{if } x \notin \{q_0, \dots, q_n\}. \end{cases}$ Since each f_n is a bounded function with only finitely many discontinuities, by Problem 16 in Problem Set 11, each f_n is Riemann integrable on $[0, 1]$. The pointwise limit of (f_n) is the function $f: [0, 1] \to \mathbb{R}$ defined by $f(x) = \begin{cases} 1 & \text{if } x \text{ is rational}. \\ 0 & \text{if } x \text{ is irrational}. \end{cases}$ By part 2 of Example 11.6, f is **not** Riemann integrable on $[0, 1]$. This example shows that the pointwise limit of a Riemann integrable function is **not** necessarily Riemann integrable.

Example 16.1 shows that the notion of pointwise convergence does not seem to have any reasonable "preservation properties." Therefore, it is not as useful as we might hope theoretically. In a moment, we will provide a stronger notion of convergence that is much better behaved. As a first step, let's write down an $\epsilon - K$ definition of pointwise convergence. The following theorem follows immediately from the definition of a convergent sequence.

Theorem 16.2: Let (f_n) be a sequence of functions on $A \subseteq \mathbb{R}$. Then (f_n) converges pointwise to f on $B \subseteq A$ if and only

$$\forall x \in B \, \forall \epsilon > 0 \, \exists K \in \mathbb{N} \, (n > K \to |f_n(x) - f(x)| < \epsilon).$$

Note: In words, (f_n) converges pointwise to f on B if for every $x \in B$ and every positive real number ϵ, there is a natural number K such that n greater than K implies that the distance between $f_n(x)$ and $f(x)$ is less than ϵ. Informally, for each $x \in B$, we can make the distance between $f_n(x)$ and $f(x)$ as small as we choose by deleting a finite portion of the beginning of the sequence of functions.

By simply moving the universal quantifier $\forall x \in B$ so that appears after the existential quantifier $\exists K \in \mathbb{N}$, we get a much stronger notion of convergence.

Let (f_n) be a sequence of functions on $A \subseteq \mathbb{R}$ and let $B \subseteq A$. We say that (f_n) **converges uniformly on B** (or (f_n) is **uniformly convergent** on B) to $f: B \to \mathbb{R}$ if

$$\forall \epsilon > 0 \, \exists K \in \mathbb{N} \, \forall x \in B \, (n > K \to |f_n(x) - f(x)| < \epsilon).$$

Notes: (1) In words, (f_n) converges uniformly on B to f if for every positive real number ϵ, there is a natural number K such that for all $x \in B$, n greater than K implies that the distance between $f_n(x)$ and $f(x)$ is less than ϵ. Informally, we can make the distance between $f_n(x)$ and $f(x)$ as small as we choose for all $x \in B$ simultaneously by deleting a finite portion of the beginning of the sequence.

(2) There is a subtle but important difference between the definitions of pointwise convergence and uniform convergence. In the definition of pointwise convergence, for a fixed $\epsilon > 0$, each $x \in B$ might require a different integer K to satisfy the conclusion given in the definition. On the other hand, for uniform convergence, given $\epsilon > 0$, a single integer K can be used for every $x \in B$ simultaneously.

(3) Observe that uniform convergence is **stronger** than pointwise convergence. If (f_n) converges uniformly on B, then (f_n) converges pointwise on B. We will see in Example 16.3 below that uniform convergence is **strictly stronger** than pointwise convergence. In other words, there are sequences that converge pointwise, but not uniformly on a given set.

(4) In general, if $P(x)$ is a property, then the negation of $\forall x(P(x))$ is $\exists x(\neg P(x))$ (see Note 2 before Example 1.16 for more information on the negation symbol, $\neg$). In other words, when we pass a negation symbol through a universal quantifier, the quantifier changes to an existential quantifier. So, $\neg \forall x(P(x)) \equiv \exists x(\neg P(x))$, where $\equiv$ is pronounced "is logically equivalent to." Similarly, we also have $\neg \exists x(P(x)) \equiv \forall x(\neg P(x))$. So, the negation of "$(f_n)$ converges uniformly on B to f" is equivalent to

$$\exists \epsilon > 0 \, \forall K \in \mathbb{N} \, \exists x \in B \, \neg(n > K \to |f_n(x) - f(x)| < \epsilon).$$

(5) Let's simplify the statement $\neg(n > K \to |f_n(x) - f(x)| < \epsilon)$. This statement has the form $\neg(p \to q)$, where $p = n > K$ and $q = |f_n(x) - f(x)| < \epsilon$. Now, $\neg(p \to q) \equiv \neg(\neg p \lor q) \equiv p \land \neg q$. These equivalences can be verified using basic propositional logic. The details of this fall a bit outside the scope of this book, and so, I leave this for the reader to explore independently. We now see that $\neg(n > K \to |f_n(x) - f(x)| < \epsilon)$ is logically equivalent to $n > K \land |f_n(x) - f(x)| \geq \epsilon$ (note that since $<$ is a linear ordering on $\mathbb{R}$, the negation of $|f_n(x) - f(x)| < \epsilon$ is equivalent to $|f_n(x) - f(x)| \geq \epsilon$). Our final conclusion is that the negation of "(f_n) converges uniformly on B" is equivalent to

$$\exists \epsilon > 0 \, \forall K \in \mathbb{N} \, \exists x \in B \, (n > K \land |f_n(x) - f(x)| \geq \epsilon).$$

(6) Using Note 4 above, we have the following method for proving that a sequence does not converge uniformly to a given function:

In order to prove that (f_n) does **not** converge uniformly on B to $f: B \to \mathbb{R}$, it suffices to find an $\epsilon > 0$, a subsequence (f_{n_k}) of (f_n), and a sequence (x_k) in B such that $|f_{n_k}(x_k) - f(x_k)| \geq \epsilon$ for all $k \in \mathbb{N}$.

Example 16.3:

1. For each $n \in \mathbb{N}$, let $f_n: \mathbb{R} \to \mathbb{R}$ be defined by $f_n(x) = x^n$. Let $f(x) = \begin{cases} 0 & \text{if } -1 < x < 1. \\ 1 & \text{if } x = 1. \end{cases}$ In part 1 of Example 16.1, we saw that (f_n) converges pointwise to f on $(-1, 1]$. Does (f_n) converge uniformly on $(-1, 1]$ to f?

   The answer is no! To see this, let $\epsilon = \frac{1}{2}$ and for each $k \in \mathbb{N}$, let $n_k = k$ and $x_k = \left(\frac{1}{2}\right)^{\frac{1}{k}}$. First observe that since $0 < \frac{1}{2} < 1$, for each $k \in \mathbb{N}$, we have $0^{\frac{1}{k}} < \left(\frac{1}{2}\right)^{\frac{1}{k}} < 1^{\frac{1}{k}}$, or $0 < \left(\frac{1}{2}\right)^{\frac{1}{k}} < 1$. So, for each $k \in \mathbb{N}$, $x_k \in (0, 1) \subseteq (-1, 1]$. Then for each $k \in \mathbb{N}$, we have

   $$\left|f_{n_k}(x_k) - f(x_k)\right| = \left|f_k\left(\left(\frac{1}{2}\right)^{\frac{1}{k}}\right) - f\left(\left(\frac{1}{2}\right)^{\frac{1}{k}}\right)\right| = \left|\left(\left(\frac{1}{2}\right)^{\frac{1}{k}}\right)^k - 0\right| = \left|\frac{1}{2}\right| = \frac{1}{2}.$$

   By Note 6 above, (f_n) does **not** converge uniformly on $(-1, 1]$ to f.

2. Let's restrict the domain of the functions given in part 1 above to $\left[-\frac{1}{2}, \frac{1}{2}\right]$. In other words, for each $n \in \mathbb{N}$, we define $f_n: \left[-\frac{1}{2}, \frac{1}{2}\right] \to \mathbb{R}$ by $f_n(x) = x^n$. We then define $f: \left[-\frac{1}{2}, \frac{1}{2}\right] \to \mathbb{R}$ by $f(x) = 0$ (so, f is the constant function that is identically 0 on $\left[-\frac{1}{2}, \frac{1}{2}\right]$). With these restricted domains, the sequence (f_n) does converge uniformly to f. To see this, let $\epsilon > 0$. Since $\lim_{n \to \infty} \left(\frac{1}{2}\right)^n = 0$ (by part 2 of Example 14.1), we can choose $K \in \mathbb{N}$ such that $n > K \to \left(\frac{1}{2}\right)^n < \epsilon$. Then for any $x \in \left[-\frac{1}{2}, \frac{1}{2}\right]$, we have $|x| \leq \frac{1}{2}$, and therefore, $n > K \to |x^n| = |x|^n \leq \left(\frac{1}{2}\right)^n < \epsilon$.

   Note that the same argument works if we replace $\left[-\frac{1}{2}, \frac{1}{2}\right]$ by $[-c, c]$ for any $c \in (0, 1)$.

3. As in part 2 of Example 16.1, let $\mathbb{Q} \cap [0, 1] = \{q_0, q_1, q_2, \ldots\}$ be the set of rational numbers such that each rational number in $[0, 1]$ appears exactly once and for each $n \in \mathbb{N}$, define $f_n: [0, 1] \to \mathbb{R}$ by $f_n(x) = \begin{cases} 1 & \text{if } x \in \{q_0, \ldots, q_n\}. \\ 0 & \text{if } x \notin \{q_0, \ldots, q_n\}. \end{cases}$ In part 2 of Example 16.1, we saw that (f_n) converges pointwise to the function $f: [0, 1] \to \mathbb{R}$ defined by $f(x) = \begin{cases} 1 & \text{if } x \text{ is rational.} \\ 0 & \text{if } x \text{ is irrational.} \end{cases}$

   In this case, (f_n) does **not** converge uniformly on $[0, 1]$ to f. Once again, we will use Note 6 above to show this. Let $\epsilon = 1$ and for each $k \in \mathbb{N}$, let $n_k = k + 1$ and $x_k = q_k$. Then for each $k \in \mathbb{N}$, we have

   $$\left|f_{n_k}(x_k) - f(x_k)\right| = |f_{k+1}(q_k) - f(q_k)| = |0 - 1| = |-1| = 1 \geq \epsilon.$$

4. For each $n \in \mathbb{N}$, define $f_n: \mathbb{R} \to \mathbb{R}$ by $f_n(x) = \frac{\sin nx}{n}$ and let $f: \mathbb{R} \to \mathbb{R}$ be the constant 0 function (that is $f(x) = 0$ for all $x \in \mathbb{R}$). Then (f_n) converges uniformly on $\mathbb{R}$ to f. Indeed, if $\epsilon > 0$, by the Archimedean Property of $\mathbb{R}$, we can choose $K \in \mathbb{N}$ so that $K > \frac{1}{\epsilon}$, or equivalently, $\frac{1}{K} < \epsilon$. Then for any $x \in \mathbb{R}$, if $n > K$, then $|f_n(x) - f(x)| = \frac{|\sin nx|}{n} \leq \frac{1}{n} < \frac{1}{K} < \epsilon$.

The Uniform Norm

Recall that a function $f: A \to \mathbb{R}$ is **bounded** on A if $\{f(x) \mid x \in A\}$ is bounded. If $f: A \to \mathbb{R}$ is a bounded function, we define the **uniform norm of f on A**, written $|f|_A$ by

$$|f|_A = \sup\{|f(x)| \mid x \in A\}.$$

The uniform norm provides us with a convenient method for testing if a given sequence of functions converges uniformly on a given set, as we now show.

Theorem 16.4: Let $B \subseteq A \subseteq \mathbb{R}$, let (f_n) be a sequence of functions that are bounded on A and let $f: B \to \mathbb{R}$. Then (f_n) converges uniformly on B to f if and only if $\lim_{n \to \infty} |f_n - f|_B = 0$.

Proof: Suppose that (f_n) converges uniformly on B to f and let $\epsilon > 0$. Then there is $K \in \mathbb{N}$ such that for all $x \in B, n > K \to |f_n(x) - f(x)| < \frac{\epsilon}{2}$. Then for all $n > K$, we have

$$|f_n - f|_B = \sup\{|f_n(x) - f(x)| \mid x \in A\} \leq \frac{\epsilon}{2} < \epsilon.$$

Since $\epsilon > 0$ was arbitrary, $\lim_{n \to \infty} |f_n - f|_B = 0$.

Conversely, suppose that $\lim_{n \to \infty} |f_n - f|_B = 0$ and let $\epsilon > 0$. Then there is $K \in \mathbb{N}$ such that

$$n > K \to |f_n - f|_B < \epsilon.$$

Since $x \in B$ implies $|f_n(x) - f(x)| \leq \sup\{|f_n(x) - f(x)| \mid x \in A\} = |f_n - f|_B$, we see that for all $x \in B$, we have $n > K \to |f_n(x) - f(x)| < \epsilon$. Therefore, (f_n) converges uniformly on B to f. □

Example 16.5:

1. For each $n \in \mathbb{N}$, let $f_n: \mathbb{R} \to \mathbb{R}$ be defined by $f_n(x) = x^n$. Let $f(x) = \begin{cases} 0 & \text{if } -1 < x < 1 \\ 1 & \text{if } x = 1 \end{cases}$. For $n > 0$, f_n is **not** bounded on $\mathbb{R}$, and therefore, $|f_n|_{\mathbb{R}}$ is not defined. So, let's restrict the domain of each f_n to $[-1, 1]$. Then for each $n \in \mathbb{N}^+$, $|f_n|_{[-1,1]} = 1$.

   If we let $f(x) = \begin{cases} 0 & \text{if } -1 < x < 1 \\ 1 & \text{if } x = 1 \end{cases}$, then (f_n) converges pointwise to f. Now, $f_n(x) - f(x)$ is bounded on $[-1, 1]$ and for all $n \in \mathbb{N}$, $|f_n - f|_{[-1,1]} = \sup\begin{Bmatrix} x^n & \text{if } -1 < x < 1 \\ 0 & \text{if } x = 1 \end{Bmatrix} = 1$. Therefore, $\lim_{n \to \infty} |f_n - f|_{[-1,1]} = 1$. So, by Theorem 16.4, (f_n) does **not** converge uniformly on $[-1, 1]$ to f.

2. Let's restrict the domain of the functions given in part 1 above to $\left[-\frac{1}{2}, \frac{1}{2}\right]$, as we did in part 2 of Example 16.3. In other words, for each $n \in \mathbb{N}$, we define $f_n: \left[-\frac{1}{2}, \frac{1}{2}\right] \to \mathbb{R}$ by $f_n(x) = x^n$. We then define $f: \left[-\frac{1}{2}, \frac{1}{2}\right] \to \mathbb{R}$ by $f(x) = 0$ (so, f is the constant function that is identically 0 on $\left[-\frac{1}{2}, \frac{1}{2}\right]$). With these restricted domains, $\lim_{n \to \infty} |f_n - f|_{\left[-\frac{1}{2}, \frac{1}{2}\right]} = \lim_{n \to \infty} \left(\frac{1}{2}\right)^n = 0$ (by part 2 of Example 14.1). So, by Theorem 16.4, (f_n) converges uniformly on $\left[-\frac{1}{2}, \frac{1}{2}\right]$ to f.

As in part 2 of Example 16.3, the same argument works if we replace $\left[-\frac{1}{2},\frac{1}{2}\right]$ by $[-c,c]$ for any $c \in (0,1)$.

3. For each $n \in \mathbb{N}$, let $f_n: [0,1] \to \mathbb{R}$ be defined by $f_n(x) = \frac{nx^2 + \cos(nx+1)}{n}$ and let $f: [0,1] \to \mathbb{R}$ be defined by $f(x) = x^2$. Then (f_n) converges uniformly on $[0,1]$ to f. This follows from Theorem 16.4 and the following computation:

$$\lim_{n\to\infty} |f_n - f|_{[0,1]} = \lim_{n\to\infty}\left(\sup\left\{\left|\frac{nx^2 + \cos(nx+1)}{n} - x^2\right| \,\Big|\, x \in [0,1]\right\}\right)$$

$$= \lim_{n\to\infty}\left(\sup\left\{\frac{|\cos(nx+1)|}{n} \,\Big|\, x \in [0,1]\right\}\right) \leq \lim_{n\to\infty}\left(\sup\left\{\frac{1}{n} \,\Big|\, x \in [0,1]\right\}\right) = \lim_{n\to\infty} \frac{1}{n} = 0.$$

Let $A \subseteq \mathbb{R}$ and let (f_n) be a sequence of functions that are bounded on A. We say that (f_n) is **uniformly Cauchy** (or **Cauchy in the uniform norm**) on A if

$$\forall \epsilon > 0 \, \exists K \in \mathbb{N} \, (m \geq n > K \to |f_m - f_n|_A < \epsilon).$$

Notes: (1) In words, (f_n) is uniformly Cauchy if for every positive real number ϵ, there is a natural number K such that $m \geq n > K$ implies that the "distance" between f_m and f_n is less than ϵ. By distance, we mean the supremum of all distances between $f_m(x)$ and $f_n(x)$, as x ranges over the elements of A. Informally, we can make the distance between $f_m(x)$ and $f_n(x)$ as small as we choose for all $x \in A$ simultaneously by deleting a finite portion of the beginning of the sequence.

(2) The negation of the statement $\forall \epsilon > 0 \, \exists K \in \mathbb{N} \, (m \geq n > K \to |f_m - f_n|_A < \epsilon)$ is the following:

$$\exists \epsilon > 0 \, \forall K \in \mathbb{N} \, (m \geq n > K \wedge |f_m - f_n|_A \geq \epsilon).$$

To see this, we can use reasoning similar to that used in Notes 4 and 5 before Example 16.3.

Theorem 16.6: Let $B \subseteq A \subseteq \mathbb{R}$ and let (f_n) be a sequence of functions that are bounded on A. Then (f_n) converges uniformly on B to a bounded function $f: B \to \mathbb{R}$ if and only if (f_n) is uniformly Cauchy on B.

Proof: Suppose that (f_n) converges uniformly on B to f and let $\epsilon > 0$. Then there is $K \in \mathbb{N}$ such that for all $x \in B$, $n > K \to |f_n(x) - f(x)| < \frac{\epsilon}{4}$. So, for all $m \geq n > K$, we have

$$|f_m(x) - f_n(x)| = |f_m(x) - f(x) + f(x) - f_n(x)| \leq |f_m(x) - f(x)| + |f(x) - f_n(x)| < \frac{\epsilon}{4} + \frac{\epsilon}{4} = \frac{\epsilon}{2}.$$

Therefore, $|f_m - f_n|_B = \sup\{|f_m(x) - f_n(x)| \,|\, x \in B\} \leq \frac{\epsilon}{2} < \epsilon$. So, (f_n) is uniformly Cauchy on B.

Conversely, suppose that (f_n) is uniformly Cauchy on B and let $b \in B$. Given $\epsilon > 0$, there is $K \in \mathbb{N}$ such that $m, n > K \to \sup\{|f_m(x) - f_n(x)| \,|\, x \in B\} = |f_m - f_n|_B < \epsilon$. Therefore, for $m, n > K$, we have

$$|f_m(b) - f_n(b)| \leq \sup\{|f_m(x) - f_n(x)| \,|\, x \in B\} < \epsilon.$$

It follows that for each $b \in B$, $(f_n(b))$ is a Cauchy sequence in $\mathbb{R}$. By Corollary 8.29, for each $b \in B$, there is $x_b \in \mathbb{R}$ such that $f_n(b) \to x_b$.

Define $f: B \to \mathbb{R}$ by $f(b) = x_b$. Then (f_n) converges pointwise to f. We will now show that the convergence is uniform. To this end, let $\epsilon > 0$. Since (f_n) is uniformly Cauchy on B, there is $K \in \mathbb{N}$ such that $m \geq n > K$ implies $|f_m - f_n|_B < \frac{\epsilon}{2}$. Therefore, if $x \in B$ and $m \geq n > K$, then

$$|f_m(x) - f_n(x)| \leq |f_m - f_n|_B < \frac{\epsilon}{2}.$$

By the Squeeze Theorem (Problem 9 in Problem Set 14), for each $x \in B$, we have

$$m > K \to \lim_{n \to \infty} |f_m(x) - f_n(x)| \leq \lim_{n \to \infty} |f_m - f_n|_B \leq \lim_{n \to \infty} \frac{\epsilon}{2},$$

Therefore, for each $x \in B$,

$$m > K \to |f_m(x) - f(x)| \leq |f_m - f|_B \leq \frac{\epsilon}{2} < \epsilon.$$

So, (f_n) converges uniformly on B to f. □

Preservation Properties of Uniform Convergence

Example 16.1 showed us that the pointwise limit of continuous functions need not be continuous, the pointwise limit of differentiable functions need not be differentiable, and the pointwise limit of Riemann integrable functions need not be Riemann integrable. In this section, we will see that uniform convergence is much more well-behaved.

Theorem 16.7: Let $B \subseteq A \subseteq \mathbb{R}$ and for each $n \in \mathbb{N}$, let $f_n: A \to \mathbb{R}$ be continuous on B. Suppose that (f_n) converges uniformly on B to f. Then $f: B \to \mathbb{R}$ is continuous on B.

Proof: Let $x \in B$ and let $\epsilon > 0$. Since (f_n) converges uniformly on B to f, there is $K \in \mathbb{N}$ such that for all $y \in B$, $n > K$ implies $|f_n(y) - f(y)| < \frac{\epsilon}{3}$. In particular, for all $y \in B$, $|f_{K+1}(y) - f(y)| < \frac{\epsilon}{3}$. Since f_{K+1} is continuous on B, it is continuous at x. So, there is $\delta > 0$ such that $|x - y| < \delta$ implies $|f_{K+1}(x) - f_{K+1}(y)| < \frac{\epsilon}{3}$. Now, assume that $|x - y| < \delta$. By the Triangle Inequality (and SACT),

$$|f(x) - f(y)| = |(f(x) - f_{K+1}(x)) + (f_{K+1}(x) - f_{K+1}(y)) + (f_{K+1}(y) - f(y))|$$
$$\leq |f(x) - f_{K+1}(x)| + |f_{K+1}(x) - f_{K+1}(y)| + |f_{K+1}(y) - f(y)| < \frac{\epsilon}{3} + \frac{\epsilon}{3} + \frac{\epsilon}{3} = \epsilon.$$

Therefore, f is continuous at x. Since $x \in B$ was arbitrary, f is continuous on B. □

Theorem 16.8: For each $n \in \mathbb{N}$, let $f_n: [a, b] \to \mathbb{R}$ be a bounded function on $[a, b]$ that is Riemann integrable on $[a, b]$. Suppose that (f_n) converges uniformly on $[a, b]$ to f. Then $f: [a, b] \to \mathbb{R}$ is Riemann integrable on $[a, b]$ and

$$\int_a^b f(x)\, dx = \lim_{n \to \infty} \int_a^b f_n(x)\, dx.$$

Proof: Let $\epsilon > 0$. Since (f_n) converges uniformly on $[a, b]$ to f, there is $K \in \mathbb{N}$ such that $n > K$ implies $|f_n - f|_{[a,b]} < \frac{\epsilon}{2(b-a)}$. So, for all $x \in [a, b]$, $n < K$ implies $|f_n(x) - f(x)| \leq |f_n - f|_{[a,b]} < \frac{\epsilon}{2(b-a)}$.

Let $x \in [a, b]$. By the Triangle Inequality (and SACT),
$$|f(x)| = |f(x) - f_{K+1}(x) + f_{K+1}(x)| \leq |f(x) - f_{K+1}(x)| + |f_{K+1}(x)| < \frac{\epsilon}{2(b-a)} + |f_{K+1}(x)|.$$

Since f_{K+1} is bounded on $[a, b]$, there is $M \in \mathbb{R}^+$ such that for all $x \in [a, b]$, $|f_{K+1}(x)| \leq M$. If we let $M^* = \frac{\epsilon}{2(b-a)} + M$, then $x \in [a, b]$ implies $|f(x)| \leq M^*$. So, f is bounded on $[a, b]$.

Since f_{K+1} is Riemann integrable on $[a, b]$, by Note 1 following the proof of Theorem 11.9 and Problem 11 from Problem Set 11, we have

$$\overline{\int_a^b} f(x)\, dx - \underline{\int_a^b} f(x)\, dx$$

$$= \overline{\int_a^b} \left(f(x) - f_{K+1}(x) + f_{K+1}(x)\right) dx - \underline{\int_a^b} \left(f(x) - f_{K+1}(x) + f_{K+1}(x)\right) dx$$

$$\leq \overline{\int_a^b} \left(f(x) - f_{K+1}(x)\right) dx + \overline{\int_a^b} f_{K+1}(x)\, dx - \underline{\int_a^b} \left(f(x) - f_{K+1}(x)\right) dx - \underline{\int_a^b} f_{K+1}(x)\, dx$$

$$= \overline{\int_a^b} \left(f(x) - f_{K+1}(x)\right) dx + \int_a^b f_{K+1}(x)\, dx - \underline{\int_a^b} \left(f(x) - f_{K+1}(x)\right) dx - \int_a^b f_{K+1}(x)\, dx$$

$$= \overline{\int_a^b} \left(f(x) - f_{K+1}(x)\right) dx - \underline{\int_a^b} \left(f(x) - f_{K+1}(x)\right) dx$$

$$\leq \frac{\epsilon}{2(b-a)}(b-a) + \frac{\epsilon}{2(b-a)}(b-a) = \epsilon.$$

Since $\epsilon > 0$ was arbitrary, by Problem 15 from Problem Set 6, $\overline{\int_a^b} f(x)\, dx - \underline{\int_a^b} f(x)\, dx = 0$. Therefore, $\overline{\int_a^b} f(x)\, dx = \underline{\int_a^b} f(x)\, dx$, and so, f is Riemann integrable on $[a, b]$.

Finally, by Theorem 11.9 and Problem 11 from Problem Set 11, for $n > K$, we have

$$\left|\int_a^b f_n(x)\, dx - \int_a^b f(x)\, dx\right| = \left|\int_a^b \left(f_n(x) - f(x)\right) dx\right| \leq \frac{\epsilon}{2(b-a)}(b-a) = \frac{\epsilon}{2} < \epsilon.$$

So, it follows that
$$\lim_{n \to \infty} \int_a^b f_n(x)\, dx = \int_a^b f(x)\, dx. \qquad \square$$

We have just seen that uniform convergence preserves continuity and Riemann integrability. What about differentiability? Unfortunately, differentiability is not preserved, as the next example will show.

Example 16.9: For each $n \in \mathbb{N}$, define $f_n: \mathbb{R} \to \mathbb{R}$ by $f_n(x) = \frac{\sin nx}{n}$ and let $f: \mathbb{R} \to \mathbb{R}$ be the constant 0 function (that is $f(x) = 0$ for all $x \in \mathbb{R}$). In part 4 of Example 16.3, we saw that (f_n) converges uniformly on $\mathbb{R}$ to f. For each $n \in \mathbb{N}$, by the chain rule (Theorem 10.20), $f_n'(x) = \frac{n \cos nx}{n} = \cos nx$. If $x \neq 0$, then $\lim_{n \to \infty} \cos nx$ does **not** converge. So, the sequence (f_n') does not converge.

Although differentiability is not preserved under uniform convergence of the original sequence, we do have the following useful result.

Theorem 16.10: Let I be a bounded interval and for each $n \in \mathbb{N}$, let $f_n: I \to \mathbb{R}$ have a continuous derivative on I. Suppose that (f_n') converges uniformly on I to $g: I \to \mathbb{R}$ and assume that there is $c \in I$ such that $(f_n(c))$ is a convergent sequence. Then (f_n) converges uniformly on I to a function $f: I \to \mathbb{R}$ with a continuous derivative on I and $f'(x) = g(x)$ for all $x \in I$.

Note: If we delete both instances of the word "continuous" in Theorem 16.9, we get a statement that is also true. However, this statement is much harder to prove. I leave it as a challenging exercise (see Problem 14 below).

Proof of Theorem 16.10: Since $(f_n(c))$ is a convergent sequence, there is $L \in \mathbb{R}$ such that $\lim_{n \to \infty} f_n(c) = L$.

Fix $n \in \mathbb{N}$. Since f_n' is continuous on I, by Theorem 11.7, f_n' is Riemann integrable on any closed interval contained in I. By the second form of the Fundamental Theorem of Calculus (Theorem 11.14), for each $x \in I$, we have

$$\int_c^x f_n'(t)\, dt = f_n(x) - f_n(c),$$

or equivalently,

$$f_n(x) = f_n(c) + \int_c^x f_n'(t)\, dt.$$

Since (f_n') converges uniformly on I, it converges uniformly on $[c, x]$ (if $c \leq x$) or $[x, c]$ (if $c > x$). By Theorem 16.8, we have $\lim_{n \to \infty} f_n(x) = \lim_{n \to \infty} f_n(c) + \lim_{n \to \infty} \int_c^x f_n'(t)\, dt = L + \int_c^x g(t)\, dt$.

By Theorem 16.7, g is continuous on I.

Define $f: I \to \mathbb{R}$ by $f(x) = L + \int_c^x g(t)\, dt$. By the first form of the Fundamental Theorem of Calculus (Theorem 11.12), $f'(x) = g(x)$ for all $x \in I$.

Let $a = \inf I$, let $b = \sup I$, and let $\epsilon > 0$. Since $(f_n(c))$ converges to $f(c) = L$, there is $K_1 \in \mathbb{N}$ such that $n > K_1$ implies $|f_n(c) - L| < \frac{\epsilon}{2}$. Since (f_n') converges uniformly on I to g, there is $K_2 \in \mathbb{N}$ such that for all $x \in I$, $n > K_2$ implies $|f_n'(x) - g(x)| < \frac{\epsilon}{2(b-a)}$. Let $K = \max\{K_1, K_2\}$. Let $x \in I$ and $n > K$. Then we have

$$|f_n(x) - f(x)| = \left| f_n(c) + \int_c^x f_n'(t)\,dt - L - \int_c^x g(t)\,dt \right|$$

$$\leq |f_n(c) - L| + \left| \int_c^x f_n'(t)\,dt - \int_c^x g(t)\,dt \right| \quad \text{(by the Triangle Inequality)}$$

$$= |f_n(c) - L| + \left| \int_c^x \left(f_n'(t) - g(t)\right) dt \right| \quad \text{(by Theorem 11.9)}$$

$$\leq |f_n(c) - L| + \left| \int_c^x |f_n'(t) - g(t)|\,dt \right| \quad \text{(by Problem 6 from Problem Set 11)}$$

$$< \frac{\epsilon}{2} + \left| \frac{\epsilon}{2(b-a)}(b-a) \right| \quad \text{(by Problem 11 from Problem Set 11)}$$

$$= \frac{\epsilon}{2} + \frac{\epsilon}{2} = \epsilon.$$

Therefore, (f_n) converges uniformly on I to f. □

Series of Functions

Let (f_n) be a sequence of functions on a set $A \subseteq \mathbb{R}$. The **kth partial sum** of the sequence is the sequence (g_k), where $g_k: A \to \mathbb{R}$ is defined as

$$g_k(x) = \sum_{n=0}^{k} f_n(x) = f_0(x) + f_1(x) + \cdots + f_k(x).$$

So, the 0th partial sum at x is $g_0(x) = f_0(x)$, the 1st partial sum at x is $g_1(x) = f_0(x) + f_1(x)$, the 2nd partial sum at x is $g_2(x) = f_0(x) + f_1(x) + f_2(x)$, and so on.

The sequence of functions (g_k) is called a **series of functions**.

The functions $f_0, f_1, \ldots, f_n, \ldots$ are called the **terms** of the series (whereas the functions $g_0, g_1, \ldots, g_n, \ldots$ are the **partial sums** of the series). So, f_n is the **nth term** of the series (whereas $g_n = f_0 + f_1 + \cdots + f_n$ is the **nth partial sum** of the series).

The following notation is used to represent both the series (g_n) and the sum of the series $\lim_{n \to \infty} g_n$:

$$\sum_{n=0}^{\infty} f_n = f_0 + f_1 + \cdots$$

If $B \subseteq A$ and (g_n) converges pointwise to $g: B \to \mathbb{R}$, then we say that the series $\sum f_n$ **converges to g on B** (or the series $\sum f_n$ **converges pointwise to g on B**).

Similarly, if (g_n) converges uniformly on B to g, then we say that the series $\sum f_n$ **converges uniformly on B to g** (or that $\sum f_n$ is **uniformly convergent** on B).

Example 16.11: For each $n \in \mathbb{N}$, define $f_n \colon \mathbb{R} \to \mathbb{R}$ by $f_n(x) = x^n$. The series with terms (f_n) is

$$\sum_{n=0}^{\infty} x^n = 1 + x + x^2 + x^3 + \cdots$$

For each $x \in \mathbb{R}$, this is a geometric series with first term 1 and common ratio x.

(Note that we adopt the convention here that $0^0 = 1$, so that the above formula holds when $x = 0$.)

The kth partial sum of this series is the function $g_k \colon \mathbb{R} \to \mathbb{R}$ defined by $g_k(x) = 1 + x + x^2 + \cdots + x^k$. If $x \neq 1$, then by part 3 of Example 15.1, $g_k(x) = \frac{1-x^{k+1}}{1-x}$. By the same example, for $|x| < 1$, the series converges to $\frac{1}{1-x}$ and for $|x| \geq 1$, the series diverges. Thus, for $|x| < 1$, we see that the series converges pointwise and we have

$$\sum_{n=0}^{\infty} x^n = \frac{1}{1-x}, \qquad x \in (-1, 1).$$

Since $\frac{1}{1-x}$ is unbounded on $(-1, 1)$, by part (i) of Problem 8 below, the series does **not** converge uniformly on $(-1, 1)$.

However, if $c \in [0, 1)$, then the series **does** converge uniformly on $[-c, c]$. To see this, let $x \in [-c, c]$. Then for $n \in \mathbb{N}$, we have the following:

$$\left| g_n(x) - \frac{1}{1-x} \right| = \left| \frac{1-x^{n+1}}{1-x} - \frac{1}{1-x} \right| = \left| \frac{-x^{n+1}}{1-x} \right| = \frac{|x|^{n+1}}{1-x}.$$

It follows from part 2 of Example 14.1 that

$$\lim_{n \to \infty} \left| g_n - \frac{1}{1-x} \right|_{[-c,c]} = \lim_{n \to \infty} \frac{c^{n+1}}{1-c} = \frac{1}{1-c} \lim_{n \to \infty} c^{n+1} = \frac{1}{1-c} \cdot 0 = 0.$$

So, by Theorem 16.4, the series converges uniformly on $[-c, c]$ to $\frac{1}{1-x}$.

The following test for uniform convergence of a series of functions is extremely useful.

Theorem 16.12 (Weierstrass M-test): Let (f_n) be a sequence of functions on $A \subseteq \mathbb{R}$ and suppose that for each $n \in \mathbb{N}$, there is $M_n \in \mathbb{R}^+$ such that for all $x \in A$ and all $n \in \mathbb{N}$, $|f_n(x)| \leq M_n$. If $\sum M_n$ converges, then $\sum f_n$ converges uniformly on A.

To prove Theorem 16.12, we will first prove a preliminary lemma. This lemma is essentially just the Cauchy Convergence Criterion for Series (Lemma 15.8) reformulated in terms of series of functions.

Lemma 16.13 (Cauchy Convergence Criterion for Series of Functions): Let (f_n) be a sequence of functions that are bounded on $A \subseteq \mathbb{R}$. The series $\sum f_n$ converges uniformly on A if and only if for every $\epsilon > 0$, there is $K \in \mathbb{N}$ such that $x \in A$ and $m \geq n > K$ implies

$$\left| \sum_{k=n+1}^{m} f_k(x) \right| = |f_{n+1}(x) + f_{n+2}(x) + \cdots + f_m(x)| < \epsilon.$$

Proof: Let (g_n) be the sequence of partial sums of the given series. $\sum f_n$ converges uniformly on A if and only if (g_n) converges uniformly on A if and only if (g_n) is uniformly Cauchy on A (by Theorem 16.6) if and only if for every $\epsilon > 0$, there is $K \in \mathbb{N}$ such that $m \geq n > K$ implies $|g_m - g_n|_A < \epsilon$.

Now, first suppose that $\sum f_n$ converges uniformly on A and let $\epsilon > 0$. By the first paragraph above, there is $K \in \mathbb{N}$ such that $m \geq n > K$ implies $|f_{n+1} + f_{n+2} + \cdots + f_m|_A = |g_m - g_n|_A < \epsilon$. So, for all $x \in A$,

$$\left| \sum_{k=n+1}^{m} f_k(x) \right| = |f_{n+1}(x) + f_{n+2}(x) + \cdots + f_m(x)| \leq |f_{n+1} + f_{n+2} + \cdots + f_m|_A < \epsilon.$$

Conversely, assume that for every $\epsilon > 0$, there is $K \in \mathbb{N}$ such that $x \in A$ and $m \geq n > K$ implies

$$\left| \sum_{k=n+1}^{m} f_k(x) \right| = |f_{n+1}(x) + f_{n+2}(x) + \cdots + f_m(x)| < \epsilon.$$

Let $\epsilon > 0$ be given. By our assumption, there is $K \in \mathbb{N}$ such that $x \in A$ and $m \geq n > K$ implies

$$\left| \sum_{k=n+1}^{m} f_k(x) \right| = |f_{n+1}(x) + f_{n+2}(x) + \cdots + f_m(x)| < \frac{\epsilon}{2}.$$

It follows that for $m \geq n > K$, we have

$$|f_{n+1} + f_{n+2} + \cdots + f_m|_A = \sup\{|f_{n+1}(x) + f_{n+2}(x) + \cdots + f_m(x)| \mid x \in A\} \leq \frac{\epsilon}{2} < \epsilon.$$

Therefore, by the first paragraph above, $\sum f_n$ converges uniformly on A. □

Proof of Theorem 16.12: First note that each f_n is bounded on A (by M_n). Let $\epsilon > 0$. Since $\sum M_n$ converges, by the Cauchy Convergence Criterion for Series (Lemma 15.8), there is $K \in \mathbb{N}$ such that $m \geq n > K$ implies

$$\left| \sum_{k=n+1}^{m} M_k \right| = |M_{n+1} + M_{n+2} + \cdots + M_m| < \epsilon.$$

So, if $x \in A$ and $m \geq n > K$, then we have

$$\left| \sum_{k=n+1}^{m} f_k(x) \right| \leq \left| \sum_{k=n+1}^{m} |f_k(x)| \right| \leq \left| \sum_{k=n+1}^{m} M_k \right| < \epsilon.$$

By Lemma 16.13, $\sum f_n$ converges uniformly on A. □

Example 16.14:

1. Let's restrict the domain of the functions given in Example 16.11 to $\left[-\frac{1}{2},\frac{1}{2}\right]$. That is, for each $n \in \mathbb{N}$, define $f_n: \left[-\frac{1}{2},\frac{1}{2}\right] \to \mathbb{R}$ by $f_n(x) = x^n$. Once again, the series with terms (f_n) is

$$\sum_{n=0}^{\infty} x^n = 1 + x + x^2 + x^3 + \cdots$$

For each $n \in \mathbb{N}$, let $M_n = \left(\frac{1}{2}\right)^n$. For $x \in \left[-\frac{1}{2},\frac{1}{2}\right]$ and $n \in \mathbb{N}$, we have $|f_n(x)| = |x^n| \leq M_n$. By part 3 of Example 15.1, $\sum M_n = \sum \left(\frac{1}{2}\right)^n$ is a convergent geometric series. Therefore, by the Weierstrass M-test (Theorem 16.12), $\sum x^n$ converges uniformly on $\left[-\frac{1}{2},\frac{1}{2}\right]$.

Note that the same argument works if we replace $\left[-\frac{1}{2},\frac{1}{2}\right]$ by $[-c,c]$ for any $c \in (0,1)$.

2. For each $n \in \mathbb{N}^+$, define $f_n: \mathbb{R} \to \mathbb{R}$ by $f_n(x) = \frac{\cos nx}{n^2}$. The series with terms (f_n) is

$$\sum_{n=1}^{\infty} \frac{\cos nx}{n^2} = \frac{\cos x}{1^2} + \frac{\cos 2x}{2^2} + \frac{\cos 3x}{3^2} + \frac{\cos 4x}{4^2} + \cdots$$

For each $n \in \mathbb{N}$, let $M_n = \frac{1}{n^2}$. For all $x \in \mathbb{R}$ and $n \in \mathbb{N}$, we have $|f_n(x)| = \frac{|\cos nx|}{n^2} \leq M_n$. By part 1 of Example 15.3, $\sum M_n = \sum \frac{1}{n^2}$ is a convergent p-series ($p = 2$). Therefore, by the Weierstrass M-test, $\sum \frac{\cos nx}{n^2}$ converges uniformly on $\mathbb{R}$.

Power Series

A **power series centered at $x = a$**, where $a \in \mathbb{R}$, is a series of the form

$$\sum_{n=0}^{\infty} c_n(x-a)^n = c_0 + c_1(x-a) + c_2(x-a)^2 + \cdots,$$

where $c_n \in \mathbb{R}$ for each $n \in \mathbb{N}$.

Note: We have adopted the convention that whenever the expression 0^0 appears in a term of a power series that we set $0^0 = 1$. This happens when $x = a$ and $n = 0$. Without this convention, we would need to write the above power series in the following unabbreviated form:

$$c_0 + \sum_{n=1}^{\infty} c_n(x-a)^n$$

It is worthwhile to keep this unabbreviated version in mind to avoid errors when doing computations. For example, naively substituting $x = a$ into the given power series might lead one to believe that the result is zero. However, we can see from the unabbreviated form that the result is actually c_0.

Example 16.15:

1. Consider the power series

$$\sum_{n=0}^{\infty} x^n = 1 + x + x^2 + x^3 + \cdots$$

If we let $s_n = x^n$, then we have

$$\lim_{n \to \infty} \left| \frac{s_{n+1}}{s_n} \right| = \lim_{n \to \infty} \left| \frac{x^{n+1}}{x^n} \right| = \lim_{n \to \infty} |x| = |x|.$$

By the Ratio Test (Theorem 15.21), if $|x| < 1$, the series is absolutely convergent and if $|x| > 1$, the series is divergent. We call the number 1 here the **radius of convergence** of the power series. Notice that the Ratio Test does **not** tell us if the series converges at $x = 1$ and $x = -1$. We will need to check these separately. In both cases, the Divergence Test (see part 2 of Example 15.1) tells us that the series diverges. It follows that the series converges absolutely on the interval $(-1, 1)$ and diverges outside of that interval. The interval $(-1, 1)$ is called the **interval of convergence** of the power series.

Note that the power series in this example is a geometric series for each $x \in \mathbb{R}$. So, we could have found the ratio of convergence and interval of converge using part 3 of Example 15.1 instead of the Ratio Test.

2. Consider the power series

$$\sum_{n=0}^{\infty} \frac{x^n}{n!} = \frac{1}{0!} + \frac{x}{1!} + \frac{x^2}{2!} + \frac{x^3}{3!} + \cdots = 1 + x + \frac{x^2}{2} + \frac{x^3}{3!} + \frac{x^4}{4!} + \cdots$$

Note that the exclamation point is pronounced "**factorial**" and by definition, $0! = 1$ and $n! = 1 \cdot 2 \cdots n$ for all natural numbers $n \geq 1$.

If we let $s_n = \frac{x^n}{n!}$, then we have

$$\lim_{n \to \infty} \left| \frac{s_{n+1}}{s_n} \right| = \lim_{n \to \infty} \left| \frac{\frac{x^{n+1}}{(n+1)!}}{\frac{x^n}{n!}} \right| = \lim_{n \to \infty} \left| \frac{x^{n+1}}{(n+1)!} \cdot \frac{n!}{x^n} \right| = \lim_{n \to \infty} \left| \frac{x^{n+1}}{x^n} \cdot \frac{n!}{(n+1)n!} \right| = \left| \frac{x}{n+1} \right| = 0.$$

Since $0 < 1$, by the Ratio Test (Theorem 15.21), the series is absolutely convergent for all $x \in \mathbb{R}$. In this case, we say that the radius of convergence is ∞ and the interval of convergence is the interval $(-\infty, \infty)$. In part 2 of Example 16.25 below, we will see that for each $x \in \mathbb{R}$, this power series converges to e^x.

3. Consider the power series

$$\sum_{n=1}^{\infty} \frac{x^n}{n} = x + \frac{x^2}{2} + \frac{x^3}{3} + \cdots$$

305

If we let $s_n = \frac{x^n}{n}$, then we have

$$\lim_{n\to\infty}\left|\frac{s_{n+1}}{s_n}\right| = \lim_{n\to\infty}\left|\frac{\frac{x^{n+1}}{n+1}}{\frac{x^n}{n}}\right| = \lim_{n\to\infty}\left|\frac{x^{n+1}}{n+1}\cdot\frac{n}{x^n}\right| = |x|\lim_{n\to\infty}\frac{n}{n+1} = |x|.$$

By the Ratio Test (Theorem 15.21), if $|x| < 1$, the series is absolutely convergent and if $|x| > 1$, the series is divergent. So, the radius of convergence of this power series is 1. Once again, the Ratio Test does **not** tell us if the series converges at $x = 1$ and $x = -1$. We will need to check these separately. For $x = 1$, we get the divergent harmonic series $\sum \frac{x^n}{n}$ (see part 2 of Example 15.3). For $x = -1$, we get a conditionally convergent alternating series $\sum \frac{(-1)^n}{n}$ (by the Alternating Series Test—see Theorem 15.5). So, the interval of convergence of this power series is the half-open interval $[-1, 1)$.

4. Consider the power series

$$\sum_{n=0}^{\infty} n!\, x^n = 1 + x + 2x^2 + 3!\, x^3 + 4!\, x^4 + \cdots$$

If we let $s_n = n!\, x^n$, then we have

$$\lim_{n\to\infty}\left|\frac{s_{n+1}}{s_n}\right| = \lim_{n\to\infty}\left|\frac{(n+1)!\, x^{n+1}}{n!\, x^n}\right| = \lim_{n\to\infty}|(n+1)x| = \begin{cases}\infty & \text{if } x \neq 0.\\ 0 & \text{if } x = 0.\end{cases}$$

By the Ratio Test (Theorem 15.21), this power series converges only at $x = 0$. In this case, we say that the radius of convergence is 0. There is no interval of convergence.

5. Consider the power series

$$\sum_{n=1}^{\infty}\left(\frac{x}{n}\right)^n = x + \left(\frac{x}{2}\right)^2 + \left(\frac{x}{3}\right)^3 + \cdots$$

If we let $s_n = \left(\frac{x}{n}\right)^n$, then we have

$$\limsup |s_n|^{\frac{1}{n}} = \limsup \left|\left(\frac{x}{n}\right)^n\right|^{\frac{1}{n}} = \limsup \frac{|x|}{n} = |x|\limsup \frac{1}{n} = 0.$$

By the Root Test (Theorem 15.25), the series is absolutely convergent for all $x \in \mathbb{R}$. So, the radius of convergence of this power series is ∞ and the interval of convergence of this power series is the interval $(-\infty, \infty)$.

6. Consider the power series

$$\sum_{n=1}^{\infty}\frac{(-1)^n\, 3^n}{n}(x-2)^n$$

This power series is centered at $x = 2$.

If we let $s_n = \frac{(-1)^n 3^n}{n}(x-2)^n$, then we have

$$\lim_{n\to\infty}\left|\frac{s_{n+1}}{s_n}\right| = \lim_{n\to\infty}\left|\frac{\frac{(-1)^{n+1} 3^{n+1}}{n+1}(x-2)^{n+1}}{\frac{(-1)^n 3^n}{n}(x-2)^n}\right| = \lim_{n\to\infty}\left|\frac{(-1)^{n+1} 3^{n+1}(x-2)^{n+1}}{n+1} \cdot \frac{n}{(-1)^n 3^n(x-2)^n}\right|$$

$$= \lim_{n\to\infty}\left(\left|\frac{(-1)^{n+1}}{(-1)^n}\right|\left|\frac{3^{n+1}}{3^n}\right|\left|\frac{n}{n+1}\right|\left|\frac{(x-2)^{n+1}}{(x-2)^n}\right|\right) = 3|x-2|\lim_{n\to\infty}\frac{n}{n+1} = 3|x-2|.$$

By the Ratio Test (Theorem 15.21), if $3|x-2| < 1$, or equivalently, $|x-2| < \frac{1}{3}$, the series is absolutely convergent and if $|x-2| > \frac{1}{3}$, the series is divergent. So, the radius of convergence of this power series is $\frac{1}{3}$.

Now, $|x-2| < \frac{1}{3}$ if and only if $-\frac{1}{3} < x - 2 < \frac{1}{3}$ if and only if $\frac{5}{3} < x < \frac{7}{3}$.

We still need to check the endpoints. When $x = \frac{7}{3}$, we get the convergent alternating series $\sum_{n=1}^{\infty}\frac{(-1)^n}{n}$. When $x = \frac{5}{3}$ we get the divergent harmonic series $\sum_{n=1}^{\infty}\frac{1}{n}$. So, the interval of convergence of the power series is $\left(\frac{5}{3}, \frac{7}{3}\right]$.

Let's now define the **radius of convergence** and **interval of convergence** of a power series more formally. To this end, consider the following general power series:

$$\sum_{n=0}^{\infty} c_n(x-a)^n = c_0 + c_1(x-a) + c_2(x-a)^2 + \cdots.$$

If we let $s_n = c_n(x-a)^n$, then we have

$$\limsup |s_n|^{\frac{1}{n}} = \limsup |c_n(x-a)^n|^{\frac{1}{n}} = |x-a|\limsup |c_n|^{\frac{1}{n}}.$$

If $\limsup |c_n|^{\frac{1}{n}} = 0$, then by the Root Test (Theorem 15.25), the series is absolutely convergent for all $x \in \mathbb{R}$. In this case, we say that the **radius of convergence** is ∞ and the **interval of convergence** is the interval $(-\infty, \infty)$.

If $\limsup |c_n|^{\frac{1}{n}} = L$ for some $L \in \mathbb{R}^+$, then by the Root Test (Theorem 15.25), the series is absolutely convergent whenever $|x-a|L < 1$, or equivalently, $|x-a| < \frac{1}{L}$ and divergent whenever $|x-a| > \frac{1}{L}$. In this case, we say that the **radius of convergence** is $\frac{1}{L}$. Now, $|x-a| < \frac{1}{L}$ if and only if $-\frac{1}{L} < x - a < \frac{1}{L}$ if and only if $a - \frac{1}{L} < x < a + \frac{1}{L}$. The **interval of convergence** is one of the four intervals $\left(a - \frac{1}{L}, a + \frac{1}{L}\right)$, $\left(a - \frac{1}{L}, a + \frac{1}{L}\right]$, $\left[a - \frac{1}{L}, a + \frac{1}{L}\right)$, or $\left[a - \frac{1}{L}, a + \frac{1}{L}\right]$, depending on whether the power series converges at each of the endpoints $x = a - \frac{1}{L}$ and $x = a + \frac{1}{L}$. Since the Root Test provides no information about the convergence of the power series at these values, they need to be tested separately using other methods.

Finally, if $\limsup |c_n|^{\frac{1}{n}} = \infty$, then by the Root Test (Theorem 15.25), the series is divergent for all $x \in \mathbb{R}$ with the exception of $x = c$. In this case, we say that the **radius of convergence** is 0 and there is no interval of convergence.

Note: By Problem 14 in Problem Set 15, we can often use the Ratio Test in place of the Root Test to determine the radius of convergence of a power series. In this case, we would compute $\lim\limits_{n \to \infty} \left|\frac{c_{n+1}}{c_n}\right|$ instead of $\limsup |c_n|^{\frac{1}{n}}$. However, there are instances when $\lim\limits_{n \to \infty} \left|\frac{c_{n+1}}{c_n}\right|$ fails to exist (see Problem 4 in Problem Set 15). In these instances, we would need to use the Root Test.

We now apply our previous results on uniform convergence to power series.

Theorem 16.16: Let $R > 0$ be the radius of convergence of a power series $\sum c_n(x-a)^n$. Then the power series converges uniformly on any closed and bounded interval contained in $(a-R, a+R)$.

Notes: (1) If $R = \infty$, then we define $(a-R, a+R)$ to be $(-\infty, \infty)$.

(2) We will be using the Weierstrass M-test (Theorem 16.12) to prove Theorem 16.16. It would be worthwhile to make sure you have a good understanding of Example 16.14 before going through the following proof.

Proof: Let $I \subseteq (a-R, a+R)$ be a closed bounded interval. Then there is $k \in \mathbb{R}^+$ with $I \subseteq [a-k, a+k] \subseteq (a-R, a+R)$. Let m be a real number such that $a+k < m < a+R$. Then $\sum c_n(m-a)^n$ converges, and so, by the Divergence Test (see part 2 of Example 15.1), $\lim\limits_{n \to \infty} c_n(m-a)^n = 0$. By Problem 2 in Problem Set 14, $(c_n(m-a)^n)$ is a bounded sequence. So, there is $M \in \mathbb{R}^+$ such that for all $n \in \mathbb{N}$, $|c_n(m-a)^n| \leq M$.

Now, if $x \in I$, then $x \in [a-k, a+k]$. So, $|x-a| \leq k$. Therefore, for each $n \in \mathbb{N}$,

$$|c_n(x-a)^n| = |c_n||(x-a)^n| \leq |c_n||k^n| = |c_n||(m-a)^n|\frac{|k^n|}{|(m-a)^n|} \leq M \left|\frac{k}{m-a}\right|^n.$$

Since $k < m-a$, we have $\left|\frac{k}{m-a}\right| < 1$. It follows that $\sum M \left|\frac{k}{m-a}\right|^n$ is a convergent geometric series. So, by the Weierstrass M-test (Theorem 16.12), $\sum c_n(x-a)^n$ converges uniformly on I. □

Corollary 16.17: Let $R > 0$ be the radius of convergence of a power series $\sum c_n(x-a)^n$. Then the power series is continuous on $(a-R, a+R)$.

Proof: Let $x \in (a-R, a+R)$. Then there is a closed bounded interval $I \subseteq (a-R, a+R)$ with $x \in I$. By Theorem 16.16, the power series converges uniformly on I. Since the partial sums of the series are polynomials, they are continuous on I. So, by Theorem 16.7, the power series is continuous on I. Since $x \in I$, the power series is continuous at x. Since $x \in (a-R, a+R)$ was arbitrary, x is continuous on $(a-R, a+R)$. □

The next Corollary essentially says that we can integrate a power series term by term.

Corollary 16.18: Let $R > 0$ be the radius of convergence of a power series $\sum c_n(x-a)^n$. Then the power series is locally integrable on $(a-R, a+R)$ and for any closed and bounded interval $[d, e]$ contained in $(a-R, a+R)$, we have

$$\int_d^e \sum_{n=0}^{\infty} c_n(x-a)^n \, dx = \sum_{n=0}^{\infty} \int_d^e c_n(x-a)^n \, dx.$$

Proof: Let $[d, e] \subseteq (a-R, a+R)$. By Theorem 16.16, the power series converges uniformly on $[d, e]$. Since the partial sums of the series are polynomials, they are Riemann integrable on $[d, e]$. The result now follows from Theorem 16.8. $\square$

In Example 16.9, we saw that uniform convergence does **not** preserve differentiability. In Problem 16 below, you will see an example of a series with continuous partial sums that converges uniformly on $\mathbb{R}$ to a function that is not differentiable anywhere. So, in general, we cannot differentiate a uniformly convergent series term by term. However, it turns out that power series are much more well-behaved than general uniformly convergent series. We can in fact differentiate a power series term by term. We will prove this in two steps. We first prove that if we differentiate a power series term by term, then the resulting power series has the same radius of convergence as the original power series.

Theorem 16.19: Let $R > 0$ be the radius of convergence of the power series

$$\sum_{n=0}^{\infty} c_n(x-a)^n = c_0 + c_1(x-a) + c_2(x-a)^2 + \cdots.$$

Then the power series

$$\sum_{n=1}^{\infty} nc_n(x-a)^{n-1} = c_1 + 2c_2(x-a) + 3c_3(x-a)^2 + \cdots.$$

also has radius of convergence R.

Note: Observe that the second series in the statement of Theorem 16.19 begins with an index of 1. This is due to the fact that the derivative of the constant c_0 is zero. When $n = 0$, we have $nc_n(x-a)^{n-1} = 0c_0(x-a)^{0-1} = \frac{0c_0}{x-a}$. This expression is undefined when $x = a$. Therefore, it should **not** be included as part of the series.

Proof: By Problem 11 in Problem Set 14, we have

$$\lim_{n \to \infty} n^{\frac{1}{n}} = 1.$$

So, by Problem 19 in Problem Set 14,

$$\limsup |nc_n|^{\frac{1}{n}} = \left(\lim_{n \to \infty} n^{\frac{1}{n}}\right)\left(\limsup |c_n|^{\frac{1}{n}}\right) = 1 \cdot \limsup |c_n|^{\frac{1}{n}} = \limsup |c_n|^{\frac{1}{n}}.$$

Therefore, $\sum_{n=1}^{\infty} nc_n(x-a)^{n-1}$ has the same radius of convergence as $\sum_{n=0}^{\infty} c_n(x-a)^n$. $\square$

Theorem 16.20: Let $R > 0$ be the radius of convergence of the power series $\sum_{n=0}^{\infty} c_n(x-a)^n$ and define $f: (a - R, a + R) \to \mathbb{R}$ by

$$f(x) = \sum_{n=0}^{\infty} c_n(x-a)^n = c_0 + c_1(x-a) + c_2(x-a)^2 + \cdots.$$

Then f is differentiable on $(a - R, a + R)$ and for all $x \in (a - R, a + R)$,

$$f'(x) = \sum_{n=1}^{\infty} nc_n(x-a)^{n-1} = c_1 + 2c_2(x-a) + 3c_3(x-a)^2 + \cdots..$$

Proof: Define $g: (a - R, a + R) \to \mathbb{R}$ by

$$g(x) = \sum_{n=1}^{\infty} nc_n(x-a)^{n-1}.$$

By Theorem 16.19, for all $x \in (a - R, a + R)$, $g(x)$ is a finite real number. Let I be a closed bounded interval contained in $(a - R, a + R)$. By Theorem 16.16, both f and g converge uniformly on I. By Theorem 16.10 (applied to the sequence of partial sums of f and g), we see that f is differentiable on I and $f'(x) = g(x)$ for all $x \in I$. Since I was an arbitrary closed bounded interval contained in $(a - R, a + R)$, f is differentiable on $(a - R, a + R)$ and $f'(x) = g(x)$ for all $x \in (a - R, a + R)$. □

Not only does Theorem 16.20 tell us that a power series $f(x) = \sum_{n=0}^{\infty} c_n(x-a)^n$ is differentiable on $(a - R, a + R)$, where R is its radius of convergence, but it also says that the derivative $f'(x)$ is a power series centered at $x = a$ with radius of convergence R (with a little help from Theorem 16.19). Therefore, we can apply Theorem 16.20 to $f'(x)$ to see that $f'(x)$ is differentiable on $(a - R, a + R)$ and $f''(x)$ is a power series centered at $x = a$ with radius of convergence R. Continuing in this fashion, we see that we can keep taking derivatives to get new power series whose radii of convergence are all R. Since the nth derivative of f, $f^{(n)}(x)$, is defined for all $n \in \mathbb{N}$ (with $f^{(0)} = f$), we say that f is **infinitely differentiable** on $(a - R, a + R)$.

The coefficients $c_0, c_1, c_2, \ldots$ are called the **Taylor coefficients** of the function f. It turns out that we can write the Taylor coefficients in terms of the original function f, as we now show.

Theorem 16.21: Let $R > 0$ be the radius of convergence of the power series $\sum_{n=0}^{\infty} c_n(x-a)^n$ and define $f: (a - R, a + R) \to \mathbb{R}$ by

$$f(x) = \sum_{n=0}^{\infty} c_n(x-a)^n = c_0 + c_1(x-a) + c_2(x-a)^2 + \cdots.$$

Then f is infinitely differentiable on $(a - R, a + R)$ and for each $n \in \mathbb{N}$,

$$c_n = \frac{f^{(n)}(a)}{n!}.$$

Proof: We first prove by induction on $k \in \mathbb{N}$ that

$$f^{(k)}(x) = \sum_{n=k}^{\infty} \frac{n!}{(n-k)!} c_n (x-a)^{n-k} = k! c_k + \frac{k!}{1!} c_{k+1}(x-a) + \frac{k!}{2!} c_{k+2}(x-a)^2 + \cdots$$

and $f^{(k)}(x)$ has radius of convergence R.

Base case ($k = 0$): $f^0(x) = f(x) = \sum_{n=0}^{\infty} c_n(x-a)^n = \sum_{n=0}^{\infty} \frac{n!}{n!} c_n(x-a)^n$.

Inductive step: Assume that $k \in \mathbb{N}$,

$$f^{(k)}(x) = \sum_{n=k}^{\infty} \frac{n!}{(n-k)!} c_n (x-a)^{n-k}$$

and $f^{(k)}(x)$ has radius of convergence R. Then by Theorem 6.20, we have

$$f^{(k+1)}(x) = \left(f^{(k)}\right)'(x) = \sum_{n=k+1}^{\infty} \frac{n!}{(n-k)!} (n-k) c_n (x-a)^{n-k-1}$$

$$= \sum_{n=k+1}^{\infty} \frac{n!}{(n-k)(n-k-1)!} (n-k) c_n (x-a)^{n-k-1} = \sum_{n=k+1}^{\infty} \frac{n!}{(n-(k+1))!} c_n (x-a)^{n-(k+1)}.$$

By Theorem 6.19, $f^{(k+1)}(x)$ has radius of convergence R.

By the Principle of Mathematical Induction, for all $k \in \mathbb{N}$, we have

$$f^{(k)}(x) = \sum_{n=k}^{\infty} \frac{n!}{(n-k)!} c_n (x-a)^{n-k}$$

and the radius of convergence of $f^{(k)}(x)$ is R.

Substituting a for x yields

$$f^{(k)}(a) = \sum_{n=k}^{\infty} \frac{n!}{(n-k)!} c_n (a-a)^{n-k} = k! c_k.$$

Therefore, for each $k \in \mathbb{N}$, we have

$$c_k = \frac{f^{(k)}(a)}{k!}.$$ $\square$

Note: When substituting a for x, observe that we used our convention that $0^0 = 1$ mentioned in the Note before Example 16.15. In other words, the first term of the series with a substituted for x is

$$\frac{k!}{(k-k)!} c_k (a-a)^{k-k} = \frac{k!}{0!} c_k \cdot 0^0 = k! c_k \cdot 1 = k! c_k.$$

Theorem 16.22 (Uniqueness Theorem): If two power series with center $x = a$, say $\sum_{n=0}^{\infty} c_n(x - a)^n$ and $\sum_{n=0}^{\infty} d_n(x - a)^n$, converge on some interval $(a - R, a + R)$ with $R > 0$ to the same function f, then for all $n \in \mathbb{N}$, $c_n = d_n$.

Proof: By Theorem 16.21, for each $n \in \mathbb{N}$, we have $c_n = \frac{f^{(n)}(a)}{n!}$ and $d_n = \frac{f^{(n)}(a)}{n!}$. Therefore, for each $n \in \mathbb{N}$, we have $c_n = d_n$. □

Taylor Series

Let $A \subseteq \mathbb{R}$, let $f: A \to \mathbb{R}$ and suppose that the nth derivative of f exists at an interior point a of A (in other words, A contains a neighborhood of a). Then the following polynomial is called the **kth Taylor polynomial for f at $x = a$**:

$$p_k(x) = \sum_{n=0}^{k} \frac{f^{(n)}(a)}{n!}(x - a)^n = f(a) + f'(a)(x - a) + \frac{f''(a)}{2!}(x - a)^2 + \cdots + \frac{f^{(k)}(a)}{k!}(x - a)^k$$

The first few Taylor polynomials for f are as follows:

$p_0(x) = f(a)$ 0th Taylor polynomial for f at $x = a$

$p_1(x) = f(a) + f'(a)(x - a)$ 1st Taylor polynomial for f at $x = a$

$p_2(x) = f(a) + f'(a)(x - a) + \frac{f''(a)}{2!}(x - a)^2$ 2nd Taylor polynomial for f at $x = a$

Notes: (1) In many cases, Taylor polynomials can be used to approximate the function f "near a." Taylor's Theorem (Theorem 16.24) can be used to determine how good these approximations are.

(2) $p_1(x)$ is also called the **linear approximation** or **tangent line approximation** for f at $x = a$. The graph of this function is the line with slope $f'(a)$ passing through the point $(a, f(a))$.

(3) $p_2(x)$ is also called the **quadratic approximation** for f at $x = a$. The graph of this function is a parabola passing through the point $(a, f(a))$. In many cases, the quadratic approximation for a function f at $x = a$ will be more accurate than the linear approximation "near a," and more generally, if $j > k$, then the jth Taylor polynomial will provide a better approximation than the kth Taylor polynomial for f near $x = a$.

(4) Note 3 above is a bit vague. What do we mean by "in many cases?" The accuracy of Note 3 needs to be checked on a case-by-case basis, often with the help of Taylor's Theorem (Theorem 16.24) below.

Example 16.23:

1. Let $f(x) = 3x^2 + 2x - 1$. Let's compute the Taylor polynomials for f at $x = 0$. We have

$f(x) = 3x^2 + 2x - 1$ $f(0) = -1$

$f'(x) = 6x + 2$ $f'(0) = 2$

$f''(x) = 6$ $f''(0) = 6$

$f^{(n)}(x) = 0, \quad n \geq 3$ $f^{(n)}(0) = 0, \quad n \geq 3$

So, we have the following:

$$p_0(x) = f(0) = -1$$
$$p_1(x) = f(0) + f'(0)x = -1 + 2x$$
$$p_2(x) = f(0) + f'(0)x + \frac{f''(0)}{2!}x^2 = -1 + 2x + \frac{6}{2}x^2 = -1 + 2x + 3x^2$$

Observe that for $k \geq 2$, the kth Taylor polynomial for f at $x = 0$ is equal to the function itself.

In fact, if f is any polynomial function of degree n, then for $k \geq n$, the kth Taylor polynomial for f at $x = 0$ will always be equal to the polynomial f.

2. Let $g(x) = e^x$. Let's compute the Taylor polynomials for g at $x = 0$. For all $n \in \mathbb{N}$, we have $g^{(n)}(x) = e^x$. So, for all $n \in \mathbb{N}$, we have $g^{(n)}(0) = e^0 = 1$. So, we have the following:

$$p_0(x) = g(0) = 1$$
$$p_1(x) = g(0) + g'(0)x = 1 + x$$
$$p_2(x) = g(0) + g'(0)x + \frac{g''(0)}{2!}x^2 = 1 + x + \frac{x^2}{2!}$$
$$p_3(x) = g(0) + g'(0)x + \frac{g''(0)}{2!} + \frac{g'''(0)}{3!} = x^2 = 1 + x + \frac{x^2}{2!} + \frac{x^3}{3!}$$

In general, for $k \in \mathbb{N}$, we see that the kth Taylor polynomial for $g(x) = e^x$ at $x = 0$ is

$$p_k(x) = \sum_{n=0}^{k} \frac{x^n}{n!} = 1 + x + \frac{x^2}{2!} + \frac{x^3}{3!} + \cdots + \frac{x^k}{k!}.$$

3. Let $h(x) = \frac{1}{x}$. Let's compute the Taylor polynomials for h at $x = 1$. We have

$h(x) = \frac{1}{x} = x^{-1}$	$h(1) = 1$
$h'(x) = -x^{-2} = -\frac{1}{x^2}$	$h'(1) = -1$
$h''(x) = 2x^{-3} = \frac{2!}{x^3}$	$h''(1) = 2!$
$h'''(x) = -2 \cdot 3x^{-4} = -\frac{3!}{x^4}$	$h'''(1) = -3!$
$\vdots$	$\vdots$
$h^{(k)}(x) = (-1)^k \frac{k!}{x^{k+1}}$	$h^{(k)}(1) = (-1)^k k!$

So, we have the following:

$$p_0(x) = h(1) = 1$$
$$p_1(x) = h(1) + h'(1)(x - 1) = 1 - (x - 1)$$
$$p_2(x) = h(1) + h'(1)(x - 1) + \frac{h''(1)}{2!}(x - 1)^2 = 1 - (x - 1) + (x - 1)^2$$

In general, for $k \in \mathbb{N}$, we see that the kth Taylor polynomial for $h(x) = \frac{1}{x}$ at $x = 1$ is

$$p_k(x) = \sum_{n=0}^{k} (-1)^n (x-1)^n = 1 - (x-1) + (x-1)^2 - (x-1)^3 + \cdots + (-1)^k (x-1)^k.$$

4. Let $b(x) = \ln x$. Let's compute the Taylor polynomials for b at $x = 1$. We have

$$b(x) = \ln x \qquad\qquad b(1) = 0$$
$$b'(x) = \frac{1}{x} = x^{-1} \qquad\qquad b'(1) = 1$$
$$b''(x) = -x^{-2} = -\frac{1}{x^2} \qquad\qquad b''(1) = -1$$
$$b'''(x) = 2x^{-3} = \frac{2!}{x^3} \qquad\qquad b'''(1) = 2!$$
$$b^{(4)}(x) = -2 \cdot 3 x^{-4} = -\frac{3!}{x^4} \qquad\qquad b^{(4)}(1) = -3!$$
$$\vdots \qquad\qquad \vdots$$
$$b^{(k)}(x) = (-1)^{k-1} \frac{(k-1)!}{x^k} \qquad\qquad b^{(k)}(1) = (-1)^{k-1}(k-1)!$$

So, we have the following:

$$p_0(x) = b(1) = 0$$
$$p_1(x) = b(1) + b'(1)(x-1) = x - 1$$
$$p_2(x) = b(1) + b'(1)(x-1) + \frac{b''(1)}{2!}(x-1)^2 = (x-1) - \frac{(x-1)^2}{2}$$

In general, for $k \in \mathbb{N}$, we see that the kth Taylor polynomial for $b(x) = \ln x$ at $x = 1$ is

$$p_k(x) = \sum_{n=1}^{k} \frac{(-1)^{n-1}(x-1)^n}{n} = (x-1) - \frac{(x-1)^2}{2} + \frac{(x-1)^3}{3} - \cdots + (-1)^{k-1} \frac{(x-1)^k}{k}.$$

In the Notes before Example 16.23, we mentioned that in certain situations the Taylor polynomials for a function f at $x = c$ provide reasonable approximations for the function f near c. The following theorem can sometimes be useful for determining if these approximations are in fact reasonable for a specific function.

Theorem 16.24 (Taylor's Theorem): Let $k \in \mathbb{N}$ and suppose that $[a,b] \subseteq A \subseteq \mathbb{R}$. Let $f : A \to \mathbb{R}$ be a function such that $f, f', f'', \ldots, f^{(k)}$ exist and are continuous on $[a,b]$ and such that $f^{(k+1)}$ exists on (a,b). If $c \in [a,b]$, then for any $x \in [a,b]$ with $x \neq c$, there is d between x and c such that

$$f(x) = \left(\sum_{n=0}^{k} \frac{f^{(n)}(c)}{n!} (x-c)^n \right) + \frac{f^{(k+1)}(d)}{(k+1)!} (x-c)^{k+1}$$

$$= f(c) + f'(c)(x-c) + \frac{f''(c)}{2!}(x-c)^2 + \cdots + \frac{f^{(k)}(c)}{k!}(x-c)^k + \frac{f^{(k+1)}(d)}{(k+1)!}(x-c)^{k+1}.$$

Notes: (1) If we let $k = 0$ in Taylor's Theorem, we see that if f is continuous on $[a, b]$ and differentiable on (a, b), then for any $c, x \in [a, b]$, there is d between x and c such that $f(x) = f(c) + f'(d)(x - c)$. In particular, if we let $c = a$ and $x = b$, we get a $d \in (a, b)$ such that $f(b) = f(a) + f'(d)(b - a)$, or equivalently,

$$f'(d) = \frac{f(b) - f(a)}{b - a}.$$

This is just the statement of the Mean Value Theorem (Theorem 10.25). So, we see that Taylor's Theorem is a generalization of the Mean Value Theorem. The proof of Taylor's Theorem mirrors the proof of the Mean Value Theorem, although it is technically more complex.

(2) Let's also look at the case $k = 1$. Here, we are given that f and f' are continuous on $[a, b]$ and f'' exists on (a, b). Fix $c \in [a, b]$. Taylor's Theorem guarantees us that if $x \in [a, b]$ with $x \neq c$, then there is d between x and c such that

$$f(x) = f(c) + f'(c)(x - c) + \frac{f''(d)}{2!}(x - c)^2.$$

In other words, we can estimate $f(x)$ with the linear approximation (or 1st Taylor polynomial for f at $x = c$) $p_1(x) = f(c) + f'(c)(x - c)$ and the error in this approximation is

$$R_1(x) = \frac{f''(d)}{2!}(x - c)^2,$$

where d is some real number between x and c. We will refer to $R_1(x)$ as the **remainder term for $p_1(x)$**.

(3) Given $k \in \mathbb{N}$, $[a, b] \subseteq A \subseteq \mathbb{R}$, and a function $f: A \to \mathbb{R}$ all satisfying the hypotheses of Taylor's Theorem, we see that given $c \in [a, b]$, we can write $f(x) = p_k(x) + R_k(x)$, where $p_k(x)$ is the kth Taylor polynomial for f at $x = c$ and $R_k(x)$ is the following **remainder term for $p_k(x)$**:

$$R_k(x) = \frac{f^{(k+1)}(d)}{(k + 1)!}(x - c)^{k+1}.$$

Finding a bound on the remainder term R_k provides us with the maximum possible error in the approximation of $f(x)$ by the Taylor polynomial $p_k(x)$.

Proof of Theorem 16.24: Let $c, x \in [a, b]$ and let I be the closed interval with endpoints c and x (so $I = [c, x]$ if $c < x$ and $I = [x, c]$ if $x < c$). Define $g: I \to \mathbb{R}$ by

$$g(t) = f(x) - f(t) - f'(t)(x - t) - f''(t)\frac{(x - t)^2}{2!} - \cdots - f^{(k)}(t)\frac{(x - t)^k}{k!}.$$

Then using the sum rule, product rule, and chain rule for derivatives, we have

$$g'(t) = -f'(t) - [f''(t)(x - t) - f'(t)] - \left[f'''(t)\frac{(x - t)^2}{2!} - f''(t)(x - t)\right] - \cdots$$
$$- \left[f^{(k+1)}(t)\frac{(x - t)^k}{k!} - f^{(k)}(t)\frac{(x - t)^{k-1}}{(k - 1)!}\right].$$

Simplifying the above expression yields

$$g'(t) = -f^{(k+1)}(t)\frac{(x-t)^k}{k!}.$$

Define $h: I \to \mathbb{R}$ by

$$h(t) = g(t) - g(c)\left(\frac{x-t}{x-c}\right)^{k+1}.$$

Observe that h is continuous on I and differentiable in the interior of I, and we have

$$h(c) = g(c) - g(c)\left(\frac{x-c}{x-c}\right)^{k+1} = g(c) - g(c) = 0 \quad \text{and} \quad h(x) = g(x) = 0.$$

By Rolle's Theorem (Theorem 10.24), there is d between c and x such that $h'(d) = 0$, or equivalently,

$$-f^{(k+1)}(d)\frac{(x-d)^k}{k!} = g'(d) = g(c)(k+1)\left(\frac{x-d}{x-c}\right)^k\left(-\frac{1}{x-c}\right) = -g(c)(k+1)\frac{(x-d)^k}{(x-c)^{k+1}}.$$

So, we have

$$g(c) = f^{(k+1)}(d)\frac{(x-c)^{k+1}}{(k+1)!},$$

or equivalently,

$$f(x) - f(c) - f'(c)(x-c) - f''(c)\frac{(x-c)^2}{2!} - \cdots - f^{(k)}(c)\frac{(x-c)^k}{k!} = f^{(k+1)}(d)\frac{(x-c)^{k+1}}{(k+1)!}.$$

Solving for $f(x)$ yields the desired result. $\square$

If f is infinitely differentiable, then we can form the **Taylor series (or Taylor expansion) for f at $x = a$**:

$$T_f(x) = \sum_{n=0}^{\infty} \frac{f^{(n)}(a)}{n!}(x-a)^n$$

$$= f(a) + f'(a)(x-a) + \frac{f''(a)}{2!}(x-a)^2 + \cdots + \frac{f^{(k)}(a)}{k!}(x-a)^k + \cdots$$

Notes: (1) $T_f(x)$ is a power series whose partial sums are the Taylor polynomials for f at $x = a$. More specifically, the kth partial sum of $T_f(x)$ is the kth Taylor polynomial for f at $x = a$.

(2) Given an arbitrary function f, it is **not** always true that $T_f(x)$ converges to f in some neighborhood of a (in Problem 12 below, you will see an example of this). One standard method for determining if $T_f(x)$ does converge to f in some neighborhood of a is by applying Taylor's Theorem (Theorem 16.24) and then showing that $R_n \to 0$ (the remainders approach 0 as n goes to ∞). In this case, we will write $f(x) = T_f(x)$ for $x \in (a - R, a + R)$, where R is the radius of convergence of the Taylor series.

(3) If $a = 0$, then the Taylor series for f at $x = 0$ is often called the **Maclaurin series (or Maclaurin expansion) for f**.

Example 16.25:

1. In part 1 of Example 16.23, we saw that if f is any polynomial function of degree n, then for $k \geq n$, the kth degree Taylor polynomial for f at $x = 0$ will be equal to the polynomial f. It follows that the Maclaurin series for a polynomial f is equal to the polynomial itself. As a specific example, the Maclaurin series for $f(x) = 17x^{320} + 5x - 6$ is $17x^{320} + 5x - 6$.

2. In part 2 of Example 16.23, we saw that for each $k \in \mathbb{N}$, the kth Taylor polynomial for $g(x) = e^x$ at $x = 0$ is

$$p_k(x) = \sum_{n=0}^{k} \frac{x^n}{n!} = 1 + x + \frac{x^2}{2!} + \frac{x^3}{3!} + \cdots + \frac{x^k}{k!}.$$

So, the Maclaurin expansion for g is defined by

$$T_g(x) = \sum_{n=0}^{\infty} \frac{x^n}{n!} = 1 + x + \frac{x^2}{2!} + \frac{x^3}{3!} + \cdots + \frac{x^k}{k!} + \cdots.$$

By part 2 of Example 16.15, the radius of convergence of $T_g(x)$ is ∞.

Now, the remainder term for $p_k(x)$ is

$$R_k(x) = \frac{g^{(k+1)}(d)}{(k+1)!} x^{k+1} = \frac{e^d x^{k+1}}{(k+1)!},$$

where d is between 0 and x. Since e^x is strictly increasing on $\mathbb{R}$, for each $x \in \mathbb{R}$, we have $|R_k(x)| \leq \frac{e^{|x|}|x|^{k+1}}{(k+1)!}$. By the Ratio Test for Sequences (Problem 18 from Problem Set 14), $\lim_{k \to \infty} \frac{e^{|x|}|x|^{k+1}}{(k+1)!} = 0$ (Check this!). Since $0 \leq |R_k(x)|$, by the Squeeze Theorem (Problem 9 in Problem Set 14), $\lim_{k \to \infty} R_k(x) = 0$. Therefore, $g(x) = T_g(x)$ for all $x \in \mathbb{R}$ ($g(x) = e^x$ is equal to its Maclaurin expansion for all $x \in \mathbb{R}$). So, we can write

$$e^x = \sum_{n=0}^{\infty} \frac{x^n}{n!} = 1 + x + \frac{x^2}{2!} + \frac{x^3}{3!} + \cdots + \frac{x^k}{k!} + \cdots \quad \text{for all } x \in \mathbb{R}.$$

3. In part 3 of Example 16.23, we saw that for each $k \in \mathbb{N}$, the kth Taylor polynomial for $h(x) = \frac{1}{x}$ at $x = 1$ is

$$p_k(x) = \sum_{n=0}^{k} (-1)^n (x-1)^n = 1 - (x-1) + (x-1)^2 - (x-1)^3 + \cdots + (-1)^k (x-1)^k.$$

So, the Taylor expansion for h at $x = 1$ is

$$T_h(x) = \sum_{n=0}^{\infty} (-1)^n (x-1)^n = 1 - (x-1) + (x-1)^2 - \cdots + (-1)^k (x-1)^k + \cdots.$$

Let's use the Ratio Test (Theorem 15.21) to find the radius of convergence of $T_g(x)$.

If we let $s_n = (-1)^n (x-1)^n$, then we have

$$\lim_{n \to \infty} \left| \frac{s_{n+1}}{s_n} \right| = \lim_{n \to \infty} \left| \frac{(-1)^{n+1}(x-1)^{n+1}}{(-1)^n(x-1)^n} \right| = |x-1|.$$

By the Ratio Test, if $|x-1| < 1$, the series is absolutely convergent and if $|x-1| > 1$, the series is divergent. So, the radius of convergence of $T_h(x)$ is 1. The Divergence Test shows that $T_h(x)$ diverges at the endpoints 0 and 2. So, the interval of convergence of $T_h(x)$ is $(0, 2)$.

Therefore, we can restrict our attention to $x \in (0, 2)$. Does Taylor's Theorem tell us that $h(x) = T_h(x)$ for all $x \in (0, 2)$? Let's check.

The remainder term for $p_k(x)$ is

$$R_k(x) = \frac{h^{(k+1)}(d)}{(k+1)!}(x-1)^{k+1} = \frac{(-1)^{k+1}(k+1)!(x-1)^{k+1}}{(k+1)!\, d^{k+1}} = \frac{(-1)^{k+1}}{d^{k+1}}(x-1)^{k+1},$$

where d is between 1 and x.

Now, if $x \geq 1$, we have $|R_k(x)| \leq \frac{1|x-1|^{k+1}}{1^{k+1}} = |x-1|^{k+1}$. If $1 < x < 2$, then $0 < x - 1 < 1$, and so, by part 2 of Example 14.1, $\lim_{k \to \infty} |x-1|^{k+1} = 0$. It follows that for $1 < x < 2$, we have $\lim_{k \to \infty} R_k(x) = 0$. Therefore, $h(x) = T_h(x)$ for all $x \in [1, 2)$ ($h(x) = \frac{1}{x}$ is equal to its Taylor expansion at $x = 1$ for all $x \in [1, 2)$).

Next, if $0 < x < 1$, then $|R_k(x)| \leq \frac{1|x-1|^{k+1}}{x^{k+1}} = \frac{(1-x)^{k+1}}{x^{k+1}} = \left(\frac{1-x}{x}\right)^{k+1} = \left(\frac{1}{x} - 1\right)^{k+1}$.

If $\frac{1}{2} < x < 1$, then $1 < \frac{1}{x} < 2$, and so, $0 < \frac{1}{x} - 1 < 1$. It follows from part 2 of Example 14.1 that $\lim_{k \to \infty} \left(\frac{1}{x} - 1\right)^{k+1} = 0$. So, $h(x) = T_h(x)$ for all $x \in \left(\frac{1}{2}, 1\right)$. Putting this together with the result from the previous paragraph, we see that $h(x) = T_h(x)$ for all $x \in \left(\frac{1}{2}, 2\right)$.

Unfortunately, if $x \leq \frac{1}{2}$, then $\frac{1}{x} - 1 \geq 2 - 1 = 1$, and so $\lim_{k \to \infty} \left(\frac{1}{x} - 1\right)^{k+1} = \infty \neq 0$. So, Taylor's Theorem does not tell us if $h(x) = T_h(x)$ for $x \in \left(0, \frac{1}{2}\right]$.

Luckily, $T_h(x)$ is a geometric series with first term 1 and common ratio $r = -(x-1)$. It follows that $T_h(x)$ converges to $\frac{1}{1-(-x+1)} = \frac{1}{x}$ whenever $|x-1| = |r| < 1$. So, despite the fact that Taylor's Theorem did not provide us with a complete answer, it is true that $h(x) = T_h(x)$ for all $x \in (0, 2)$. Therefore, we can write the following:

$$\frac{1}{x} = \sum_{n=0}^{\infty} (-1)^n (x-1)^n = 1 - (x-1) + (x-1)^2 - \cdots + (-1)^k (x-1)^k + \cdots, \quad x \in (0, 2)$$

Problem Set 16

Full solutions to these problems are available for free download here:
www.SATPrepGet800.com/RABQXZ

LEVEL 1

1. Let $f, g: A \to \mathbb{R}$ be bounded on A. Prove that $|fg|_A \leq |f|_A \cdot |g|_A$.

2. Let $f, g: A \to \mathbb{R}$ be bounded on A. Prove that $|f + g|_A \leq |f|_A + |g|_A$.

LEVEL 2

3. A sequence of functions (f_n) is said to be **decreasing** on A if for all $x \in A$ and $n \in \mathbb{N}$, $f_n(x) \geq f_{n+1}(x)$. Find a decreasing sequence (f_n) of continuous functions on $[0, 1)$ that converges to a continuous function $f: [0, 1) \to \mathbb{R}$, but the convergence is **not** uniform on $[0, 1)$.

4. Determine the radius of convergence and the interval of convergence of each of the following power series:

   (i) $\sum_{n=1}^{\infty} \frac{(-1)^n x^{2n}}{2^{2n}(n!)^2}$

   (ii) $\sum_{n=1}^{\infty} \left(\frac{n+1}{n}\right)^n x^n$

   (iii) $\sum_{n=1}^{\infty} \frac{5^n(x+6)^n}{\sqrt{n}}$

LEVEL 3

5. For each $n \in \mathbb{N}^+$, let $f_n: [0, \infty) \to \mathbb{R}$ be defined by $f_n(x) = \frac{x}{n+x}$. Prove each of the following:

   (i) f_n converges to 0 pointwise on $[0, \infty)$.

   (ii) f_n does **not** converge uniformly on $[0, \infty)$ to 0.

   (iii) For each $c \in \mathbb{R}^+$, f_n converges uniformly on $[0, c]$ to 0.

6. Let (f_n) and (g_n) be sequences of functions on $A \subseteq \mathbb{R}$, let $B \subseteq A$, and suppose that (f_n) and (g_n) converge uniformly on B to f and g, respectively. Determine if each of the following statements is true or false. If true, provide a proof. If false, provide a counterexample.

   (i) $(f_n + g_n)$ converges uniformly on B to $f + g$.

   (ii) $(f_n g_n)$ converges uniformly on B to fg.

7. Find the Maclaurin series $T_f(x)$ for $f(x) = \sin x$ and show that $f(x) = T_f(x)$ for all $x \in \mathbb{R}$.

Level 4

8. Let $B \subseteq A \subseteq \mathbb{R}$ and for each $n \in \mathbb{N}$, let $f_n: A \to \mathbb{R}$ be a function that is bounded on B. We say that (f_n) is **uniformly bounded on B** if there is $M \in \mathbb{R}^+$ such that $|f_n(x)| \le M$ for all $n \in \mathbb{N}$ and all $x \in B$. Prove each of the following:

    (i) If (f_n) converges uniformly on B to $f: B \to \mathbb{R}$, then f is bounded on B.

    (ii) If (f_n) converges uniformly on B, then (f_n) is uniformly bounded on B.

    (iii) If (f_n) is uniformly bounded on B and (f_n) converges pointwise to $f: B \to \mathbb{R}$, then f is bounded on B.

9. Find the Taylor series $T_f(x)$ for $f(x) = \ln x$ at $x = 1$ and show that $f(x) = T_f(x)$ for all x inside the radius of convergence of $T_f(x)$.

10. Let (f_n) be a sequence of continuous functions on $A \subseteq \mathbb{R}$, let $B \subseteq A$, and suppose that (f_n) converges uniformly on B to f. For each $n \in \mathbb{N}$, let $x_n \in B$ and let $x \in B$ with $x_n \to x$. Prove that $\lim_{k \to \infty} \left(\lim_{n \to \infty} f_n(x_k) \right) = \lim_{n \to \infty} \left(\lim_{k \to \infty} f_n(x_k) \right)$. Is the result still true if we replace the word "uniformly" by the word "pointwise?"

11. Define $f_0: \mathbb{R} \to \mathbb{R}$ by $f_0(x) = 1$ and for each $n \in \mathbb{N}$, define $f_{n+1}: \mathbb{R} \to \mathbb{R}$ recursively by $f_{n+1}(x) = 1 + \int_0^x f_n(t)\, dt$. Prove that the sequence (f_n) converges to a function $f: \mathbb{R} \to \mathbb{R}$ such that the convergence is uniform on any closed interval and $f(x) = e^x$ for all $x \in \mathbb{R}$.

Level 5

12. Define $f: \mathbb{R} \to \mathbb{R}$ by $f(x) = \begin{cases} e^{-\frac{1}{x^2}} & \text{if } x \ne 0 \\ 0 & \text{if } x = 0. \end{cases}$ Prove that f is infinitely differentiable at every $x \in \mathbb{R}$, but $f \ne T_f$, where $T_f(x)$ is the Maclaurin expansion for f.

13. A sequence of functions (f_n) is said to be **increasing** on A if for all $x \in A$ and $n \in \mathbb{N}$, $f_n(x) \le f_{n+1}(x)$. Similarly, a sequence of functions (f_n) is said to be **decreasing** on A if for all $x \in A$ and $n \in \mathbb{N}$, $f_n(x) \ge f_{n+1}(x)$. Finally, a sequence of functions (f_n) is said to be **monotone** on A if the sequence is increasing or decreasing on A.

    Let (f_n) be a monotone sequence of continuous functions on $A \subseteq \mathbb{R}$, let $[a, b] \subseteq A$, and suppose that (f_n) converges on $[a, b]$ to a continuous function $f: [a, b] \to \mathbb{R}$. Prove that (f_n) converges uniformly on $[a, b]$ to f. This result is known as **Dini's Theorem**.

Challenge Problems

14. Let I be a bounded interval and for each $n \in \mathbb{N}$, let $f_n: I \to \mathbb{R}$ be differentiable on I. Suppose that (f_n') converges uniformly on I to $g: I \to \mathbb{R}$ and assume that there is $c \in I$ such that $(f_n(c))$ is a convergent sequence. Then (f_n) converges uniformly on I to a differentiable function $f: I \to \mathbb{R}$ and $f'(x) = g(x)$ for all $x \in I$.

15. For each $n \in \mathbb{N}$, let $f_n: [a, b] \to \mathbb{R}$ be a bounded function on $[a, b]$ that is Riemann integrable on $[a, b]$. Suppose that (f_n) converges to a function $f: [a, b] \to \mathbb{R}$ that is Riemann integrable on $[a, b]$ and that there is $M \in \mathbb{R}^+$ such that for all $n \in \mathbb{N}$ and $x \in [a, b]$, $|f_n(x)| \leq M$. Prove that

$$\int_a^b f(x)\, dx = \lim_{n \to \infty} \int_a^b f_n(x)\, dx.$$

This result is known as the **Bounded Convergence Theorem**.

16. Consider the following series:

$$\sum_{n=0}^{\infty} \frac{\cos(3^n x)}{2^n}$$

Prove that this series converges uniformly on $\mathbb{R}$ to a function that is continuous on $\mathbb{R}$ and is not differentiable at any real number.

INDEX

Abelian, 103
Abelian group, 103
Abel's Lemma, 276
Abel's Test, 279
Absolute maximum, 166
Absolute minimum, 166
Absolute extremum, 166
Absolute value, 30, 85, 122
Absolutely convergent integral, 251
Absolutely convergent series, 283
Absolutely integrable, 251
Accumulation point, 130
Additive identity, 79
Algebraic function, 65
Alternating, 272
Alternating harmonic series, 275
Alternating p-series, 275
Alternating series, 272
Alternating Series Test, 273
Angle in standard position, 67
Antiderivative, 220
Antireflexive relation, 36
Antisymmetric relation, 36
Arbitrary but specific element, 15
Archimedean Property, 117
Argument, 16
Associative, 80
Associativity, 27, 80, 103
Asymptote, 64
Axiom of Extensionality, 21
Base case, 78
Biconditional, 20
Bijection, 52
Bijective function, 52
Binary expansion, 97
Binary relation, 35
Binary operation, 79, 101

Blackboard bold, 10
Bolzano-Weierstrass Theorem, 158
Boundary point, 137
Bounded above, 114
Bounded below, 114
Bounded function, 166
Bounded set, 114
Bounded sequence, 158, 258
Boundedness Theorem, 166
Bounding set, 11
Cantor's Theorem, 93
Cantor-Schroeder-Bernstein Theorem, 95
Cantor set, 129
Cardinality, 13, 92
Cartesian plane, 23
Cartesian product, 22, 23
Cauchy Convergence Criterion for Series, 276
Cauchy Convergence Criterion for Series of Functions, 303
Cauchy Criterion for Convergence, 158
Cauchy in the uniform norm, 297
Cauchy sequence, 61, 85, 157, 256
Center, 66
Chain rule, 182
Change of variables, 221
Circumference, 67
Closed, 79, 101, 103
Closed ball, 124
Closed disk, 124
Closed interval, 41, 123
Closed set, 128
Closing statement, 16
Closure, 80, 103
Closure of a set, 130
Codomain, 50

Coefficient, 64
Common ratio, 59
Commutative, 80, 103
Commutative group, 103
Commutative monoid, 80
Commutativity, 27, 80
Compact, 132
Comparability condition, 39
Comparable, 39
Comparison Test (for integrals), 249
Comparison Test (for series), 280
Complement, 128
Complete ordered field, 116
Completeness, 114
Completeness property, 116
Complex numbers, 13, 87
Complex Plane, 13
Complex-valued function, 58
Complex-valued sequence, 58
Composite function, 55
Composition, 55
Conditional, 16
Conditionally convergent integral, 251
Conditionally convergent series, 283
Conditionally integrable, 251
Congruence modulo 2, 43
Congruence modulo n, 43
Conjunction, 20
Connected set, 134
Constant function, 50, 62
Constant sequence, 58, 86, 156
Constant term, 64
Continuity, 143, 163
Continuous function, 143, 163
Contradiction, 38, 78, 94
Contrapositive, 27

Convergent improper integral, 241
Convergent sequence, 60, 156, 256
Convergent sequence of functions, 293
Convergent series, 264
Converges, 156, 256, 264, 301
Converges absolutely, 251, 283
Converges conditionally, 251, 283
Converges pointwise, 293, 301
Converges uniformly, 294, 301
Countable set, 93
Countably infinite set, 93
Counterexample 14
Cover, 132
Covering, 132
Cubic function, 64
Cubic polynomial, 64
Curve, 195
Darboux integrable, 209
Darboux integral, 209
Darboux's Theorem, 194
De Morgan's laws, 27
Decimal, 12
Decimal expansion, 98
Decimal point, 12
Decreasing function, 187
Decreasing sequence, 258
Dedekind cut, 121
Degree of a polynomial, 64
Deleted neighborhood, 124
Density Theorem, 117
Denumerable, set, 93
Dependent variable, 234
Derivative, 178
Derivative from the left, 218
Derivative from the right, 218
Diameter, 123
Dictionary order, 38
Difference, 80, 85
Difference quotient, 172

Differentiable, 178
Differential, 227
Digit, 12
Dirichlet's Test (for integrals), 252
Dirichlet's Test (for series), 276
Disconnected, 134
Disconnected, set, 134
Disconnection, 134
Disjoint, 26, 30
Disjunction, 20
Distance, 85, 122
Distributive, 105
Distributivity, 27, 80, 105
Divergence test, 265
Divergent improper integral, 241
Divergent sequence, 61, 156, 256
Divergent series, 264
Diverges, 156, 256, 264
Divisible, 43, 89
Division, 108
Domain of a function, 50
Domain of a relation, 36
Doublestruck, 10
Element, 9
Ellipses, 9
Empty set, 13
$\epsilon - \delta$ definition of limit, 147
Equal set, 19
Equality relation, 43
Equinumerosity, 92
Equinumerous, 92
Equivalence class, 44
Equivalence relation, 43
Even, 112
Even natural numbers, 80
Even integers, 10, 112
Even natural numbers, 10
Eventually decreasing, 274
Existential quantifier, 36
Explicit function, 234

Exponential function, 66
Factorial, 305
Fence-post formula, 14
Field (algebraic structure), 105
Field axioms, 105, 106
Field of a relation, 36
Finite sequence, 57
Finite set, 93
First derivative, 191
Fourth derivative, 191
Fraction, 12
Function, 50
Fundamental Theorem of Calculus- first form, 216
Fundamental Theorem of Calculus- first form, 219
Gamma function, 253
General exponential function, 233
General logarithmic function, 234
Generalized Mean Value Theorem, 188
Geometric sequence, 59, 258
Geometric series, 266
Greatest lower bound, 114
Group, 103
Group axioms, 103, 104
Half-open interval, 41
Half plane, 153
Harmonic series, 267
Heine-Borel Theorem, 133
Homeomorphic, 167
Homeomorphism, 151, 167
Horizontal asymptote, 64
Horizontal strip, 138, 153
Hypotenuse, 111
Idempotent laws, 27
Identity, 79, 80, 103
Identity element, 79, 103
Identity function, 56
Image, 57
Imaginary axis, 13

Imaginary part, 87
Implication, 15
Implicit differentiation, 235
Implicit function, 235
Improper integral, 241
Inclusion map, 96
Increasing function, 187
Increasing sequence, 258
Independent variable, 234
Indeterminate form, 188, 192
Index, 264
Indices, 264
Indirect proof, 27
Inductive hypothesis, 78
Inductive step, 78
Infimum, 115
Infinite closed interval, 41
Infinite interval, 41
Infinite open interval, 41
Infinite sequence, 57
Infinite set, 9, 93
Infinite slope, 171
Infinitely differentiable, 310
Initial ray, 67
Injection, 52
Injective function, 52
Integer-valued function, 58
Integer-valued sequence, 58
Integers, 9, 83
Integral Test, 267
Integration by parts, 244
Intercept, 62
Interior point, 178
Intermediate Value Theorem, 167
Interior point, 137
Intersection, 24, 29
Interval, 40
Interval of convergence, 305, 307
Interval notation, 41
Inverse, 103
Inverse function, 53

Inverse image, 57
Irrational numbers, 13
Isomorphism, 110
ith term, 64
kth partial sum, 264, 301
kth Taylor polynomial, 312
Leading coefficient, 64
Leading term, 64
Least element, 77
Least upper bound, 114
Left Riemann rectangle, 195
Left Riemann sum, 200
Leg, 111
Lemma, 81
Length of a sequence, 57
Length of an interval, 197
L'Hôpital's rule, 188
Lim inf, 260
Lim sup, 260
Limit, 141, 145
Limit Comparison Test (for integrals), 250
Limit Comparison Test (for series), 281
Limit from the right, 155
Limit inferior, 260
Limits involving infinity, 153
Limit of a sequence, 156
Limit superior, 260
Line, 170
Linear approximation, 312
Linear function, 62, 170
Linearly ordered set, 39
Local extremum, 185
Local maximum, 185
Local minimum, 185
Locally integrable, 241
Logarithmic differentiation, 235
Logarithmic function, 66
Logical connective, 16
Logically equivalent, 16, 20
Lower bound, 114

Lower Darboux integral, 209
Lower Darboux sum, 207
Lower Riemann integral, 209
Lower Riemann rectangle, 195
Lower Riemann sum, 201
Maclaurin expansion, 317
Maclaurin series, 317
Maximum-Minimum Theorem, 166
Mean Value Theorem, 187
Member, 9
Membership relation, 9, 35
Monoid, 80, 104
Monotone Convergence Theorem, 259
Monotone function, 187
Monotone sequence, 258
Monotonic function, 187
Monotonic sequence, 258
Mutually exclusive, 26
n-ary relation, 37
Natural exponential function, 230
Natural logarithmic function, 225
Natural numbers, 9, 75
Negation, 20
Negative exponents, 182
Negative identities, 70
Negative slope, 171
Neighborhood, 123
Nondecreasing function, 187
Nonincreasing function, 187
Norm of a partition, 223
nth derivative, 191
nth partial sum, 264
nth root, 65
nth term of a sequence, 57, 256
nth term of a series, 264, 301
nth Term Test, 265
Odd, 112
Odd natural numbers, 10

Odd integers, 112
One-sided limits, 155
One-to-one function, 52
One-to-one correspondence, 34
Open covering, 132
Open ball, 123
Open disk, 123
Open interval, 41, 123
Open rectangle, 139
Open set, 124
Onto, 52
Opening statement, 16
Ordered field, 108
Ordered k-tuple, 22
Ordered pair, 21
Ordered ring, 108
Ordered triple, 22
Ordered tuple, 22
Origin, 13, 23
p-integral, 243
p-series, 267
Pairwise disjoint, 30
Parabola, 63
Parity, 43
Partial binary operation, 103
Partial ordering, 36, 37
Partial sum, 264, 301
Partially ordered set, 37
Partition of a closed interval, 197
Partition of a set, 45
Pointwise convergence, 293, 301
Pointwise convergent series, 301
Polynomial degree, 64
Polynomial function, 64
POMI, 77
Poset, 37
Positive integers, 10
Positive slope, 171
Positive square root, 65, 122

Power rule, 182
Power series, 304
Power set, 18
Predecessor, 75
Prime number, 94
Principle of Mathematical Induction, 77
Product rule, 181
Proof by contradiction, 38, 78, 94
Proof by contrapositive, 27
Proper subset, 15
Property, 10
Propositional logic, 15
Propositional variables, 15
Punctured open interval, 124
Pure imaginary numbers, 13, 87
Pythagorean identity, 70
Pythagorean Theorem, 111
Quadrantal angles, 69
Quadratic approximation, 312
Quadratic function, 63
Quotient rule, 182
Raabe's Test, 292
Radian measure, 67
Radius, 66
Range of a function, 50
Range of a relation, 36
Radius of convergence, 305, 307
Ratio Test, 286
Ratio Test for Sequences, 263
Rational-valued function, 58
Rational-valued sequence, 58
Rational numbers, 12, 84
Rational function, 64
Ray, 67
Real axis, 13
Real line, 12
Real numbers, 12, 85
Real part, 87
Real-valued function, 58

Real-valued sequence, 58, 256
Rearrangement, 284
Rearrangement Theorem, 285
Recursive definitions, 76
Refinement, 209
Reflexive relation, 36
Regrouping, 291
Reindexed, 264
Relative extremum, 185
Relative maximum, 185
Relative minimum, 185
Remainder term, 315
Repeating decimal, 13
Representative, 47
Riemann integrable function, 209
Riemann integral, 209
Riemann rectangle, 195
Riemann-Stieltjes integrable, 224
Riemann-Stieltjes integral, 224
Riemann sum, 197, 199
Right Riemann rectangle, 195
Right Riemann sum, 200
Ring, 106
Rolle's Theorem, 186
Root Test, 288
Roster method, 9
SACT, 81
Secant line, 172
Second derivative, 191
Selfish set, 91
Semigroup, 104
Semiring, 106
Separation, 134
Sequence, 57, 256
Sequence of functions, 293
Sequence of real numbers, 58
Series, 264
Set, 9
Set-builder notation, 10
Set difference, 25
Sigma notation, 198

Slope, 62, 170
Slope-intercept form of the equation of a line, 170
Square root, 65, 122
Squeeze Theorem, 161, 262
Standard Advanced Calculus Trick, 81
Statement, 15
Strict linearly ordered set, 39
Strict partial ordering, 37
Strict partially ordered set, 37
Strict poset, 37
Strictly decreasing function, 169, 187
Strictly increasing function, 59, 169, 187
Strictly monotone function, 187
Strictly monotonic function, 187
Strip, 138
Strip game, 144
Subcover, 132
Subinterval, 197
Subsequence, 59
Subset, 14
Subtraction, 108
Successor, 75
Sum identities, 71
Sum of a series, 264
Summation by parts, 276
Supremum, 115
Surjection, 52
Surjective function, 52
Symmetric relation, 36
Symmetric difference, 25
Tabular integration by parts, 245
Tag, 200
Tail of 9's, 12
Tail of a sequence, 116
Tail of a series, 268
Tangent line, 176
Tangent line approximation, 312
Tautology, 17
Taylor coefficients, 310
Taylor expansion, 316
Taylor polynomial, 312
Taylor series, 316
Taylor's Theorem, 314
Telescoping sum, 266
Term of a polynomial, 64
Term of a sequence, 57
Term of a series, 264, 301
Terminal ray, 67
Terminating decimal, 13
Ternary expansion, 98
Ternary relation, 37
Third derivative, 191
Tietze Extension Theorem, 169
Topologically equivalent, 167
Totally ordered set, 39
Transitive relation, 14, 19, 36
Transitive set, 90
Transitivity, 18, 19
Tree diagram, 18
Triangle Inequality, 86, 123
Trichotomy, 39
Trigonometric function, 66, 69
Trivial equivalence relation, 43
Truth assignment, 15
Unary relation, 37
Unbounded set, 114
Uncountable set, 93
Uniform continuity, 150
Uniform norm, 296
Uniformly Cauchy, 297
Uniformly continuous, 150
Uniformly convergent, 294, 301
Uniformly differentiable, 194
Uniqueness Theorem, 311
Union, 24, 29
Unit circle, 67
Universal set, 15
Universal quantifier, 15
Unordered pair, 21
Upper bound, 114
Upper Darboux integral, 209
Upper Darboux sum, 207
Upper Riemann integral, 209
Upper Riemann rectangle, 195
Upper Riemann sum, 201
Urysohn's Lemma, 169
Vacuously true, 17
Variable, 10
Vector, 87
Venn diagram, 15
Vertex, 63
Vertical asymptote, 64
Vertical strip, 138, 153
Weierstrass M-test, 302
Well-defined, 47
Well Ordering Principle, 77
Without loss of generality, 54
WOP, 77
Wrapping function, 68
y-intercept, 62, 170
Zero slope, 171

About the Author

Dr. Steve Warner, a New York native, earned his Ph.D. at Rutgers University in Pure Mathematics in May 2001. While a graduate student, Dr. Warner won the TA Teaching Excellence Award.

After Rutgers, Dr. Warner joined the Penn State Mathematics Department as an Assistant Professor and in September 2002, he returned to New York to accept an Assistant Professor position at Hofstra University. By September 2007, Dr. Warner had received tenure and was promoted to Associate Professor. He has taught undergraduate and graduate courses in Precalculus, Calculus, Linear Algebra, Differential Equations, Real Analysis, Complex Analysis, Mathematical Logic, Set Theory, and Abstract Algebra.

From 2003 – 2008, Dr. Warner participated in a five-year NSF grant, "The MSTP Project," to study and improve mathematics and science curriculum in poorly performing junior high schools. He also published several articles in scholarly journals, specifically on Mathematical Logic.

Dr. Warner has nearly two decades of experience in general math tutoring and tutoring for standardized tests such as the SAT, ACT, GRE, GMAT, and AP Calculus exams. He has tutored students both individually and in group settings.

In February 2010, Dr. Warner released his first SAT prep book "The 32 Most Effective SAT Math Strategies," and in 2012 founded Get 800 Test Prep. Since then Dr. Warner has written books for the SAT, ACT, SAT Math Subject Tests, AP Calculus exams, and GRE. In September 2018, Dr. Warner released his first advanced math book "Pure Mathematics for Beginners." Since then, he has released several more books in advanced mathematics.

Dr. Steve Warner can be reached at

steve@SATPrepGet800.com

BOOKS BY DR. STEVE WARNER